CURRENT RESEARCH IN HEAT AND MASS TRANSFER

CURRENT RESEARCH IN HEAT AND MASS TRANSFER

A Compendium and a Festschrift for Professor Arcot Ramachandran

Edited by

M. V. Krishna Murthy
V. M. Krishna Sastri
P. K. Sarma
S. P. Sukhatme

HEMISPHERE PUBLISHING CORPORATION

New York Washington Philadelphia London

DISTRIBUTION OUTSIDE NORTH AMERICA

SPRINGER–VERLAG

Berlin Heidelberg New York London Paris Tokyo

CURRENT RESEARCH IN HEAT AND MASS TRANSFER

1 2 3 4 5 6 7 8 9 0 EBEB 8 9 8

Library of Congress Cataloging in Publication Data

Current research in heat and mass transfer.

Bibliography: p.
Includes index.
1. Heat—Transmission. 2. Mass transfer.
3. Ramachandran, Arcot, date. I. Ramachandran, Arcot, date. II. Krishna Murthy, M. V.
TJ260.C87 1986 621.402′2 86-7707
ISBN 0-89116-578-9 Hemisphere Publishing Corporation

DISTRIBUTION OUTSIDE NORTH AMERICA:
ISBN 3-540-16827-3 Springer-Verlag Berlin

Professor Arcot Ramachandran

Contents

Preface

This Festschrift is published as a tribute to Professor Arcot Ramachandran on the occasion of his sixty-second birthday. It is a very happy occasion not only for the members of his family but also for his former students, his colleagues and his friends. Every one of them would think of him according to the circumstances of their privileged association with him—as an affectionate family man, an educator, an accomplished scientist, or a committed administrator.

Professor Ramachandran was born in Madras on 6th April 1923. He obtained his Honours degree in Mechanical Engineering from the University of Madras in 1943. He pursued his graduate work in heat transfer at Purdue University under the Late Professor Max Jakob and Professor George Hawkins during the period of 1945–49. On his return to India, he joined as a faculty member of the Mechanical Engineering Department at the Indian Institute of Science, Bangalore which had already built up an international reputation for research in sciences. At that time, however, very little research work was being done in India in the field of heat transfer since there were hardly any graduate programmes in engineering. During the period 1950–67, Professor Ramachandran established a number of graduate programmes in Mechanical Engineering and Industrial Management. In particular, he built up a school of research in heat transfer which attracted a number of visiting scholars and professors. On the basic research side, he worked in the areas of pool and film boiling heat transfer, heat transfer from vibrating and rotating surfaces, free convection heat transfer, thermal entry length studies in annuli, and hydrodynamics and heat transfer in non-circular passages. The needs of a developing nation led him to focus his attention also on problems of thermal power such as the use of lignite as a fuel for power generation. He also conducted extensive research in heat transfer problems related to production technology such as solidification, and thermal properties and characteristics of metals and non-metals which gained him international recognition.

After his illustrious professional career, he was called upon to take over as the Director of the Indian Institute of Technology at Madras in 1967. During his tenure there, he gave a major thrust to research by initiating and encouraging basic and applied research in many fields of science and engineering. With his charming personality and capacity to take quick and incisive decisions, he was directly responsible for the institute's growth as a leading research institute in a brief span of five years. In his own field of heat transfer, a new school of research was established by him. This school, now, is a recognised centre of excellence for research in heat transfer problems related to food processing, fluidized bed combustion, and passive and active solar thermal systems.

Apart from being a great academician, a researcher and an excellent administrator, Professor Ramachandran is a builder of institutions. As the founder-President of the Indian Society for Heat and Mass Transfer since 1974, he has shaped the society into an effective vehicle of interaction among the heat transfer community in India and abroad. With his insight into the similarity of problems of the developing countries, he was instrumental in founding the Regional Centre for Energy, Heat and Mass Transfer for Asia and the Pacific which is the regional centre of excellence in the field to be established in cooperation with UNESCO, International Centre of Heat and Mass Transfer and national governments. He has been a delegate to the International Assembly for Heat Transfer Conferences and is a member of Scientific Council of the International Centre for Heat and Mass Transfer in Belgrade.

His achievements attracted the attention of international organization and the national government. In 1973, he was invited to assume the position of Secretary in the newly established Department of Science and Technology of the Government of India and was later concurrently the Director-General of the Scientific and Industrial Research. This department was then responsible for planning and direction of research and development in science and technology through a network of research laboratories, academic institutions and industrial organisations both in the private and public sectors. During his tenure, the first Science and Technology Plan was approved. Other noteworthy achievements are the establishment of National Remote Sensing Agency, Ocean Science and Technology Agency, Science and Engineering Research Council and the Central Electronics Limited - an industrial undertaking for manufacturing ferrites, opto-electronic instruments and solar cells. He was the architect of the National Programme for Research and Development in Renewable Sources of Energy. It was also during his tenure that fiscal incentives for research and development in industry were introduced by the Government of India.

Professor Ramachandran played a leading role in engineering education at the undergraduate and graduate levels as Chairman and member of various boards. He was the Chairman of the Preparatory Committee for the United Nations Conference on Science and Technology for Development (1977–78). He was Chairman of UNESCO International Conference on Education and Training for Engineers and Technicians, the Association for Engineering Education in South and Central Asia, the UNESCO Expert Group on Environmental Aspects of Engineering Education and ESCAP Experts Meeting on the Regional Centre for Transfer of Technology. Based on the recommendations of this group, the ESCAP Regional Centre for Transfer of Technology was established in Bangalore in July 1978.

In October 1978, he assumed his present position as Under-Secretary General and Executive Director of the newly established United Nations Centre for Human Settlements (Habitat) with headquarters at Nairobi. In the short span of six years, the Centre has become fully operational and renders assistance to several countries in Africa, Asia, West Asia, Latin America and the Carribean.

Professor Ramachandran has received many national and international honours in recognition of his outstanding achievements in professional and related fields of activity. He was honoured by Purdue University with an honorary degree of Doctor of Engineering and the Distinguished Alumnus Award. He was awarded Honorary Doctorates by Stuttgart University (Federal Republic of Germany), University of Roorkee, Andhra University, Jawaharlal Nehru Technological University and the Indian Institute of Technology, Madras. The Institute of Engineers (India) elected him as Honorary Life Fellow and the Indian Institute of Science bestowed the Honorary Fellowship of the Institute on him.

He is a fellow of the American Society of Mechanical Engineers, Institution Mechanical Engineers (London) and Indian National Science Academy. He is also a member of the Editorial Boards of several technical journals including the International Journal of Heat and Mass Transfer. He is the Editor-in-chief of the Regional Journal of Energy, Heat and Mass Transfer.

Professor Ramachandran has been blessed with a charming wife, Susila Ramachandran. With her suave and gentle manners. she has contributed a lot in an unobtrusive and self-effacing way to Professor Ramachandran's success. The Ramachandrans have two accomplished sons, Dr. Balakrishnan and Mr. Mahendra.

For some of us who have been fortunate enough to have had a long association with him, it is a great pleasure to wish Professor Ramachandran and his family a long life of health, happiness and further accomplishment. We are sure that his students, colleagues and friends from all over the world will join the members of the Indian Heat and Mass Transfer community in paying our tributes to him.

This Festschrift is a small token of appreciation and a tribute to the foresight, vision and creative imagination of Professor Ramachandran. The Editors are indeed grateful to all the distinguished authors of the papers published in this volume for readily agreeing to send their contributions. Deliberately, no attempt has been made to arrange the papers in any topical sequence. It is our hope that this collection of papers will serve as a useful source book for research workers in the field of heat transfer, thermodynamics and energy.

M. V. Krishna Murthy
V. M. Krishna Sastri
P. K. Sarma
S. P. Sukhatme

December 29, 1985

Contributors

Adachi, K., Department of Chemical Engineering, Nagoya University, Furocho, Chikusaku, Nagoya 464, Japan.

Bergles, A. E., Department of Mechanical Engineering, Iowa State University, Ames, Iowa, USA.

Bhavnani, S. H., Department of Mechanical Engineering, Iowa State University, Ames, Iowa, USA.

Bose, T. K., Department of Aeronautical Engineering, Indian Institute of Technology, Madras 600 0036, India.

Chang, T. D., Department of Mechanical Engineering, State University of New York, Stony Brook, New York 11794, USA.

Clark, J. A., Department of Mechanical Engineering and Applied Mechanics, University of Michigan, Ann Arbor, Michigan 48109-2125, USA.

Fisch, N. M., Institut für Thermodynamik, Universität Stuttgart, PO BOX 801140, D-7000, Stuttgart 80, FRG.

Gopichand, T., Department of Chemical Engineering, Indian Institute of Technology, Madras 600 036, India.

Hahne, E. W. P., Institut für Thermodynamik and Wärmetechnik, Universität Stuttgart, PO BOX 801140, D-7000 Stuttgart 80, FRG.

Harada, T., Department of Chemical Engineering, Kobe University, Kobe 657, Japan.

Hasan, M., Department of Chemical Engineering, McGill University, Montreal, PQ, Canada H3A2A7.

Hasatani, M., Department of Chemical Engineering, Nagoya University, Furocho, Chikusaku, Nagoya 464, Japan.

Himasekhar, K., Department of Mechanical Engineering, Andhra University, Visakhapatnam 530 003, India.

Itaya, Y., Department of Chemical Engineering, Nagoya University, Furocho, Chikusaku, Nagoya 464, Japan.

Irvine, Jr., T. F., Department of Mechanical Engineering, State University, of New York, Stony Brook, New York 11794, USA.

Kataoka, K., Department of Chemical Engineering, Kobe University, Kobe 657, Japan.

Khabakhpasheva, Ye. M., Institute of Thermophysics of the Siberian Branch, of USSR Academy of Sciences, 1 Lavrentyev Avenue, Novosibirsk-90, 630090, USSR.

Kandlikar, S. G., Mechanical Engineering Department, Rochester, N.Y.
New York 14623, USA.

Korovkin, V. N., Luikov Heat and Mass Transfer Institute, Byelorussin Academy of Sciences, Minsk 22078, USSR.

Krishna Murthy, M. V., Department of Mechanical Engineering, Indian Institute of Technology, Madras 600 036, India.

Kutataladze, S. S., Institute of Thermophysics of the Siberian Branch of the USSR Academy of Sciences, 1 Lavrentyev Avenue, Novosibirsk-90, 630090, USSR.

Laddha, G. S., Department of Chemical Engineering, A. C. College of Technology, University of Madras, Madras 600 025, India.

Lienhard, J. H., Department of Mechanical Engineering, University of Houston, Texas 77004, USA.

Martynenko, O. G., Luikov Heat and Mass Transfer Institute, Byelorussian Academy of Sciences, Minsk, 220728, USSR.

Mathur, A., Department of Chemical Engineering, University of Illinois at Chicago, Box 4348, Chicago, Illinois 60680, USA.

Mayinger, F., Institut für Thermodynamik-A, Universität München, FRG.

Mehta, S. K., Reactor Group, Bhabha Atomic Research Centre, Trombay, Bombay-400085, India.

Mohanty, A. K., Department of Mechanical Engineering, Indian Institute of Technology, Kharagpur, India.

Mujumdar, A. S., Department of Chemical Engineering, McGill University, Montreal P.Q., Canada H3A 2A7.

Rao, N. S., Department of Mechanical Engineering, Indian Institute of Technology, Madras 600 036, India.

Reddy, B. S., Faculty of Engineering, Memorial University of Newfoundland, St. John's Newfoundland, A1B BX5, Canada.

Sahara, R., Department of Chemical Engineering, Kobe University, Kobe 657, Japan.

Sastri, V. M. K., Department of Mechanical Engineering, Indian Institute of Technology, Madras 600 036, India.

Saxena, S. C., Department of Chemical Engineering, University of Illinois at Chicago, Box 4348, Chicago, Illinois 60680, USA.

Schlünder, E. U., Institut für Thermische Verfahrenstechnik der Universität Karlsruhe, Kaiserst 12, 7500 Karlsruhe 1, FRG.

Schnittger, R., Institut für Thermodynamik-A, Universität München, München, FRG.

Shah, R. K., Harrison Radiator Division, General Motors Corporation Lockport, New York 14094, USA.

Sharan, A. M., Faculty of Engineering, Memorial University of Newfoundland, St. John's, Newfoundland, A1BBX5, Canada.

Sarma, P. K., Department of Mechanical Engineering, Andhra University, Visakhapatnam 533003, India.

Sokovischin, Yu. A., Luikov Heat and Mass Transfer Institute, Byelorussian Academy of Sciences, Minsk 220728, USSR.

Srinivasan, S., Department of Chemical Engineering, A. C. College of Technology, University of Madras, Madras 600 025, India.

Sukhatme, S. P., Department of Mechanical Engineering, Indian Institute of Technology, Powai, Bombay 400 076, India.

Swamy, B., Department of Mechanical Engineering, S. V. University College of Engineering, Tirupati 517 502, India.

Tsotsas, E., Institut für Thermische Verfahrenstechnik der Universität Karlsruhe, Kaiserst 12, 7500 Karlsruhe 1, FRG.

Venkat Raj, V., Reactor Group, Bhabha Atomic Research Centre, Trombay, Bombay 400 0085, India.

Interferometric Study of Laminar Natural Convection from an Isothermal Vertical Plate with Transverse Roughness Elements

S. H. BHAVNANI and A. E. BERGLES
Heat Transfer Laboratory
Department of Mechanical Engineering
Iowa State University
Ames, Iowa

A Mach-Zehnder interferometer was used to experimentally evaluate the local natural convection-heat transfer characteristics of several enhanced vertically oriented surfaces. The surface configurations studied were of two types--repeated ribs and stepped. Air was the working fluid and a constant wall temperature boundary condition was applied. Laminar flow up to a Grashof number of 2×10^7 was studied. A general objective of this study was to provide a basic understanding of how natural convection boundary layers are affected by large scale surface roughness elements. The results obtained indicate that, while the ribbed configuration, in general, caused a reduction in average heat transfer coefficient, the stepped plate served to enhance the heat transfer. A plane flat plate was tested to provide the basis for comparison. Several interesting features of the variation of the temperature field in the vicinity of the wall and of the local Nusselt number are described.

1. INTRODUCTION

Since the pioneering experimental work of Ray [1] in 1920, natural or free convection has developed into one of the most studied topics in heat transfer. However, relatively little information is available on the effect of complex geometries on natural convection. The problem of natural convection from vertical surfaces with some sort of large scale surface roughness elements is encountered in several technological applications. Of particular interest is the dissipation of heat from electronic circuits, where component performance and reliability are strongly dependent on operating temperature. Natural convection represents an inherently reliable cooling process. Further, this mode of heat transfer is often designed as a backup in the event of the failure, due to fan breakdown, of a forced convection system.

Circuit boards represent a naturally enhanced free convection situation. In other applications where the heat dissipating surface is normally smooth, it may be necessary to enhance the surface to achieve the desired temperature level. The traditional solution is to add fins; however, roughening the surface would be a more attractive solution if the heat transfer coefficient increase is substantial. Few studies have been carried out to determine the effect of surface roughness elements on free convection heat transfer. The results obtained seem to conflict with each other as to the increase of the heat transfer coefficient.

Several studies have examined the effects of roughness on average heat transfer coefficients for vertical isothermal surfaces in a large enclosure (unbounded free convection) Jofre and Barron [2] obtained data for a vertical surface, roughened with triangular grooves, to air. A boundary layer solution, based on forced flow over a rough plate, was presented to explain the experimental results. At $Ra_L \sim 10^9$ they quoted an improvement in the average Nusselt number of about 200% relative to the turbulent predictions of Eckert and Jackson [3]. However, the flow is probably not turbulent over the whole smooth plate. Compared to other conditions, the improvement is closer to 100%.

Furthermore, it appears that the reported Nusselt numbers are too high due to underestimation of the radiation correction [4].

Ramakrishna et al. [5] developed an analogy correlation of heat transfer data as obtained by Sastry et al. [6] for a vertical cylinder. Sastry et al. roughened the cylinder by wrapping 0.45mm to 1.45mm diameter wire around it; the wrapping was done with a pitch equal to the wire diameter (i.e., no gap between wires). The enhancement was typically about 50%. These studies were also made with air.

Heya et al. [7] conducted experiments on horizontal cylinders of 35-mm and 63-mm diameter with dense pyramid, streak-type and check-type roughness elements of heights varying from 0.15 mm to 0.72 mm and spacing varying from 0.76 mm to 2.00 mm. The tests were conducted using both water and air. No increase in average heat transfer coefficients was reported for the range of $4 \times 10^4 < Ra < 10^7$ for air and $3 \times 10^6 < Ra < 2 \times 10^8$ for water.

Fujii et al. [8] roughened large-diameter vertical cylinders with repeated ribs, dispersed protrusions, and closely spaced pyramids. Heat transfer to water and spindle oil was studied. Maximum increases in the average Nusselt number of 10% with water were observed relative to their previous data for a smooth cylinder. The latter data are slightly higher than McAdams' correlation [9] for smooth vertical plates.

Prasolov [10] presented some results of an experimental investigation of the influence of machined roughness of a horizontal cylinder on heat transfer to air. The type of roughness elements used were densely packed pyramids of heights varying from 0.08 mm to 0.36 mm at spacings varying from 1 mm to 2 mm. A very significant increase in the average Nusselt number was observed in the Rayleigh number range $5 \times 10^4 < Ra_D < 5 \times 10^6$. For lower or higher values of the Rayleigh number, little enhancement was observed. A qualitative explanation is offered in terms of intensification of turbulence in the transition region. It is difficult to accept this, however, as transition is normally considered to occur at $Ra_D \sim 10^9$.

Yao [11] used a transformation method to numerically analyze heat transfer from a vertical sinusoidal surface at uniform temperature. The results show a decrease in the Nusselt number when compared to a plane flat plate. Some explanation is offered in terms of the thickening of the boundary layer near the nodes of the sinusoidal pattern.

There are also several related studies of bounded convection. Bohn and Anderson [12] recently studied free convection heat transfer from large machine-roughened vertical surfaces in an enclosure. They concluded that for an isothermal surface the rough texture produced fully turbulent behavior at Ra_L about half that characterizing turbulence for a smooth surface. This produced increases in section-average heat transfer of up to 40% and in the surface average heat transfer of about 16%. The results were complicated by recirculating flow in the enclosure. They found no enhancement in the laminar regime, stated to be $Ra_L < 2 \times 10^{10}$.

Forced convection duct flow with a roughened wall simulating a circuit board has been studied by several investigators. Lehmann and Wirtz [13] used an interferometer to determine local heat transfer coefficients around protuberances on one of the duct walls. This type of flow situation, although of potential interest, is different from the one presently studied because of the recirculatory nature of the flow as a consequence of higher Reynolds numbers. Such separation and reattachment are not expected in natural-convection driven flows; however, this point does not appear to have been confirmed.

In this study two different types of large-scale surface roughness elements were studied--namely, repeated ribs and stepped surfaces. Typical schematics of the two types of surfaces are shown in Fig. 1. The study was carried out using a Mach-Zehnder interferometer in order to nonintrusively evaluate local phenomena. To the authors' knowledge, this kind of study has not been carried out before with vertical plates having large-scale roughness elements.

2. EXPERIMENTAL APPARATUS AND PROCEDURE

2.1 Interferometer

A Mach-Zehnder interferometer (MZI) was used in the study. Interferometry is an ideal technique for taking local measurements in free convection flows. The MZI was configured as shown in Fig. 2. The basic interferometer consisted of two splitter plates and two mirrors. The associated optical equipment was employed merely to obtain a good quality interferogram. The light source used was a 2 mW helium-neon laser whose wavelength is 6328 A°. Besides producing monochromatic light, the laser's chief advantage is its long coherence length that eliminates the need for a compensating chamber. The paths of the light beam from the laser to the camera are indicated in Fig. 2.

The principle of operation and the design of the MZI has been described in detail in Hauf and Grigull [14] and Eckert and Goldstein [15], and the reader is referred to these texts for further information.

Interferograms were recorded on Kodak Panatomic-X fine grain, 35-mm black and white film using a Canon TX camera equipped with a 135-mm lens along with a set of close-up rings. This was done to enlarge the image to fill the film frame. All optical components were placed on an optical table with air bag suspension, with the exception of paraboloidal mirror PM1, which was placed on a separate table.

PM1 was vibrated gently using a low power D.C. motor. This served to eliminate background noise fringes so as to give a clear picture of the true fringes. The splitter plates and mirrors were positioned such that the light beam was incident on them at an angle of 30^0. This differs from the conventional angle of 45^0 and was done in order to obtain a wider beam of light.

The experimental rig was housed in a climate-controlled room with filtered ventilation air. The walls of the room were well insulated.

2.2 Test Section

A cross-sectional view of a typical test section itself is shown in Fig. 3. Nichrome-wire heater assemblies were custom fabricated and pasted to the rear surface of the plate using Glyptal Red enamel. The test sections were made from 1/4-in. thick aluminum plate and were 0.127 m x 0.178 m in size. The ribbed test sections had ribs that were press fit ensuring adequate thermal contact. Twelve 30-gauge copper-constantan thermocouples were located in a grid in the test section as shown in Fig. 3; care was taken to locate them as close to the front surface as possible. The rear surface was insulated with glass-wool to minimize heat loss. An isothermal boundary condition was established with the four heater assemblies and was monitored by the thermocouples. The variation in temperature over the surface of the plate was $\pm$ o.3^0 C.

Various aspect ratios, p/s and p/q, were studied for both ribbed and stepped surfaces.

2.3 Procedure

The experimental data collection procedure involved a careful alignment of the MZI to provide a good infinite fringe pattern. For more details on alignment, the reader is referred to Tolpadi [16]. Care was taken to ensure that the optical table was horizontal by adjusting the pressure in the four air bag suspension units. The test section was suspended in the measuring leg of the MZI and carefully aligned so that the surface was parallel to the light beam. Power to the heaters was then switched on. Steady state was reached after 3 to 4 hours. A polyethylene sheet was wrapped around the MZI to minimize the disturbance due to stray air currents.

The studies made were essentially two dimensional in nature. Several interferograms had to be made to cover the entire length of the plate. The vertical movement desired was facilitated by the use of a lab jack, which was moved 0.0254 m at a time. Composites of several interferograms are shown in Figs. 4 and 5 for the cases studied. The pins shown are for the purposes of indication only and are not in the flow field being studied.

The plates were tested at temperatures that varied from plate to plate between 48^0 C and 75^0 C. Note that the plate temperature was not used as a variable, the variation coming about merely to facilitate obtaining a conveniently readable interferogram.

The data reduction procedure involved reading the interferograms at several different x locations. This was done using a tool-maker's microscope. The fringe shift information obtained was then used in conjunction with equations for the variation of index of refraction of air to evaluate both a predicted wall temperature, T_W, using the ambient temperature as a reference, and the slope at the wall, dT/dy, using a curve fitting program.

Note that for the ribbed surface, one measurement was made at the middle of the upstream side, the top, and the downstream side of the rib. Each of these measurements was considered to be a representative average for that surface. Fringe shift measurements were made perpendicularly outward from the respective surfaces.

The local heat transfer coefficients and Nusselt numbers were then calculated as follows:

$$h_x \equiv - k \cdot \frac{dT}{dy} \quad \frac{1}{(T_w - T_a)}$$

and

$$Nu_x \equiv - \frac{dT}{dy} \cdot \frac{1}{(T_w - T_a)} \cdot x$$

The local Grashof numbers were evaluated at the film temperature,

$$T_f = \frac{T_w + T_a}{2}$$

3. RESULTS AND DISCUSSION

Figures 6 and 7 show the local heat transfer coefficient plotted against distance from the leading edge and local Nusselt number plotted against local Grashof number for the plane flat plate. The agreement with the usual boundary layer analytical results is good, indicating that the experimental method is adequate.

Figures 8 and 9 show the variation of local coefficients and Nusselt numbers for the ribbed test section with aspect ratios p/s = 8:1 and p/q = 8:1. The solid line represents the boundary layer solution for an isothermal vertical plane flat plate. It can be seen that most of the data fall below the prediction with the exception of points located on the top of the rib. The local coefficients are especially low just downstream and upstream of the rib. This is attributed to flow stagnation and separation thickening of the boundary layer at these locations. This, in turn, is probably due to the retardation of the buoyant forces caused by the projection. A second set of clearly defined peaks in the heat transfer coefficient, of lesser magnitude, occurs between ribs. This is where the boundary layer reattaches itself to the surface; this is clearly brought out in the composite interferogram (see Fig. 4). The stagnation zone effect was more pronounced in the test section with larger ribs with p/s = 16:5 and p/q = 4:1 than it was in the smaller ribbed test section with p/s = 8:1 and p/q = 8:1.

An analysis made in Yao [11] seems to agree well in principle with the above thoughts on thickening of boundary layers. Yao found that the boundary layer thickness in his numerical results was thicker at the nodes (corresponding to the regions just downstream and upstream of the ribs in the present study) than near the crest (face of the rib) and the trough (point of reattachment.) For the portion of the surface parallel to the buoyant driving force, the flow velocity is larger and so is the heat transfer rate.

Figures 10 through 15 illustrate the variation of local coefficients and Nusselt numbers for the stepped test sections with aspect ratios, p/q = 8:1, 16:1, 32:1, respectively. It is readily apparent that the heat transfer performance of this type of surface is far superior to that of the ribbed plates studied. One obvious reason is the complete elimination of the downstream stagnation zone caused by this configuration. For the test sections with aspect ratios 16:1 and 32:1, almost all the data points lie above the exact solution for a plane flat plate. This results in an increase in the mean integrated heat transfer coefficient, h of 23.8% and 11.5% respectively, using the projected area as basis for the calculations. As before, the peaks occur just after the steps as was the case with the ribs. Also, there is no readily observable second set of peaks, as is to be expected.

The data for the test section with aspect ratio 8:1 exhibit wide fluctuations similar to the variations for the ribbed test sections, with many data points falling well below the exact solution for a plane flat plate. This suggests that the steps may be too frequent, causing a retardation of the buoyancy force to such an extent that it completely offsets the local increase in heat transfer coefficient obtained at the step location. The percentage increase in integrated, mean heat transfer coefficient for this case was 11.2%.

The increase in heat transfer coefficient is more pronounced as the upper edge of the plate is approached. This is reasonable as the ribs or steps should have a greater influence as the transition

Grashof number ($Ra_L \sim 10^9$) is approached. This suggests that both kinds of transverse roughness elements may be useful in improving heat transfer performance in the transition regime, in accordance with the results presented in Ref.12.

It would seem that both the height of the roughness and the spacing between elements are of great importance. It is interesting to qualitatively examine the previously reported work with regard to this point. The increase in heat transfer coefficient reported by Jofre and Barron [2] may be attributed to the effect the triangular grooved surface (height 0.76 mm and spacing 0.89 mm) had on promoting an early transition to turbulence, since the data are at Rayleigh numbers of about 10^9.

Fujii et al. [8] do not report significant increases in the Nusselt number in their experiments, which can be attributed to the fact that the dense pyramid-type roughness used is not conducive to increasing heat transfer in the laminar region. This is because individual roughness elements would lie in the downstream dead zone caused by the preceding roughness element. For the Fujii et al. tests conducted using repeated ribs, it is likely that their rib height of 0.5 mm was insufficient to disturb the boundary layer but enough to cause a dead zone and, therefore, a thickening of the boundary layer.

Sastry et al. [6] conducted experiments in the transition region, where their roughness element heights (0.45 mm to 1.45 mm) were probably sufficient to trip the boundary layer into turbulence, leading to the 50% increase in Nusselt number reported.

The data of Heya et al [7] show no increase in the Nusselt number using dense pyramid, streak-type and check-type roughness elements, since the testing was done in the laminar region; this again supports the hypothesis that this type of roughness is not effective in this flow regime, where the boundary layer is stable enough to resist tripping.

Bohn and Anderson's [12] experiments were performed using a machine-roughened plate consisting of two sets of grooves 1-mm deep and 1-mm wide cut at right angles to each other. This type of surface, for reasons similar to those expressed for Fujii's data, may not be best suited for laminar natural convection enhancement. This is evidenced by results reported in [12], where increases in Nusselt number in the laminar region are of the order of 4% and those in the transition region are about 16%.

Studies conducted with forced convection seem to suggest the existence of an optimum protuberance spacing below which the dead zone is large enough to more than compensate for the increase in heat transfer caused by the accelerated flow at reattachment and above which the dead zones and accelerated flow zones are too small to produce significant changes in the heat transfer coefficient.

Thus, it is evident that with proper sizing and shape selection, enhancement of heat transfer from a vertical plate is possible in natural convection even in the laminar region. However, before optimum parameters can be prescribed a more detailed quantitative analysis needs to be carried out.

4. CONCLUSIONS

The effect on local heat transfer performance of two types of transverse roughness elements was studied. It was shown that aspect ratio and protuberance spacing are both parameters of considerable importance. The stepped surfaces helped to improve the performance. The ribbed surfaces, on the other hand, served to decrease the heat transfer performance overall because of the formation of stagnation zones both upstream and downstream of the ribs. A qualitative explanation in terms of the retardation of buoyant forces by the roughness elements has been suggested.

The results indicate that it is possible to enhance the heat transfer performance in the laminar region. An even greater effect is expected in the transition region.

NOMENCLATURE

Gr_x Local Grashof number = $g\beta\Delta Tx^3/\nu^2$

g Acceleration of gravity, m/s^2

h_x Local heat transfer coefficient, W/m^2-K

k thermal conductivity, W/m-K

Nu_x Local Nusselt number = $h_x x/K$

p Rib-to-rib pitch (also step-to-step pitch), m

q Height of rib or step, m

Ra Rayleigh number

s Width of rib, distance from leading edge along profile, m

T Temperature, ^0C

T_a Ambient temperature, ^0C.

T_f Film temperature, ^0C

T_w Wall temperature, ^0C

x Longitudinal coordinate, also distance from leading edge, m

y Transverse coordinate

Greek letter symbols

β Volumetric coefficient of thermal expansion, K^{-1}

Δ Designates a difference when used as a prefix

ν Kinematic viscosity, m^2/s

REFERENCES

1. Ray, B.B., Proc. Indian Assoc. Cultivation Sci. 6 (1920) 95.

2. Jofre, R.J., and Barron, R.F., "Free Convection Heat Transfer to a Rough Plate," ASME Paper No.WA 67/HT-38, 1967.

3. Eckert, E.R.G., and Jackson, T.W., "Analysis of Turbulent Free Convection Boundary Layer on Flat Plate," NACA Report, (1951) 1015.

4. Bergles, A.E., and Junkhan, G.J., "Energy Conservation Via Heat Transfer Enhancement," Quarterly Progress Report, Jan.1 1979 - Mar.31 1979, COO-4649-5, (1979).

5. Ramakrishna, K., Seetharamu, K.N., and Sarma, P.K., "Turbulent Heat Transfer from a Rough Surface," J. Heat Transfer, 100 (1978) 727-729.

6. Sastry, C.V.S.N., Murthy, V.N., and Sarma, P.K., "Effect of Discrete Wall Roughness on Free Convection Heat Transfer from a Vertical Tube," Presented at International Centre for Heat and Mass Transfer, Dubrovnik, Yugoslavia, (1976).

7. Heya, N., Takeuchi, M., and Fujii, T., "Influence of Surface Roughness on Free Convection Heat Transfer from a Horizontal Cylinder," Chem. Eng. J., 23 (1982) 185-192.

8. Fujji, T., Fujii, M., and Takeuchi, M., "Influence Various Surface Roughness on the Natural Convection," Int. J. Heat Mass Transfer, 16 (1973) 629-640.

9. McAdams, W.H., Heat Transmission, 3rd Ed., McGraw Hill, New York, (1954) 173-176.

10. Prasolov, R.S., "On the Effects of Surface Roughness on Natural Convection Heat Transfer from Horizontal Cylinders to Air," Inzh. Fiz. Zh. (Russian), No.5, 4 (1961) 3-7.

11. Yao, L.S., "Natural Convection Along a Vertical Wavy Surface," J. Heat Transfer, 105 (1983) 465-468.

12. Bohn, M., and Anderson, R., "Heat Transfer Enhancement in Natural Convection Enclosure Flows", SERI Report Number SERI/TR-252-2103, Aug.(1984).

13. Lehmann, G.L., and Wirtz, R.A., "Convection from Surface Mounted Repeating Ribs in a Channel Flow," ASME Paper No.84-WA/HT-88, (1984).

14. Hauf, W., and Grigull, U., "Optical Methods in Heat Transfer," in Advances in Heat Transfer, New York, Academic Press, 6 (1960) 133-366.

15. Eckert, E.R.G., and Goldsterin, R.J., Measurements in Heat Transfer, 2nd ed., McGraw Hill, New York, (1976).

16. Tolpadi, A.K., "Experimental and Numerical Study of Conjugate Natural Convection Heat Transfer from Fins," M.S. Thesis in Mechanical Engineering, Iowa State University (1983).

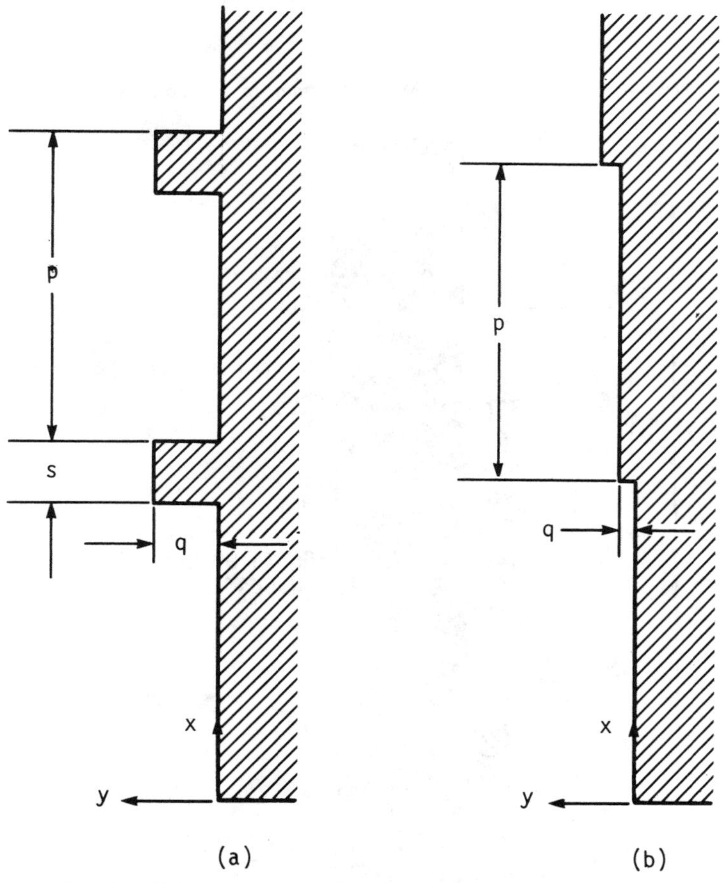

Fig.1 Schematic of flow geometry; (a) ribbed, and (b) stepped.

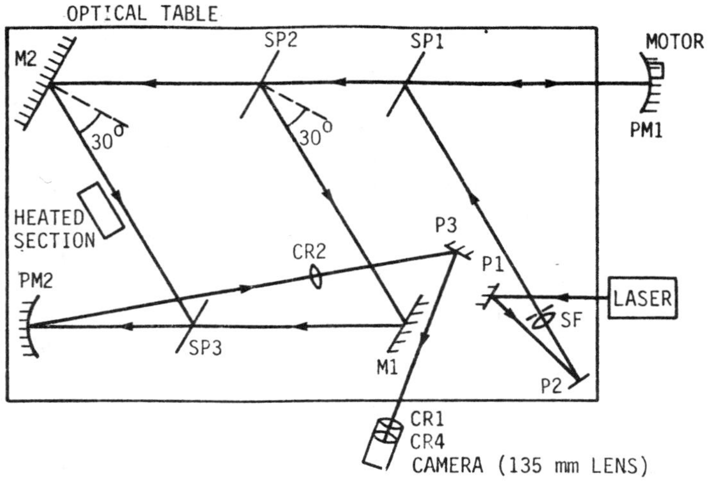

P1, P2, P3 - SMALL PLANE MIRRORS

SP1, SP2, SP3 - SPLITTER PLATES

PM1, PM2 - PARABOLOIDAL MIRRORS

SF - SPATIAL FILTER

M1 - 0.102 m MIRROR

M2 - 0.152 m MIRROR

CR1, CR2, CR4 - CLOSE-UP RINGS #1, 2, AND 4

Fig.2 Plan view of the Mach-Zehnder interferometer

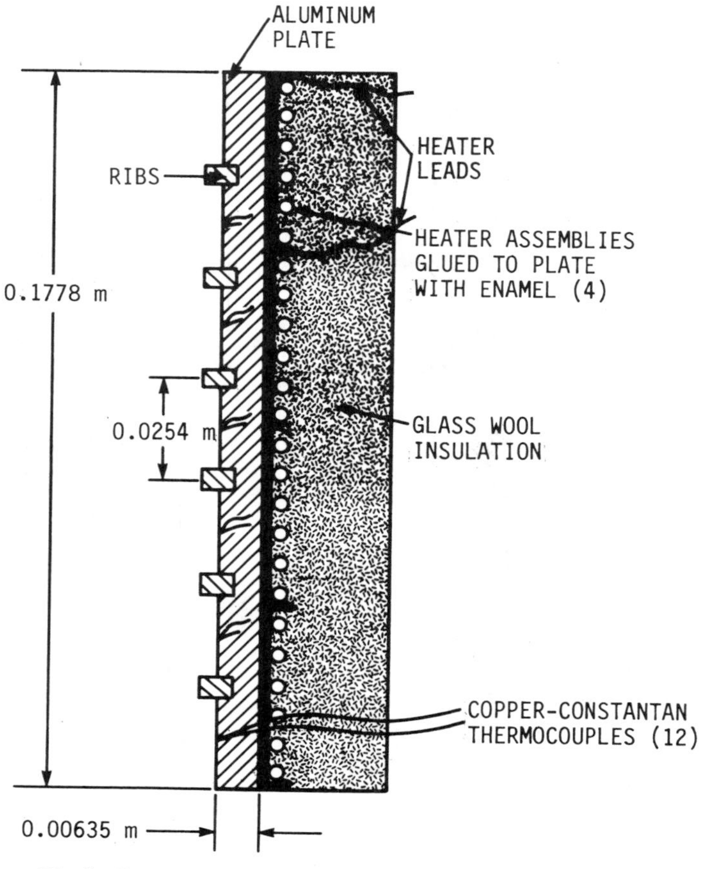

Fig.3. Cross sectional view of ribbed test section

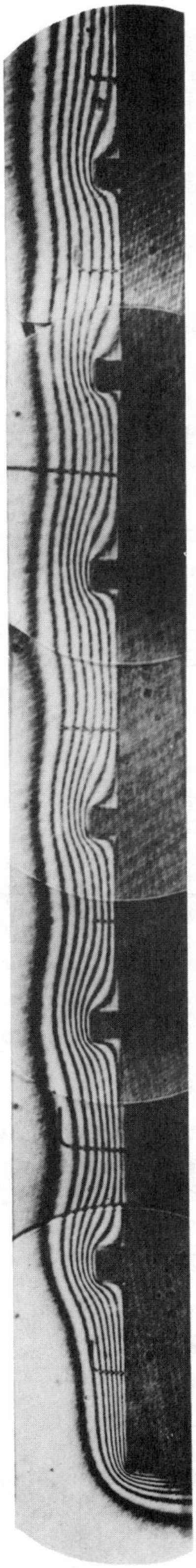

Fig.4 Composite interferogram of the ribbed test section with p/s = 8.1 and p/q = 8.1

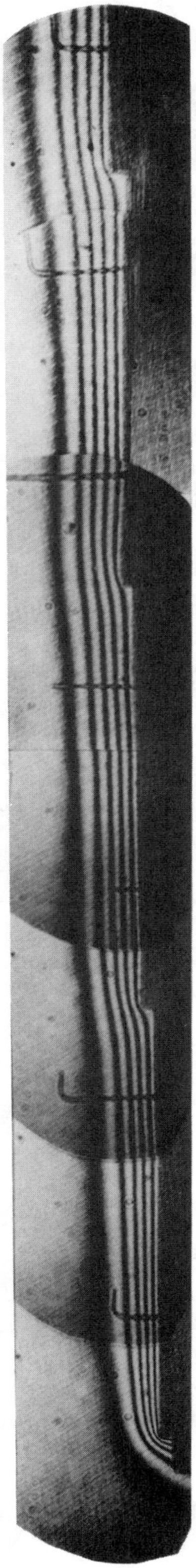

Fig.5 Composite interferogram of the stepped test section with p/q = 32:1

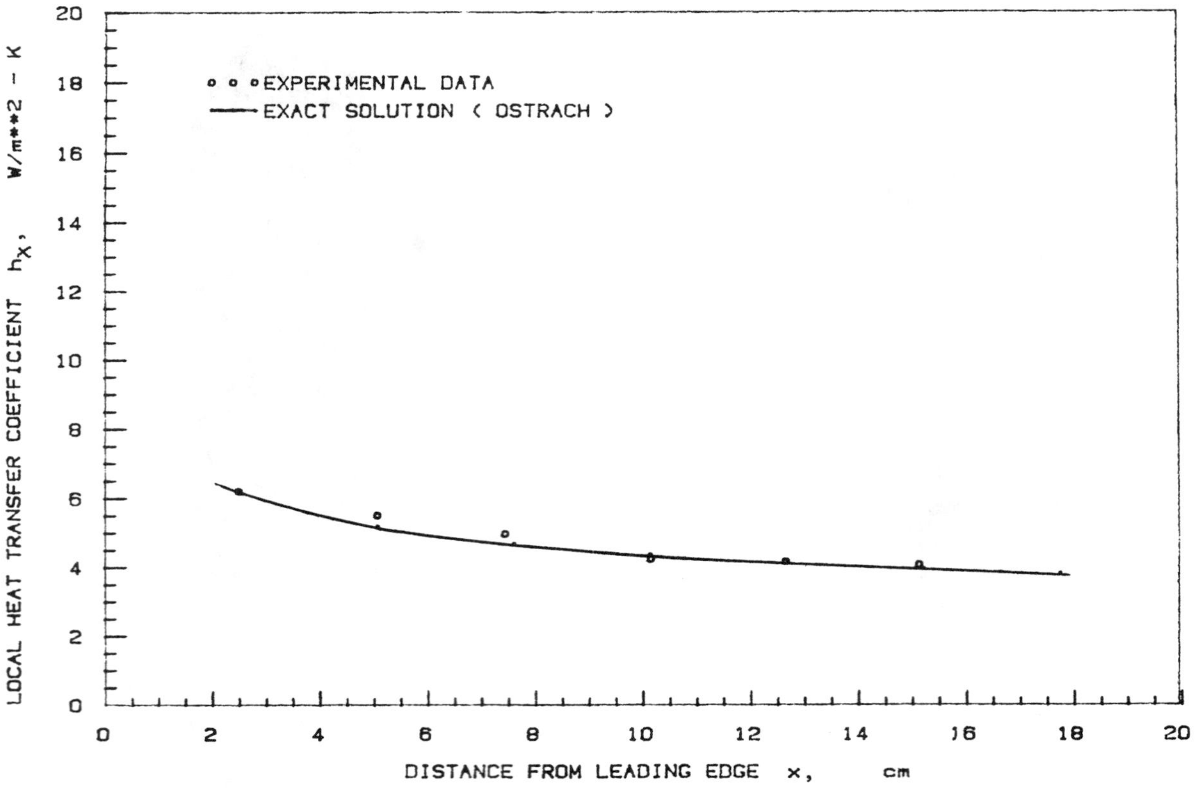

Fig.6 Local heat transfer coefficient for plane flat plate

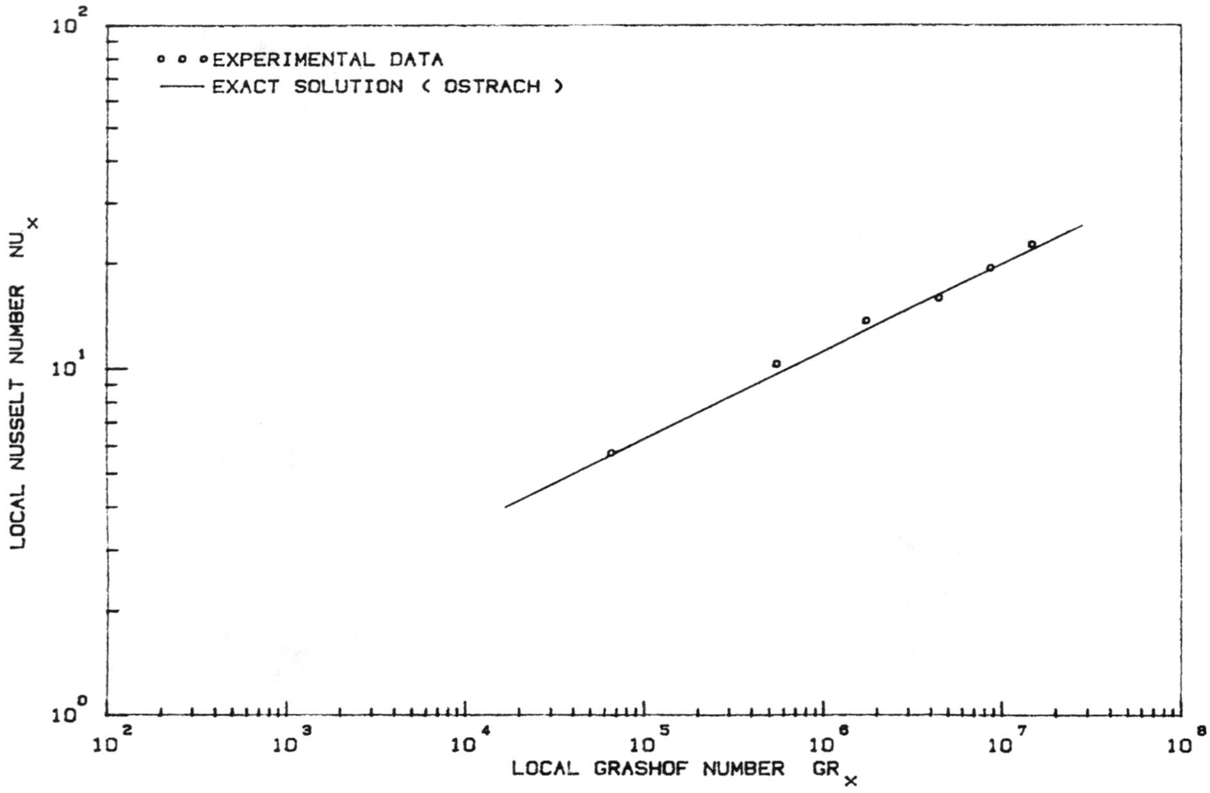

Fig.7 Local Nusselt number for plane flat plate.

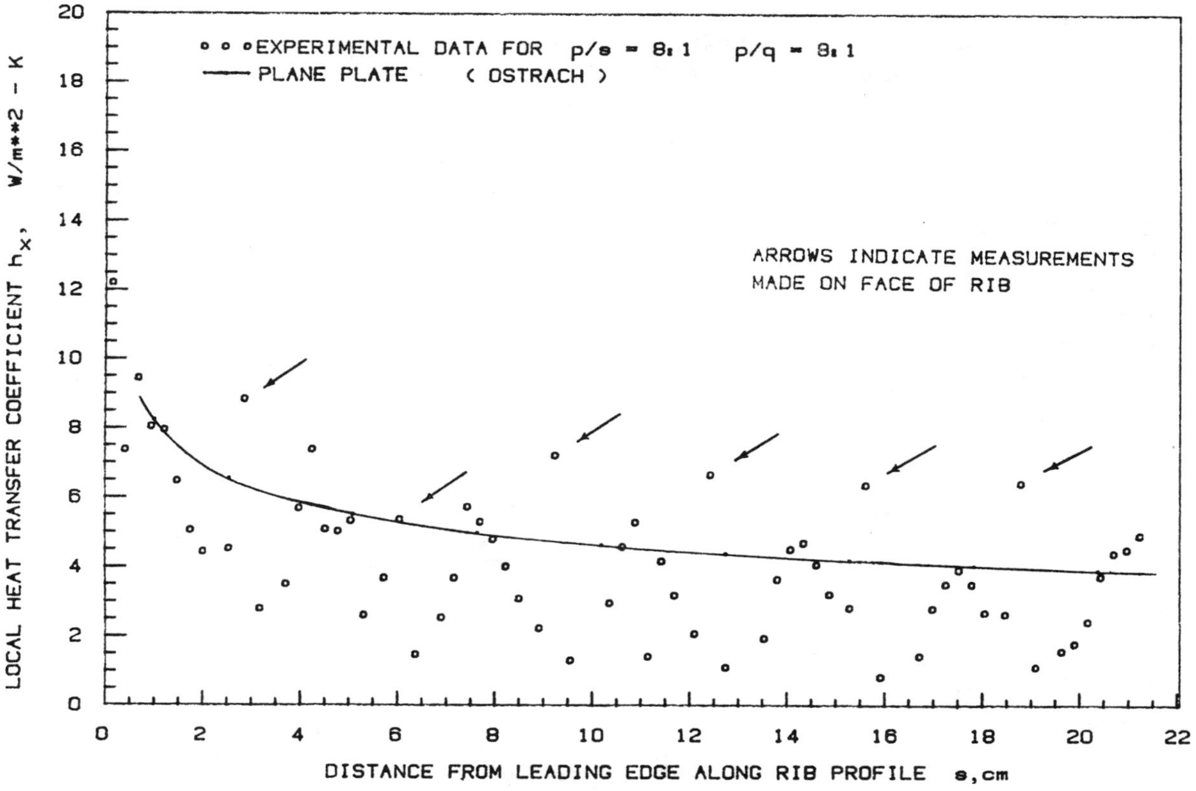

Fig.8 Local heat transfer coefficient for ribbed plate with p/s = 8:1 and p/q = 8:1

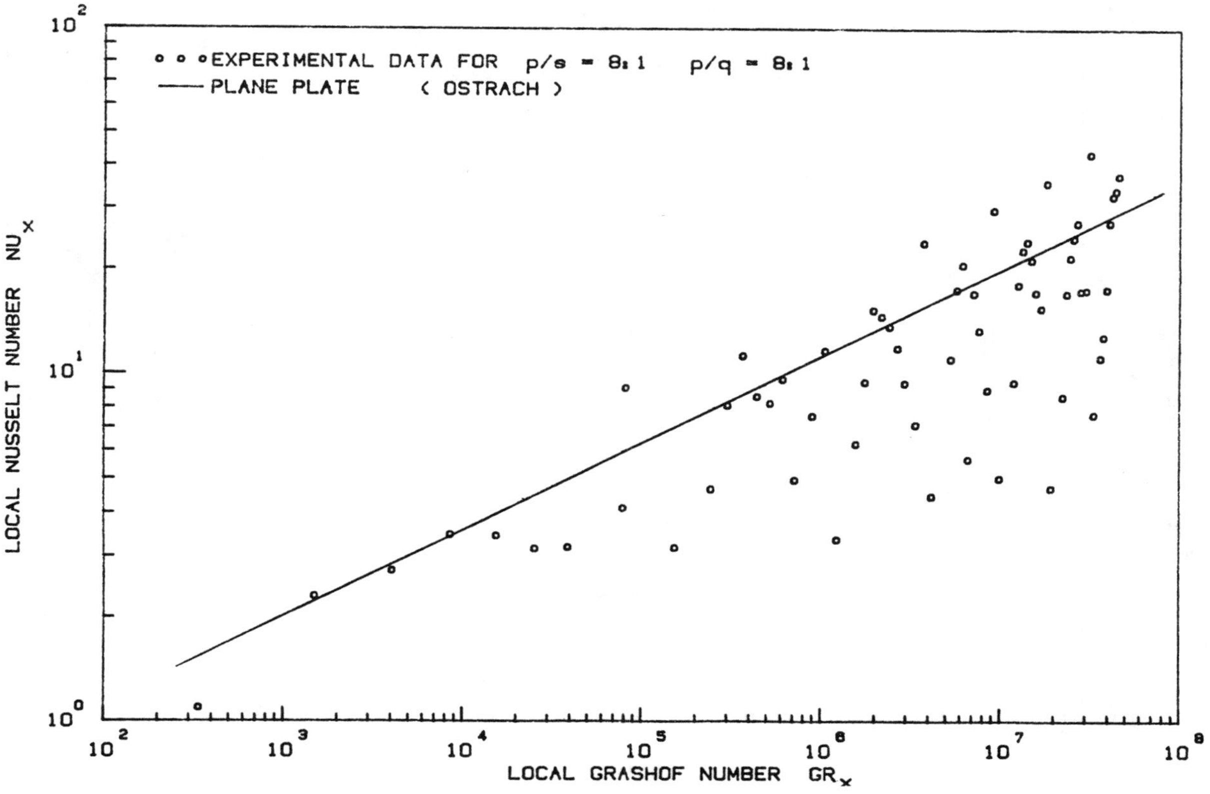

Fig.9 Local Nusselt number for ribbed plate with p/s = 8:1 and p/q = 8:1

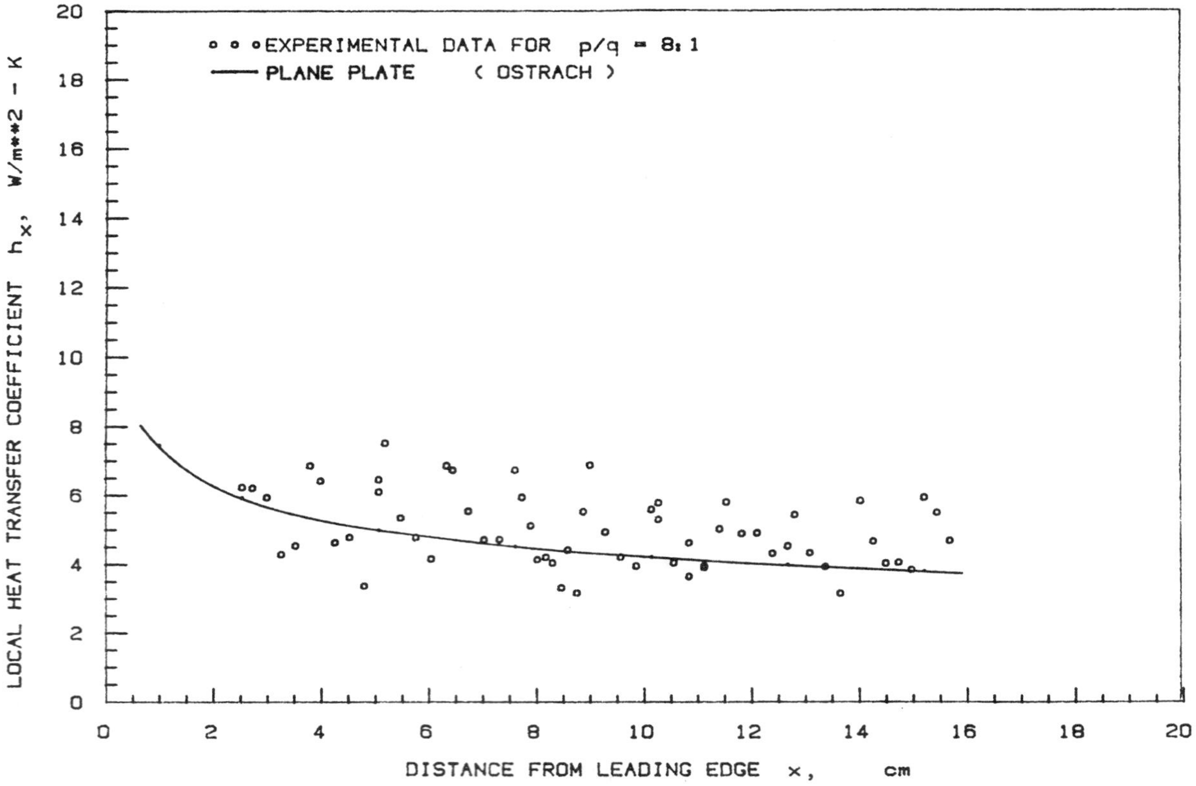

Fig.10 Local heat transfer coefficient for stepped plate with p/q = 8:1

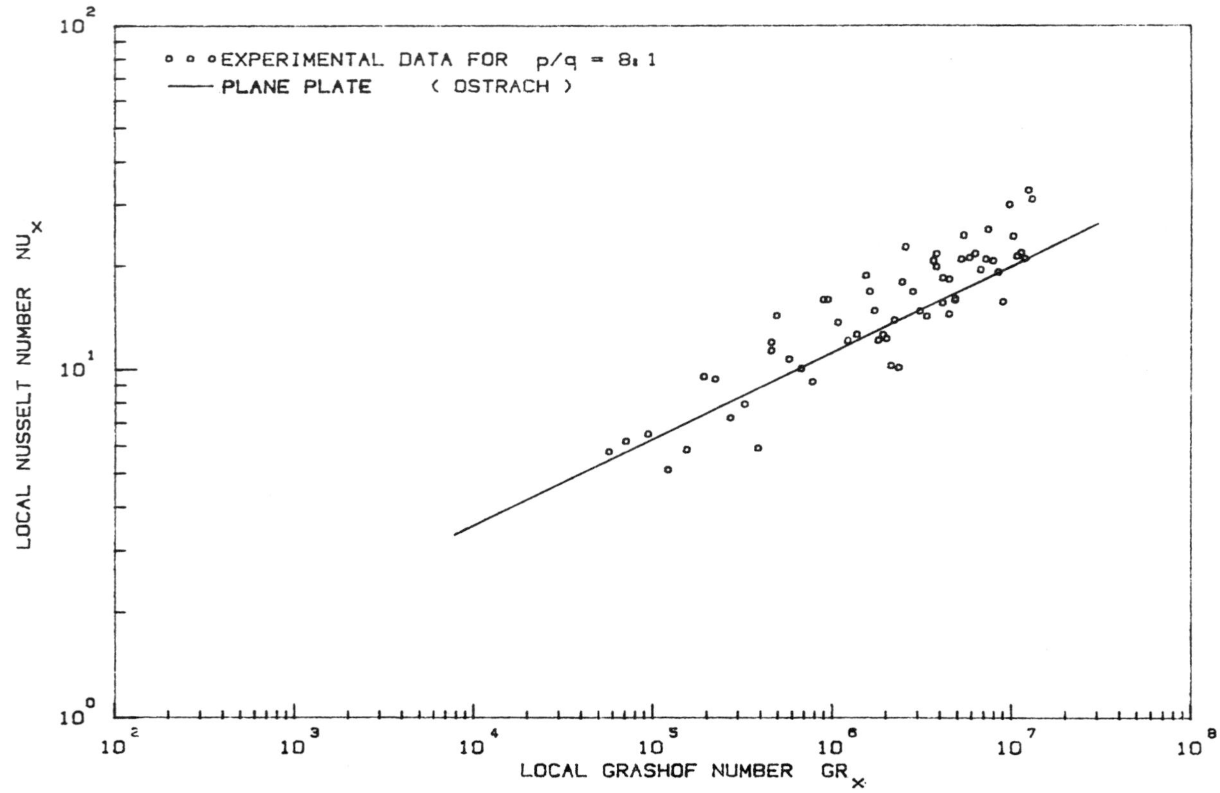

Fig.11 Local Nusselt number for stepped plate with p/q = 8:1

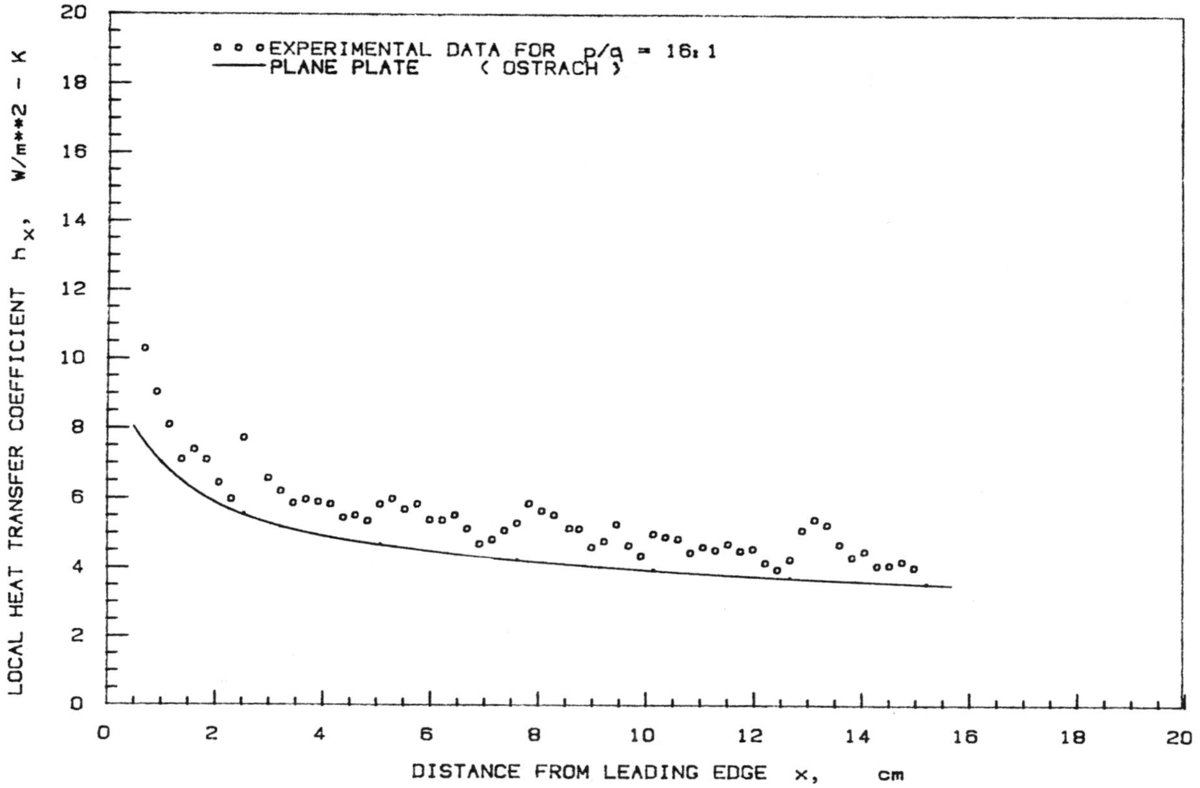

Fig.12 Local heat transfer coefficient for stepped plate with p/q = 16:1

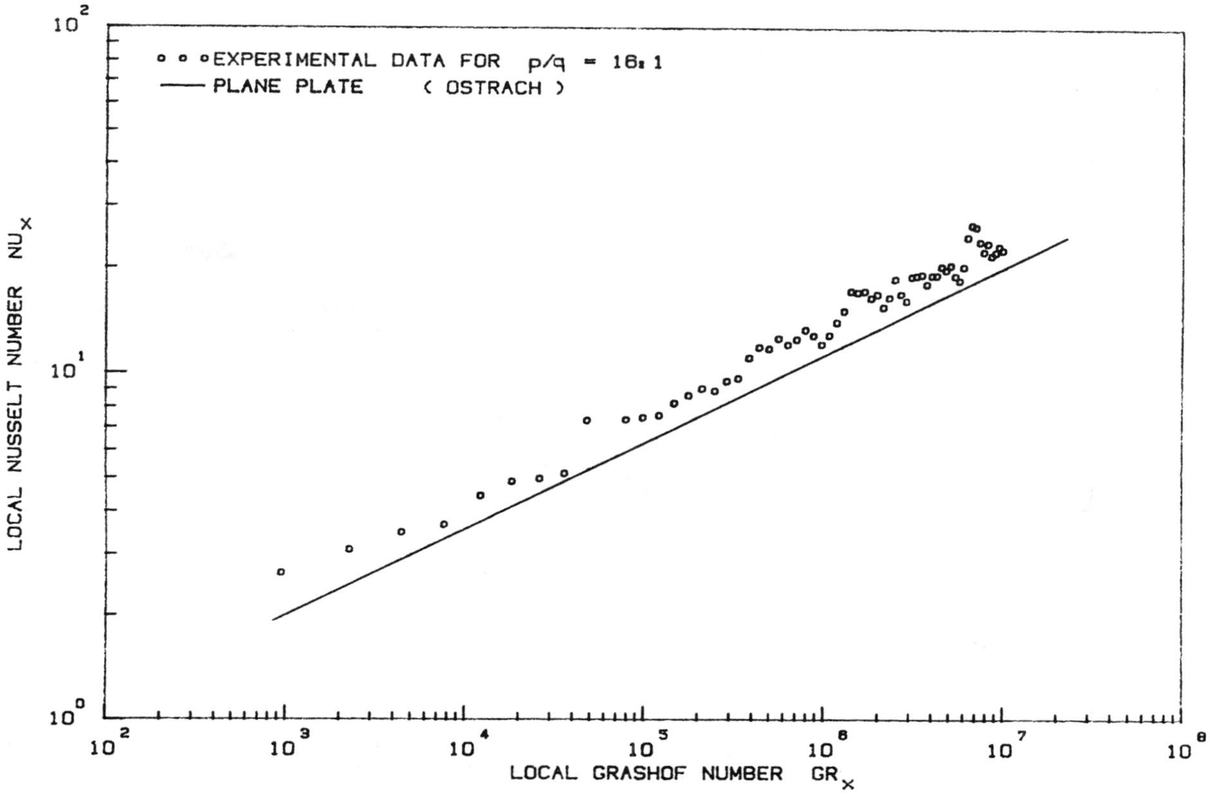

Fig.13 Local Nusselt number for stepped plate with p/q = 16:1

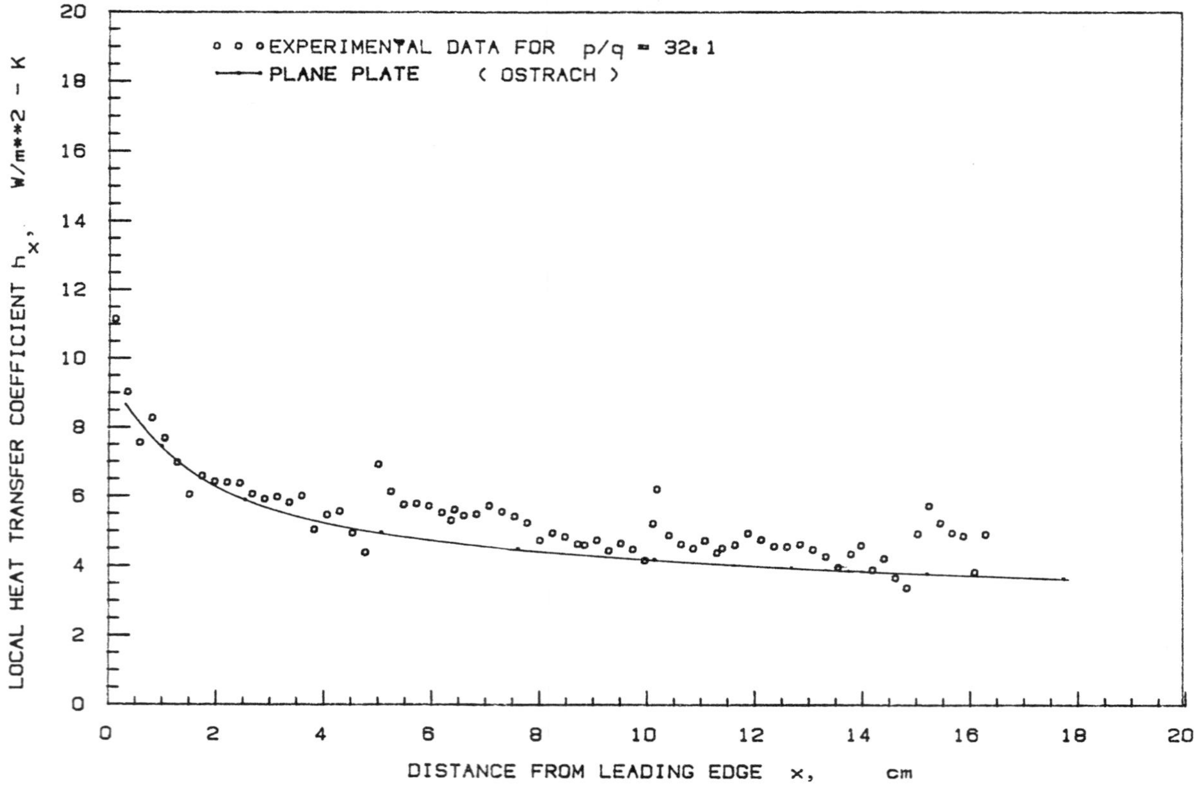

Fig.14 Local heat transfer coefficient for stepped plate with p/q = 32:1

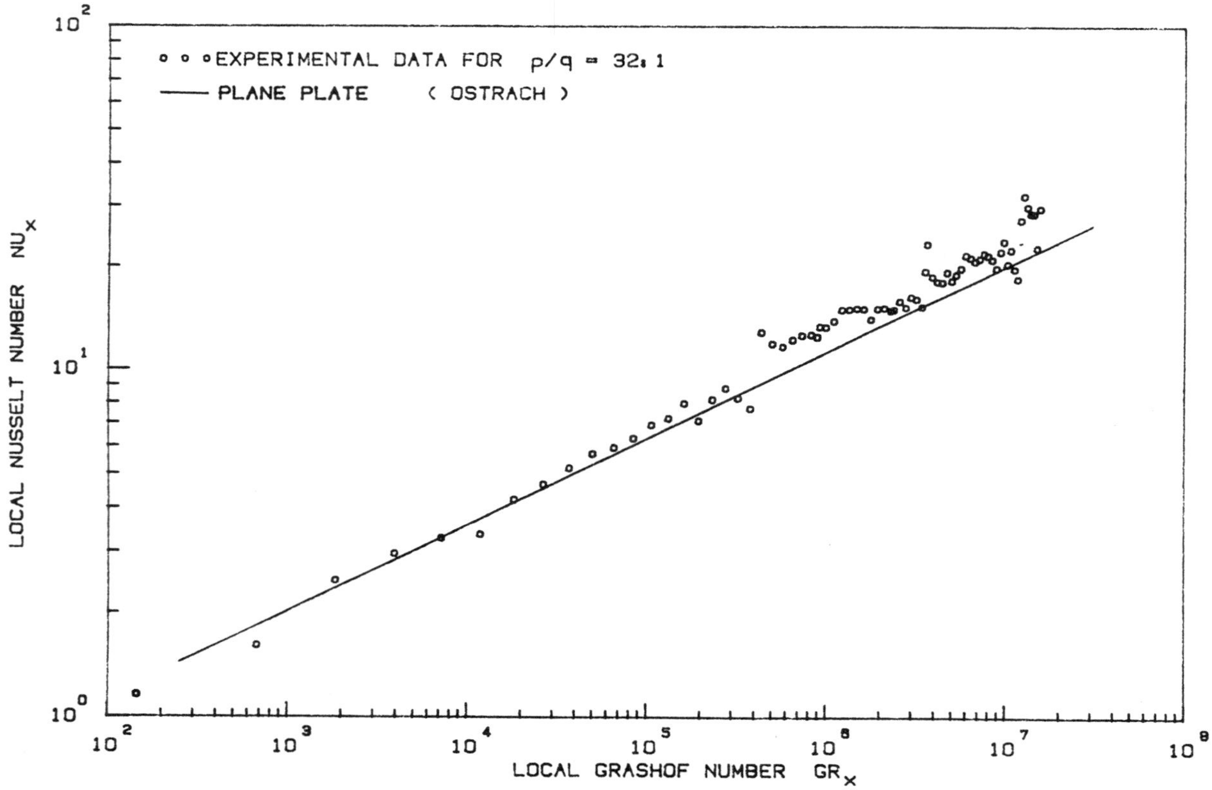

Fig.15 Local Nusselt number for stepped plate with p/q = 32:1

Thermodynamic Analysis for a MGD Power Plant

TARIT KUMAR BOSE
Indian Institute of Technology
Madras, India

A Faraday type magnetogasdynamic (MGD) generator using a potassium seeded water gas burning with 50 percent oxygen enriched air is studied with the help of an enthalpy-entropy Mollier chart for equilibrium combustion product. For this purpose equilibrium composition is evaluated for temperature in the range 1500 to 3000°K and pressure in the range 0.1 to 10 bars by considering 14 specie in the mixture, and electrical conductivity is computed with the help of rigorous kinetic theory and plotted in the Mollier chart. Limitations in the expansion process and in the power generation was discussed.

1. INTRODUCTION

A high temperature ratio is known to give a high thermodynamic efficiency in a thermodynamic cycle. Since a conventional gas turbine or steam turbine can work only at a fairly low temperature in comparison to the temperature of the combustion product, a big loss occurs due to irreversible nature of energy transfer from the higher temperature of the combustion product to the comparatively lower temperature of the turbine working medium. This loss can somewhat be prevented by topping the gas turbine cycle with the help of a seeded combustion product driven magnetogasdynamic (MGD) power generator. However, the current, generated due to the induced electric field, can not flow without any resistance, and a loss due to the irreversible nature of the Joule heating occurs, which can be studied best with the help of entropy change. For this purpose, equilibrium composition, mixture enthalpy, mixture entropy and electrical conductivity are calculated and taken into the analysis. Such an analysis including thermophysical and transport properties of a seeded combustion product may not have been analyzed before. The channel flow being considered is strictly one-dimensional[1-2], and for the combustion product a combustion of water gas burning with 50 percent oxygen enriched air is considered. Such a combustion gas is of importance in view of the already operational Indian magnetogasdynamic pilot power plant using the same or similar combustion gas [3-5].

2. ANALYSIS

One-dimensional channel flow geometry and direction of the magnetic induction for a Faraday type magnetogasdynamic (MGD) generator are shown schematically in Fig.1. In this type of generator, externally applied magnetic induction B is applied perpendicular to the flow direction and the current is drawn in a direction perpendicular to both the flow and magnetic induction directions. The electrodes are connected individually to external loads and may be shown schematically with the help of resistors R. In case the opposite electrodes are short-circuited, a current density that will flow in the z-direction would be $j = \sigma u B$ and the external electric field is $E = 0$. On the other hand, the opencircuit field (for $j = 0$), that would be induced is given by the relation $E = UB$. With an external load, the unit works as a generator, as if an electric field is induced by the load, and the current flows in the direction indicated in the figure so long $UB - E \geq 0$. Thus under a load condition the current density in the z-direction is

$$j = \sigma (UB - E) = \sigma UB (1 - K) \quad (1)$$

where

$$K = E/(UB) < 1 \quad (1a)$$

is the electro-magnetic field ratio.

Now the external power to be generated is

$$P = jE = \sigma U^2 B^2 K (1 - K) \quad (2)$$

For open-circuit case, $K = 1$ and $P = 0$, whereas with short circuited electrodes, $K = 0$ and $P = 0$. The maximum power that can be obtained for given U and B is of course, if $K = 1/2$ (optimum power).

Now the differential equations for an one-dimensional, steady, inviscid magnetogasdynamic channel flow are [1,2].

$$\rho U \frac{dU}{dx} = -\frac{dp}{dx} - \sigma UB^2 (1 - K) \quad (3a)$$

$$\rho U \frac{dh}{dx} = U\frac{dp}{dx} + \sigma U^2B^2(1 - K)^2 \quad (3b)$$

$$\rho U \frac{dh^o}{dx} = -\sigma U^2 B^2K (1 - K) = - P \quad (3c)$$

Thus extracted power is obtained by conversion of the stagnation enthalpy into the electrical energy, and rate of this conversion per unit channel length is dependent on the electrical conductivity. For a seeded combustion type of plasma discussed later in this paper, the temperature range of magnetogasdynamic operation is found, therefore, only upwards of 1900°K. However in addition to the consideration of the extracted power per unit channel length, we have to consider also the loss due to the irreversible nature of the Joule heating by considering the entropy change. From the laws of thermodynamics,

$$T\frac{ds}{dx} = \frac{dh}{dx} - \frac{1}{\rho}\frac{dp}{dx} = \frac{\delta uB}{\rho}(1 - K)^2 \quad (4)$$

From Eqs. (3c) and (4), we can write, therefore,

$$\frac{ds}{dh^{o}} = - \frac{1-K}{KT} = - \frac{1}{T} \text{ (for } K = 1/2) \tag{5}$$

Thus for the open-circuit situation, K = 1 and ds = 0, which is evident, since no current flows. On the otherhand for short-circuit, K = 0, dh° = 0 and ds/dx can be calculated from Eq. (4). This entropy change is positive. In addition for a given K, the entropy change depends on the temperature, and higher the temperature, lower is the entropy change. This positive entropy change is due to the heating of the fluid because of Joule heating.

Further to the general statement made above regarding the entropy change, two special cases are now considered, namely U remains constant along the channel and p remains constant. For the first case, that is, dU/dx = 0,

$$\frac{dp}{dx} = - \sigma UB^2 (1 - K) \tag{6a}$$

and

$$\rho U \frac{dh}{dx} = - \sigma U^2B^2K (1-K) = \rho U \frac{dh^{o}}{dx} \tag{6b}$$

and hence, both the static pressure and the static enthalpy decrease in the flow direction. On the otherhand for the case of dp/dx = 0, we get

$$\rho U \frac{dU}{dx} = - \sigma UB^2 (1 - K) \tag{7a}$$

and

$$\rho U \frac{dh}{dx} = \sigma U^2B^2 (1 - K)^2 \tag{7b}$$

and hence, while the flow velocity decreases in the flow direction, the static enthalpy increases. These two results are shown schematically in Fig.2 for an ideal non-reacting gas, in which point 2 is the state reaching for constant velocity channel and 2 for constant pressure channel. It can be seen, that in the constant pressure case the extracted power has limitation as U tends to zero. On the other hand the average operating static temperature in the constant pressure case is higher and hence, for the same total enthalpy change, the process is less irreversible than the constant velocity case. Further, while in the constant pressure case the extracted power per unit channel length becomes smaller and smaller as U tends to zero, in the constant velocity case this happens due to tending towards zero as the static temperature is decreased inspite of decreasing static pressure.

We would now examine what happens to the channel cross-section for the above two special cases. From the condition of conservation of mass and equation of state for an ideal gas, we write

$$\frac{1}{U}\frac{dU}{dx} + \frac{1}{\rho}\frac{d\rho}{dx} + \frac{1}{A}\frac{dA}{dx} = 0 \tag{8a}$$

and

$$-\frac{1}{p}\frac{dp}{dx} + \frac{1}{\rho}\frac{d\rho}{dx} + \frac{1}{T}\frac{dT}{dx} = 0 \tag{8b}$$

Thus,

$$\rho U \frac{dh}{dx} = \frac{pU}{RT} c_p \frac{dT}{dx} = \frac{\gamma}{\gamma - 1} pU\left[\frac{1}{p}\frac{dp}{dx} + \frac{1}{U}\frac{dU}{dx} + \frac{1}{A}\frac{dA}{dx}\right]$$
$$= U\frac{dp}{dx} + \sigma U^2B^2 (1 - K)^2 \tag{9}$$

For the first case, that is, dU/dx = 0, we get

$$\frac{1}{A}\frac{dA}{dx} = \frac{\gamma\sigma UB^2 (1 - K)}{p} \tag{10a}$$

Similarly for the second case, that is, dp/dx = 0, we get

$$\frac{1}{A}\frac{dA}{dx} = \frac{\sigma B^2 (1 - K)}{\rho U}[1 + (\gamma - 1) M^2] \tag{10b}$$

We find, therefore, that in both cases dA > 0, and we require a divergent magnetogasdynamic channel.

While the above analysis is for an ideal gas, an estimate is done for a real gas. For the Indian magnetogasdynamic power plant various gaseous products of coal, that is, blue water gas, Lurgi, Koppers-Totzek and producer gas, and their variations have been considered[3] and in order to attain atleast a channel temperature of 2500°K, oxygen enriched air was proposed [4, 5]. We have, therefore, considered water gas with the following composition: CO_2 7%, N_2 5.5%, CO 37.5%, H_2 49.5% and CH_4 0.5%. This water gas is mixed with 50% oxygen enriched air in stoichiometric ratio, and 1% potassium by weight is added. Starting from the tabulated values of molar enthalpy and entropy of 14 gas components (CO_2, O_2, N_2, H_2, K^+, e^-, CO, O, N, NO, H, H_2O and CH_4), equilibrium composition, mixture specific enthalpy and specific entropy of the combustion product, as well as transport properties like the viscosity coefficient, the heat conductivity coefficient and the electrical conductivity are calculated[6]. Some of these results, namely the enthalpy, the entropy, and the electrical conductivity are shown in Fig.3, on the basis of which a few interesting conclusions can be drawn as follows:

Firstly, since the isobaric lines in certain temperature range is almost parallel to the isentropic lines, an expansion due to an externally forced pressure gradient and without magnetic field can not be isentropic, since the heat that is released due to shifting equilibrium composition increases the entropy. Secondly, in the temperature range around 2500°K, a comparative small change in the enthalpy and pressure can lead to a fairly large gas flow velocity. At this temperature range the electrical conductivity is only about 10 A/Vm, which is about two orders of magnitude smaller than that for a noble gas fully-ionized plasma, and even this electrical conductivity is reduced by an order of magnitude, if the temperature is reduced to 1900°K. In addition the loss due to irreversibility becomes excessive below 1900°K also. Thus the conclusion that can be drawn from the above analysis is that for a seeded magnetogasdynamic plasma the power that can be extracted from the seeded combustion plasma is at best only marginal, although an improvement in the thermodynamic efficiency is expected.

NOMENCLATURE

A	=	Channel cross-section area
B	=	magnetic induction
h,h°	=	Specific static and stagnation enthalpy, respectively.
j	=	(electric) current density
K	=	electro-magnetic field ratio
P	=	extracted power
p,p°	=	static and stagnation pressure, respectively
R	=	gas constant
s	=	specific entropy
T,T°	=	static and stagnation temperature respectively
U	=	gas velocity
x	=	co-ordinate in the flow direction
y	=	co-ordinate in the (external) magnetic field direction
z	=	co-ordinate in the (external) electric current direction
γ	=	specific heat ratio
ρ	=	gas density
σ	=	electrical conductivity

REFERENCES

1. Rosa, R.A., Magnetogasdynamic Energy Conversion, (1968) McGraw Hill.
2. Bose, T.K., High Temperature Gasdynamics, (1979) Macmillan (India).
3. Das, A.K., Study of thermodynamic and transport properties of combustion plasma, Ph.D. Thesis, Bombay Univ., (1975)
4. Sathyamurthy, P., Venkataramani, N., and Rohatgi V.K., Analysis of combustion MHD generator using a coupled core-boundary layer model, Energy Conversion Management, (1984)
5. Anathapadmanabhan, P.V., and 14 Indian co-authors and 15 Russian co-authors, An Indo-soviet experiment on an MHD generator test section at the soviet U-02 facility, Proc. Indian Acad. of Sciences (Engineering Sciences), Part 3, 5 (1982) 169-195.
6. Bose, T.K., Thermophysical and transport properties of multicomponent gas plasma at multiple temperatures, Progress in Aerospace Sciences, to be published.

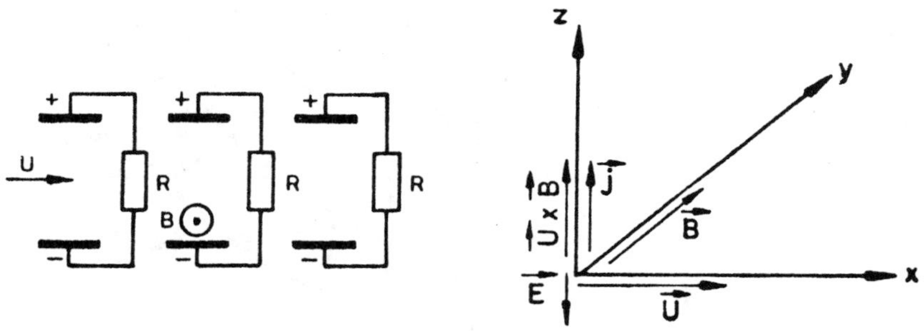

Fig.1 Schematic Sketch of a Faraday MGD Generator

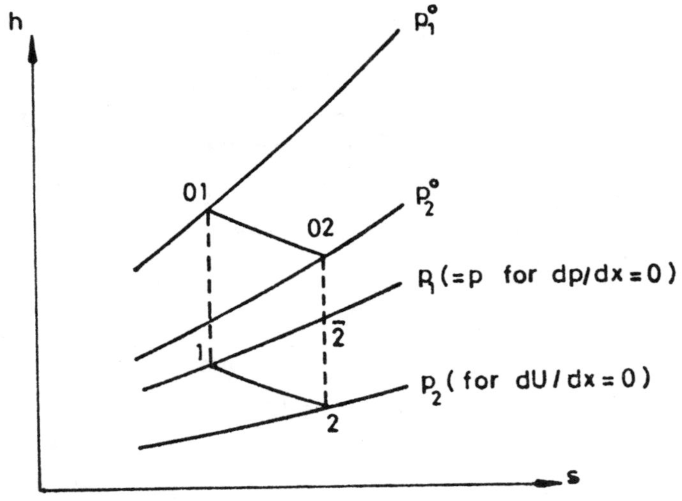

Fig.2 Schematic Sketch of a (h,s) Mollier Chart for an Ideal Gas

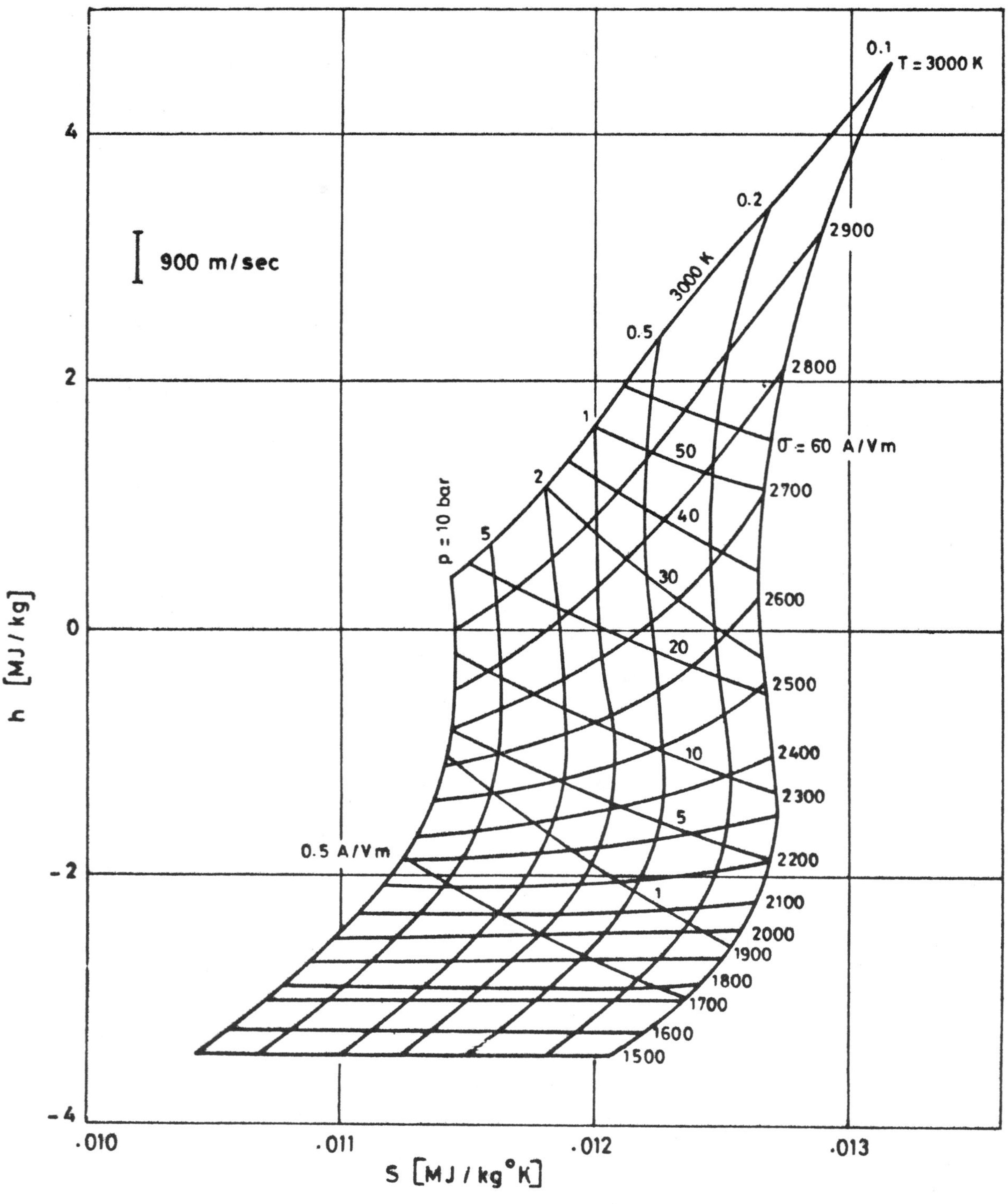

Fig.3 Mollier Chart Seeded Combustion Gas (see text)

Fully Developed Turbulent Pressure Drops in a Triangular Duct for a Power Law Fluid

T. D. CHANG and T. F. IRVINE, Jr.
Mechanical Engineering Department
State University of New York at Stony Brook
New York 11794, USA

A flow model was developed and applied to the prediction of fully developed turbulent pressure drops in triangular ducts for Newtonian and power law fluids. The numerical calculations for the predicted turbulent velocity profiles and friction factors were carried out by utilizing a modified integral transformation and using the known laminar velocity solutions for power law fluids.

Friction factors for Newtonian and power law fluids were measured over a Reynolds number range of 500 to 40,000 in an equilaterial triangular duct. Solutions of carbopol-934 in concentrations of 2500, 5000 and 5600 wppm were prepared for the power law studies.

Comparisons were made between the theoretical and experimental results. The model was successful in pedicting friction factor-Reynolds number relations for power law fluids in the index range $0.844 < n \leq 1.0$.

1. INTRODUCTION

Many non-Newtonian fluids are characterized by large "apparent viscosities" and thus lead to laminar flow situations more often than Newtonian fluids. For this reason, a large number of theoretical and experimental investigations have been carried out on the laminar flow characteristics of purely viscous rheological fluids.

There are many occations however because of large pipe sizes or flow velocities when non-Newtonian turbulent flows occur. It is therefore a fundamental engineering problem to be able to predict the turbulent pressure drop for a non-Newtonian fluid flowing in a duct or passage.

Quite naturally, the emphasis of investigations up to the present time has been on flow in circular ducts because of their wide availability and use. However, there are many situations, especially in the design of compact heat exchangers, when non-circular passages are more appropriate.

With only several exceptions, there have been no investigations, either theoretical or experimental, on the turbulent flow pressure drop for non-Newtonian fluids in non-circular ducts. To the author's knowledge, only a recent paper by Kostic and Hartnett [1] considers this subject in the light of experimental data. They proposed a method of using a geometrical correction factor applied to a circular tube friction factor-Reynolds number relation to account for the effects of noncircularity. More recently, Irvine [2], has suggested a modification of the Kostic-Hartnett technique, by utilizing a simplified circular tube relation but in either case the methods are empirical.

The purpose of the present investigation was to carry out a fundamental theoretical and experimental study on the turbulent flow of a power law fluid in an isosceles triangular duct. A fluid model has been constructed based on a modified mixing length concept which can predict velocity profiles in triangular ducts and thus through integration obtain friction-Reynolds number relations.

Pressure drop experiments have been carried out to test the validity of the flow model and to determine its range of applicability.

2. ANALYSIS

Following the "turbulent viscosity" concept introduced by Boussinesq and considering that the fluid shear stress is a linear function of the distance from the wall for fully developed flow, the equation of motion for a Newtonian fluid can be written in dimensionless form as (see nomenclature):

$$\frac{du^+}{dy^+} = \frac{1 - y^+/R^+}{1 + \varepsilon^+} \tag{1}$$

With his analogy to kinetic theory, Prandtl related the dimensionless eddy viscosity, ε^+, to the velocity field through the relation

$$\varepsilon^+ = \ell^+ \frac{du^+}{dy^+} \tag{2}$$

Where ℓ^+ is the dimensionless mixing length. He then assumed that ℓ^+ was a simple linear relation of y^+ and could then solve Eq.(1) for the velocity field. Subsequent measurements by Nikuradse indicated that a more realistic relationship between ℓ^+ and y^+ could be written as

$$\ell^+ = R^+[0.14 - 0.008\,(1-y^+/R^+)^2 - 0.06(1-y^+/R^+)^y]\,[1 - \exp(-y^+/C_i^+)] \tag{3}$$

Where C_i^+ varies with Reynolds number as shown in Table 1

C_i^+	38.0	38.0	26.6	26	26
Re	$3x10^3$	$5x10^3$	10^4	$8x10^4$	10^5

Table 1

Thus, Eqs. (1), (2) and (3) can now be solved for the dimensionless velocity field and through integration, to find the average velocity, the friction factor-Reynolds number relation can be obtained.

This solution is relatively straight forward in the case of a circular pipe with its axial symmetry. For a non-circular duct, however, the question arises of how to utilize Eq. (3) to relate the mixing length to the distance from the wall. Diessler and Taylor [3] assumed that Eq. (3) or its equivalent could be applied in non-circular shapes along lines from the wall to the geometric center of the duct that followed paths always being perpendicular to the isovels. This scheme however leads to a time consuming trial and error solution and has not been often used. In a later paper, Krajewski [4] proposed an integral transformation of the form

$$w^+ = \int_0^{u^+} (1 + \varepsilon^+)\, dy^+ \tag{3a}$$

where ε^+ is an explicit function of y^+ only, i.e.

$$\frac{dw^+}{dy^+} = (1 + \varepsilon^+) \frac{du^+}{dy^+} \tag{4}$$

Introducing Eq. (4) into Eq. (1) yields the equation

$$\frac{dw^+}{dy^+} = 1 - y^+/R^+ \tag{5}$$

which is similar in form to the laminar equation of motion cosidering w^+ as the dependent velocity.

If Eq. (5) is solved for the w^+ field, then using the $\varepsilon^+ = \varepsilon^+(u^+)$ relation for a circular tube, the u^+ field can be recovered without explicity identifying the y^+ path. Figure 1 shows the relation between u^+, ε^+ and w^+ for a circular tube which allows this transformation. More detailed discussion of this transformation scheme are in the Ph.D. thesis of Chang [5].

So far, only the situation with respect to Newtonian fluids has been considered. For power law fluids, several complications arise. First, the equation of motion becomes non-linear due to the nature of the shear stress-velocity field relation and only numerical solutions are practical. Second, the equation of motion becomes more complicated because of the cross coupling that appears between the shear terms in the different coordinate directions. For example, the laminar equation of motion written for fully developed duct flow in dimensionless form is [6]

$$\frac{\partial}{\partial y^+} \left\{ \left| \left(\frac{\partial u^+}{\partial y^+}\right)^2 + \left(\frac{\partial u^+}{\partial z^+}\right)^2 \right|^{\frac{n-1}{2}} \frac{\partial u^+}{\partial y^+} \right\} + \frac{\partial}{\partial z^+} \left\{ \left| \left(\frac{\partial u^+}{\partial z^+}\right)^2 + \left(\frac{\partial u^+}{\partial y^+}\right)^2 \right|^{\frac{n-1}{2}} \frac{\partial u^+}{\partial z^+} \right\} = - \frac{1}{D_n^+} \tag{6}$$

where it can be seen that the cross-coupling terms drop out when n = 1.

In spite of these difficulties, it is possible to construct a reasonable model for fully developed turbulent flow of a power law fluid leading to equations similar to Eqs. (1), (2) and (3). The following assumptions are made.

1. The shear law for turbulent flow is taken as

$$\frac{\tau}{\rho} = (\nu + \varepsilon_n) \left(\frac{du}{dy}\right)^n \tag{7}$$

2. In analogy to Pra'ndtl's reasoning, the dimensionless mixing length is defined through the equation

$$\varepsilon_n^+ = (\ell^+)^2 \left(\frac{du^+}{dy_n^+}\right)^{2-n} \tag{8}$$

3. The dimensionless eddy viscosity, ε_n^+, is taken to be the same for Newtonian and power law fluids except that it is now defined as $\varepsilon_n^+ = \rho\varepsilon_n/K$, where K is the fluid consistency.

4. Although the dimensionless velocity is defined the same way for both fluids, the coordinate distance, y_n^+ requires a new definition from dimensional considerations such that

$$y_n^+ = y(u*)^{\frac{2-n}{n}} / \nu_k^{1/n}$$

5. The Newtonian integral transformation must be modified from Eq. (3a) to

$$w_n^+ = \int_0^{u^+} (1 + \varepsilon_n^+)^{1/n}\, du^+ \tag{9}$$

With this model, the equation of motion now becomes:

$$\left(\frac{du^+}{dy_n^+}\right)^{2-n} = \frac{1 - y_n^+/R_n^+}{(1 + \varepsilon_n^+)} \tag{10}$$

If Eq. (10) is "closed" by using Eqs. (8) and (3) with y^+ and R^+ replaced by y_n^+ and R_n^+ respectively, solutions can be obtained both for Newtonian fluids (n=1) and power law fluids ($n \neq 1$). In order to use the integral transform technique, the power law laminar flow solutions of Cheng [7] were utilized. The results will be discussed when the experimental data are presented.

3. EXPERIMENT

A closed flow loop was constructed as illustrated in fig. 2. The details of construction and the measurement techniques are described in detail in Ref. 5. Of particular importance was that six pressure taps were installed along the duct so that it was possible to determine that the flow was fully developed in each experimental run. Also, in the cases of the runs with carbopol, the rheological properties, n and K were measured with a capillary tube viscometer before and after each experimental run to insure that they were not affected by fluid degradation. Figure 3 shows an example of typical flow curves taken in this manner.

Since the rheological properties of a power law fluid can be a function of the shear rate (which is related to the Reynolds number) it is important that the properties be measured in the shear rate range which will be encountered in the actual duct flows. In the present experiments, this was done as nearly as possible. Table 2 shows a comparison of the shear rate ranges of the capillary tube viscometer and the duct flows. The table indicates that there was not a complete overlap, especially at the higher regions of the shear rate ranges. However, other measurements indicate that carbopol has constant rheological properties over a considerably larger shear rate range than measured with the present capillary tube viscometer.

Solution	Concent. (ppm)	$\dot{\gamma}$ Range of Capillary Tubes (s^{-1})	$\dot{\gamma}$ Range of Duct Flow (s^{-1})
Carbopol - 934	2500	1951.9-8939.8	317.1-49771.9
"	5000	3305.9-7764.3	1703.4-21072.9
"	5600	2203.5-9557.2	1087.5-38402.8

Table 2

4. RESULTS AND DISCUSSION

The first calculations using the model described in the analysis section were for the purpose of comparing the calculation with other known results either analytical or experimental. Figure 4 shows a comparison of the calculated Newtonian velocity profile in a circular tube with the experimental velocity measurements of Laufer [9] . Also shown in the figure is the simplified two zone velocity profile and the distribution of the dimensionless eddy viscosity, ε^+ . Good agreement is seen with the experimental data and the calculations of the present model.

Next, a series of calculations was made for a power law fluid flowing in a circular pipe. These are shown in fig. 5 in comparison with the semi-empirical relations of Dodge and Metzner [10]. It is apparent that there is some discrepancy between the two results that increases as the flow index, n, decreases. On the other hand, there are only a limited amount of experimental data at low values of n and the differences do not appear to be excessive.

Figure 6 shows a comparison of the calculations for a Newtonian fluid in an equilateral triangular duct along with the experimental correlation by Leonhardt and Irvine [11], the measurements of Aly et al. [12], calculations by Diessler and Taylor [3] and the experimental data from the present study. The calculations agree well with both the present experiments, the experimental correlation of Leonhardt and Irvine and the data of Aly et al. The calculations by Diessler and Taylor are about 15% lower.

A series of experimental runs was made on the triangular duct with carbopol solutions having n values of 0.930 and 0.844. Since the lower the n value, the greater the test of the fluid model, only the results for n = 0.844 will be presented and discussed here. the results for the other runs are essentially the same.

Figure 7 shows a comparison of the measured and calculated factors against the generalized Reynolds number, Re_g. The solid line represents the present calculations. The agreement is seen to be reasonable although the predictions are somewhat high at the larger Reynolds numbers. In the laminar range, there is good agreement between the data and the theoretical value given in the figure. It should also be noted that when using the generalized Reynolds number that transition occurs at approximately $Re_g = 2000$.

The dashed line in Fig. 7 shows the results of the approximate calculation as proposed in Ref. [2] which is a modification of the Kostic and Hartnett procedure [1]. The agreement in this case with the experimental data is essentially as good as the more lengthy numerical calculation.

Finally, to look at one more possible correlation technique, Fig. 8 shows both the water and n = 0.844 data ploted against a Reynolds number, Re_g^+, which is defined by

$$Re_g^+ = \frac{Re_g \, 8^{1-n}}{(a+b)\left(\frac{a+bn}{n}\right)^n} \tag{11}$$

Where a and b are geometric constants after Kozicki et al. [13]. This Reynolds number was used by Chang and Irvine [14] to correlate the pressure drop of a power law fluid in laminar flow through a triangular duct. It is related to the Reynolds number used by Kostic and Hartnett and it also correlates the data well using the numerical solution. Figure 8 can thus be used as a summary of the numerical solution for an equilateral triangular duct for $0.5 \leq n \leq 1$ in a Reynolds number range $2x10^3 \leq Re_g^+ \leq 10^5$.

Clearly, it is desireable that additional experimental data be obtained.

5. SUMMARY AND CONCLUSIONS

Using a power law modified mixing length flow model, calculations were made of the friction factor-Reynolds number relation for fully developed turbulent flow in an equilateral triangular duct. The predictions were in reasonable agreement with the experimental data obtained in the flow index $0.844 \leq n \leq 1.0$. An approximate correlation technique which is simple to use was also in agreement with the experimental data.

NOMENCLATURE

C_i^+	constant in Eq. (3)
D_h, D_h^+	dimension and dimensionless hydraulic diameter $D_h^+ = D_h u^{*\frac{2-n}{n}} / \nu_k^{1/n}$
f	Darcy friction factor, defined by $f = \frac{4\tau_w}{\rho \bar{u}^2/2}$
K	fluid consistency
ℓ, ℓ^+	dimension and dimensionless mixing length, $\ell^+ = \ell u^{*\frac{2-n}{n}} / \nu_k^{1/n}$
n	flow index
p	pressure
R, R^+	dimension and dimensionless radius of a circular pipe $R^+ = R u^{*\frac{2-n}{n}} / \nu_k^{1/n}$
Re	Reynolds number
Re_g	generalized Reynolds number $\frac{\rho \bar{u}^{2-n} D_h^n}{K}$
Re_g^+	generalized Reynolds number,defined by Eq.(11)
u	time average velocity in flow direction
u*	friction velocity, $\sqrt{\tau_w/\rho}$
u^+	dimensionless velocity, u/u*
W	integral transformation, defined by Eq. (3a)
$\bar{W}$	average value of W
W^+	$W/\bar{W}$
y	distance from the wall; y-axis in coordinate system
y^+	dimensionsless distance, $y u^{*\frac{2-n}{n}} / \nu_k^{1/n}$
z^+, z^{++}	same as y^+ and y^{++}, in z direction
$\dot{\gamma}$	shear rate
τ	effective shear stress
τ_w	wall shear stress
ρ	desity
μ	dynamic viscosity
ν	kinematic viscosity
ν_k	K/ρ
ε	eddy viscosity - Newtonian fluid
ε_n	eddy viscosity-Power law fluid
ε_n^+	dimensionless eddy viscosity, ε / ν_k

REFERENCES

1. Kostic,M., and Hartnett,J.P., Predicting Turbulent Friction Factors of Non-Newtonian Fluids in Non-Circular Ducts, Int. Comm. Heat and Mass Transfer, 11 (1984).
2. Irvine,T.F., Jr., A Generalized Blasius Equation for Power Law Fluids, submitted for publication, (1985).
3. Deissler,R.G., and Taylor,M.F., Analysis of Turbulent Flow and Heat Transfer in Noncircular Passages, N.A.C.A. TN 4384, (1958).
4. Krajewski,B., Determination of Turbulent Velocity Field in a Rectilinear Duct with Non-Circular Cross Section, Int. J. Heat and Mass Transfer. 13, (1970).
5. Chang,T.D., The Prediction of Fully Developed pressure Drops in a Triangular Duct for a Power Law Fluid, Ph.D. thesis (Mech. Engr. Dept) State University of New York, Stony Brook, New York, (1985).
6. Fredrickson,A.G., Principles and Applications of Rheology, Prentice-Hall, New Jersey, USA, (1964).
7. Cheng,J.A., Laminar Forced Covective Heat Transfer of Power Law Fluids in Isosceles Triangular Ducts with Peripheral Wall Conduction, Ph.D thesis (Mech. Engr. Dept.) State University of New York, Stony Brook, N.Y., (1985).
8. Park,N.A., Measurement at the Rheological Properties of Non-Newtonian Fluids with the Falling Needle Viscometer, Ph.D. Thesis (Mech. Engr. Dept.) State University of New York, Stony Brook, N.Y., (1984).
9. Laufer,J., The Structure of turbulence in Fully Developed Pipe Flow, N.A.C.A. TN 2954,(1953).
10. Dodge,D.W., and Metzner, A.B., Turbulent Flow in Non-Newtonian Systems,A.I.Ch.E.J., 5,(1959).
11. Leonhardt,W.J., and Irvine,T.F., Jr., Experimental Friction Factors for Fully Developed Flow in Dilute Aqueous Polyethylene-Oxide Solutions in smooth Wall Triangular Ducts, Heat and Mass Transfer Sourcebook, Scripta Publishing Co., Washington, D.C.,(1977).
12. Aly,A.M., Trupp,A.C., and Gerrard,A.D., Measurement, and Prediction of Fully Developed Turbulent Flow in an Equilateral Triangular Duct, J.Fluid Mechs., 85 (1978).
13. Kozicki,W., Chou,C.H., and Tiu,C., Non-Newtonian Flow in Ducts of Arbitrary Cross-Sectional Shape, Chem. Eng. Sc., 21 (1966).
14. Chang,T.D., and Irvine,T.F., Jr. , Fully Developed Laminar Pressure Drop in a Triangular Duct for a Power Law Fluid, proc. VII All-Union Heat and Mass Transfer Conferenc, Minsk, USSR, (1984).

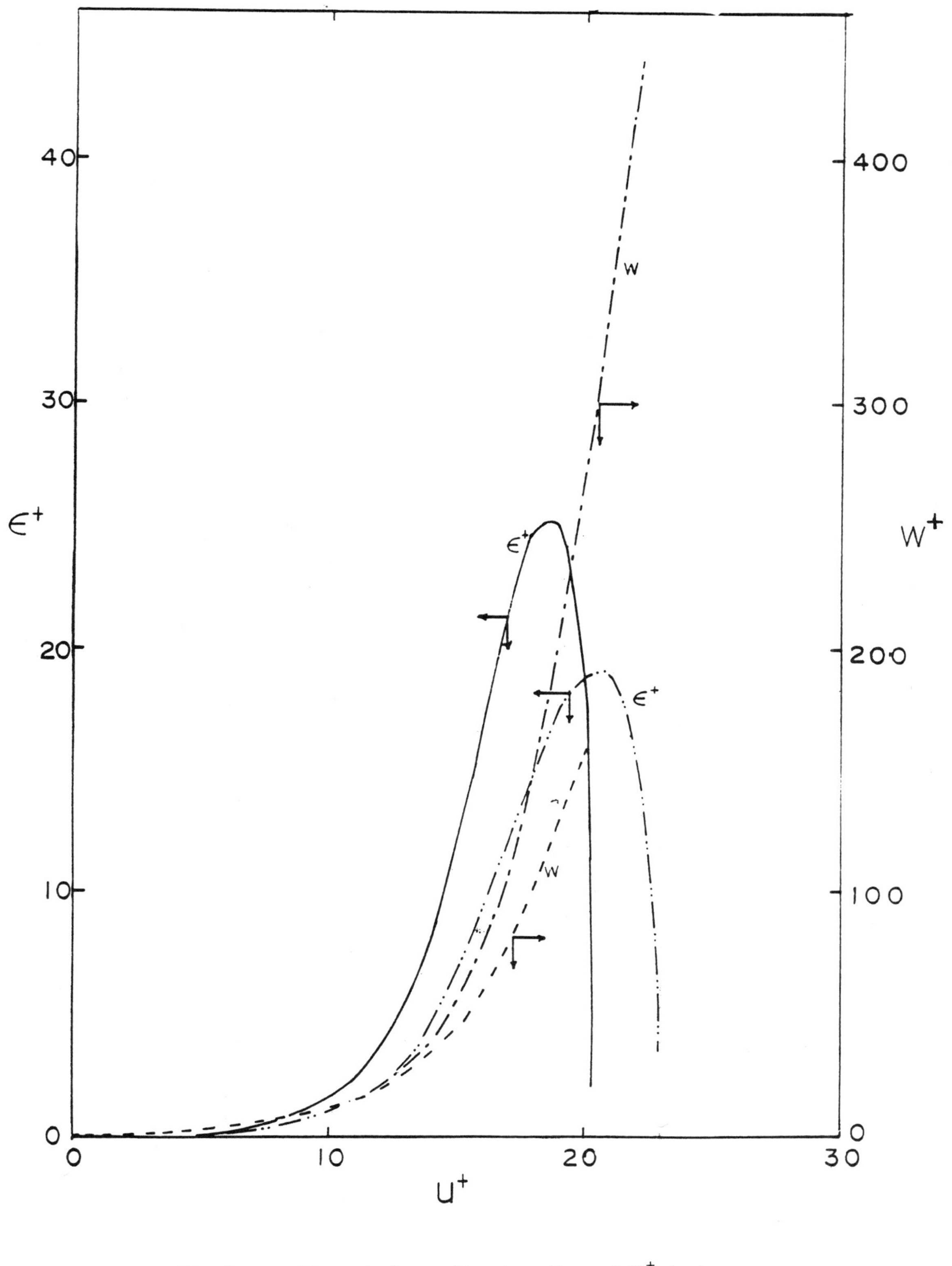

Fig. 1 - Dimensionless eddy viscosity and W^+ realtions with u^+ in a circular pipe flow.

ε^+ { ———— : n = 1, Re = 10^4
- -. . - - : n = 0.7, Re_g = 10^4

W { -------- : n = 1, Re = 10^4
-- .. -- : n = 0.7, Re_g = 10^4

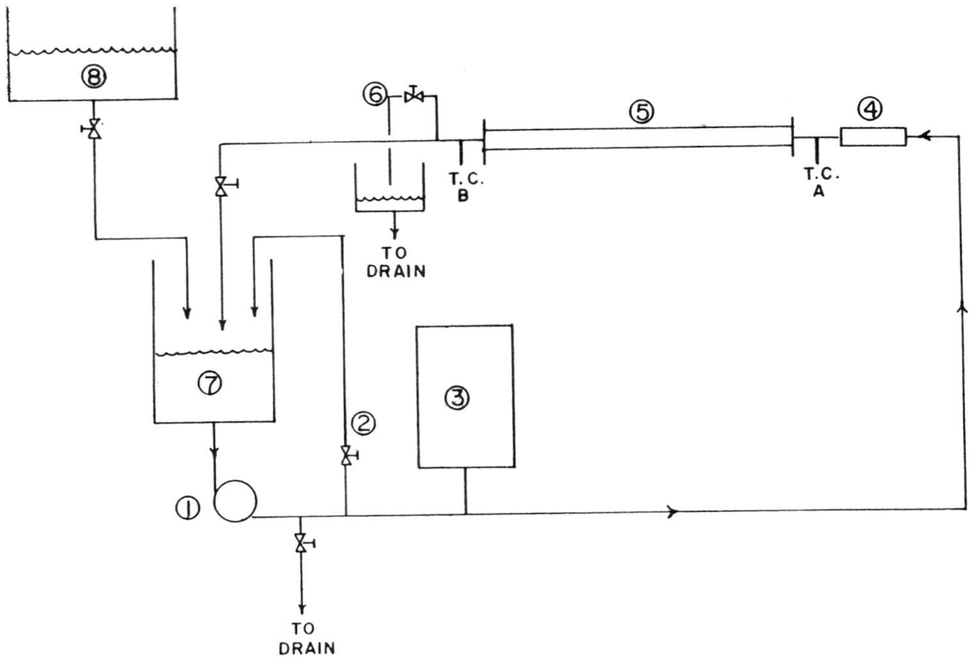

Fig. 2 - Schematic diagram of the experimental system

1. pump
2. by-pass valve
3. surge tank
4. flow straightener
5. test section
6. drain valve
7. reservoir
8. open tank

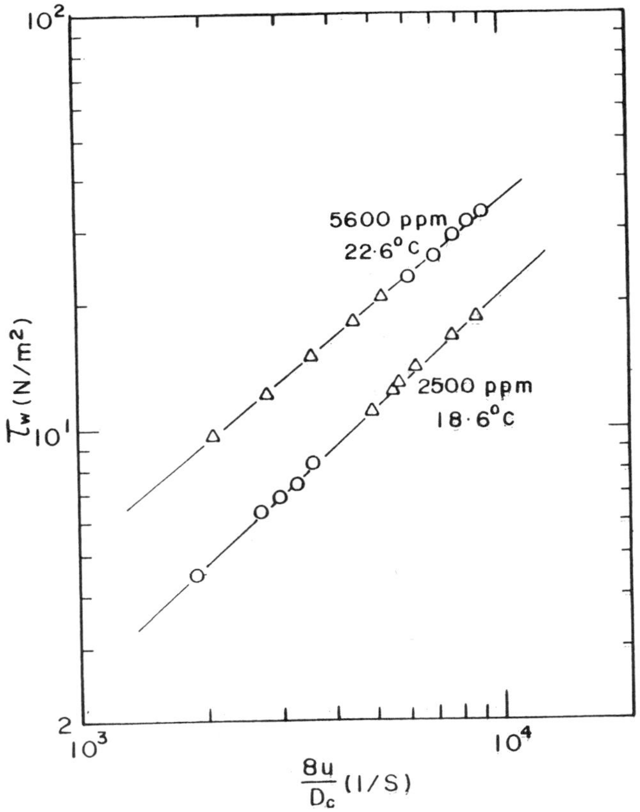

Fig. 3 Flow diagram of Carbopol solutions,

O : before the experiment

Δ : after the experiment

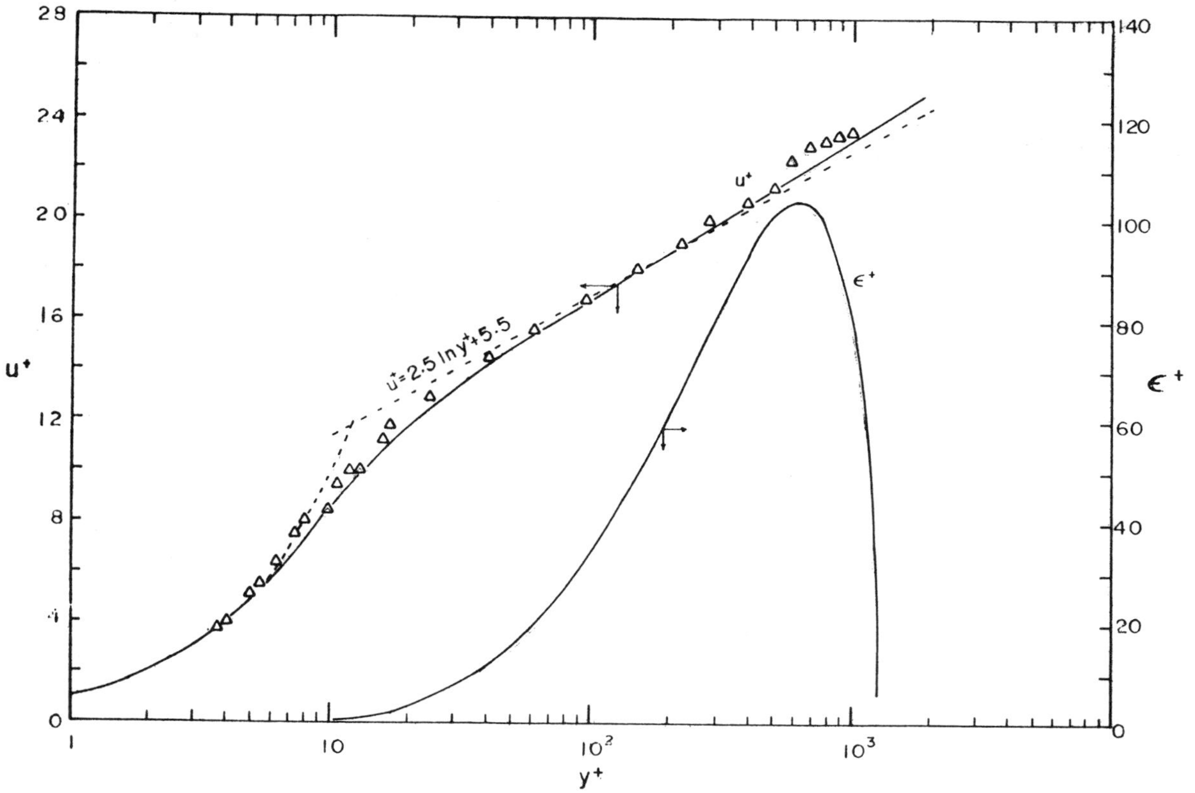

Fig. 4 - Velocity profile and dimensionless eddy viscosity distribution in a circular pipe flow for water at Re = 50,000.

________ : predicted velocity profile

Δ : Laufer's experimental data, replotted from ref. [9].

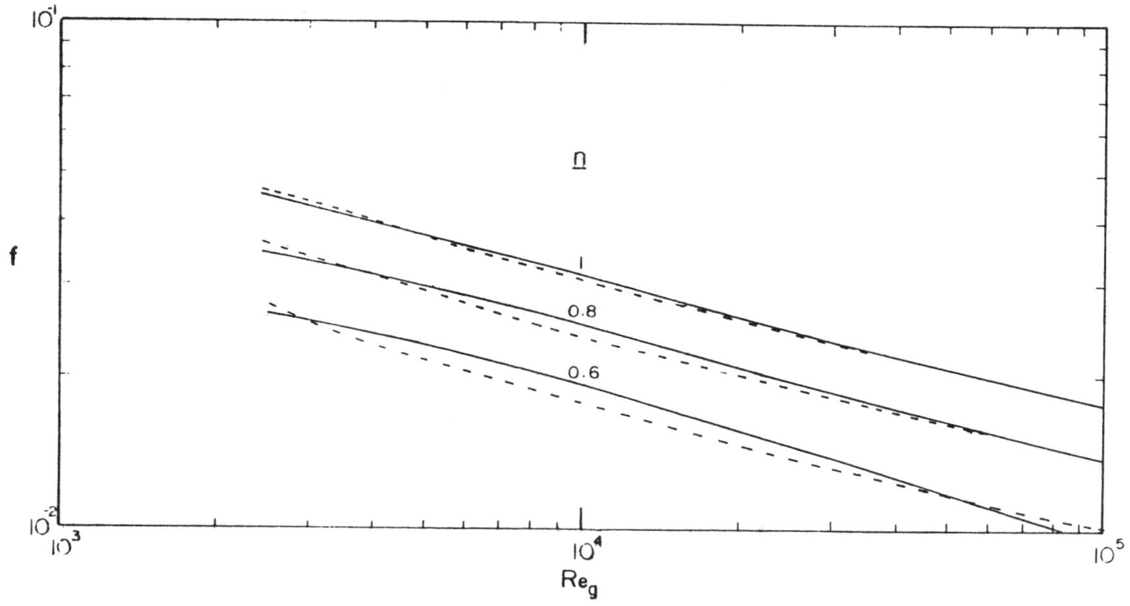

Fig. 5 - Friction factors in a circular pipe when n = 1, 0.8 and 0.6.

________ : predicted relations

-------- : Dodge and Metzner's relations [10]

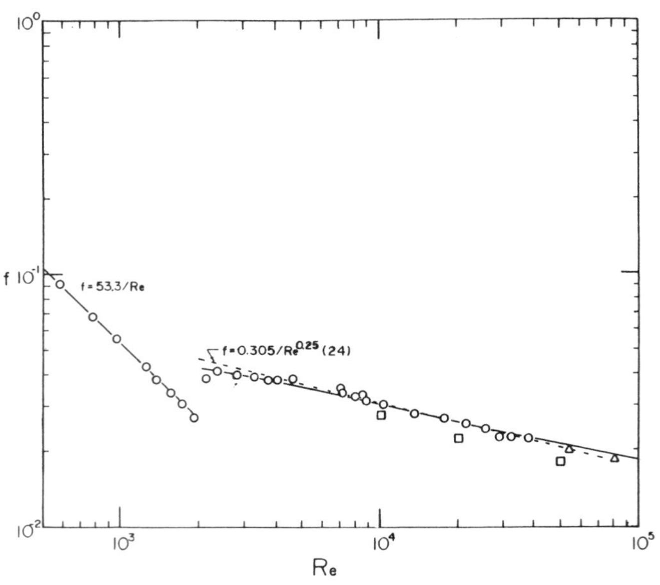

Fig. 6 - Friction factors in 60° duct for water, f vs. Re

______ : predicted relation

-------- : Leonhardt and Irvine [11]

O : present measurements

Δ : Aly et al. [12] experimental data

(: Deissler and Taylor [3] calculated points

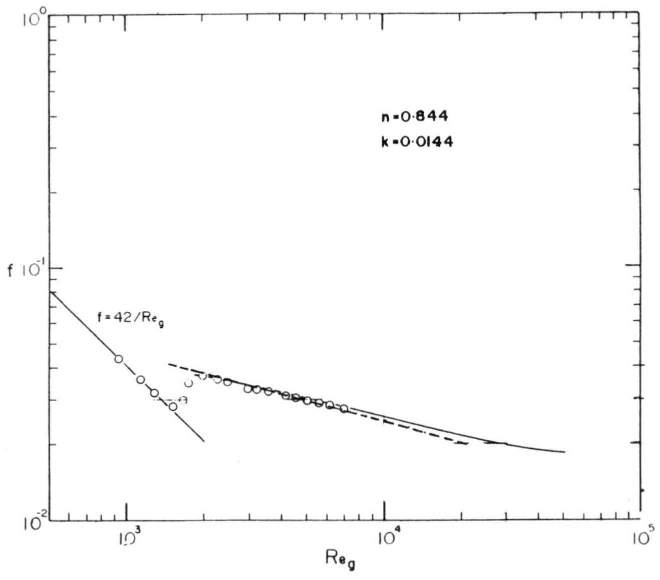

Fig. 7 - Friction factors in 60° duct for 5600 ppm Carbopol-034 solution, f vs. Re_g

______ : predicted relation

O : present measurements

--------- : method of Irvine [2]

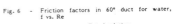

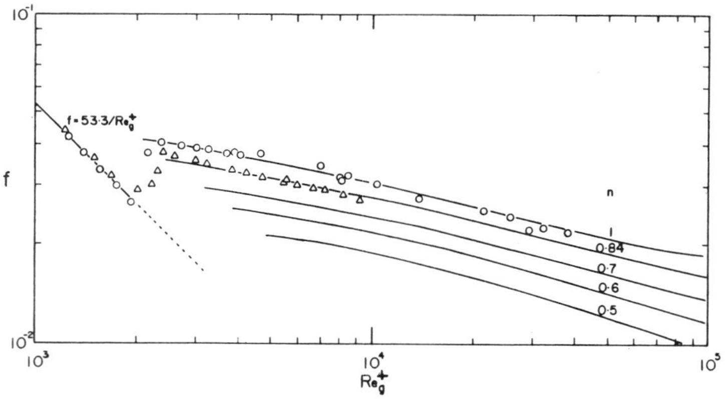

Fig. 8 - Friction factors in 60° duct when $0.5 \leq n < 1$, f vs. Re_g^+

______ : present predictions

O : present measurements, n = 1

Δ : present measurements, n = 0.844

Analysis of the Thermal Performance of Boiling Collectors

JOHN A. CLARK
Professor of Mechanical Engineering, Solar Energy Laboratory
Department of Mechanical Engineering and Applied Mechanics
University of Michigan
Ann Arbor, Michigan 48109–2125 USA

The analysis of the thermal performance of a boiling collector is presented. Comparison is made with the performance of non-boiling collectors. A generalized heat removal factor is developed and it is demostrated that the conventional heat removal factor for non-boiling collectors is a limiting case of a more generalized result. The influence of inlet sub-cooling is evaluated and shown to be consistent with recent experimental observations.

1. INTRODUCTION

The analysis of the thermal performance of flat-plate solar collectors containing a single-phase fluid, developed by Hottel and Woertz [1] and Hottel and Whillier [2], is well established for forced flow systems. Stuides of single-phase collector systems operating in a thermosiphon mode have been reported by Shitzer, et al [3. 4], Close [5], Gupta and Garg [6] and Zvirin, et al [7], among others. Morrison and Braun [8] have recently summarized both the analytic modeling and the thermal performance of these systems.

The solar collector containing a boiling fluid operating in a thermosiphon mode has only recently been given attention. Soin, et al [9] investigated the thermal performance of a thermosiphon collector containing boiling acetone and petroleum ether and correlated their experimental results with a modified form of the Hottel-Whillier-Bliss equation (HWB) to account for the fraction of liquid level in the collector. Downing and Waldin [10] stuided the boiling heat transfer process in solar water heating using R-11 and R-114. The thermal analysis of a boiling collector has been presented by Al-Tamimi and Clark [11] and the thermal performance of this collector is given by the same authors in [12] The present paper extends the discussion of the thermal analysis of boiling collectors and provides additional insights into its performance characteristics.

The boiling collector possesses several important and desirable advantages over non-boiling collectors. For equivalent operating parameters the conversion efficiency of a boiling collector is inherently greater than that of a non-boiling collector. This is a consequence of its lower thermal losses as, under boiling condition, the mean absorber plate temperature is lower than that of a non-boiling collector. Further, in a boiling mode a thermosiphon condition is very effective in producing flow because of the inherently large density differences generated between the boiling fluid in the collector flow channels and the liquid return line. In many installations this eliminates the need for a collector circulating pump, thus simplifying the system. With the use of a refrigerant in the collector freeze protection is achieved with certainity which avoids the need for draining values or flow circulation provisions and their associated controls and further simplified the system. Boiling collectors operating in a thermosiphon mode begin to produce energy at very low levels of incident sunlight. This provides excellent thermal response. Corrosion in the collector loop is virtually eliminated by the chemical inertness of the common refrigerants (R-11, R-114) with steel and copper, the normal materials used in construction. A test standard has been developed by ASHRAE for the thermal performance testing of boiling collectors [13].

A typical thermosiphon system [14] for a boiling collector is shown in Figure 1. The collector cross-section showing the tube plate configuration used for analytic modeling [11] is given in Figure 2.

2. BACKGROUND ANALYSIS ON CONVERSION EFFICIENCY, η_B

The useful energy produced by a boiling collector in a steady operating condition is

$$q_u = \eta_B \, I A_c \tag{1}$$

The thermal conversion efficiency, η_B has been developed in (11) as

$$\eta_B = \mathbf{F}_R \left[(\alpha \tau)_e - \frac{U_L(T_1 - T_a)}{I} \right] \tag{2}$$

where $\mathbf{F}_R$ is a Generalized Heat Removal Factor [11], valid for both boiling conditions (in which the collector fluid is in a saturated exit state) and non-boiling conditions. The factor $\mathbf{F}_R$ is shown [11] to be

$$\mathbf{F}_R = F_R \left[\frac{1 - \exp(-az^*)}{1 - \exp(-a)} + \frac{(1-z^*)\exp(-az^*)}{F_R / F_B'} \right]$$

$$= F_R F_B = f(a, z^*, F_B'/F', F') \tag{3}$$

In this result F_R ($F_R = F'F''(a)$, Equation (4) below), is the conventional (single phase flow) heat removal factor [15], z^* is the ratio of the non-boiling (collector) channel length to the total channel length, L_{NB}/L, a^{-1} is $(w_c/Ac)C_{p\ell}/F'U_L$ and F_B' is the conventional collector efficiency factor, F' [15], in which the channel heat transfer coefficient is that for the boiling condition. The other variables are those in conventional usage as given by Duffie and Beckman (15). F_B is defined by equation (3) as the bracketed term and F_R is defined [15] as

$$F_R = F'F''(a) \quad (4)$$

where,

$$F''(a) = a^{-1}(1-e^{-a}), \quad (5)$$

and

$$a = \frac{F' U_L}{(w_c/Ac)C_{p\ell}} \quad (6)$$

Since the conversion efficiency, η, for a non-boiling collector is [15]

$$\eta = F_R \left[(\alpha\tau)_e - \frac{U_L(T_1-T_a)}{I} \right] \quad (7)$$

it is evidence that the ratio of the conversion efficiency of a boiling collector to that of a non-boiling collector, η_B/η, for fixed operating parameters, i.e., fixed values of a, F^1, $(\alpha\tau)_e$, U_L, and $(T_1-T_a)/I$, is

$$\frac{\eta_B}{\eta} = F_B , \quad (8)$$

or, as shown in [12] this ratio may also be written

$$\frac{\eta_B}{\eta} = \frac{1 - \exp(-az^*) + a(1-z^*)\,(F_B'/F')\exp(-az^*)}{1 - \exp(-a)} \quad (9)$$

$$= F_B \geq 1.0, \; (0 \leq z^* \leq 1.0). \quad (10)$$

Hence,

$$\frac{\eta_B}{\eta} = f(a, z^*, F_B'/F') \quad (11)$$

The ratio η_B/η is shown in Fig.3 for values of F_B'/F' of 1.00 and 1.10, which represent limits of this quantity that may reasonably be expected under normal operating conditions and values of z^* from 0.0 to 1.0. The range for a is from 0.01 to 10 with an operating range under normal thermosiphon conditions of 1.2 to 5.2, as shown. These results clearly demonstrate the increased conversion efficiency of a boiling collector over that of a non-boiling collector at all values of a. This increase in efficiency is especially pronounced at large values of a, conditions that correspond, among other things, to low collector flow rates.[1]

The value of z^*, the dimensionless non-boiling length, L_{NB}/L, is given in [11] in terms of the efficiencies of the collector operating in the non-boiling mode as

$$z^* = \frac{1}{a} \log_e \frac{\eta_{T_1}}{\eta_{T_{sat}}} \quad (12)$$

where, for fixed values of a, U_L, F' and $(\alpha\tau)_e$,

$$\eta_{T_1} = F_R \left[(\alpha\tau)_e - \frac{U_L(T_1-T_a)}{I} \right] \quad (13)$$

and

$$\eta_{T_{sat}} = F_R \left[(\alpha\tau)_e - \frac{U_L(T_{sat}-T_a)}{I} \right] \quad (14)$$

or, the efficiencies of a non-boiling collector for inlet temperatures of T_1 (less than T_{sat}) and T_{sat}, respectively, where T_{sat} is the saturation temperature at the collector inlet. When the inlet temperature is equal to T_{sat}, the inlet fluid is saturated and z* is zero, as required by the physical circumstances and indicated by Equation (12). When the inlet temperature is T_1 (necessarily less than T_{sat}), the inlet fluid is in a sub-cooled state and $\eta_{T_1}/\eta_{T_{sat}}$ is greater than 1.0. Hence, z^* will be greater than zero and a finite length of the tube will be under non-boiling conditions. It is by the parameter z^* that the influence of inlet sub-cooling ($T_1 < T_{sat}$) on the thermal performance of a boiling collector is determined. This will be discussed further in connection with the analysis of the thermal efficiency of a boiling collector.

3. GENERALIZED HEAT REMOVAL FACTOR, F_R

The generalized heat removal factor, $\mathbf{F}_R$, given by equation (3), can also be expressed as

$$\frac{\mathbf{F}_R}{F_B'} = F''(a) \frac{1-\exp(-az^*) + a(1-z^*)\,(F_B'/F' \exp(-az^*)}{(F_B'/F')\,[1 - \exp(-a)\,]} \quad (15)$$

$$= F_B''\,(a, z^*, F_B'/F') \quad (16)$$

In this form the generalized nature of $\mathbf{F}_R$ becomes clear. In the limiting condition of a fully non-boiling collector ($z^* = 1.0$) the bracketed term in equation (15) becomes F'/F_B' and

$$\frac{\mathbf{F}_R}{F_B'} = \frac{F'F''(a)}{F_B'} \quad (17)$$

$$\mathbf{F}_R = F'F''(a) = F_R. \quad (18)$$

Hence, for a fully non-boiling collector the generalized heat removal factor, $\mathbf{F}_R$, and the heat removal factor, F_R, of reference (15) become identical, which they should. In this case F_B' and F' would also be identical. For the condition of a fully boiling collector ($z^* = 0$), the bracketed term in equation (15) becomes $1/F''(a)$, and for all values of F_B'/F',

$$\frac{\mathbf{F}_R}{F_B'} = 1.0 \quad (19)$$

and

$$\mathbf{F}_R = F_B' \quad (20)$$

In this case $\mathbf{F}_R$ has its maximum possible value for a fixed set of operating conditions.

Accordingly, for a fixed set of collector design parameters of $(\alpha\tau)_e$ and U_L, this condition corresponds to the maximum possible collector conversion efficiency, η_B, and is independent of the value of a, except for the small influence this parameter may have on the value of the channel boiling heat transfer coefficient, h_B, which in turn influences the magnitude of F_B'. The total effect of a on F_B' is probably third order or smaller and, in any case, is very difficult to demonstrate by a physical experiment.

For collectors operating in a partially boiling mode z* will have values intermediate between 0 and 1.0, or $0 \leq z^* \leq 1.0$. In this case, for all values of the parameter a, the range of values for the generalized heat removal factor, $\mathbf{F}_R$, will be

$$F_B' \geq \mathbf{F}_R \geq F_R, \tag{21}$$

and,

$$1.00 \geq \frac{\mathbf{F}_R}{F_B'} \geq F''(a) \tag{22}$$

or,

$$1.00 \geq F_B''(a, z^*, F_B'/F') \geq F''(a) \tag{23}$$

Implied in this last result is the equivalent roles the functions $F_B''(a, z^*, F_B'/F')$ and $F''(a)$ have in collector design analysis. In fact, the function $F_B''(a, z^*, F_B'/F')$ is the Generalized Collector Flow Factor, appropriate to both boiling and non-boiling collectors, for which the traditional collector flow factor, $F''(a)$, reference [15], becomes the limiting value for fully non-boiling conditions, i.e., $z^* = 1.0$.

The generalized nature of the function $F_B'' = \mathbf{F}_R/F_B'$ is shown in Figures 4 and 5 for values of F_B'/F' of 1.000 and 1.100 and over a range of the parameter a from 0.01 to 10.0. The lower limit of F_B'' is the traditional factor $F''(a)$ and corresponds to the fully non-boiling condition, $z^* = 1.0$. The upper limit, which is the maximum possible value for F_B'', is 1.000 and corresponds to the fully boiling condition, $z^* = 0$.

The similarity between the ratios

$$\frac{\mathbf{F}_R}{F_B'} = F_B''(a, z^*, F_B'/F') \tag{24}$$

and

$$\frac{F_R}{F'} = F''(a) \tag{25}$$

is to be noted. The ratio $\mathbf{F}_R/F_B'$ is the generalized result.

The asympotic limit of $\mathbf{F}_B''$ for $a^{-1} = \infty$ o $a = 0$ may be shown to be

$$F_B''(0, z^*, F_B'/F') = 1 - z^* \left(\frac{F_B'/F' - 1}{F_B'/F'}\right). \tag{26}$$

Thus, when $F_B'/F' = 1.000$, this limit is 1.000 for all values of z*, as indicated in Figure 4. Further, for $z^* = 0$, the limiting value of F_B'' is 1.000 for all values of F_B'/F'. The limits for F_B'' corresponding to $F_B'/F' = 1.100$ are shown in Figure 5 and are in accord with equation (27), except for the curve for $z^* = 1.000$ for which two values are given. In this case it should be recognized that when $z^* = 1.00$ the only possible value for F_B'/F' is 1.000 and the corresponding limit for $F_B''(0, 1.00, 1.00)$ is 1.000. However, equation (27) for $z^* = 1.000$ produces an asympotic value for $F_B''(0, 1.00, 1.100)$ of $(1.100)^{-1} = 0.9090$ since the equation does not include the effect of a coupled relationship between z* and F_B'/F' which does exist but is unknown at present.

Some additional insights into the performance of boiling collectors in comparison with non-boiling collectors is also provided by the asympotic limit analysis in which $a = 0$ or $a^{-1} = \infty$. For this case $\mathbf{F}_R$ is written

$$\mathbf{F}_R(a = 0) = F_B' F_B''(0, z^*, F_B'/F') \tag{27}$$

which, with equation (26), becomes

$$\mathbf{F}_R(a = 0) = F'[z^* + \frac{F_B'}{F'}(1 - z^*)]. \tag{28}$$

However, for this limiting case of $a = 0$, F_R becomes equal to F' since $F''(0)$ is exactly 1.000. Accordingly, **for the asympotic limit of a = 0,**

$$\left(\frac{\mathbf{F}_R}{F_R}\right) = z^* + \frac{F_B'}{F'}(1 - z^*) \tag{29}$$

Further, for fixed values of $(\alpha\tau)_e$, U_L and $(T_1 - T_a)/I$, this result is also the ratio of the conversion efficiencies of a boiling to a non-boiling collector, as

$$\left(\frac{\mathbf{F}_R}{F_R}\right)_{a=0} = \left(\frac{\eta_B}{\eta}\right)_{a=0}$$

$$= z^* + \frac{F_B'}{F'}(1 - z^*), \tag{30}$$

Thus, in the limit of large values of a^{-1} (i.e., $a = 0$), a boiling collector will always produce conversion efficiences greater than those of a non-boiling collector for the circumstances of equation (30). The maximum value of this ratio (η_B/η) occurs at $z^* = 0$, the fully boiling mode, and is equal to F_B'/F'. The minimum value of this ratio is 1.000 and this occurs at $z^* = 1.000$, the condition of a fully non-boiling collector.

In most tube-plate collector designs for boiling operations it is expected that the ratio F_B'/F' will be in the range 1.000 to 1.100. This suggests that at large circulation rates (small a) the boiling collector will have a conversion efficiency approximately 10% greater than a non-boiling collector of the same design and exposed to the same operating conditions.

4. THERMAL CONVERSION EFFICIENCY, η_B

4.1. General Discussion

The thermal conversion efficiency, η_B, of a boiling collector operating in the steady state is given by equation (2). With this formulation it is possible to represent η_B in the same manner as traditionally employed for non-boiling collectors [16] as a function of $(T_1 - T_a)/I$. There are, however, important differences in the representations of the efficiencies of these two types of collectors.

For a non-boiling collector the efficiency is expressed using the factor $\mathbf{F}_R$ which is primarily a function of the parameter a and, except for second order effects on U_L, is essentially independent of the solar irradiance, I. Thus, to a satisfactory degree, the efficiency of a non-boiling collector can be represented as a function of $(T_1 - T_a)/I$ by a single curve having an intercept of $F_R(\alpha\tau)_e$ and a slope of $-F_RU_L$

In the case of a boiling collector, the representation of the thermal efficiency requires the factor $\mathbf{F}_R$, which is a function of a, z*, F' and F_B'/F'. Of these variables both z* and F_B' are functions of the level of solar irradiance, I. Accordingly, unlike the non-boiling collector, the thermal efficiency of a boiling collector is implicity dependent on the solar irradiance, I, through the variables z* and F_B'. Further, in a thermosiphon system (a common mode for field operation) the rate of flow of coolant and, hence, the parameter a, is also dependent on the solar irradiance. Because of these effects the representation of the efficiency of a boiling collector is considerably more complicated than that for a non-boiling collector and requires the specification, either explicitly or implicitly, of the variables a, z* and F_B'.

4.2. Generalized Comparison with Non-Boiling Collectors Operating at Different Flow Rates per Unit Area

For boiling and non-boiling collectors each having different flow rates and, hence, different values of the parameter a, the ratios of their conversion efficiencies, for similar design and operating parameters of $(\alpha\tau)_e$, U_L and $(T_1 - T_a)/1$, may be formulated in a general way from equations (2) to (7) as

$$\frac{\eta_B(a_1)}{\eta(a_2)} = \frac{\mathbf{F}_R(a_1)}{F_R(a_2)} \qquad (31)$$

where a_1 and a_2 are the values of a, equation (6), for boiling and non-boiling conditions, respectively. This result can also be written

$$\frac{\eta_B(a_1)}{\eta(a_2)} = \frac{F'(a_1)\,F''(a_1)}{F'(a_2)\,F''(a_2)}\,F_B(a_1) \qquad (32)$$

or, using equation (9) this becomes

$$\frac{\eta_B(a_1)}{\eta(a_2)} = \frac{F'(a_1)}{F'(a_2)}\,\frac{F''(a_1)}{F''(a_2)}\left[\frac{\eta_B(a_1)}{\eta(a_1)}\right]. \qquad (33)$$

Accordingly, in a generalized sense, the ratio of the efficiencies of a boiling and non-boiling collectors is determined by the relative values of a_1 and a_2. Although, for a given value of a_1 the ratio $\eta_B(a_1)/\eta(a_1)$ will always be greater than 1.000 (as shown in Figure 3) it is certainly possible for a_2 to be sufficiently small that the ratio $F''/a_1/F''(a_2)$ would be less than 1.000 and the ratio $\eta_B(a_1)/\eta(a_2)$ equation (33) to be less than 1.000. In such cases it is likely that the small value of a_2 would be obtained by making the flow rate of collector coolant per unit area, (w_c/A_c), for the non-boiling collector considerably greater than that of the boiling collector. This, in fact, probably occurs in most non-boiling systems where the values of (w_c/A_c) may be as much as 10-times those for a boiling collector, particularly one operating in a thermosiphon mode. However, even in this case, a fully boiling collector may very well be operating in a range of a_1 that produces a value of $\eta_B(a_1)/\eta(a_1)$ of 2.0 to 4.0, in which case it still produces an efficiency significantly greater than the non-boiling collector having the larger flow rate per unit area.

4.3. Collector Efficiency, η_B, for a Fully Boiling Mode and the Influence of Inlet Sub-Cooling.

The efficiency of a boiling collector having saturated exit states and with thermal conditions in the coolant channels ranging from fully non-boiling to fully-boiling ($0 \leq z^* \leq 1.0$) is given by Equation (2)

$$\eta_B = \mathbf{F}_R\left[(\alpha\tau)_e - \frac{U_L(T_1 - T_a)}{I}\right] \qquad (2)$$

or, using Equation (3) this may be written

$$\eta_B = f\left(a,\ z^*, F_B'/F', F', (\alpha\tau)_e, U_L, \frac{T_1 - T_a}{I}\right) \qquad (34)$$

The quantities $(\alpha\tau)_e$ and U_L are design parameters and $(T_1 - T_a)/I$ is an operation parameter, all of which are independent of the thermodynamic conditions in the channel. The remaining quantities in equation (34) are determined by the thermodynamic conditions in the coolant and the design of the absorber plate. The quantity z*, defined by equation (12), is determined by the parameter a and the efficiency of a fully non-boiling collector (z* = 1.0), which also depends on a.

Hence, for a set of fixed values of a, F_B'/F', F', $(\alpha\tau)_e$ and U_L, the efficiency of a boiling collector may be expressed as

$$\eta_B = f\left(z^*, \frac{T_1 - T_a}{I}\right) \qquad (35)$$

Using representative values for these fixed parameters

η_B is shown in Figures 6 through 11 as a function of $(T_1 - T_a)/I$ with z* as a parameter having a range of 0. to 1.0. These results are for values of F_B'/F' of 1.000 and 1.1000 and values of a of 0.500, 1,000, 5.000 and 10.000. These ranges are those which should encompass most operating conditions for refrigerant charged boiling collectors operating in a thermosiphon mode. The values of z* are indicated on Figs. 6 and 7 and are the same for all Figures 6 - 11.

The efficiency of a non-boiling collector, corresponding to the particular value of a, is given in Figures 6 - 11 as the curve for z* = 1.0 and that of the fully-boiling collector as the curve for z* = 0. The collector efficiency for collectors that are partially boiling, that is, those with finite inlet subcooling correspond to the curves z* between 0 and 1.0. The y-axis intercept is, in general

$$F_R(\alpha\tau)_e \tag{36}$$

which for z* = 1.0 is

$$F_R(\alpha\tau)_e, \tag{37}$$

and for z* = 0.0 is

$$F_R'(\alpha\tau)_e. \tag{38}$$

The slopes of all the efficiency curves is

$$-F_R U_L, \tag{39}$$

which for z* = 0.0 is

$$-F_R U_L, \tag{40}$$

and for z* = 0.0 is

$$-F_B' U_L. \tag{41}$$

The x-axis intercept (not shown in Figures 6 - 11) is common to all the efficiency curves and is

$$\frac{(\alpha\tau)_e}{U_L} \tag{42}$$

This corresponds to the "stagnation", or no flow, condition in the collector and is the same whether the collector is boiling or not boiling.

As is once again evident these figures show that the performance of a boiling collector as measured by its thermal efficiency is always greater than that of a non-boiling collector for systems having the same value of a. The influence of z* is very significant in determining the efficiency of a boiling collector but one which diminishes noticably as a is reduced, that is, as the coolant flow rate increases. The efficiency is, however, independent of a or coolant flow rate for a fully-boiling collector[2] for which z* = 0.0. Hence, the efficiency curves for z* = 0.0 are unique in Figures 6 - 11.

Another series of curves are included in the Figures which indicate the effect of sub-cooling. These are the curves for which A - A in Figures 6, 8 and 11 are typical. For these curves, only the variable $(T_1 - T_a)/I$ is changed. The point A on the z* = 0.0 curve corresponds to a saturated inlet state. Hence, the variable $(T_1 - T_a)/I$ for this point becomes

$$\frac{T_{sat} - T_a}{I}. \tag{43}$$

Since $(T_1 - T_a)/I$ can be written in general

$$\frac{T_1 - T_a}{I} = \frac{T_{sat} - T_a}{I} - \frac{\Delta T_{sc}}{I}, \tag{44}$$

where

$$\Delta T_{sc} = T_{sat} - T_1, \tag{45}$$

all points to the left of A(z* = 0.0), such as point C, Figure 6, correspond to finite inlet sub-cooling. As is shown on Figure 6 by the curve A - A, sub-cooling of the maximum practical amount, i.e., $T_1 = T_a$, produces a very small increase in the efficiency. This is the result of a relatively small effect on the non-boiling length z* of this amount of subcooling as z* increases from 0.0 to 0.045, hardly enough to affect the thermal performance for this value of a. However, for smaller values of a the effect of sub-cooling is more pronounced, as is shown in Figures 8 and 11. In all cases, the influence of sub-cooling is to increase the efficiency from the value corresponding to a saturated inlet state, such as that at point C compared with point A (z* = 0.0), Figure 6.

Various levels of inlet saturated states correspond to the points, such as B, Figure 6, along the z* = 0.0. curve. In an operating system these states will be established by the conditions in the condenser, Figure 1. To a good approximation the saturation conditions in the condenser are determined by the inlet temperature of the water entering the coil in a well designed condenser. Because this temperature is essentially the same as the temperature of the water in the storage tank (not shown in Figure 1), the saturation state in the condenser is essentially that corresponding to storage tank temperature. Because of this the various points B, Figure 6, represent ranges of storage tank conditions, B corresponding to a storage tank at a lower temperature than that for A and hence, a higher efficiency of the collector.

The curves A - A in Figures 6, 8 and 11 each correspond to the values of the parameter a given. Since coolant flow rate is the principal variable in a, the curves A - A can be considered lines of constant coolant flow rate. In a thermosiphon system the coolant flow rate is primarily governed by the solar irradiance, I. Accordingly, the curves A - A can also be considered, to a close approximation, as lines of constant I, each such curve being associated with a particular saturation temperature (Point A for z* = 0.0) in the condenser (or storage tank temperature)

These influences of sub-cooling at constant solar irradiance have been confirmed in some initial experimental investigations on the thermal performance of boiling collectors (11, 12). In operating systems the magnitute of the inlet subcooling is not large, being in the range 0 to 30 F. Thus, the values of $\Delta T_{sc}/I$ in equation (44) will be generally less than 0.20.

The range of possible subcooling shown in Figure 6 - 11 is therefore significantly greater (by a factor of 2 - 3) than that which will be found in practical systems.

The effect on the thermal performance of an

increase in the solar irradiance, I, can also be established by these results. For a situation in which T_l and T_a are fixed, an increase in the solar irradiance increases the coolant flow rate and, hence, also a^{-1}, and decreases $\Delta T/I$. Taking Point A ($z^* = 0.0$), Figure 6, as a reference point and using coolant flow rate data from (11), an increase in the solar irradiance from 127 BTU/Hr-Ft^2 (400 w/m^2) to 206 BTU/Hr Ft^2 (650 w/m^2) will double the coolant flow rate. This corresponds to an increase in a^{-1} from 0.100, Figure 6 to 0.200, the conditions of Figure 7. At the same time the value of $\Delta T/I$ decreases from 0.40 (point A, Figure 6) to 0.246 (Figure 7). Thus, using Figure 7 it may be seen that this increase in solar irradiance produces an increase in the efficiency (For $z^* = 0.0$) from 0.52 to 0.64. Accordingly, increases in solar irradiance produce upward shifting in the lines A - A. Also, it should be noted, the thermal performance will increase by a larger amount, which in this case is almost a factor of 2, since the useful energy from the collector is the product of I, η_B and the net collector absorber area.

ACKNOWLEDGEMENT

The author wishes to acknowledge the contribution to this research of his graduate assistant, Mohammad Iftekhar Ahmed, who developed the computer programs that generated the Figures in this paper.

REFERENCES

1. Hottel, H.C., and Woertz, B.B., "The Performance of Flat-Plate Solar Collector," Trans. ASME, 64 (1942) 91

2. Hottel, H.C., and Whillier, A., "Evaluation of Flat-Plate Solar Collector Performance," Trans. of the conference on the use of Solar Energy, Univ. of Arizona, (1955) 74-104.

3. Shitzer, A., D. Kalmanoviz, Y. Zvirin and G. Grossman, "Experiments with a Flat-Plate Solar Water Heating System in Thermosyphonic Flow," Solar Energy, 22 (1979) 27.

4. Grossman, G., A. Shitzer and Y. Zvirin, "Heat Transfer Analysis of a Flat-Plate Solar Energy Collector," Solar Energy , 19, (1977) 493.

5. Close, D.J., "The Performance of Solar Water Heaters with Natural Circulation," Solar Energy , 6, (1962) 33.

6. Gupta, C.L. and H.P. Garg, "System Design in Solar Water Heaters with Natural Circulation," Solar Energy , 12 (1968) 163.

7. Zvirin, Y., A. Shitzer and G. Grossman, "The Natural Solar Heater-Models with Linear and Non-linear Temperature Distributions," Int. J. Heat Mass Trans., 20 (1977) 997.

8. Morrison, G.L. and J.E. Braun, "System Modelling and Operation Characteristics of Thermosyphon Solar Water Heaters," Solar Energy , 34 (1985) 4/5.

9. Soin, R.S., K. Sangameswar Rao, D.P. Rao and K.S. Rao, "Performance of Flat-Plate Solar Collector with Fluid Undergoing Phase Change," Solar Energy , 23 (1979) 69.

10. Downing, R.C. and V.M. Waldin, "Phase-Change Heat Transfer in Solar Hot Water Heating using R-11 and R-114," ASHRAE Trans., Part 1, 86 (1980).

11. Al-Tamimi, A.I. and J.A. Clark, "Thermal Analysis of a Solar Collector Containing a Boiling Fluid" 1983 Annual Meeting, Proceedings, American Solar Energy Society, Minneapolis, MN, June 1-3, (1983).

12. Al-Tamimi, A.I. and J.A. Clark, "Thermal Performance of a Solar Collector Containing a Boiling Fluid," AT-84-13, No.1, ASHRAE Symposium on Boiling Collectors, Atlanta, GA January-February (1984).

13. Standard 109-P, ASHRAE, Atlanta, GA (1985)

14. Refrigeration Research, Inc. Solar Research Division, Brighton, Mich., USA (1985)

15. Duffie, J.A. and W.A. Beckman, Solar Engineering of Thermal Processes, John Wiley & Sons, (1980).

16. ASHRAE standard 93-77, "Methods of Testing to Determine the Thermal Performance of Solar Collectors," American Society of Heating, Refrigerating and Air-Conditioning Engineers, New York, (1978)

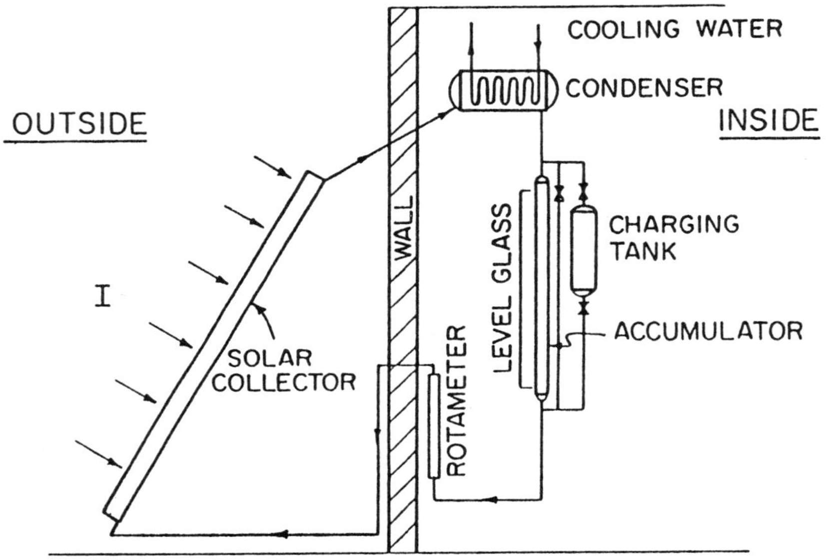

Figure 1. Loop Assembly

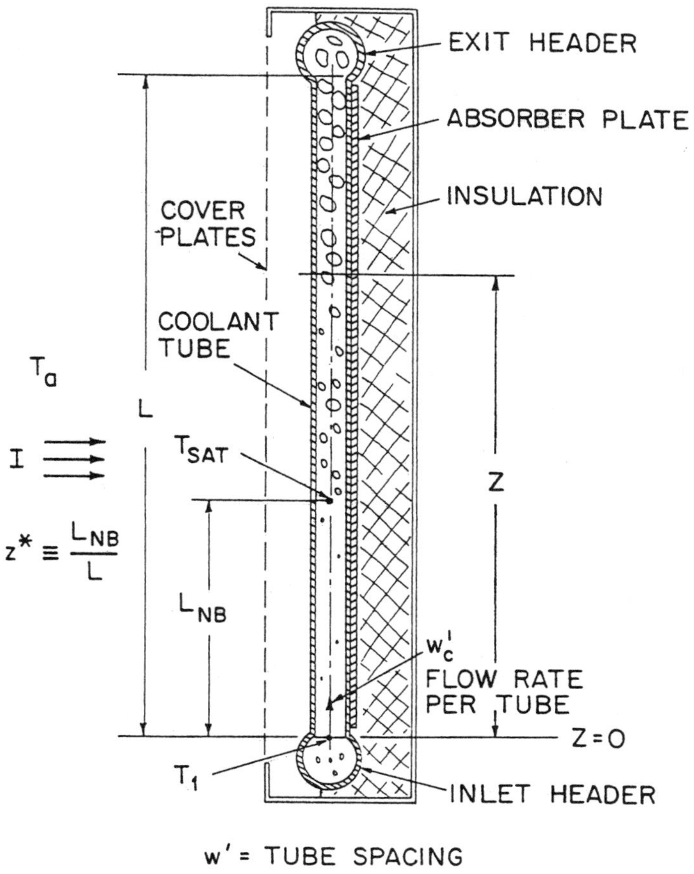

Figure 2. Typical Tube-Plate Crossection (Not to Scale)

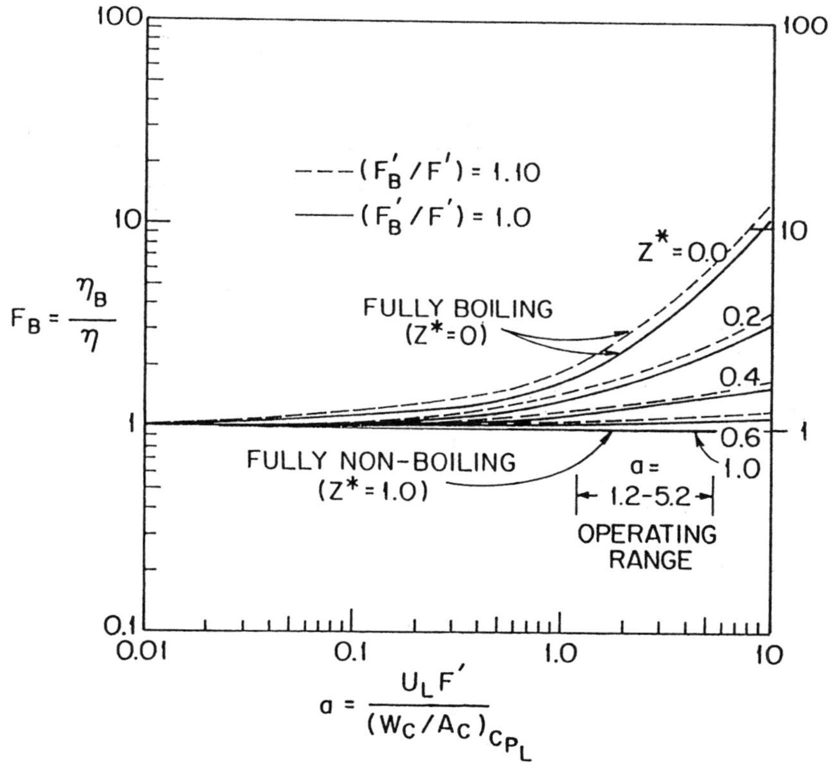

Figure 3. Ratio of Efficiencies of a Boiling Collector to a Non-Boiling Collector

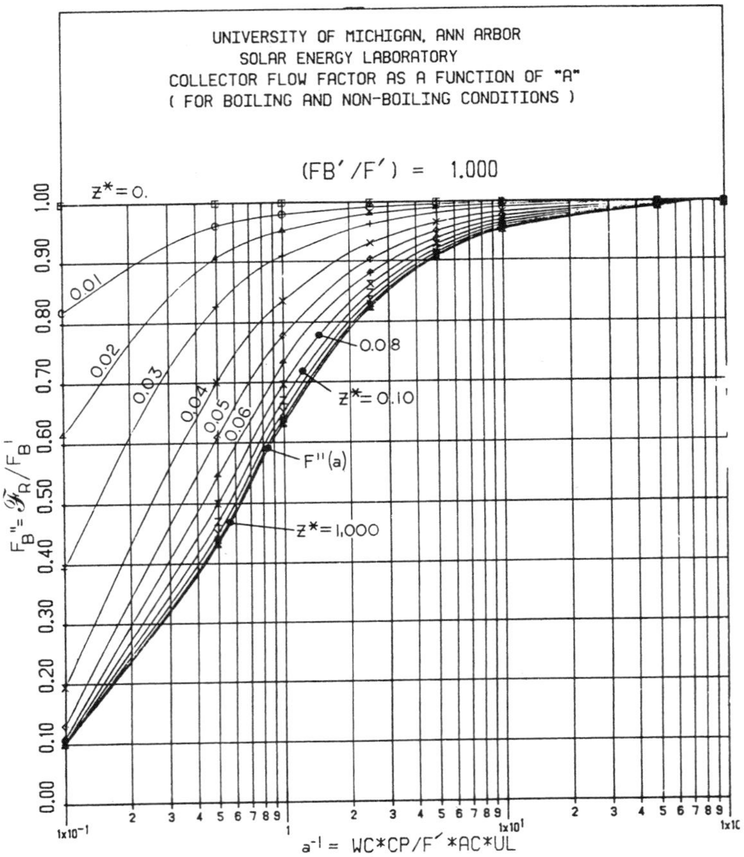

Figure 4. Generalized Collector Flow Factor F''_B for F'_B/F' = 1.000

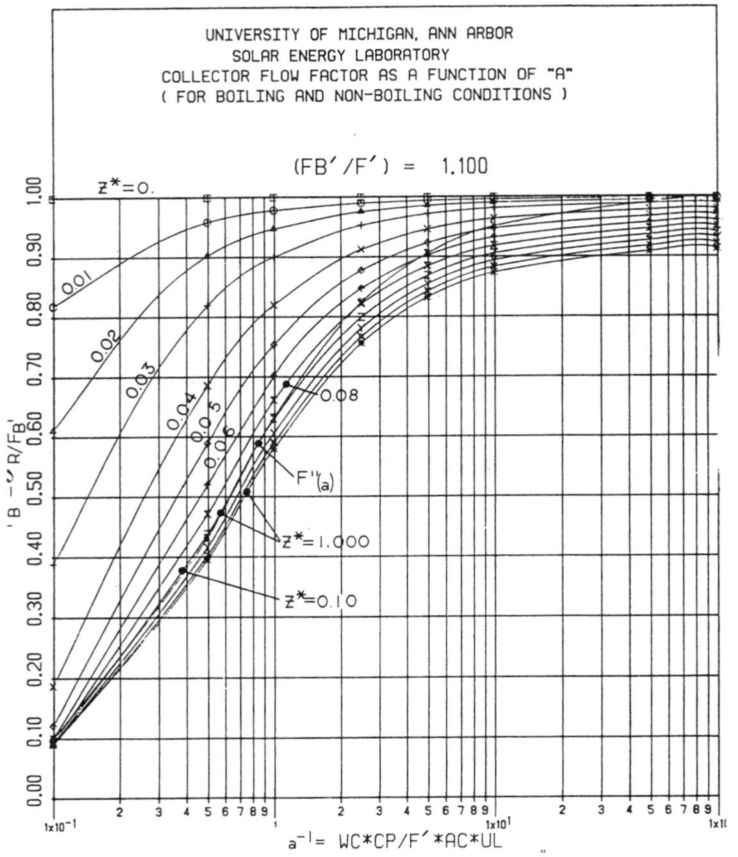

Figure 5. Generalized Collector Flow Factor F''_B for F'_B/F' = 1.100

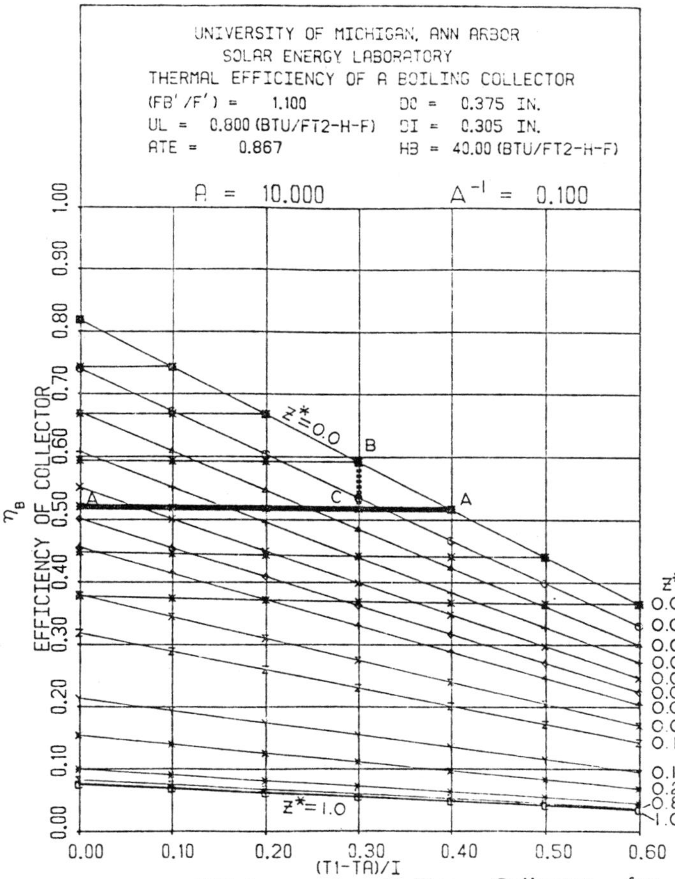

Figure 6. Efficiency of a Boiling Collector for $a = 10.0$, $F'_B/F' = 1.100$

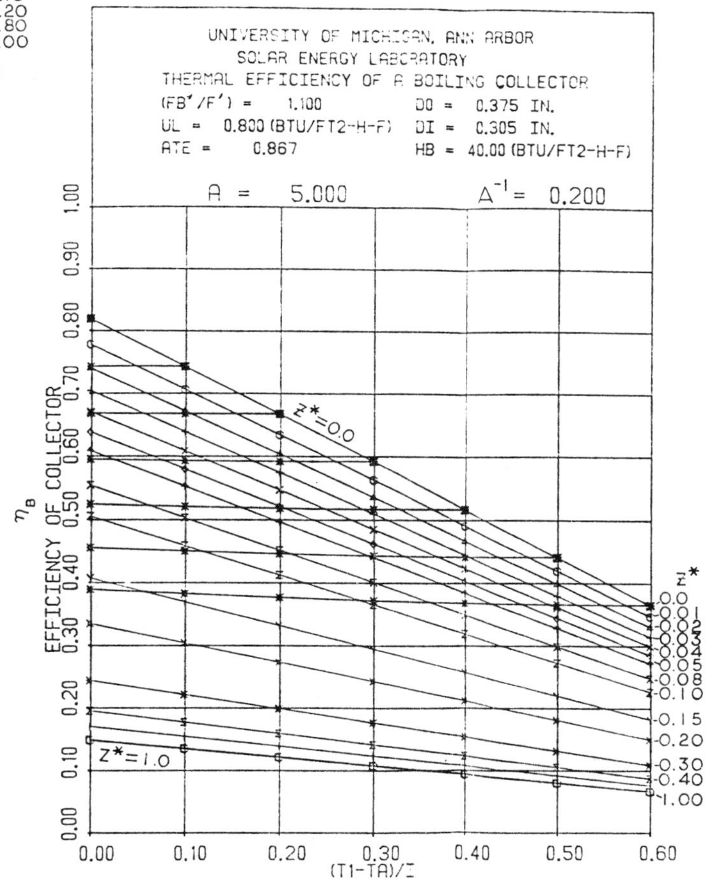

Figure 7. Efficiency of a Boiling Collector for $a = 5.0$, $F'_B/F' = 1.100$

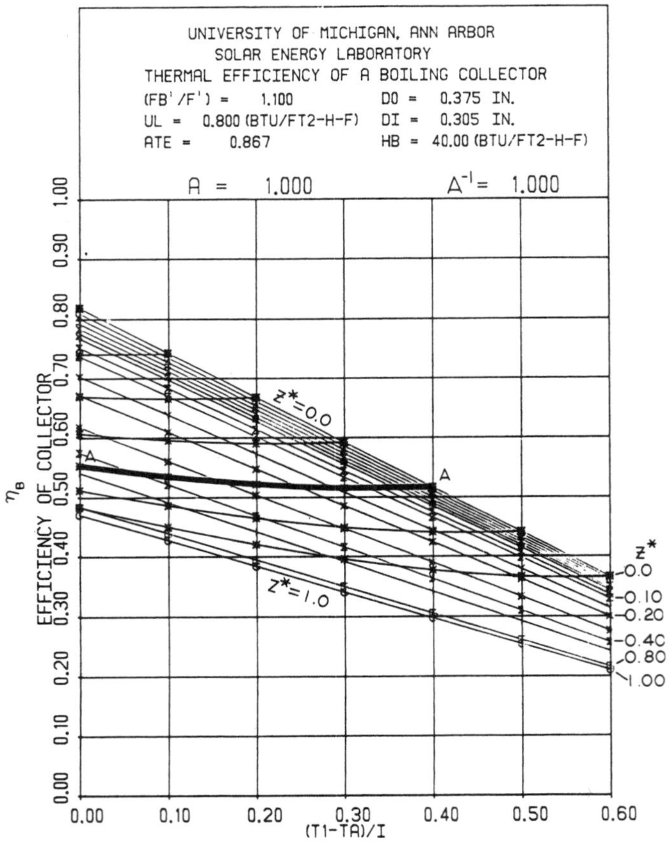

Figure 8. Efficiency of a Boiling Collector for a = 1.000, F'_B/F' = 1.100

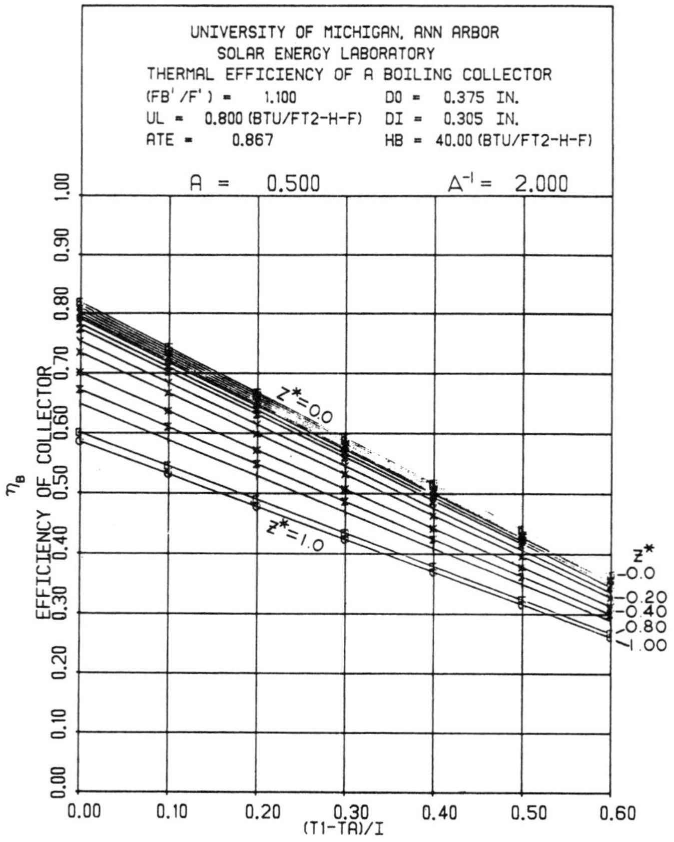

Figure 9. Efficiency of a Boiling Collector for a = 0.500, F'_B/F' = 1.100

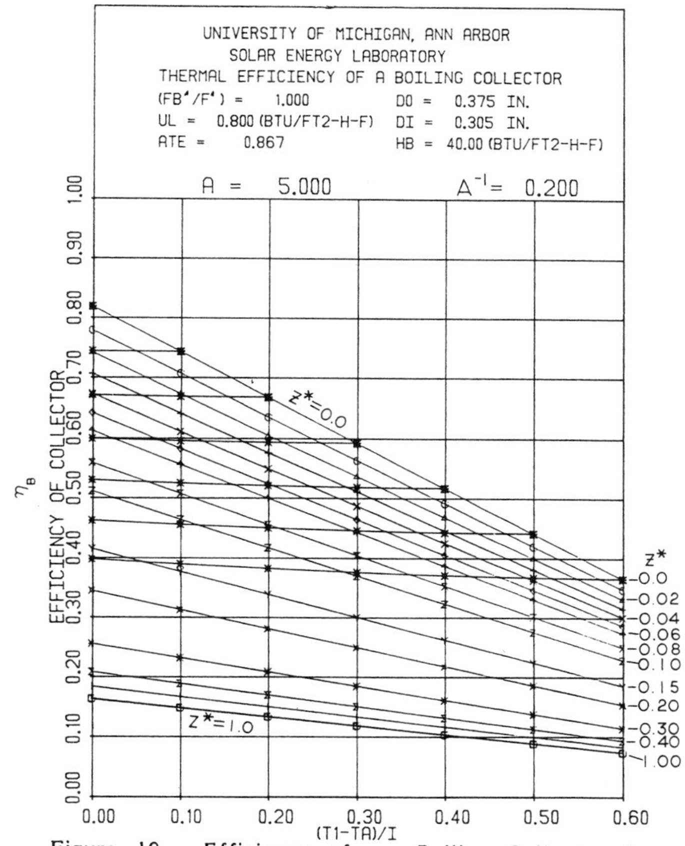

Figure 10. Efficiency of a Boiling Collector for $a = 5.000$, $F'_B/F' = 1.000$

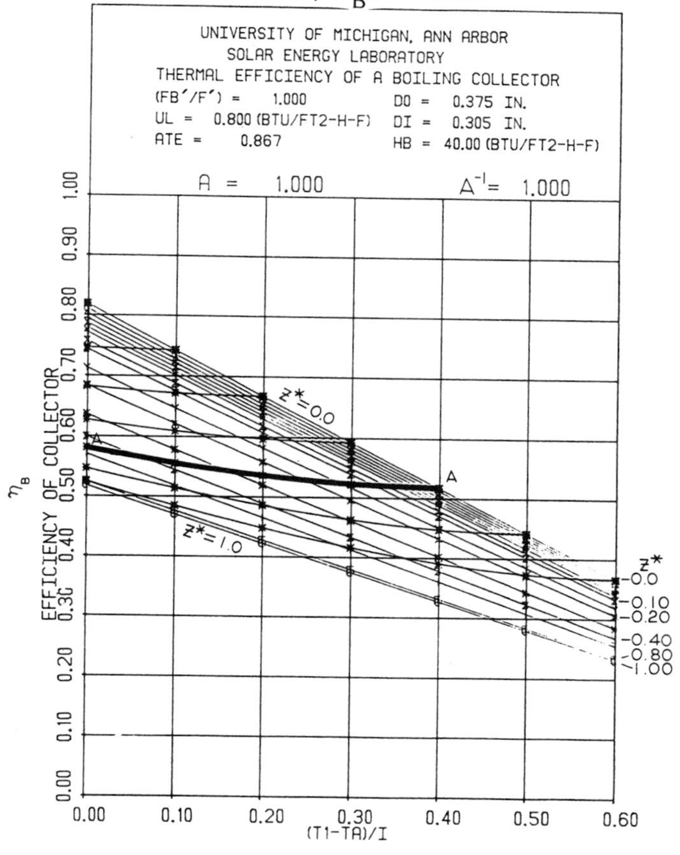

Figure 11 Efficiency of a Boiling Collector for $a = 1.000$, $F'_B/F' = 1.000$

Systems Approach for Modeling Natural Convection: One Dimensional Systems

T. GOPICHAND
Department of Chemical Engineering
Indian Institute of Technology
Madras 600 036, India

Entropy generation of natural convective flow field is considered as a source, to augment the conductive mode of heat transfer and for evolution of dissipatory structures. The non-linearity of equations, of a simple one dimensional steady state heat transfer between two vertical plates is analysed. For a systems view point, possible lumped parameter elements and their connectivity is postulated for further investigations.

1. INTRODUCTION

Natural convection phenomena, has at its root, a thermal gradient, driving or creating a convective flow, augmenting the conductive mechanism, which in the first place provided the thermal gradient. This self excited loop, further gets reinforced as thermal flux increases, leading to formation of what are being currently called dissipative structures. A general class of phenomena called Dis-Order to Order Transformations, has come into vogue, of which the above is considered to be an example. In so far as Natural convection phenomena is considered, the Dis-Order at molecular level, is said to be transformed into an Order of geometrical patterns at macroscopic level. For instance, Bernard Cells observed experimentally around 1900 (1), when a thin horizontal layer is subjected to a thermal gradient, develops a hexagonal pattern of cells. Such patterns were latter subjected to careful experimental and theoretical studies (2,3,4,5). Buoyancy effects in fluids, Hydrodynamic stability and dissipative structures received considerable attention (6,7,8). Interest in this problem, has been further highlighted in this decade (9,10,11,12), since the Dis-Order to Order transformations, as a class of phenomena, have been considered to be of basic importance encompassing not only non-living and living systems but also human systems.

In models of Natural Convection used in engineering practice, one makes a common assumption, for effecting considerable simplification of analysis. This assumption is stated in two parts as (13,14):

1) A non-isothermal system would be either in a state of forced or free convection

2) Forced and free convection represent two limiting conditions - one in which buoyancy forces are negligible and the other in which the effects of pressure and gravitational forces can be expressed entirely in terms of buoyancy forces.

It is felt desirable, to examine the limitation imposed by such an ideal conceptual frame work.

The main objective of this study is, to carefully examine the simplest one dimensional models of natural convection, used in engineering practice, to modify them and to identify the Elements (lumped) and their Connectivity ; to be more specific, the steady state heat transfer between two vertical plates separated by a gap (at y = -b and y = +b, a gravity acting along negative "Z" direction). A long term objective of this study and others to follow in the sequel is to understand the net-work of Dir-Order to Order Transformations of Natural convection phenomena.

2. PROBLEM STATEMENT

The simplest one dimensional steady state model from reference 14 (pp. 297-300) is shown in Fig.1 and the salient features of the solution are shown in Fig.2. Examination of this rudimentary model and the modifications necessary thereof, forms the subject matter of the present investigation.

In the rudimentary model shown in Fig.2, the one dimensional energy and momentum equations are decoupled and simultaneously solved for temperature and velocity profiles. The latter is a typical natural convective flow field. However, the net result of this model, indicates that the steady state heat transfer coefficient for the gap is independent of ΔT, since

$$q = -k\frac{dT}{dy} = -k\frac{\Delta T}{2b} \equiv h\,\Delta T \tag{1}$$

Hence for a substantive study of system elements

and their connectivity, the model indicates, it is still a conductive mode and that the element is just a thermal dissipator with no specific role assigned to velocity profile in the heat transfer mechanism.

If one considers, the viscous heat dissipation of the ϕ - profile of the rudimentary model shown in Fig.2, one gets, for the integral average,

$$q_{irr} = \frac{\bar{\rho}^2 \bar{\beta} g^2 (\Delta T)^2 b}{90 \mu}$$

$$= \frac{Gr}{90} \left[\frac{\bar{\beta} g (\Delta T) \mu}{b^2 g_c} \right] \qquad (2)$$

If one writes,

$$q + q_{irr} \equiv h_1 \Delta T \qquad (3)$$

One gets,

$$h_1 = \frac{k}{2b} + \frac{Gr}{90} \left[\frac{\bar{\beta} g \mu}{b^2 g_c} \right] \qquad (4)$$

indicating that at best one has $h_1 \alpha (\Delta T)$ only.

Next higher level of modelling taken from Reference 14 (pp. 330-333) is shown in Fig.3. The flow field considered is $\vee$ $v_z(y,z)$ and $v_y(y,z)$. The latter considered is small. Continuity, and energy equation considered along with partial derivatives neglected for facilitating decoupling are shown in Fig.4 along with the salient results. The final solution does indicate that the heat transfer coefficient is proportional to $(Gr.Pr)^{1/4}$. The typical steady state heat transfer coefficient for a vertical plate is shown in Fig.5. [References 15,16 and 17 are cited in (14) as relevant to this model]. In so far as the present study is concerned, the simple model is first analysed and the implications of the second model is discussed towards the end.

In what follows, first the modifications of the simple model necessary to understand the system elements are presented followed by proposing a connectivity of the elements identified and discussing its implications for understanding the next level models.

3. MODIFICATION OF RUDIMENTARY MODEL

As discussed in the earlier section, even though the rudimentary model gives a flow field, ϕ, its contribution to enhancement of the conductive mode is neglected. To account for it, one should have to solve the following set of one dimensional energy and momentum equations:

$$- k \frac{d^2 T}{d y^2} = \frac{\mu}{g_c} \left[\frac{d v_z}{d y} \right]^2 \qquad (5)$$

$$\mu \frac{d^2 v_z}{d y^2} = - \bar{\rho} \bar{\beta} g (T - \bar{T}) \qquad (6)$$

wherein, the following approximate equation of state is used,

$$\rho = \bar{\rho} [1 - \bar{\beta} (T - \bar{T})] \qquad (7)$$

The boundary conditions are,

$$T = T_2 \quad \text{at} \quad y = -b \qquad (8)$$

$$T = T_1 \quad \text{at} \quad y = +b \qquad (9)$$

$$v_z = 0 \quad \text{at} \quad y = \pm b \qquad (10)$$

Let the following dimensionless numbers be defined,

$$\eta = \frac{y}{b} \qquad (11)$$

$$\Theta = \frac{T - T_1}{T_2 - T_1} \qquad (12)$$

$$\phi = \frac{b v_z \bar{\rho}}{\mu} \qquad (13)$$

Hence equations (5) and (6) become,

$$\frac{d^2 \Theta}{d \eta^2} = - \frac{\mu^3}{b^2 \bar{\rho}^2 g_c k \Delta T} \left(\frac{d \phi}{d \eta} \right)^2 \qquad (14)$$

$$\frac{d^2 \phi}{d \eta^2} = - \frac{b^3 \bar{\rho}^2 \bar{\beta} g \Delta T}{\mu^2} \left[\Theta - \frac{1}{2} \right] \qquad (15)$$

In writing Eq. (14), it is assumed that $\bar{T} = T_m = 1/2 (T_2 + T_1)$. Equations (14) and (15) in terms of conventional Grashof number, becomes,

$$\frac{d^2 \Theta}{d \eta^2} = - \frac{1}{Gr} \left[\frac{\mu b \bar{\beta} g}{k g_c} \right] \left(\frac{d\phi}{d\eta} \right)^2 \qquad (16)$$

$$\frac{d^2 \phi}{d\eta^2} = - Gr \left[\Theta - \frac{1}{2} \right] \qquad (17)$$

The dimensionless quantity in the square brackets is called a modified Brinkmann number and will be designated as K. Its significance will be discussed towards the end. In terms of a new variable, $\theta = -Gr\,[\textcircled{H} - 1/2]$, one can write, equations (16) and (17) as,

$$\frac{d^2\theta}{d\eta^2} = K\left(\frac{d\phi}{d\eta}\right)^2 \quad (18)$$

$$\frac{d^2\phi}{d\eta^2} = \theta \quad (19)$$

The two equations have to be simultaneously solved for $\theta(\eta)$ and $\phi(\eta)$, for the following boundary conditions,

$$\theta = -\frac{Gr}{2} \quad \text{for } \eta = -1 \quad (20)$$

$$\theta = +\frac{Gr}{2} \quad \text{for } \eta = +1 \quad (21)$$

$$\phi = 0 \text{ for } \eta = \pm 1 \quad (22)$$

If one neglects the non-linear term in equation (18), one gets the simplest solutions shown in Fig.2. This non-linear term is considered as the primitive root of natural convection phenomena and its systematic interpretation is the subject matter of this study.

4. DECOUPLING AND SOLUTION GOVERNING EQUATIONS

The equations to be solved are,

$$\theta'' = K(\phi')^2 \quad (23)$$

$$\phi'' = \theta \quad (24)$$

where the primes are differentiation with respect to η. Rigorous simultaneous solution of the above set requires, the solution of,

$$\phi^{IV} = K(\phi')^2 \quad (25)$$

which is fourth order non-linear equation. It is proposed, to understand solutions of a simpler second order non-linear equation obtained by subtracting (24) from (23), and rewriting as,

$$\theta'' - \theta = K(\phi')^2 - \phi'' \quad (26)$$

Since both sides are functions of η only, it is written as,

$$\theta'' - \theta = K(\phi')^2 - \phi'' = -\left(G_0 + G_1\eta + G_2\frac{\eta^2}{2!} + \ldots\right) \quad (27)$$

Considering the set, involving first term of the G-series,

$$\theta_o'' - \theta_o = -G_o \quad (28)$$

$$\phi_o'' - K(\phi_o')^2 = G_o \quad (29)$$

Equation (28) is linear and equation (29) has an integrating factor. The solutions are,

$$\theta_o = C_1 \sin(\eta) + C_2 \cos(\eta) + G_o \quad (30)$$

$$\phi_o = -\frac{1}{K}\log\cos\left[(G_0 K)^{1/2}.\,\eta\right] + C_3\eta + C_4 \quad (31)$$

Further, for a general G-series consisting of N terms equation (27) becomes,

$$\phi'' - K(\phi')^2 = \left[G_o + G_1\eta + G_2\frac{\eta^2}{2!} + \ldots + G_N\frac{\eta^N}{N!}\right] = G_{(N)} \quad (32)$$

which reduces to,

$$-G(N)\cdot\frac{f}{(f')^2}\left[f'' - \frac{G'(N).}{G(N)}f' + K\,G_{(N)}.f\right] = 0 \quad (33)$$

if $\phi' = G_{(N)} \cdot \frac{f}{f'}$ and f, f' are functions of η. Further,

equation (31) indicates that $f_0 = \sin[(G_o K)^{1/2}.\,\eta]$ with a possibility to find appropriate combination of sine functions for f_1, f_2 etc., which satisfy the d.E on f within the square brackets in eq. 33 going to zero. However, the log cos function in eq. (31) to start with gives rise to problems in evaluating C_3 and C_4 and to get a velocity profile with a positive velocity in the interval, $-1 < \eta < 0$ and a negative velocity in the interval $0 < \eta < 1$.

As pointed out in the last paragraph of previous section, neglecting $K(\phi^1)^2$ in eq.(18) which is same as (23) gives the simplest solutions of Fig.2. Hence other method of decoupling could be to postulate,

$$K(\phi)^2 = \left[H_0 + H_1\eta + H_2\frac{\eta^2}{2!} + \ldots\right] = \left[\sum_{i=0}^{\infty} H_i\frac{\eta^i}{i!}\right] \quad (34)$$

Using equation (23), one can integrate twice and obtain θ and using equation (24), one can obtain ϕ. It was found necessary, while evaluating constants after substituting boundary conditions to split the H-series into odd and even terms. One gets, after substitution of boundary conditions,

$$\theta_i = \left[\sum_{i=0}^{\infty}\frac{H_{2i}}{(2i+2)!}(\eta^{2i+2}-1)\right] + \left[\sum_{i=0}^{\infty}\frac{H_{2i+1}}{(2i+3)!}(\eta^{2i+3}-\eta)\right] + \frac{Gr}{2}\eta \quad (35)$$

$$\phi_i = \left[\sum_{i=0}^{\infty} \frac{H_{2i}}{(2i+4)!} (\eta^{2i+4}-1) - \left(\frac{1}{2} \sum_{i=0}^{\infty} \frac{H_{2i}}{(2i+2)!}\right) (\eta^2 - 1) \right] + \left[\sum_{i=0}^{\infty} \frac{H_{2i+1}}{(2i+5)!} (\eta^{2i+5} - \eta) - \left(\frac{1}{6} \sum_{i=0}^{\infty} \frac{H_{2i+1}}{(2i+3)!}\right) (\eta^3 - \eta) \right] + \frac{G\eta}{12} (\eta^3 - \eta) \tag{36}$$

When all the H's are zero, one gets the simple solutions of Fig.2 ($\theta_.$; $\phi_.$). For $i = 0$ there are two cases, $H_0 \neq 0$, $H_1 = 0$ gives, (θ_0 ; ϕ_0) and $H_0 \neq 0, H_1 \neq 0$ gives (θ_1 ; ϕ_1). These three solution sets are given below:

$$\theta_. = \frac{Gr}{2} \eta \tag{37}$$

$$\phi_. = \frac{Gr}{12} (\eta^3 - \eta) \tag{38}$$

$$\theta_0 = \left[\frac{H_0}{2} (\eta^2 - 1)\right] + \frac{Gr}{2} \eta \tag{39}$$

$$\phi_0 = \left[\frac{H_0}{24}(\eta^4 - 1) - \frac{H_0}{4}(\eta^2 - 1) + \frac{Gr}{12} (\eta^3 - \eta)\right. \tag{40}$$

$$\theta_1 = \left[\frac{H_1}{6} (\eta^3 - \eta)\right] + \frac{H_0}{2} (\eta^2 - 1) + \frac{Gr}{2} \eta \tag{41}$$

$$\phi_1 = \left[\frac{H_1}{120} (\eta^5 - \eta) - \frac{H_1}{36} (\eta^3 - \eta)\right] + \frac{H_0}{24} (\eta^4 - 1) - \frac{H_0}{4} (\eta^2 - 1) + \frac{Gr}{12} (\eta^3 - \eta) \tag{42}$$

Other solution sets can all be derived using the general equations (35) and (36).

No matter how many terms are evaluated, each solution set is approximate in the sense what one assumes for $K(\phi')^2$ by equation (34) to get the solution started and the ϕ' of the solution set differs from what one assumes in the beginning. Hence even if the McLaurin series expansion of series in equation (34) converges, the solution set at any stage is only an approximation of the exact solution.

Hence the first method of decoupling gives a formal closed form solution (at least for G_0 and G_1 which can be easily computed), the difficulty involved in fitting the boundary conditions is not yet surmounted. In the second method one can calculate profiles to any accuracy one desires but still the solution is only approximate. Equations (35) and (36) are natural extensions of solution set of Fig.2. However, for the preliminary attempt of this study they can be used in a meaningful way to identify the elements and postulate connectivity.

In the next section, physical meaning for the constant H_0, H_1 is discussed and will be followed-up by identification of elements and connectivity.

5. PHYSICAL SIGNIFICANCE OF H_0, H_1.

Heat flux entering at $\eta = -1$ and leaving at $\eta = +1$, can be calculated from the temperature profiles given in equations 37-42, by,

$$q = -k \frac{dT}{dy} = \frac{k(T_2 - T_1)}{b.\,Gr} \frac{d\theta}{d\eta} \tag{43}$$

Hence one gets,

$$q_.|_{y=-b} = \frac{k(T_2 - T_1)}{b.\,Gr} \times \frac{Gr}{2} \tag{44}$$

$$q_.|_{y=+b} = \frac{k(T_2 - T_1)}{b.\,Gr} \times \frac{Gr}{2} \tag{45}$$

$$q_0|_{y=-b} = \frac{k(T_2 - T_1)}{b.\,Gr} \left[\frac{Gr}{2} + \frac{H_0}{2}(-2)\right] \ldots \tag{46}$$

$$q_0|_{y=+b} = \frac{k(T_2 - T_1)}{b.\,Gr} \left[\frac{Gr}{2} + \frac{H_0}{2}(+2)\right] \tag{47}$$

$$q_1|_{y=-b} = \frac{k(T_2 - T_1)}{b.\,Gr} \left[\frac{Gr}{2} + \frac{H_0}{2}(-2) + \frac{H_1}{3}\right. \tag{48}$$

$$q_1|_{y=+b} = \frac{k(T_2 - T_1)}{b.\,Gr} \left[\frac{Gr}{2} + \frac{H_0}{2}(+2) + \frac{H_1}{3}\right. \tag{49}$$

If the one dimensional system is well insulated on all faces except the conducting surfaces at $\eta = \pm 1$, viscous heat dissipation will show up as difference between $q|_{y=-b}$ and $q|_{y=+b}$. The system is at steady state, from an experimental view point - T (y) and ϕ (y) being steady. If one defines a steady state heat transfer coefficient based on $q|_{y=+b}$ (i.e at $\eta = +1$), one has,

$$h_. = \frac{k}{b.\,Gr} \times \frac{Gr}{2} = \frac{k}{2b} \tag{50}$$

$$h_0 = \frac{k}{2b} \left[1 + \frac{2H_0}{Gr}\right] \tag{51}$$

$$h_1 = \frac{k}{2b} \left[1 + \frac{2H_0}{Gr} + \frac{2H_1}{3Gr}\right] \tag{52}$$

Or in terms of Nusselt Number, $Nu_b = 2\,bh/k$,

$$Nu_{b.} = 1 \quad (53)$$

$$Nu_{bo} = 1 + \frac{2H_0}{Gr} \quad (54)$$

$$Nu_{b1} = 1 + \frac{2H_o}{Gr} + \frac{2H_1}{3Gr} \quad (55)$$

A steady state experimental measurement, can be used to identify, the values of H_o and H_1. Experimental data such as the one shown in Fig.5 could not be used, since the characteristic dimension was taken as, L. Eckert and Drake (18) reported observations of temperature field for natural convection for a gap between two vertical surfaces made with Zehnder-Mach interferometer. An interesting correlation of Nusselt vs $Gr_b \cdot Pr$ was also presented. Bejan (19), gave an excellent summary and review of Natural convection in enclosures. It is proposed to use these two sources, for checking with the results of this study in due course .

Further, if one considers the centre line temperature, that is $\theta(0)$, one has

$$\theta_{.}(0) = 0 \quad (56)$$

$$\theta_0(0) = -\frac{H_0}{2} \quad (57)$$

$$\theta_1(0) = +\frac{H_0}{2} \quad (58)$$

In terms of Ⓗ and T, then one has,

$$Ⓗ_{.}(0) = \frac{T_{(.)}(0) - T_1}{T_2 - T_1} = \frac{1}{2} \quad (59)$$

$$Ⓗ_{(0)}(0) = \frac{T_{(0)}(0) - T_1}{T_2 - T_1} = \frac{1}{2} + \frac{H_0}{2Gr} \quad (60)$$

$$Ⓗ_{(1)}(0) = \frac{T_{(1)}(0) - T_1}{T_2 - T_1} = \frac{1}{2} \quad \frac{H_0}{2Gr} \quad (61)$$

Since the centre line temperature $T_{(\bullet)}(0)$, has to be in the interval $T_1 < T_{(\bullet)}(0) < T_2$; the following are thermodynamic restrictions on the temperature profile,

$$0 < Ⓗ(0) < 1 \quad (62)$$

$$0 < \frac{1}{2} + \frac{H_0}{2Gr} < 1 \; ; -\frac{1}{2} < \frac{H_0}{2Gr} < \frac{1}{2} \quad (63)$$

$$0 < \frac{1}{2} - \frac{H_0}{2Gr} < 1 \; ; -\frac{1}{2} < -\frac{H_0}{2Gr} < \frac{1}{2} \quad (64)$$

In other words, one has,

$$-1 < \pm\frac{H_0}{Gr} < 1 \quad (65)$$

In view of equation (34), one has,

$$(\phi_o')^2 = \frac{H_0}{K} = \frac{H_0}{Pr} \times \left[\frac{Cp}{b\beta} \cdot \frac{g_c}{g} \right] \quad (66)$$

Hence one can consider equation (65) as

$$-1 < \pm\frac{Pr\left[\frac{b\beta}{Cp} \times \frac{g}{g_c}\right]}{Gr} \cdot (\phi'_0)^2 < 1 \quad (67)$$

a thermodynamic limitation on the permissible stable oscillation of the velocity field of zero order. Or to be more relevant to the present study, the non-linear equation in terms of the H-series, gives solutions of zero order with $-1 < \frac{H_0}{2Gr} < 1$ and first order with no such restriction on H_1.

The second order non-linear equation given as equation (27), is a well studied equation (20 to 23). The approximations considered here have yet to be compared for their relevance in so far as mathematical rigor is considered.

6. SYSTEM ELEMENTS AND CONNECTIVITY

If Natural convection is to be modelled as a self-excited, oscillatory, competitive system capable of displaying characteristic limit cycles of inherent non-linear elements or generation of dissipative structures, the preliminary results of this study indicates a possible choice of elements and connectivity as presented in this section.

The positive and negative flows of the intervals, $-1 < \eta < 0$ and $0 < \eta < +1$ could be lumped as shown in Fig.6 and 7. The connectivity of the system is shown in Fig.8. The salient point of Fig.8 is a transforming transducting element. However the postulates involved in Fig.7 and 8, like the approximate solutions of the non-linear system of equations presented in the earlier section, require experimental verification on a one dimensional system involving both steady state and a dynamic response. This aspect of study is currently in progress, in our laboratories. However for completeness of the present study it is included herein.

Finally, some comments are in order for the next higher level model shown in Fig.3 and 4. For simplicity of analysis, it is a common practice in engineering to consider the y-component of flow as small and neglect y-component of Navier-Stokes equation, leading to boundary layer theory solutions. By such a method, one gets the prominant features of the system displayed in the domain of stable zone, but the features which contribute to transformations that a dynamic system undergoes such as dissipative structures cannot be grasped. It is the belief of the postulates of the connectivity of present study, that elements such as transforming transductors will enable one to systematise efforts in this direction. The transforming transductor element presented in the preceeding para is one in which the potential energy is transducted into thermal energy. In the system depicted in Fig.3 and 4, the corresponding additional element will involve v_y to v_z transforming element in addition to the transduction element in Z-direction. Such additional complexities have to wait till what is postulated herein is first established experimentally.

7. SUMMARY AND CONCLUSION

One dimensional momentum and energy equations, of steady state heat transfer between two vertical plates, with a fluid in between undergoing natural convection is analysed to study the effect of non-linearities of the governing equations. A choice of system elements and their connectivity to study the dynamic behaviour is postulated. Possibility to extend it for higher level models is also discussed.

ACKNOWLEDGEMENT

The author acknowledges with pleasure, participation in discussions by Prof. M.S. Ananth of Chemical Engineering and Prof. R. Subramanian of Mathematics of our Institute, during this study.

NOMENCLATURE

b Half width of gap, ft

Br Modified Brinkman Number, Dimensionless,

$$= K = \left[\frac{\mu b \bar{\beta}}{k} \times \frac{g}{g_c} \right] = Pr \left[\frac{b\bar{\beta}}{C_p} \times \frac{g}{g_c} \right]$$

f Arbitrary function of η

G, H Infinite Series; G_i and H_i are constants occuring in the ith term of the series ; Dimensions as per definition of series ; $G_{(N)}$ denotes the series consisting of N terms.

Gr Grashoff Number, Dimensionless

$$= \frac{b^3 \bar{\rho}^2 \bar{\beta} g \Delta T}{\mu^2}$$

g Gravitational acceleration, ft. sec^{-2}

g_c Gravitational conversion factor, dimensionless, (Lb_m/Lb_f) (ft. sec^{-2}).

h Heat transfer coefficient, B.T.U. ft^{-2} $(°F)^{-1}$. sec ; Subscripts as defined in text.

i index in infinite series

k Thermal conductivity of fluid, B.T.U. ft. ft^{-2}. $(°F)^{-1}$. sec;

K Dimensionless number, Modified Brinkman Number

N Suffix on G or H occuring on the constant of the Nth term of the series

Nu_b Nusselt Number, Dimensionless, with b as characteristic length,

$$= \frac{2bh}{k}$$

P Pressure, Lb_f. ft^{-2}. sec^{-1} ; Subscripts as defined in text.

Pr Prandtl number, Dimensionless $= C_p \mu / k$

q Heat flux, B.T.U. ft^{-2}. $sec.^{-1}$; Subscripts as defined in text.

q_{irr} Irreversible heat generation due to viscous dissipation, units as for q

T Temperature, °F ; subscripts as defined in Text; **T** - Lumped temperature in an interval of n.

ΔT Temperature difference, °F

v_z, v_y velocity components in z and y direction, ft. sec^{-1}; v - Lumped velocity in an interval n.

$\bar{\beta}$ Thermal coefficient of volumetric expansion $(°F)^{-1}$

$\bar{\rho}$ Density of fluid at $\bar{T}$, Lb_m. ft^{-3}

μ Viscosity of fluid, Lb_m. ft^{-1}. sec^{-1}

ϕ Dimensionless velocity $= b v_z \bar{\rho} / \mu$

Ⓗ θ Dimensionless temperatures as defined in text.

REFERENCES

1 Rayleigh, Scientific papers, 4 (1911-1919) 433.

2 Low, Proc. Roy. Soc. - A, 125 (1929) 180.

3 Jeffreys, Phil. Mag, 2, (1926) 833.

4 Jeffreys, Proc. Roy. Soc, Lond, CXVIII, 195 (1928)

5 Jeffreys, ibid., 118 (1928) 195

6 Turner, J.S., Buoyancy in fluids, Cambridge University Press, (1973)

7 Drazin, P.G., and Reid, H.W., Hydrodynamic stability, Cambridge University Press, (1981)

8 Glansdorff, P., and Prigogine, I., Thermodynamic theory of structure, Stability and Fluctuations, John Wiley, (1971)

9 Ahlers, G., and Welden, R., Phys. Rev. Letters 44 (1980) 445

10 Greenside, H.S., Coughran Jr., and Schreyer,N.L., ibid., 49 (1982) 726

11 Haken, H., Advanced Synergetics, Springer-Verlag, (1983)

12 Horsethemke , W.,and Lefever, R., Noise induced transitions, Springer-Verlag, (1984)

13 Acrivos, A., AIChE Journal, 4 (1958) 285-289

14 Bird,R.B., Stewart,W.E., and Lightfoot,E.N., Transport phenomena, John Wiley, (1960).

15 Lorenz, L., Weiderman's Ann. Physik, 13 (1881) 582-606

16 Schmidt,E., and Beckman,W., Tech. Mech. Thermo-dynamik, 1 (1930) 341-391.

17 Sparrow, E.M., and Grogg, J.W., Trans ASME 80 (1958) 379-386 .

18 Eckert, E.R.G., and Drake Jr., R.M., Heat and Mass Transfer, Mc Graw Hill - Koka Kusha, Tokyo, (1959) 330-331 .

19 Bejan, A., Convective Heat Transfer, Chapt.5, p.159 - 184, John Wiley, (1984).

20 Kamke, E., Differential gleischungen, Chesla pub Co., NY, (1959)

21 Korn, G.A., and Korn, T.M., Mathematical hand book for Scientists and Engineers, Mc Graw Hill, (1961) 277-286

22 Friedley, J.C., Dynamic behaviour of processes, Prentice Hall Inc., NJ, (1972), [pp.465, 466; Table 11.2-1]

23 Barnfield, S.R., and Lakshmikantham,V., in 'Non-linear phenomena in Mathematical Sciences' Edt. by V.Lakshmikantham, Academic Press, (1982) 117-122.

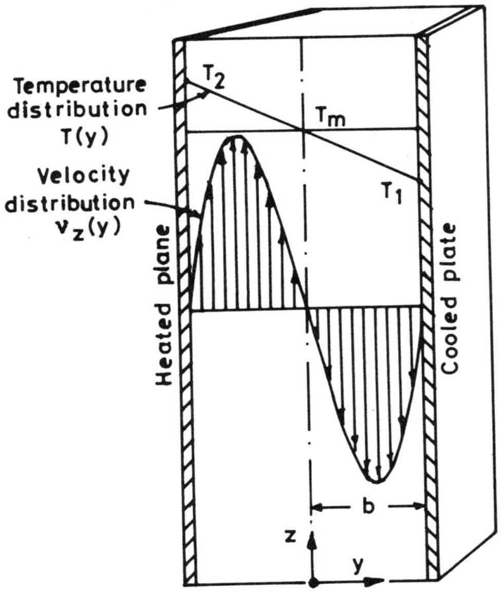

Fig.1 Rudimentary Model (Ref.14)

RUDIMENTARY MODEL

[FROM REF.(14)]

ENERGY EQN $\frac{d^2T}{dy^2} = 0$; BC: $y = -b;\ T = T_2$
$y = +b;\ T = T_1$

MOMENTUM EQN $\mu \frac{d^2 v_z}{dy^2} = -\bar{\rho}\bar{\beta} g (T - \bar{T})$; BC : $v_z = 0$ at $y = \pm b$

EQN OF STATE $\rho = \bar{\rho}[1 - \bar{\beta}(T - \bar{T})]$

TEMPERATURE PROFILE $\boxed{T = T_m - \frac{1}{2}\Delta T \left(\frac{y}{b}\right)}$ $T_m = \frac{1}{2}(T_2 - T_1)$; $\Delta T = (T_2 - T_1)$

VELOCITY PROFILE $v_z = \frac{\bar{\rho}\bar{\beta} g b^2 \Delta\bar{T}}{\mu^2}[n^3 - An^2 - n - A]$

$n = \frac{y}{b}$; $A = \frac{6(T_m - \bar{T})}{\Delta T}$

NO NET FLOW $\int_{-1}^{+1} v_z\, dn = 0$

$\therefore A = 0$; or $T_m = \bar{T}$

$\therefore v = \frac{\rho \beta g b^2 \Delta T}{12\mu}(n^3 - n)$

DIMENSIONLESS VELOCITY PROFILE $\phi = \frac{b v_z \bar{\rho}}{\mu}$

$Gr = \frac{\bar{\rho}^2 \bar{\beta} g b^3 \Delta T}{\mu^2}$

$\boxed{\phi = \frac{Gr}{12}(n^3 - n)}$

Fig.2 Rudimentary Model (From Ref.[14])

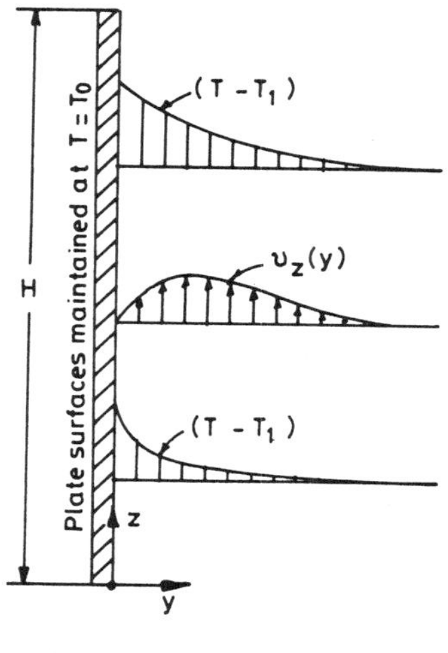

Fig.3 Next Level Model (Ref.14)

NEXT LEVEL MODEL

[FROM REFERANCE (14)]

FLOW FIELD $v_z(x,y)$; $v_y(x,y)$

CONTINUITY EQN $\frac{\partial v_y}{\partial y} + \frac{\partial v_z}{\partial z} = 0$

MOMENTUM $\rho\left(v_y\frac{\partial}{\partial y} + v_z\frac{\partial}{\partial z}\right)v_z = u\left(\frac{\partial^2 v_z}{\partial y^2} + \boxed{\frac{\partial^2 v_z}{\partial x^2}}\right) + \rho g\beta(T-T_1)$

ENERGY $\rho\hat{C}_p\left(v_y\frac{\partial}{\partial y} + v_z\frac{\partial}{\partial y}\right)(T-T_1) = k\left(\frac{\partial^2}{\partial y^2} + \boxed{\frac{\partial^2}{\partial z^2}}\right)(T-T_1)$

BC

$y = 0;\ v_y = v_z = 0;\ T = T_0$

$y = \infty;\ v_y = v_z = 0;\ T = T_1$

$z = -\infty;\ v_y = v_z = 0;\ T = T_1$

DIMENSIONLESS VARIABLES

$\textcircled{H} = \frac{T-T_1}{T_0-T_1};\ \zeta = \frac{z}{H};\ n = \left(\frac{B}{\mu\alpha H}\right)^{1/4}\cdot y$

$\phi_z = \left(\frac{\mu}{B\alpha H}\right)^{1/4}\cdot v_z$

$\phi_y = \left(\frac{\mu H}{\alpha^2 B}\right) v_y$

$\alpha = \frac{k}{\rho\hat{C}_p};\quad B = \rho g\beta(T_0 - T_1)$

DIMENSIONLESS GOVERNING EQNS

$\frac{\partial\phi_y}{\partial n} + \frac{\partial\phi_z}{\partial\zeta} = 0$

$\frac{1}{Pr}\left(\phi_y\frac{\partial}{\partial n} + \phi_z\frac{\partial}{\partial\zeta}\right)\phi_z = \frac{\partial^2\phi_z}{\partial n^2} + \textcircled{H}$

$\phi_y\frac{\partial\textcircled{H}}{\partial n} + \phi_z\frac{\partial\textcircled{H}}{\partial\zeta} = \frac{\partial^2\textcircled{H}}{\partial n^2}$

BC

$n = 0;\ \phi_y = \phi_z = 0;\ \textcircled{H} = 1$

$n = \infty;\ \phi_y = \phi_z = 0;\ \textcircled{H} = 0$

$\zeta = -\infty;\ \phi_y = \phi_z = 0;\ \textcircled{H} = 0$

AVERAGE HEAT FLUX $q_{Avg} = +\frac{k}{H}\int_0^1 -\frac{\partial T}{\partial y}\Big|_{y=0} dz$

$= k(T_0 - T_1)\left(\frac{B}{\mu\alpha H}\right)^{1/4}\boxed{\int_0^1 -\frac{\partial\textcircled{H}}{\partial n}\Big|_{n=0} d\zeta}$

$= C\cdot\frac{k}{H}(T_0 - T_1)(Gr\cdot Pr)^{1/4}$

Fig.4 FOR EVALUATION OF "C" SEE REFERANCES 15, 16 and 17

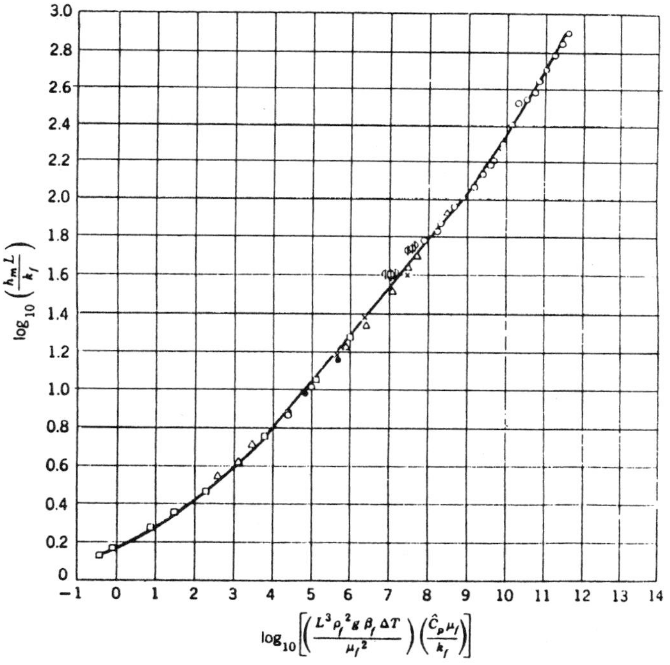

Fig.5 From Ref.(14)

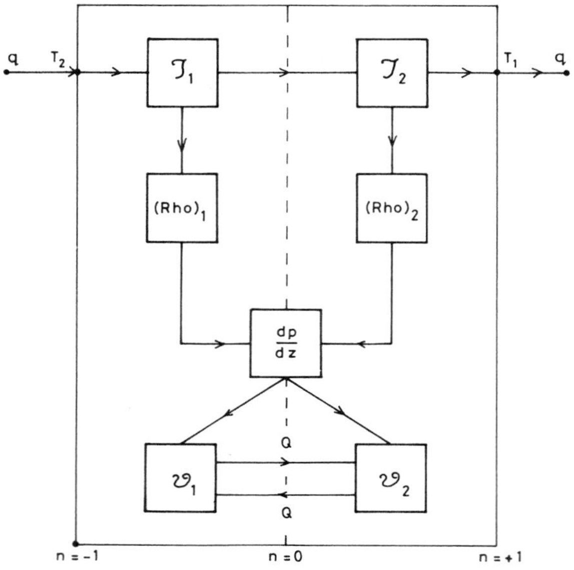

Fig.6

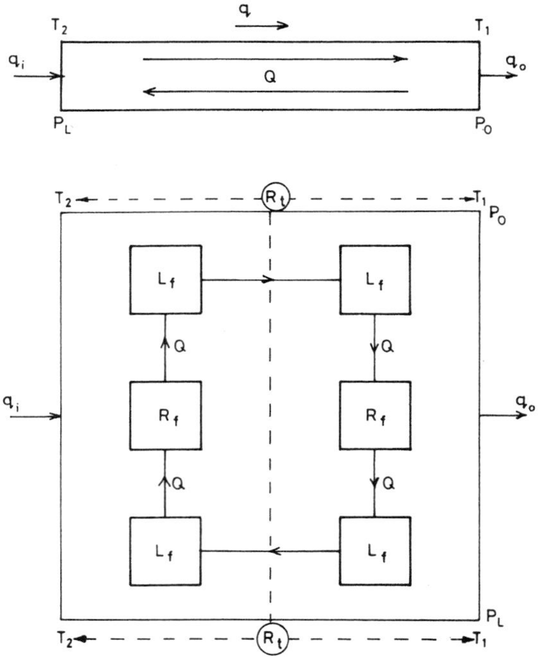

Fig.7

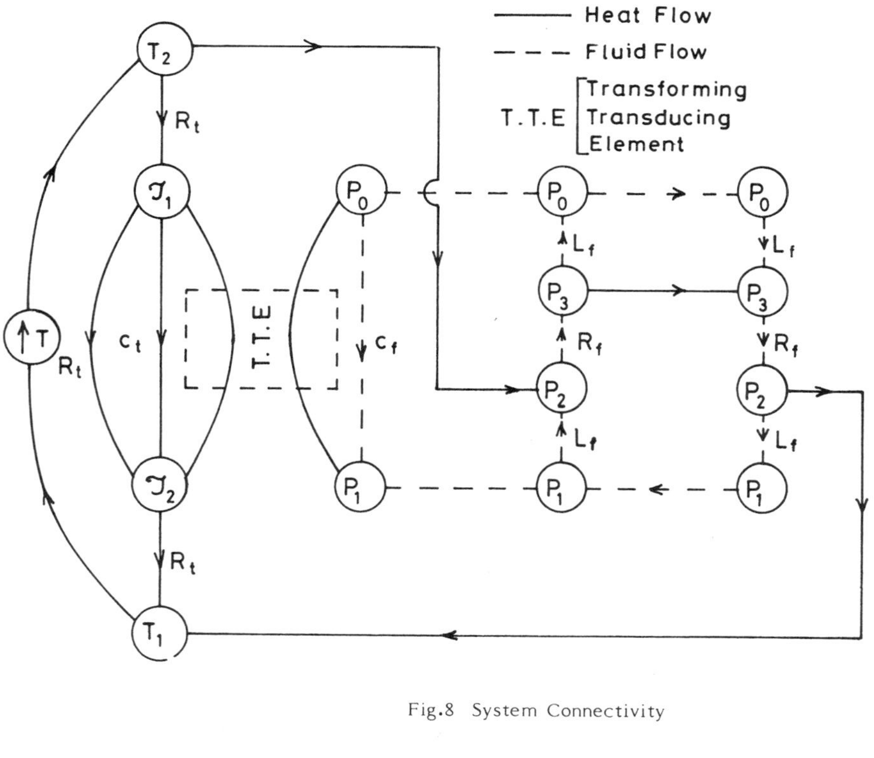

Fig.8 System Connectivity

Technical Utilization of Solar Energy— Its Effects on Heat Transfer Research

E. W. P. HAHNE and N. M. FISCH
Institut für Thermodynamik und Wärmetechnik
Universität Stuttgart
P.O. Box 80 11 40, D-7000 Stuttgart 80, FRG

1. INTRODUCTION

At the beginning of the Seventies - with the first oil crisis - the interest in solar energy as a possible resource to replace or save fossil - or nuclear sources, has spontaneously risen. Early, oversized enthusiasm could not be verified, for many reasons; one of them, and not the least one, was that laws of heat transfer had not been properly applied or were even ignored.

From the heat transfer point of view, the utilisation of solar energy means the solution of transient problems, comprising all three modes of heat transfer: radiation, convection and conduction and as a boundary condition an intermittingly acting, uncontrollable heat source of a low heat flow density (compared to electrical or fossil fuel heating).

So far, such conditions were unmet in heating technology. They called for a number of components to be combined in a system: a collector, a storage device, an auxiliary heat source and control-and regulating installations. Because of the restrictive boundary condition, the best components have to be found to interact optimally in the system. A weak component or an improper regulation strategy deteriorates the entire system. Component-and system optimisation was and is the goal in heat transfer research.

2. SOLAR COLLECTORS

The most common solar collector is the flat plate fin-tube type: a number of tubes which are connected by thin metal plates form the absorber, which collects the solar radiation, transforms it into heat and transmits to the liquid flowing in the tubes; for insulation the absorber is covered with one or more glass plates and a backside insulation. This all forms the collector.

For a heat transfer analysis, the absorber presents a fin-problem with combined radiation, convection, conduction and a variable heat flow density boundary condition. The temperature in the absorber is three - dimensional and a function of the following parameters:

$\vartheta = \vartheta$ [geometrical parameters: x,y,z; δ_A, δ_G; s.

radiation paraameters : α^*_A , α^*_G ; ε^*_A, ε^*_G, τ^*_G.

thermal parameters : k_A, k_G; c_A, c_G.

Meteorological parameters: E_{glob}, w; ϑ_a , ϑ_{sky}.

operational parameters: m, ϑ_i

number of (glass) cover plates

geometrical and thermal parameters of the backside insulation material and the frame] .

Calculations of the temperature field in the absorber can be performed numerically based on the heat balances for a symmetric element of the collector [1] as shown in figure 1. Elements of length Δy are thermally interconnected to form the entire length of the collector.

A respective nodal model for such an element is shown in figure 2. This already contains some simplifications such as uniform glass cover-and insulation-temperature.

As a result, a computer plot for the temperature field within the fin and tube (x,z plane) at the very beginning of the absorber (y = 0) is presented in figure 3. Even for the comparatively small thermal conductivity of κ = 5 W/m K, as used here, the temperature difference within the fin thickness is small enough to be neglected for more complex calculations.

The temperature distribution along the length (in flow direction of the liquid within the tubes) of the collector is shown in figure 4. The temperature rise in the first half of the absorber plate amounts to about 60% due to lower heat losses with lower absorber temperatures. With this result, further simplifications with larger Δ y can be introduced into the more complex calculations.

A large number of properties and complicated interrelations still remain. In order to allow for comparison and valuation of collectors, a collector-efficiency is introduced:

η = usable solar energy/solar energy incident on collector.

This can be measured according to

$$\eta = \dot{m}\, c_p\, (\vartheta_o - \vartheta_i) / E_{glob}$$

and defined as (1)

$$\eta = [E_{glob}\, \tau^*_G\, \alpha^*_A - k\ (\vartheta_A - \vartheta_{amb})\ ]\ /\ E_{glob}$$

$$\eta = \tau^*_G \alpha^*_A - K/E_{glob}\, (\vartheta_A - \vartheta_{amb}) \qquad (2)$$

The absorber temperature ϑ_A can hardly be determined accurately (see figs.2 and 3), but we can easily measure the working fluid inlet - and outlet temperatures ϑ_i, ϑ_0 and calculate an arithmetic mean temperature

$$\vartheta = (\vartheta_i + \vartheta_0)\ /\ 2.$$

Introducing this mean temperature instead of ϑ_A into equ. (2) we obtain

$$\eta' = \tau^*_G \alpha^*_A - k\ /\ E_{glob}\ (\vartheta_i + \vartheta_o)\ /\ 2 - \vartheta_{amb} \qquad (3)$$

The absorber temperature ϑ_A is certainly higher than the mean fluid temperature $\bar{\vartheta}$ because the heat has to be conducted through the fin under a temperature gradient. Thus we have

$$[(\vartheta_i + \vartheta_0)\ /\ 2 - \vartheta_{amb}] < (\vartheta_A - \vartheta_{amb}), \qquad (4)$$

and $$\eta' > \eta \qquad (5)$$

In order to obtain realistic values for the efficiency and still be able to use the easily measurable ϑ_i and ϑ_0, a factor, the "efficiency factor"

$$F' = \frac{\eta}{\eta'} \qquad (6)$$

can be introduced to yield

$$\eta = F'\, \tau^*_G\, \alpha_1 - F'\, k/E_{glob}\ [(\vartheta_i + \vartheta_0)\ /\ 2 - \vartheta_{amb}\ ] \qquad (7)$$

As no transients are taken into account, these equations yield instantaneous efficiencies.

For transient processes an "effective efficiency" η_{eff} can be calculated from

$$\eta_{eff} = [\ \sum_{t=0}^{t} c_F\ \dot{\bar{m}}\ (t)\ (\vartheta_0\ (t) - \vartheta_i\ (t)\ )\ \Delta t\ ]\ /\ \sum_{t=0}^{t} E_{glob}\ (t)\ \ \Delta t \qquad (8)$$

Instantaneous efficiencies are plotted advantageously in the form

$$\eta \text{ vs. } [\ (\vartheta_i + \vartheta_0)\ /\ 2 - \vartheta_{amb}\ ]\ /\ E_{glob}$$

as shown in figure 5. Parameters which are assumed constant are indicated in the figure. The number of glass cover plates is given by different symbols. In most cases, the metal materials give higher efficiencies than plastic ones.

In the hypothetical case that

$$(\vartheta_i + \vartheta_0)\ /\ 2 - \vartheta_a = \underline{0} \qquad (9)$$

equ. (7) yields

$$\eta = \eta_0 = F'\ \tau^*_G\, \alpha^*_A \qquad (10)$$

This means that the ordinate intersection η_0 provides an information on the optical properties τ^*_G and α^*_A of the collector.

The differentiation of equ. (7)

$$\frac{d\,\eta}{d\,\{\,[\,(\vartheta_i + \vartheta_0)\ /\ 2 - \vartheta_{amb}\ ]\ /\ E_{glob}\ \}} = -\,F'\,k \qquad (11)$$

gives the inclination of the curves in fig.5 and with this an information of the overall heat losses k of the collector.

These possibilities of information are often mentioned in papers and textbooks, however, they have to be handled cautiously: from fig.5 it appears that 1) the metal absorber has a much better optical quality than the plastic one, while 2) the overall heat transfer coefficient for the metal absorber is inferior to that of plastic. Both assumptions are wrong: The curves were numerically obtained with the same optical parameters (as given in the figure), no temperature effects were taken into account; thus, 1) the same ordinate intersection values η_0 should be obtained. The plastic absorber being a poor heat conductor will have higher absorber temperatures than the metal absorber, with L/d = const, consequently 2) the overall heat transfer coefficient for plastic absorbers must be larger than for metal absorbers. The results of fig. 5, however, are correct and can be interpreted correctly when the efficiency factor F' is taken into account (equs. (10) and (11)): If, in an efficiency plot, simultaneously, the ordinate intersection η_0 and the inclination of the curve is small, then, most likely the efficiency factor F' is small and we have poor collector from the beginning. In such a case, the heat flow from the fin to the tube should be improved before any other corrections are made.

Another pecularity of this type of plotting is demonstrated in figure 6. The efficiency curve splits up for different solar irradiances, wind velocities and ambient temperatures. This means that efficiency plots obtained by open air tests where all these data vary, will exhibit an unavoidable scatter in data just because of the scale of the plot.

In fig. 6, the region of scatter is presented for quite extreme conditions and steady - state calculations. Transient conditions can also be well simulated.

In fig. 7, a comparison is presented for measured and calculated outlet temperatures subject to a highly intermittent solar irradiance and a variable fluid flow rate [2]. The agreement is very good when the transient conditions are taken into account, the steady - state model calculation yields large deviations.

The interrelation of paramaters is, relatively, easy to obtain in numerical parameter studies [3]: in tables 1 and 2, the "effective efficiency" taken over the period of one day and the "warming up" time of a collector are presented.

In June, the transient - state efficiency of a collector with parameters as listed in column "CONST." is 57%. If e.g. the selectivity of the absorber would be increased from $\alpha_A^* / \varepsilon_A^* = 1$ to $\alpha_A^* / \varepsilon_A^* = 9$ (listed under c) and again only one glass cover would be applied, then the efficiency increases to 64.5%. With three glass covers, the effect of selectivity becomes quite small.

To heat up the collector from 20°C to 35°C takes 84 minutes. Not before 35°C is reached the fluid flow through the collector is switched on. When this collector is not tilted by 45°, but located flatly on the roof (with the tilt 0°) the warming time is only 36 minutes.

In winter, the efficiency of this collector is only 46.5% and warming-up time 96 minutes. With a zero inclination of the collector it would now take 198 minutes to heat from 10° C to 35° C.

Solar collectors only costitute one component of a solar heating system, another one - not less important - is a device to store thermal energy.

3. THERMAL STORAGE

An enormous amount of ideas, new concepts and applications were and are being born in the field of thermal storage. Thermal energy can be stored as sensible heat, e.g. in water or solid material or as chemical heat of fusion, or as latent heat with melting or evaporation processes. Heat transfer processes occur with charging and discharging of the store and with internal and external heat losses. Transient natural convection in enclosures, combined with forced convection and thermal conduction constitute the controlling mechanisms. Again there is the handicap that energy-density and temperatures are small and all processes are transient.

The physics behind these processes are known, problems occur when predictions are required for proper design.

For example, natural convection heat transfer from a spirally wound finned tube heat exchanger in a water tank finally was found to be best predicted according to the following model: each cylindrical coil is subdivided into two horizontal and two vertical sections for which heat transfer coefficients are calculated sectionwise and coilwise with [4]

$$Nu = \{0.60 + 0.387\, Ra^{1/6} [1 + (0.559/Pr)^{9/16}]^{8/27}\}^2 \qquad (12)$$

and taking temperature-dependent properties into account.

In figure 8, a comparison is presented between calculated and measured heat transfer for three different heat exchanger configurations [5]. The configurations are described in table 3.

Solid particle regenerators obtained revived interest in solar air-heating applications; new investigations had to be performed, because geometries, mass flow rates and exchange periods were quite different from those found with industrial regenerators. The temperature response of a cube-shaped 1.16 m^3 model pebble bed store during the charging period is shown in figure 9 [6]. This store contains granite pebbles - average diameter 40 mm - and is charged here by hot air with 68.5° C and a flow rate of 254 kg/h. Temperatures are measured in 9 points in two different levels, respectively and in the inlet and outlet duct.

In figure 10, the temperature distribution in these levels is shown, for a time shortly after the start of charging (point ⓐ in figure 9). The good heat transfer to the pebbles causes a fast, steep increase of temperatures in the upper level (see also fig.9) (the upper level is about 180 mm below the top layer of the pebble bed), while the lower level experiences only a small temperature increase. The higher temperatures in the central part are caused by the inlet duct, a 125 mm tube in the top centre.

The isotherms across the two layers at various times are presented in figure 11. The letters ⓐ to ⓓ indicate corresponding times in figure 9. A front of large temperature gradients (these are directly proportional to the number of isotherms and inversely proportional to the distance between isotherms) migrates downwards through the bed, heating the various layers consecutively. From this process it can be concluded whether the store is properly sized for a given air flow, pebble size and pebble material.

The temperature response of a single stone (d = 50 mm) was measured at its surface and its centre. The result is shown in figure 12. The temperature in the centre of the stone follows closely enough that of the surface, thus a homogeneous stone temperature may be assumed for mathematical simulations. If the effective thermal conductivity is calculated [7], values of $\lambda_{eff} = 0.4$ to 0.5 W/m k for 30° and 80° are obtained, respectively. The measured values proved to be larger by 25%.

Pebble beds flooded with water are applied as so-called "ground stores" for the storage of solar heat. The effects of ground structure, geometrical configurations, mass flow rates and temperatures during charging and discharging can be studied in model stores.

Such a store is shown in figure 13. The store can be charged or discharged by direct exchange of water or indirectly by six heat transfer tubes arranged in two levels. Temperatures are measured in 13 levels within the bed and their distributions for different times during charging and discharging are shown in figure 14 [8]. Store temperatures for charging are presented as

$$\theta_{S,c} = (\vartheta - \bar{\vartheta}_{S,init}) / (\vartheta_i - \vartheta_{S,init}) \qquad (13)$$

in the upper part of the diagram, and for discharging as

$$\theta_{S,dc} = 1 - (\vartheta - \bar{\vartheta}_{S,init}) / (\vartheta_i - \bar{\vartheta}_{S,init}) \qquad (14)$$

Temperature changes are largest within the first two hours of charging or discharging - as one would expect - because the driving temperature differences are largest then as well as the heat-transfer coefficient (α = 217 W/m^2 k, for this permeability of $B = 0.122\ 10^{-6}\ m^2$). The zone below the lower level of the heat exchanger tubes is hardly affected in either case and thus takes little part in the storing process.

If cycling times of about 12 hours are anticipated, this type of storage configuration (geometry of heat exchanger tubes and arrangement, permeability of the bed) appears proper. For solar energy utilization, however, charging temperatures and mass flow rates depend on collectors and all their interrelations to other parameters; discharging temperatures and mas flow rates depend on heating requirements and habits of inhabitants.

All such interrelations have to be taken into account, so that evaluations for solar utilization finally mean the evaluation of a system where human beings, a building, collectors, storage devices, auxillary heaters and regulators are combined.

4. HEATING SYSTEMS

Quite a number of solar heating projects have become failures, because system considerations were insufficient.

They are - no doubt - very complicated: a "simple" house as in figure 15 turns into a complex energy system [6] as in figure 16.

For the mathematical treatment of such a system, a data flow model has to be conceived: in figure 17, e.g. the meteorological data and the building construction data together with users' behaviour affect the heat load of the building; the meteorological data and the heating-system parameters affect the collected useful energy. This energy is either used directly or stored; in either case it affects - together with the heat load - the amount of auxiliary energy necessary to meet the demands. The miniature plots show that all processes are transient.

There is a huge amount of data to be handled and data compression is a must: meteorological outputs are compressed into a diurnal distribution averaged for each month of a year; thus we have 12 different curves for irradiance, ambient temperature and wind velocity versus time of the day. Users' habits are averaged in the same way. The collector behaviour is taken from efficiency curves and processes within thermal store are most often calculated from heat balance equations only (transfer processes are neglected).
Computer programs have been developed which take account of all the many detail parameters in different varieties and completness.

A property to evaluate and compare systems is the system efficiency η_{syst}, which is defined as

$$\eta_{syst} = \frac{\text{usable heat flow}}{\text{from solar energy}} \Big/ \frac{\text{solar energy incident}}{\text{on collector field}} .$$

$$\eta_{syst} = \sum_{t=0}^{t} Q_{usable}(t)\,\Delta t \Big/ A_c \sum_{t=0}^{t} E_{glob}(t)\,\Delta t \qquad (15)$$

This system efficiency (for the house shown in fig.15) is presented in figure 18 in its dependence from collector area and geographic location. For a special type of a one - family house and specific users' habits, a collector area of about 7 m^2 appears best with efficiencies between 18 and 25%. The decrease in efficiency with an increased collector area indicates that additionally collected energy is not transformed into a usable heat flow, becasue of system - or component deficiencies . The diagram of fig.18 also presents the electrical energy consumption for pumps and ventilators for the house.

The system efficiency indicates how well solar energy is used; another property - the "solar fraction" f- indicates, how much of the total heat load is covered by solar energy:

$$f = \frac{\text{usable heat flow}}{\text{from solar energy}} \Big/ \frac{\text{total heat load}}{\text{of building}}$$

$$f = \sum_{t=0}^{t} Q_{usable}(t)\,\Delta t \Big/ \sum_{t=0}^{t} Q_{required}(6)\,\Delta t \qquad (16)$$

This solar fraction for our house-heating system is plotted versus the collector area in figure 19. With a larger collector more of the required heat is collected from solar energy. This solar fraction is very sensitive to weather conditions as these affect both the usable heat flow and the required heat load simultaneously. Small collector areas with about 7 m^2 as found best for efficiency would yield very low solar fractions, so further considerations have to be made. For a given house and given users' habits, ofcourse, other system may be conceived. Eight variations are described in table 4: the air flow through the collector field may be open (once through) or closed, it may be controlled or not, so that outlet temperature is constant or not and there may be heat stores or not - water and / or pebble bed - for heating and domestic and hot water supply.

The solar fraction vs. collector area for all these different systems is shown in figure 20: a heat store clearly increases the solar fraction as for systems SYS 05 to 08, a water store in the air - heating system (SYS 07) is not as good as a pebble bed for $A_c > 30$ m^2.. for small collector field areas ($A_c < 10$m^2) all systems with stores for the air-heating system yield about the same solar fraction. The relation of collector area and storage capacity is important for the scaling of heating systems. For the system of fig.15 and a cube-shaped pebble-bed store, the solar fraction is presented in figure 21 with respect to heat capacity of the store and various collector areas.

Stores may easily be oversized; evern for collector fields larger than 100 m^2, a pebble-bed store of about 20 m^3 is large enough to provide a solar fraction of about 37%; when the store is twice as large, only an extra 2% are gained for the solar fraction.

Utilisation of solar energy has risen many questions in heat transfer research. Only a few were mentioned here many more remain unmentioned, e.g. those connected with "passive" solar utilisation, solar cooling or drying and solar power stations.

Principally, solar energy utilisation has enriched heat transfer research with problems of transient combined heat transfer in highly interacting systems.

NOMENCLATURE

A	heat transferring area
B	permeability
C	heat capacity
c_p	isobaric specific heat capacity
d	tube diameter
E_{glob}	global solar irradiance
F'	efficiency factor
f	solar fraction
h	height
k	overall heat transfer coefficient
L	length
$\dot{m}$	mass flow rate per m^2 collector area
n	number of glass covers
Q	heat flow
s	distance: absorber - glass cover
t	time
(t)	function of time
w_{amb}	wind velocity
x,y,z	coordinates
—	mean value
Nu	Nusselt number
Ra	Raleigh number
Pr	Prandtl number

Greek symbols

α	heat transfer coefficient
α^*	absorptivity
γ	collector tilt
δ	thickness
ε^*	emissivity
η	efficiency
θ	dimensionless temperature
	temperature
λ	thermal conductivity
τ^*	transmissivity

Indices

A	absorber
amb	ambient
C	collector (field)
c	charging
dc	discharging
F	fluid
G	glass cover
I	insulation
i	inlet (fluid)
o	outlet (fluid)
S	storage
sky	sky
syst	system

REFERENCES

1. Hahne, E., Fisch, N., Aráfa, A., "The flat plate solar collector, its steady-state and transient-state behaviour" Solar Energy Int. Prog. 1 (1978) 159-186 Pergamon Press.
2. Arafa, A., "Ein Beitrag zur optimalen Auslegung eines solaren Heizungssystems fur ein Wohngebaude" Doctoral thesis, Stuttgart (1982).
3. Hahne, E., "Parameter Effects on Design and Performance of Flat Plate Solar Collectors" Solar Energy (ISES) 1985 (in press)
4. Churchill, S.W., and Chu H.H.S., "Correlating equations for laminar and turbulent free convection from a horizontal cylinder". Int. J. Heat Mass Transfer 18 (1975) 1049-1053.
5. Fisch, N., et al., "Technische Nutzung solarer Energie - Solar thermische Wandlung und Warmespeicherung". BMFT-Report 03E8021 A,(1985) 1984.
6. Fisch, N., "Systemuntersuchungen zur Nutzung der Sonnenenergie bei der Beheizung von Wohngebauden mit Luft als Warmetrager" Doctoral thesis, Stuttgart (1984).
7. Schlunder, E.U., "Heat Transfer in Packed Beds" Z.Warme - und Stoffubertragung 1 (1968) 153-158.
8. Hahne, E., et al., "Zukunftsorientierte Warmeversorgung fur Institute der Energietechnik der Universitat Stuttgart" BMFT - Report 03E81 87A, (1985) 1984.

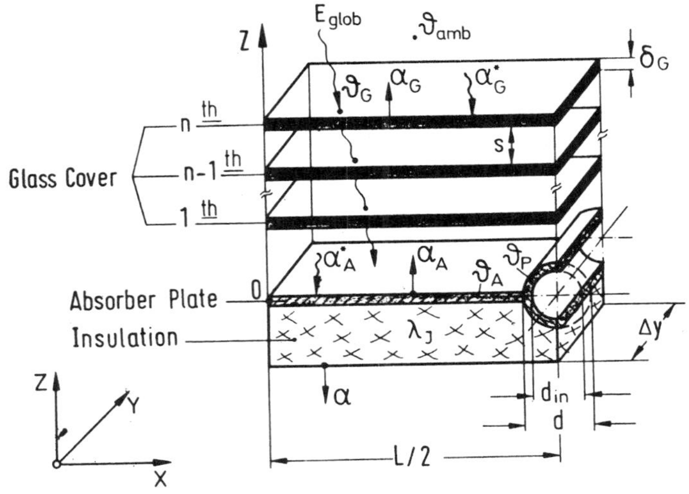

Fig.1 Schematic illustration of a symmetric element of a flat plate collector

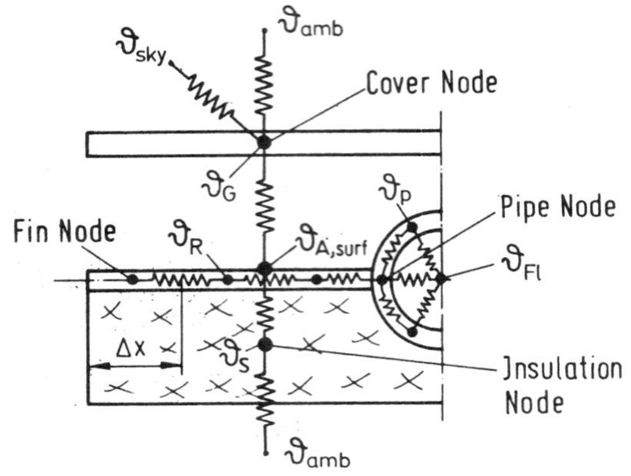

Fig. 2:
Nodal model of the symmetric element

Fig.2 Nodal model of the symmetric element

A respective nodal model for such an element is shown in figure 2. This already contains some simplifications such as uniform glass cover-and insulation-temperature.

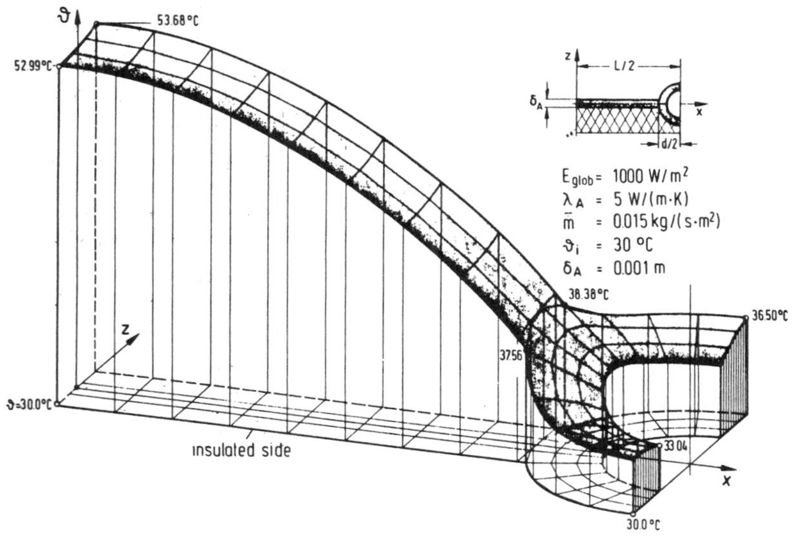

Fig.3 Temperature field in the x,z plane within the first element (y=0)

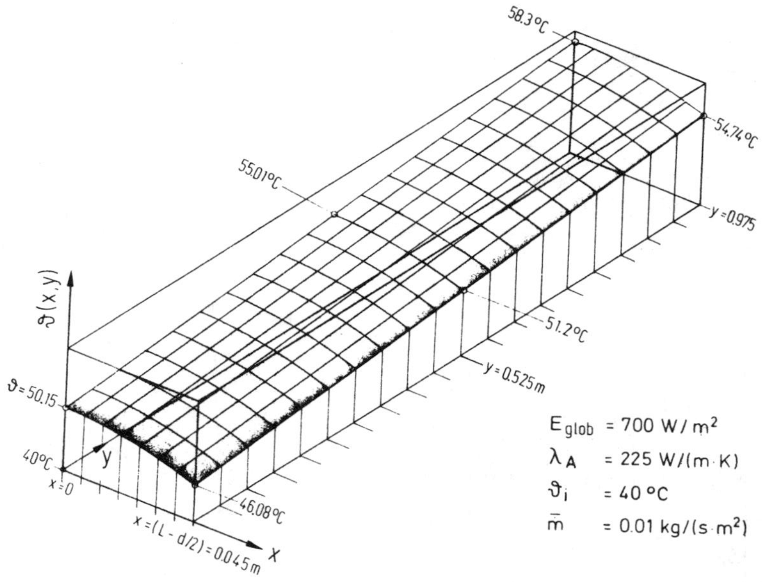

Fig.4 Temperature field in the x,y plane of the absorber plate

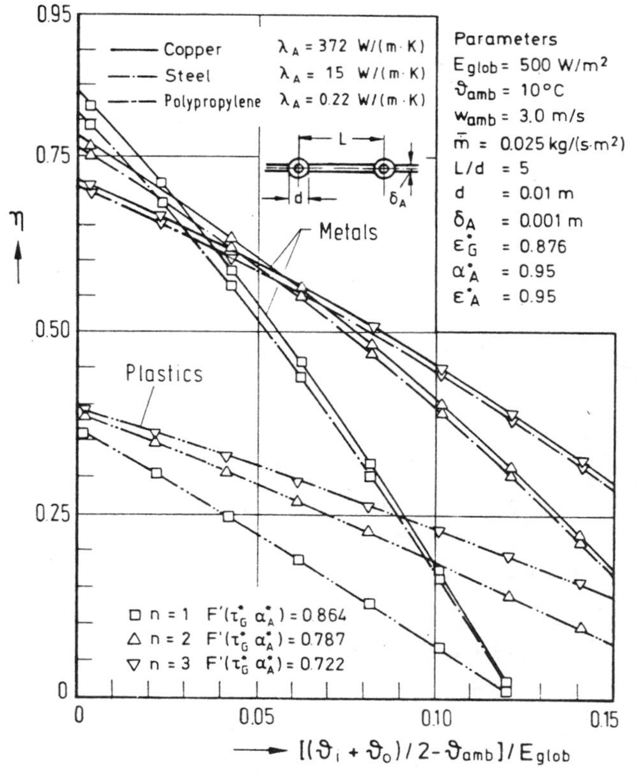

Fig.5 Efficiency plot of a flat plate collector. Effect of absorber material

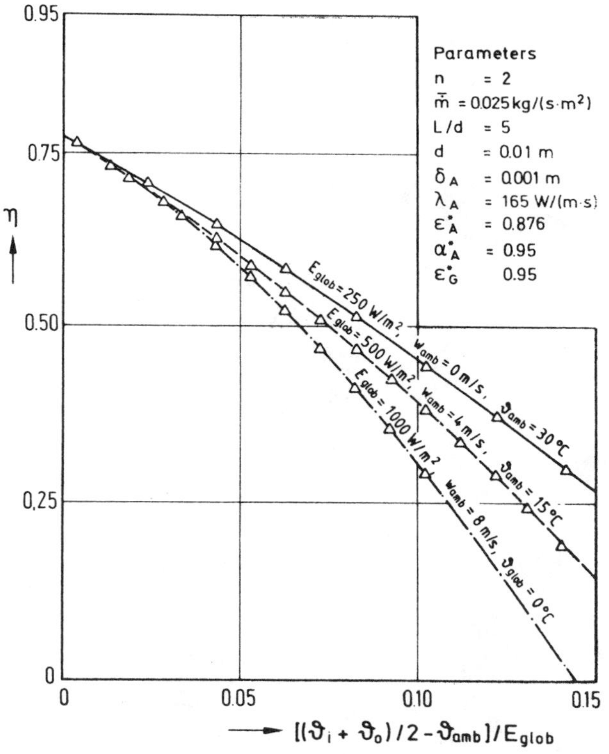

Fig. 6

Fig.6 Efficiency plot of a flat plate collector. Effect of meteorological parameters

4C

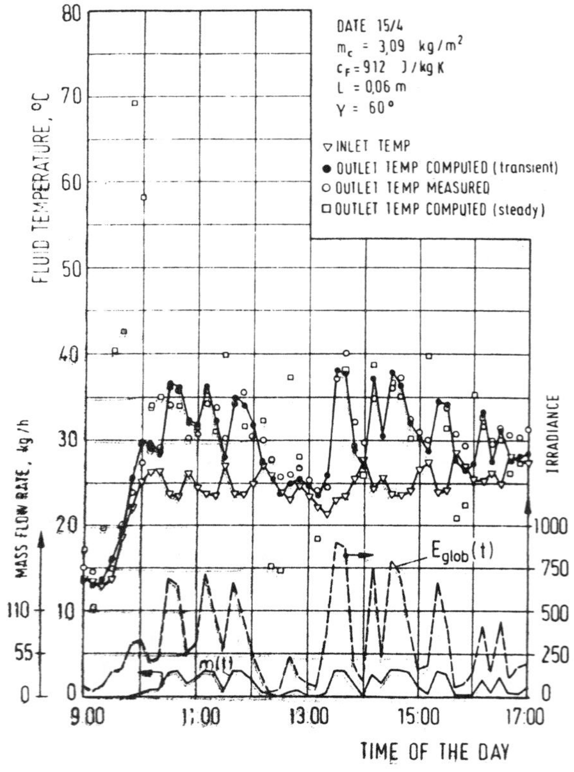

Fig.7 Comparison between measured and calculated outlet temperatures, for transient conditions

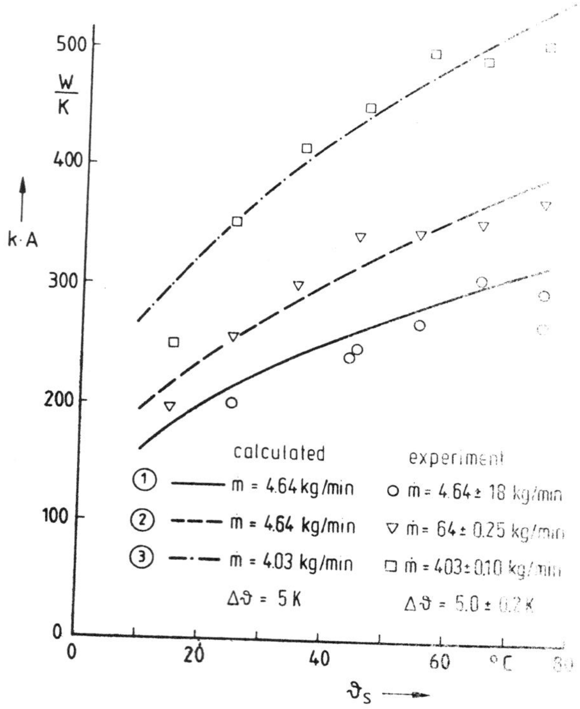

Fig.8 Comparison of calculated and measured heat transfer κ.A for three different heat exchangers

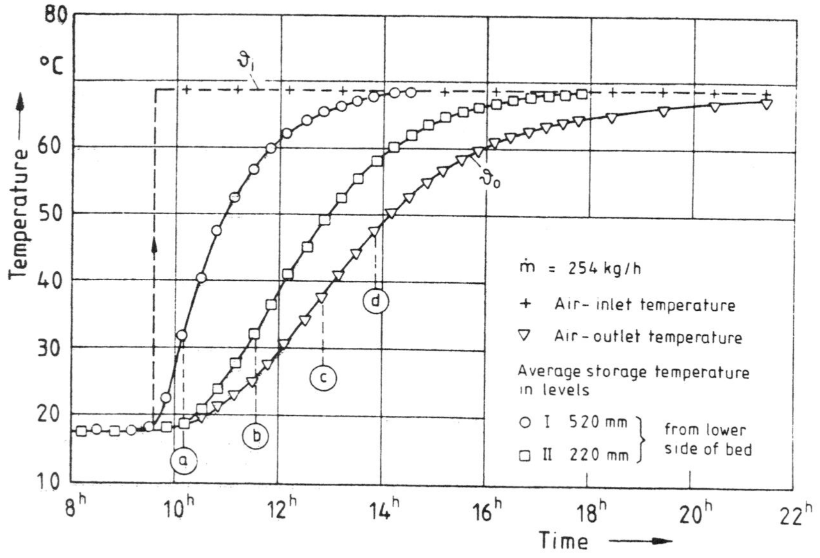

Fig.9 Temperature response of a model pebble-bed store

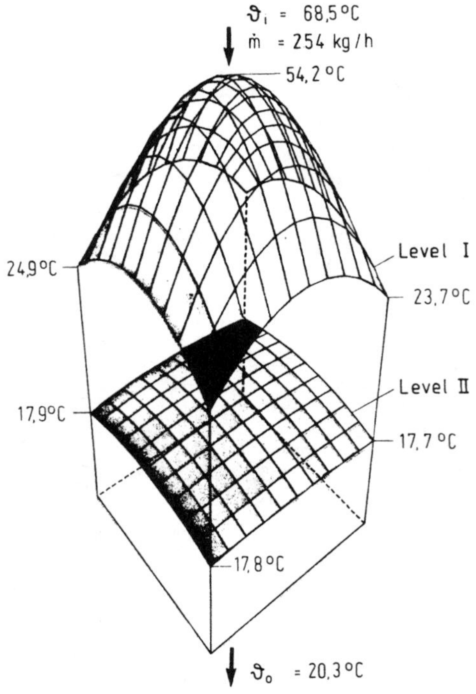

Fig.10 Temperature distribution in two levels of the pebble bed store (point (a) of figure 9)

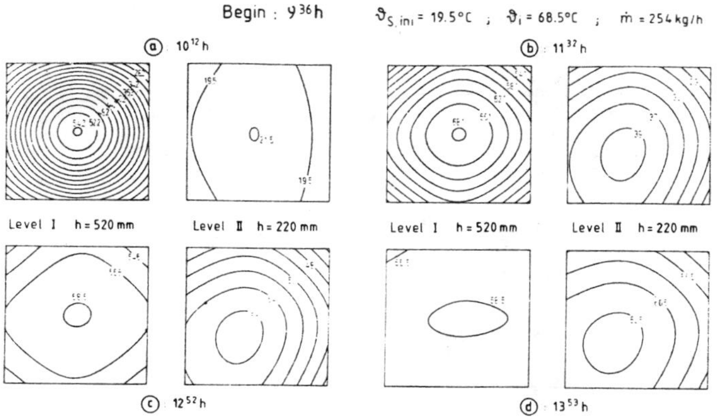

Fig. Fig.11 Isotherms across two levels in the pebble-bed store

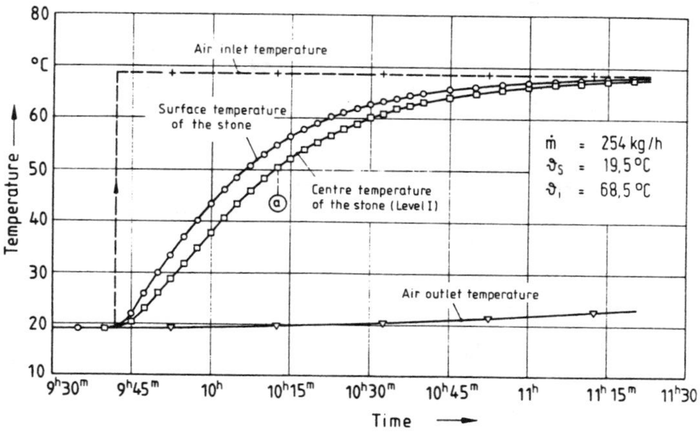

Fig.12 Temperature response of a single stone (d = 50 mm) on its surface and in its centre

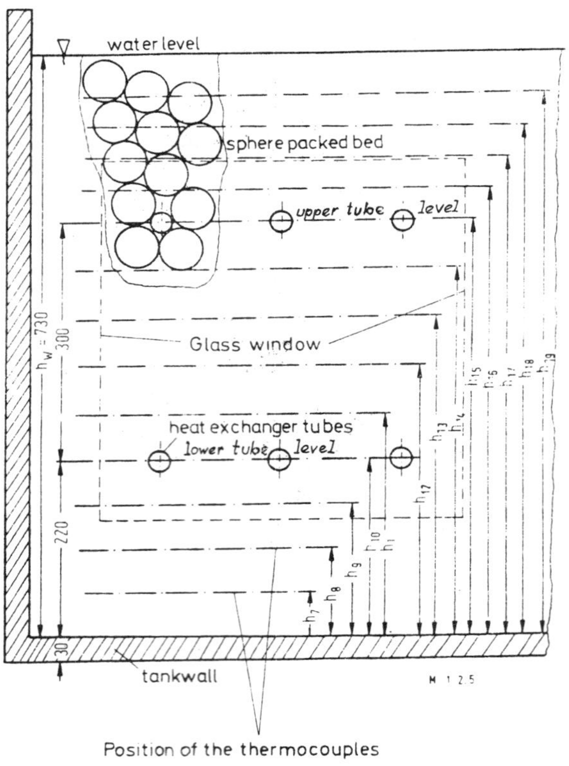

Fig.13 Model ground store filled with Steatit spheres and flooded by wate

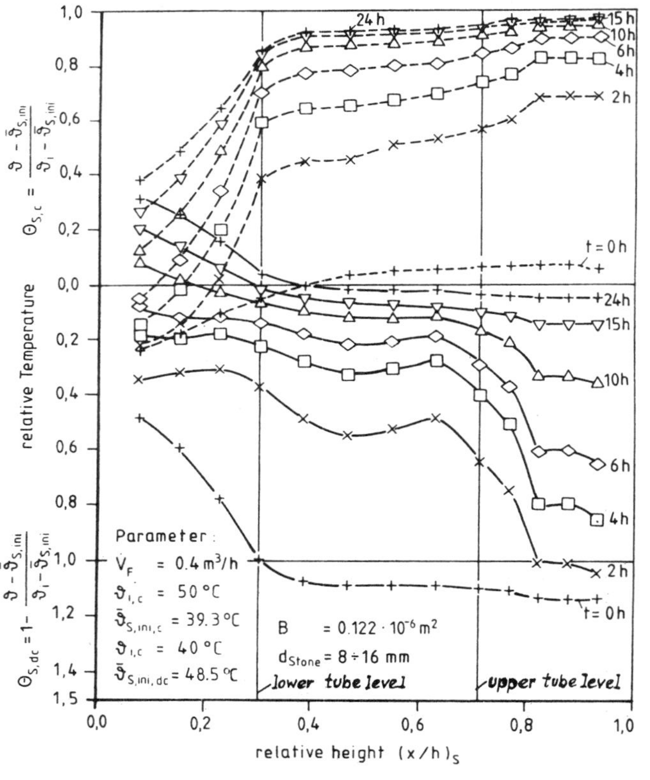

Fig.14 Time-dependent temperature distributions within the ground store during charging and discharging

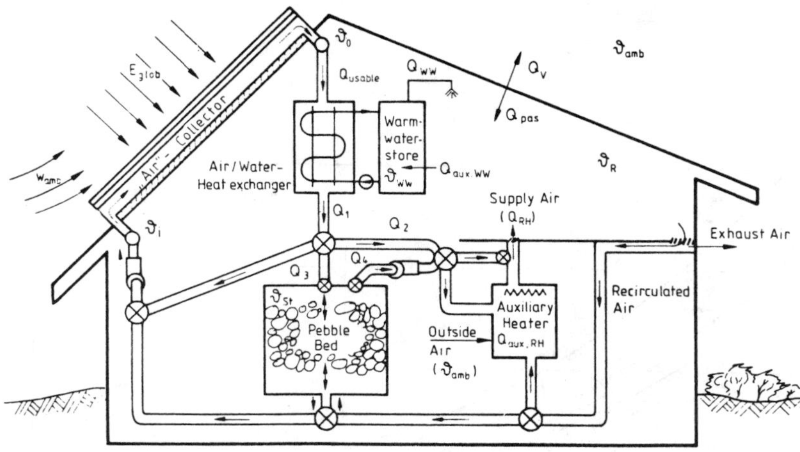

Fig.15 Solar air-and water-heating system of a house

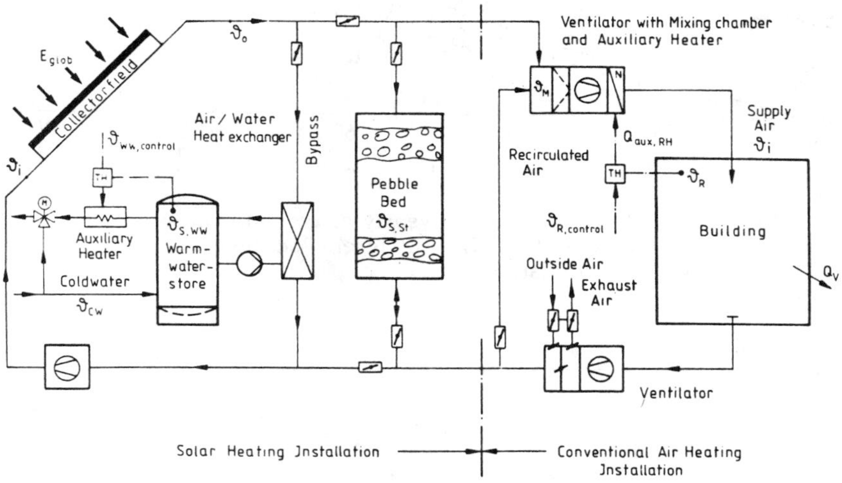

Fig.16 System configuration for the house of figure 15

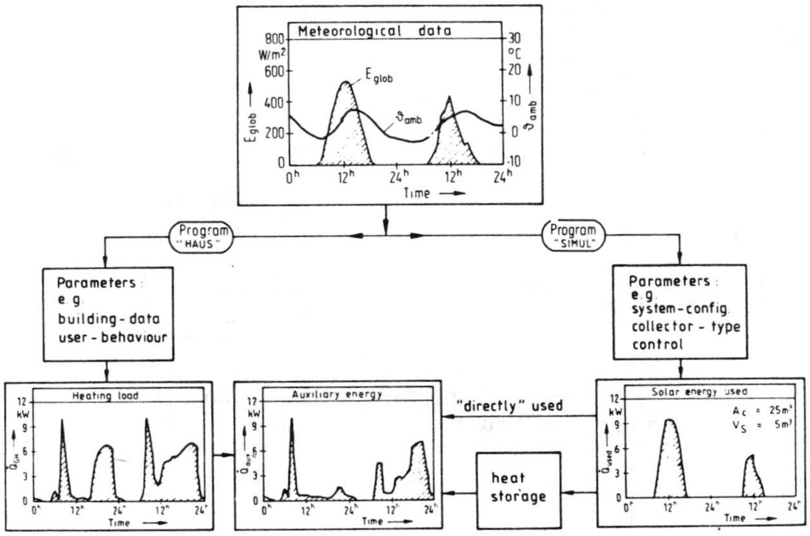

Fig.17 Data flow model of the house-heating svstem

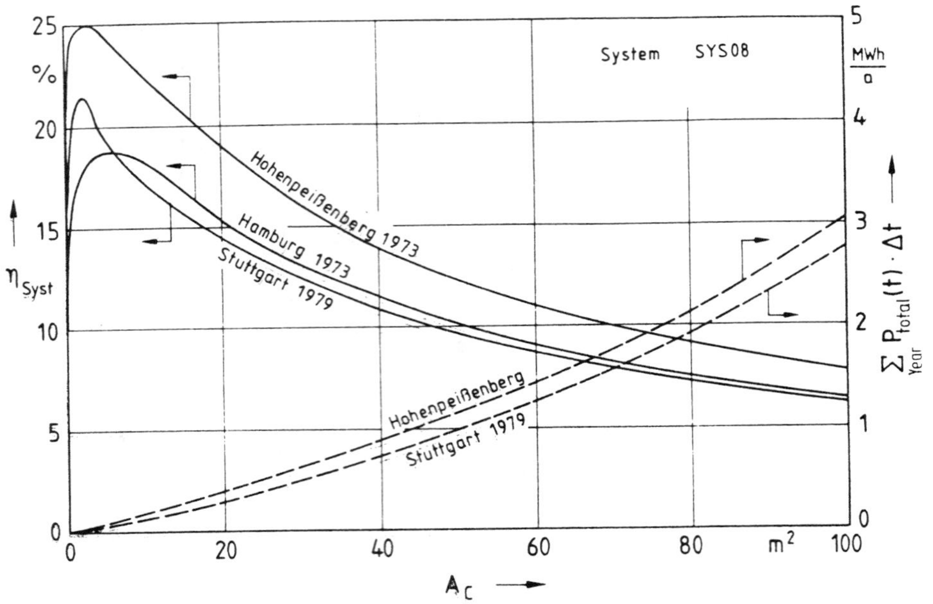

Fig.18 System efficiency and yearly electrical energy demand of the house (fig.15) at different locations vs. collector area

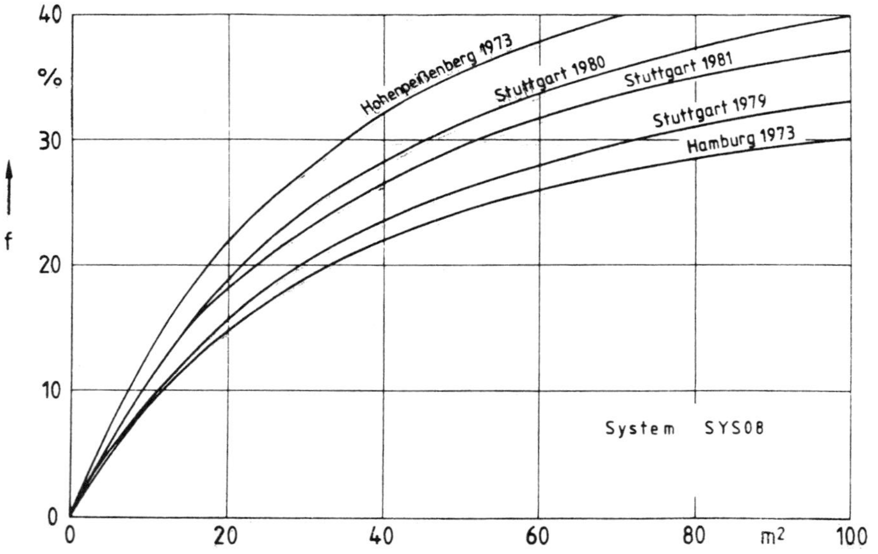

Fig.19 Solar fraction vs. collector area for different locations and weather data (Stuttgart)

Table 1

PARAM.	CONST.	VAR	$\eta_e = \left[c_F \sum_{t=0}^{t} \bar{m}(t)(\vartheta_0(t)-\vartheta_i(t))\Delta t\right] / \sum_{t=0}^{t} E_{glob}(t)\Delta t$	WARMING UP TIME of collector
			35 38 40 42 44 46 48 50 52 54 56 58 60 62 64%	36 42 48 54 60 66 72 78 84 90Min 102 108 114
Glass Covers	2	1-3		
Pipe Spacing mm	50	0-90		
Selectivity $\alpha^*_A/\varepsilon^*_A$	1	(a, b, c)		
Material $\delta_A \lambda_A$ W/K	Steel 0.02	Al Cu Polypropylene		
Collector Tilt	45°	0-60		
Inlet Temperature °C	35	25-50		

Selectivity sub-table:

	α^*_A	ε^*_A	$\alpha^*_A/\varepsilon^*_A$
a	0.95	0.95	1.0
b	0.95	0.527	1.8
c	0.95	0.105	9.0

Material sub-table:

$\delta_A \lambda_A$	
0.22	Al
0.37	Cu
0.0006	PP

June 21st
$\vartheta_{amb}=20°C$, $w_{amb}=3$ m/s, $w_F=0.01$ m/s
$d=10$ mm, $\varepsilon^*_G=0.876$, $\delta_G/\lambda_G=2.5$ m²K/W
$\vartheta_{start}=20°C$, $\vartheta_{end}=\{35°C, \vartheta_i\}$

Collector mass m_c at L/d=5

PP	Al	Steel	Cu
2.34	4.23	9.92	11.04 kg/m

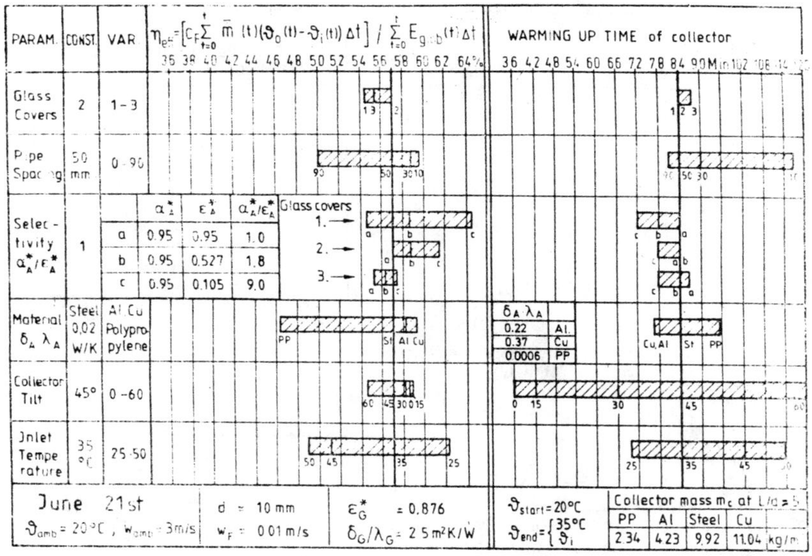

Table:1 Parameter effects in transient-state conditions: June 21st

Table 2

PARAM.	CONST.	VAR	$\eta_e = \left[c_F \sum_{t=0}^{t} \bar{m}(t)(\vartheta_0(t)-\vartheta_i(t))\Delta t\right] / \sum_{t=0}^{t} E_{glob}(t)\Delta t$	WARMING UP TIME of collector
			34 36 38 40 42 44 46 48 50 52 54 56 58 60 62%	72 78 84 90 96 102 108 114 120 126 132 138 144 150 Min
Glass Covers	2	1-3		
Pipe Spacing mm	50	10-90		
Selectivity $\alpha^*_A/\varepsilon^*_A$	1	(a, b, c)		
Material $\delta_A \lambda_A$ W/K	Steel 0.02	Al, Cu Polypropylene		
Collector Tilt	45°	0-60		
Inlet Temperature °C	35	25-50		

Pipe spacing sub-table:

L/d	m_c steel
1	18.1
9	9.01

Selectivity sub-table:

	$\alpha^*_A/\varepsilon^*_A$
a	1.0
b	1.8
c	9.0

Material sub-table:

	$\delta_A \lambda_A$
Al	0.22
Cu	0.37
PP	0.0006

December 21st
$\vartheta_{amb}=10°C$, $w_{amb}=3$ m/s, $w_F=0.01$ m/s
$d=10$ mm, $\varepsilon^*_G=0.876$, $\delta_G/\lambda_G=2.5$ m²K/W
$\vartheta_{start}=10°C$, $\vartheta_{end}=\{35°C, \vartheta_i\}$

Collector mass m_c at L/d=5

PP	Al	Steel	Cu
2.34	4.23	9.92	11.04 kg/m

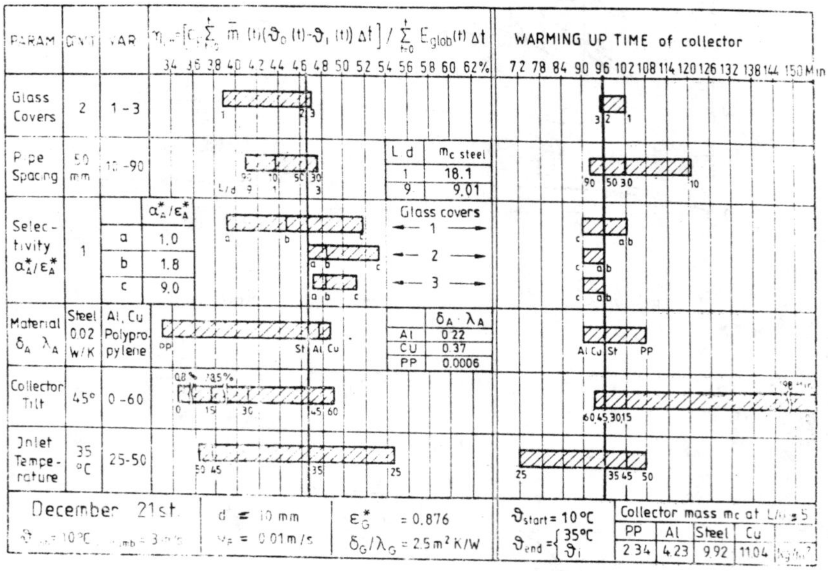

Table 2: Parameter effects in transient-state conditions: December 21st

Table 3

	①	②	③
Position			
Tube diameter d_i	10.4 mm	12.5 mm	17.0 mm
Fin height h	3.2 mm	3.25 mm	3.1 mm
Fin spacing t_f	2.2 mm	2.2 mm	2.2 mm
Outer surface A	0.971 m²	1.439 m²	2.497 m²
Inner surface A_i	0.170 m²	0.233 m²	0.482 m²
Tube length L	5.23 m	5.97 m	9.04 m

Table 3: Description of three different coiled, finned-tube heat exchangers

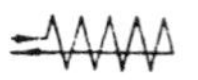

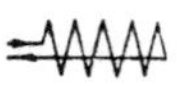

System	Air flow open	Air flow closed	Continuously Controlled flow	Heat Store Room Heating Water	Heat Store Room Heating Pebbles	Domestic Hot Water by Air
SYS01	×	–	×	–	–	–
SYS02	×	–	–	–	–	–
SYS03	–	×	×	–	–	–
SYS04	–	×	×	–	–	×
SYS05	–	×	×	–	×	–
SYS06	–	×	×	–	×	×
SYS07	–	×	×	×	–	×
SYS08	–	×	–	–	×	×

Table 4: System variations for fig. 15 treatable in the computer program "SIMUL"

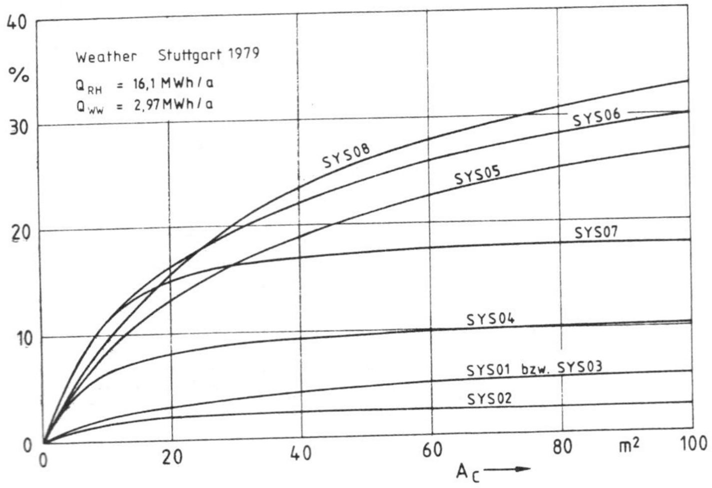

Fig.20 Solar fraction vs. collector area for the different systems listed in table 4.

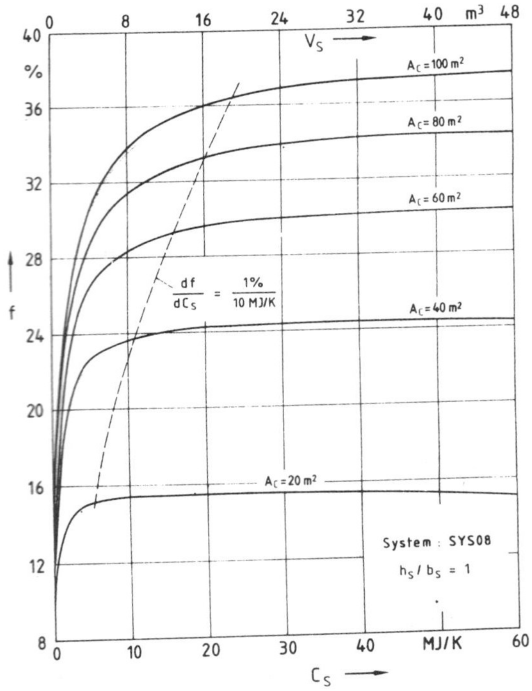

Fig.21 Solar fraction vs. thermal capacity of pebble-bed store (storage volume) for various collector areas

Heat Transfer and Thermal Storage Characteristics of Optically Semitransparent Material Packed Bed Solar Air Heater

MASANOBU HASATANI, YOSHINORI ITAYA, and KOUJI ADACHI
Department of Mechanical Engineering
Nagoya University
Furocho, Chikusaku, Nagoya 464, Japan

1. INTRODUCTION

The air at comparatively low temperature up to about 100°C is sufficient for drying of grains, foods, coals and so on. Solar air heaters would be most applicable to such low grade heating systems of air from the point of view of a Solar energy utilization. Solar air heaters are also available for air conditioning and the regeneration of desiccant and others. In flat plate type of solar air heaters, however, since the heat transfer rate between the absorber on which solar radiation can be transformed into heat and the air is usually much lower than the case of water heating, it is difficult to obtain the high efficiency of the energy collection. In order to enhance the heat transfer between the absorber and the air, for instance, the following devices have been proposed and those heat transfer characteristics have been reported since Close [1] ; setting of the absorber under the air channel [2], the use of the V-shaped absorber [3] , the multi-pass mode of air channel [4] , and packing of porous materials in the air channel [6] etc. Further, it is to be desired that a solar energy utilization system should also have heat storage effect in order to compensate for an irregular supply of solar radiation. But the air itself can not be used for heat storage because of low sensible heat of the air and then a separate heat storage system such as a rock bed is substantially necessary. This heat storage system would also decrease the overall thermal efficiency of the air heater. Few researches including the heat storage system have been seen with the exception of the report by Mishra and Sharma [5] , in which they tested the performance of a packed-bed solar air heater with iron-chips, aluminium-chips and pebbles.

In this work, an optically semitransparent materials packed-bed solar air heater, which may have two advantages to the flat plate solar air heaters was proposed ; (1) it has accelerative effects of heat transfer rate due to increases in the optical depth of the air layer and the heat transfer area, and (2) the packed material itself can be used as a heat storage material. As one of basic steps of the R&D on the solar air heater of the proposed type, the effects of the thermal and optical properties of packed materials and the air flow rate on the overall efficiency of the energy collection and thermal storage were investigated theoretically and experimentally. Then the applicability of this type of solar air heater to a practical solar system was discussed comparing with the flat plate type.

2. THEORY

2.1 Basic Equation

Consider the optically semitransparent material packed-bed solar air heater shown in Figure 1. The channel with a cover glass of width D, depth B and length L, of which an aspect ratio is comparatively small, is uniformly heated from the upper side by radiation. In the channel, an air is flowing with a constant and uniform velocity u in the direction x. Assuming that (1) the temperature distributions within each packed material and the cover glass are uniform, (2) conductive heat transfer in the flow direction is negligibly small, (3) natural convection is not generated within the channel, and (4) the physical properties of the packed material are independent of the temperature, each basic equation is given for the unsteady heat transfer in the channel as follows:

(packed material)

$$C_p \rho_p (1-\varepsilon_p) \frac{\partial T_p}{\partial \theta} = \frac{\partial}{\partial z}\left(\lambda_e \frac{\partial T_b}{\partial z} - Q_r\right) - h_p a_p (T_p - T_g) \tag{1}$$

The initial and boundary conditions for Eq.(1) are:

$$\theta = 0,\ 0 \leqq z \leqq B\ ;\ T_p = T_0 \tag{2}$$

$$\theta \geqq 0,\ z = 0\ ;\ \lambda_e \frac{dTb}{dz} + h_w (T_p - T_s) = 0 \tag{3}$$

$$\theta \geqq 0,\ z = B\ ;\ \lambda_e \frac{dT_b}{dz} - Q_r = 0 \tag{4}$$

(air)

$$C_g \rho_g \varepsilon_p \frac{\partial T_g}{\partial \theta} = -h_p a_p (T_g - T_p) - C_g \rho_g u \frac{\partial T_g}{\partial x} \tag{5}$$

For Eq. (5) the initial and boundary conditions become

$$\theta = 0,\ 0 \leqq x \leqq L\ ;\ T_g = T_0 \tag{6}$$

$$\theta \geqq 0,\ x = 0\ ;\ T_g = T_0 \tag{7}$$

(Cover glass)

$$C_s \rho_s d_s \frac{dT_s}{d\theta} = - h_g(T_s - T_a) - h_w (T_s - T_p) + Q_{r0} - Q_{r1} \tag{8}$$

and initial condition is given

$$\theta = 0 \; ; \; T_s = T_0 \tag{9}$$

where T_b is a temperature of the packed-bed and could be approximated by the average temperature of the packing and the air as:

$$T_b = \frac{T_p + T_g}{2} \tag{10}$$

2.2. Radiative Heat Transfer Equations

We make the following assumptions for the radiative heat transfer:

1) the radiative heat transfer is one-dimensional in the direction z.
2) the packed-bed is an isotropically homogeneous semitransparent layer for incident radiation.
3) the effect of scattering is negligibly small.
4) the incident radiation from the heat source is diffuse.
5) the emission in the layer is involved in the effective thermal conductivity expression of the semi-transparent material packed-bed.
6) the boundary surface and the packed-bed are gray materials, and the optical properties are independent of the wavelength.

Under these conditions, the relations between the radiosity R and the irradiation H shown in Figure 1 are given as below.

$$\begin{aligned} H_0 &= Q_1 + \sigma T_a^4 \\ R_0 &= r_s H_0 + \varepsilon_s \sigma T_a^4 + (1 - r_s - \alpha_s) H_1 \\ R_1 &= r_s H_1 + (1 - r_s - \alpha_s) \\ H_1 &= 2R_2 E_3(\tau_0) \\ H_2 &= 2R_1 E_3(\tau_0) \\ R_2 &= (1 - \varepsilon_w) H_2 \\ H &= 2R_1 E_3(\tau) \\ R &= 2R_2 E_3(\tau_0 - \tau) \end{aligned} \tag{11}$$

Then the radiative heat flux Q_r is given by:

$$Q_r = H - R \tag{12}$$

2.3 Effective Thermal Conductivity and Wall Heat Transfer Coefficient

As described in the above Assumption (5), in this work, the effect of the emission in the bed has been tentatively involved in the effective thermal conductivity and the wall heat transfer coefficient. At first, according to the previous paper [7]., the stagnant effective thermal conductivity of the semi-transparent material packed-bed has been calculated.

$$\lambda_e^0 = \frac{1 + \varepsilon_p^{1.3}}{\dfrac{1}{\lambda_p + h_{rv} d_p \varepsilon_p^{1.3}/(1 - \varepsilon_p^{1.3})} + \dfrac{1}{(\lambda_g/\phi) + h_{rs} d_p}} + \frac{4 h_{rv} \varepsilon_p^{1.3} d_p \lambda_p}{\lambda_p + h_{rv} d_p \varepsilon_p^{1.3}/(1 - \varepsilon_p^{1.3}) + (\lambda_g/\phi) + h_{rs} d_p} + \frac{16 n^2 \sigma}{3\alpha} T^3 \tag{13}$$

where h_{rv} and h_{rs} are radiative heat transfer coefficients [8] and are given by

$$h_{rv} = [0.1952/\{1 + \frac{\varepsilon_p}{2(1 - \varepsilon_p)} \frac{1 - \varepsilon}{\varepsilon}\}] (\frac{T_b}{100})^3$$

$$h_{rs} = 0.1952\{\varepsilon/(2 - \varepsilon)\} (\frac{T_b}{100})^3 \tag{14}$$

Then substituting the λ_e^0 obtained by Eq.(13) into Yagi and Kunii's equation [8], the effective thermal conductivity in the normal direction z to the airflow could be estimated. The wall heat transfer coefficient h_w has been also estimated as according to the similar idea [9] as

$$\frac{1}{h_w d_p / \lambda_g} = \frac{1}{2} \{ \frac{1}{\lambda_{ew}/\lambda_g} \quad \frac{1}{\lambda_e/\lambda_g} \} \tag{15}$$

where λ_{ew}/λ_g could be seen in the literature [9].

In order to solve the basic equations, the similar dimensionless variables to those defined in the previous paper [10] have been introduced. Then the basic equations have been solved by the numerical method [10] to obtain the theoretical time-changes of the temperature distribution in the packed-bed.

3. EXPERIMENTAL APPRATUS AND PROCEDURES

A schematic diagram of the experimental apparatus employed is shown in Figure 2. An air channel was made of glass with the test section of 1000 mm in length, 100 mm in width and 30 mm in depth. The side wall, the bottom and the other top of the channel than the test section were thermally insulated with Styrofoam. The bottom surface of the test section was painted black and the top surface was covered with a glass plate of 3 mm in thickness. Glass beads of 3.4 mm, 5.2 mm and 10.0 mm in diameter and glass tubes of 5 mm in outer-diameter and 4 mm in inner-diameter as semitransparent packings and porcelain beads of 3 mm in diameter as opaque packings were employed, respectively. As a radiative heat source model, eight infrared lamps (100V-125W) were used and were arranged in two lines to obtain a uniform heat flux along the flow direction of the channel. The time-change of temperature distribution in the packed-bed was measured by 100 μm ϕ CA-thermocouples placed at the positions shown in Figure 2(b). The pressure drop between the inlet and the outlet of the bed was also measured.

The experimental procedures are as follows: After it was confirmed that the air flow and the bed temperature were steadily uniform and constant, the channel was stepwisely heated by the lamps and then the time-change of the temperature distribution in the bed was measured. When the steady state of the temperature was confirmed under heating conditions, the lamps heating was cut off and then transient cooling temperature distribution was measured. Further, the same measurement was conducted in the case of a usual flat plate air heater, no packing in the channel, for the purpose of comparison.

The entire measuring apparatus was surrounded with an aluminium foil and a vinyl sheet to ensure diffuse irradiation-and natural convection- conditions on the cover glass of the channel.

4. RESULTS AND DISCUSSION

4.1 Characteristic Values Used in Theoretical Calculations

The other characteristic values than those described in §2.3, which were used in solving theoretically the basic equations, were estimated or measured in the following manners:

The absorption coefficient of the packed-bed was experimentally determined from the results of a stagnant experiment by a trial-and-error procedure. It was to find the absorption coefficient satisfying that the theoretical, steady state temperature distribution at the air flow velocity $u = 0$ showed a good agreement with the experimental data obtained by radiative heating of the stagnant packed-bed. The absorption coefficients thus determined are listed in Table 1.

For the optical properties of the cover glass, $\alpha_s = 0.25$, $r_s = 0.09$ and $\varepsilon_s = 0.90$ were used, which were calculated from the refractive index $n = 1.5$ and the absorption coefficient given by Neuroth [11]. $\varepsilon_w = 0.8$ was used as the emissivity of bottom wall. The void fraction of the packed bed ε_p was measured and the specific surface area a_p was calculated from the ε_p and the diameter of the packing. These values are also summarized in Table 1. The other thermal properties were quoted from the literature [14]. The heat transfer coefficient h_g on the surface of the cover glass was calculated from the authors' empirical equation [12].

4.2 Heat-Trap Characteristics

Temperature distributions. Figure 3 shows an example of the experimental results for the time-change of the temperature distribution in the bed at x = 600 mm which is compared with the results of the porcelain beads packed-bed and the empty channel. In the case of the empty, the temperatures of the cover glass and the bottom surface were remarkably higher than that of the air layer. In this case, the incident radiation is absorbed only by the cover glass and on the bottom surface and the air is heated only by convective heat transfer from the glass and the bottom. The temperature distribution in the porcelain beads packed-bed showed high temperature on the top surface at the first period of heating and became gradually uniform to the steady state ($\theta = 3.0$ h) since a large portion of the incident radiation is absorbed around the top of the bed and the heat absorbed there is transferred into the bed by thermal conduction and convection. While, in the case of the glass beads packing, the temperature distribution was remarkably different from the results of those two cases, that is, the temperature peak was observed within the bed.

Figure 4 (a) shows the temperature distribution change in the air flow direction x. The position of the temperature peak in the bed shifted from the top to the bottom with increase in x or θ. This tendency would be desirable from the viewpoint of improvement in heat collection efficiency since the top surface temperature is the lower, the heat lost from the top the lower. Figure 4 (b) and (c) show the effects of the air flow velocity and the packed glass beads diameter on the temperature distribution, respectively. The experimental temperature distribution is seen from Figure 4 (b) to become more uniform with increase in the air flow velocity. Any remarkable effect of the packed glass beads diameter on the temperature distribution was not observed from Figure 4 (c) except that the absorption coefficient of the packed-bed slightly decreases and the bottom temperature becomes slightly higher with increase in packed glass beads diameter.

Through Figure 4 (a) to (c), the experimental data are compared with the theoretical calculations. As is seen from the Figure, the agreement between the theoretical results and the experimental data was fairly good in the case of low air flow velocity. ($G < \sim 1500$ kg/m^2 h). However, fairly large differences between them appeared with increase in the air velocity. This disagreement may be due to the following reason; the air flow is not really a plug flow but is nonuniformly distributed in the packed-bed because of biased packing conditions or end effects though the plug flow is assumed in the theoretical model.

The temperature distributions in Figure 4 show similar configuration to that of the semitransparent liquid heated by radiation [13], and then the glass beads packing is considered effective to increase the optical depth of the air layer for enhancement of the heat trap.

Heat trap efficiency. Figure 5 shows the effects of the experimental parameters on the heat trap efficiency defined by the following equation:

$$E_f = \frac{C_g GB(T_{gout} - T_0)}{Q_1 L} \times 100 \quad (\%) \qquad (16)$$

In the range of the experimental conditions employed, the glass beads packing show 5 to 15% higher heat trap efficiency in an absolute value than those of the empty or the porcelain beads packings. Only little effect of the glass beads diameter was observed in the range of d_p 3.4 to 10.0 mm. In the Figure, the result in the case of glass tubes packing is also indicated. When $G <\sim 2000$ kg/m^{2}h, the efficiency was almost equal to those in the case of the glass beads packings.

4.3 Heat Storage Characteristic

Figure 6 indicates an example of the experimental results for the time-change of the outlet air mean temperature in the heating period as well as that in the cooling period. As already described, in the cooling experiment, the lamps heating was stepwisely cut off keeping the other conditions the same. The steady outlet air temperature in the case of the glass beads packing was about 8°C higher than that of no packing. The hot air of higher temperature over 10°C than the inlet was obtained for 0.3 h in no packing and for 1.7 h in the glass beads packing after cutting off the lamps. This means that the glass beads packing has a considerable heat storage effect by its heat capacity. However, in the case of the glass tubes packing, it was obtained only for 0.9 h. This may be due to larger void fraction, that is, the heat capacity decreases with increase in void fraction and then a heat storage effect decreases, too.

5. CONCLUSION

An optically semitransparent materials packed-bed solar air heater was proposed and the heat collection and storage characteristics were investigated theoretically and experimentally.

It was observed from the experimental data analysis that the solar air heater in which semitransparent materials like glass beads or a glass tubes were used for a heat collection-and storage-material had higher efficiency of the energy collection and the thermal storage than a usual flat plate collector or the collector in which opaque packings like porcelain beads were packed. Those experimental tendencies could be predicted fairly well by the numerical calculation based on a formulated model. From these results, it is likely that the proposed solar air heater has enough heat transfer and thermal storage characteristics to enable the development of a new type of solar air heater with high efficiency.

NOMENCLATURE

a_p = specific surface area

B = thickness of bed

C_g, C_p, C_s = specific heats of air, packed material and cover glass, respectively

d_p = diameter of packed material

d_s = thickness of cover glass

E_3 = Exponential integral function

G = mass flow rate of air

h_g = convective heat transfer coefficient between cover glass and surrounding

h_p = convective heat transfer coefficient between packed material and air

Q_l = radiative heat flux from lamp

Q_r, Q_{r0}, Q_{rl} = radiative heat fluxes

r_s = reflectivity of cover glass

T_a, T_g, T_p, T_s = temperature of surrounding, air, packed material and cover glass, respectively

T_0 = inlet and initial temperature

α = absorption coefficient

α_s = absorptivity

ε_p = void fraction of packed-bed

$\varepsilon_s, \varepsilon_w$ = emissivities of cover glass and bottom wall respectively

θ = time

λ_e = effective thermal conductivity

λ_g, λ_p = thermal conductivities of air and packed material, respectively

ρ_g, ρ_p, ρ_s = densities of air, packed material and cover glass, respectively

σ = Steafan-Boltzmann constant

τ = optical distance

Subscript

out = outlet

REFERENCES

1. Close, D.J., Solar Air Heaters, Solar Energy, No.3, 7 (1963) 117-124

2. Bhargava, A.K., Garg, H.P., and Sharma, V.K., Evaluation of the Performance of Air Heaters of conventional Designs, Solar Energy, No.6, 29 (1982) 523-533.

3. Parker, B.F., Derivation of Efficiency and Loss Factors for Solar Air Heaters, Solar Energy, No.1, 26 (1981) 27-32.

4. Wijieysundera, N.E., Ah, L.L., and Tjioe, L.E., Thermal Performance Study of Two-Pass Solar Air Heaters, Solar Energy, No.5, 28 (1982) 363-370.

5. Mishra, C.B., and Sharma, S.P., Performance Study of Air Heated Packed-Bed Solar-Energy Collectors, Energy, 6 (1981) 153-157.

6. Lalude, O., and Buchberg, H., Design and Application of Honeycomb Porous-Bed Solar-Air Heaters, Solar Energy, 13 (1971) 223-224.

7. Sugiyama, S., Hasatani, M., and Yada, A., Effective Thermal Conductivity of a Packed-Bed of Glass Sphere, Kagaku Kogaku, No.5, 34 (1970) 545-548.

8. Yagi, S., Kunii, D., Studies on Effective Thermal Conductivities in Packed-Beds, AICHE J., No.3,

3 (1957) 373-381.

9. Kunii, D., Suzuki, M., and Ono, N., Heat Transfer from Wall Surface to Packed Beds at High Reynolds Number, J. Chem. Eng. Japan, No.1, 1 (1968) 21-26.

10. Hasatani, M., and Arai, N., Unsteady Two-Dimensional Heat and Mass Transfer in a Packed Bed of Fine Particles with an Endothermic Process, Chem. Eng. Commun., 10 (1981) 223-242.

11. Neuroth, N., Der Einfluβ der Temperature auf die Spektrale Absorption von Glasern in Ultraroten, Glastechn. Ber., 25 (1952) 242-249.

12. Arai, N., Takahashi, S., and Sugiyama, S., Natural-Convection Heat and Mass Transfer from a Horizontal Upward-Facing Plane Surface to the Air, No.5, 5 (1979) 471-475.

13. Arai, N., Itaya, Y., and Hasatani, M., Development of a "Volume Heat-Trap" Type Solar Collector Using a Fine-Particle Semi-transparent Liquid Suspension (FPSS) as a Heat Vehicle and Heat Storage Medium, Solar Energy, No.1, 32 (1984) 49-56.

14. Kagaku Kogaku Binran, 4th Ed., Maruzen, (1978)

Table 1 Packed materials used in this experiment

Packed Materials	d_p [mm]	ε_p [-]	a_p [m^2/m^3]	α [1/m]
glass beads	3.4	0.41	1140	70
	5.2	0.43	430	60
	10.0	0.46	360	50
porcelain beads	3.0	0.43	1120	-
glass tubes	5.0	0.76	960	-

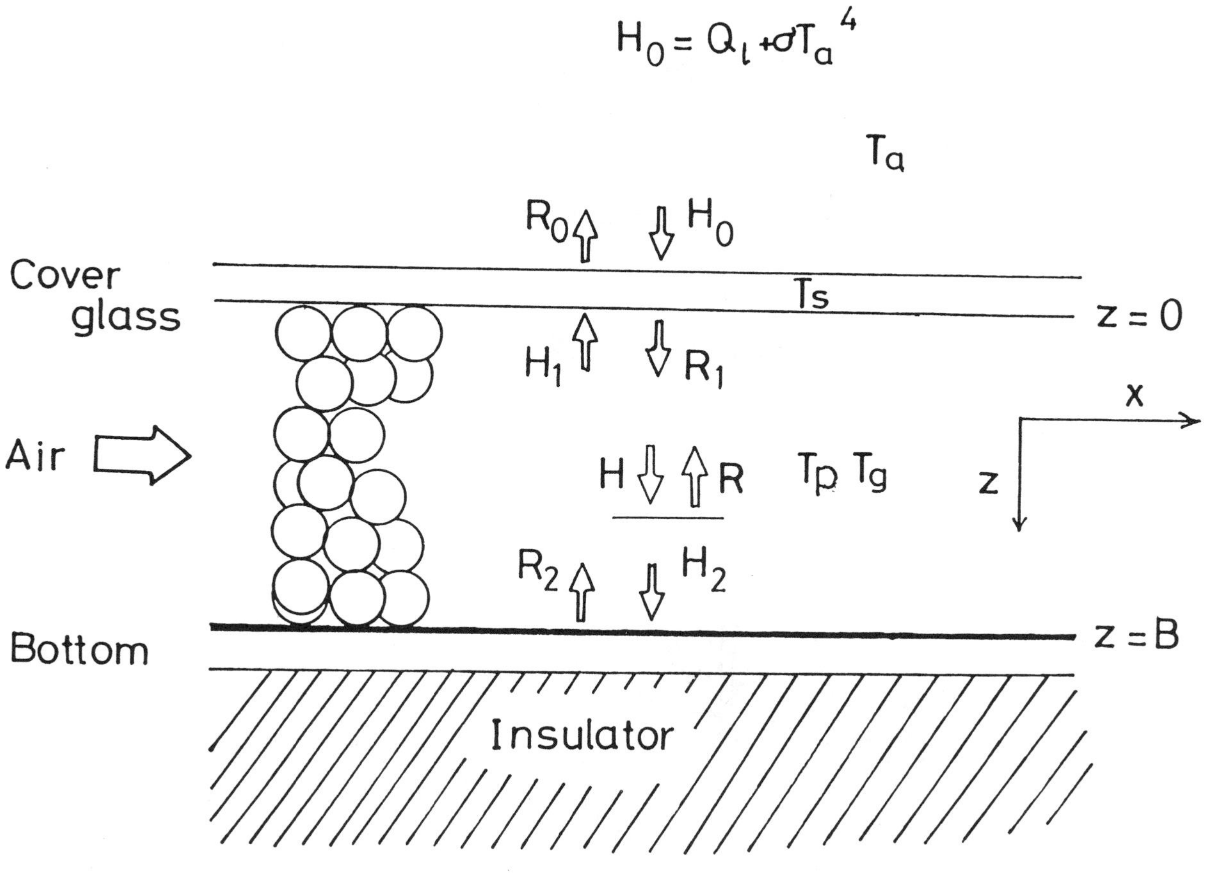

Fig. 1 Radiative heat transfer

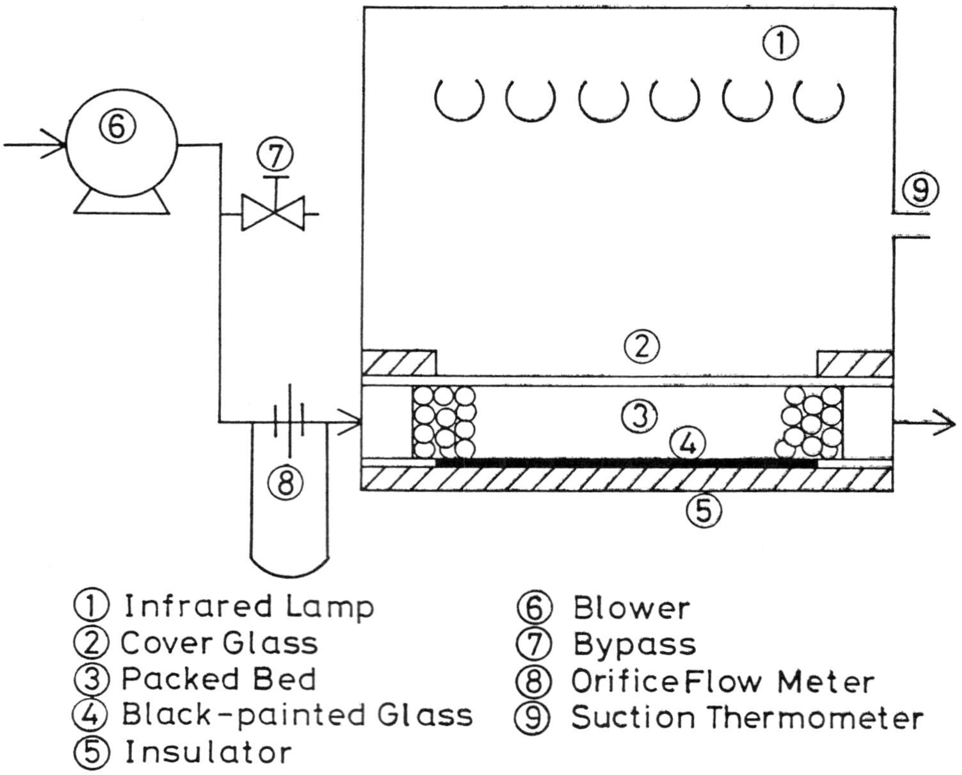

Fig.2 (a) Experimental apparatus

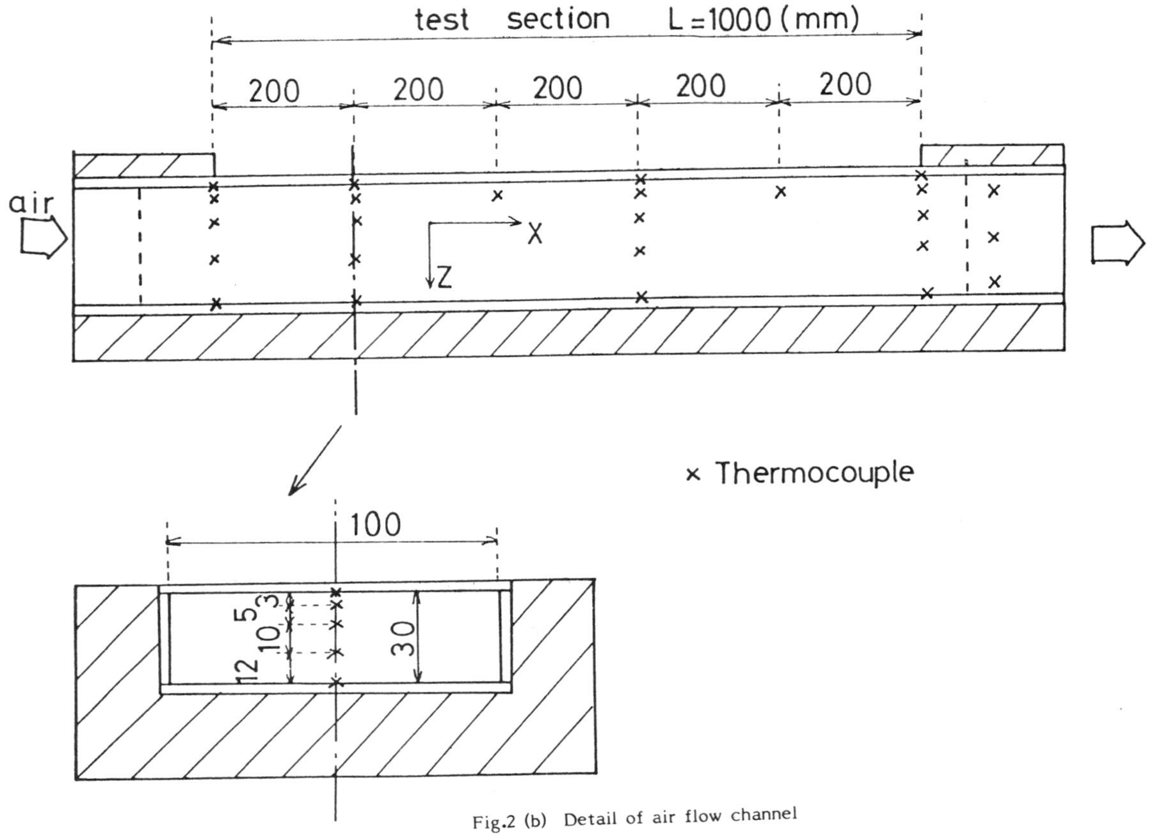

Fig.2 (b) Detail of air flow channel

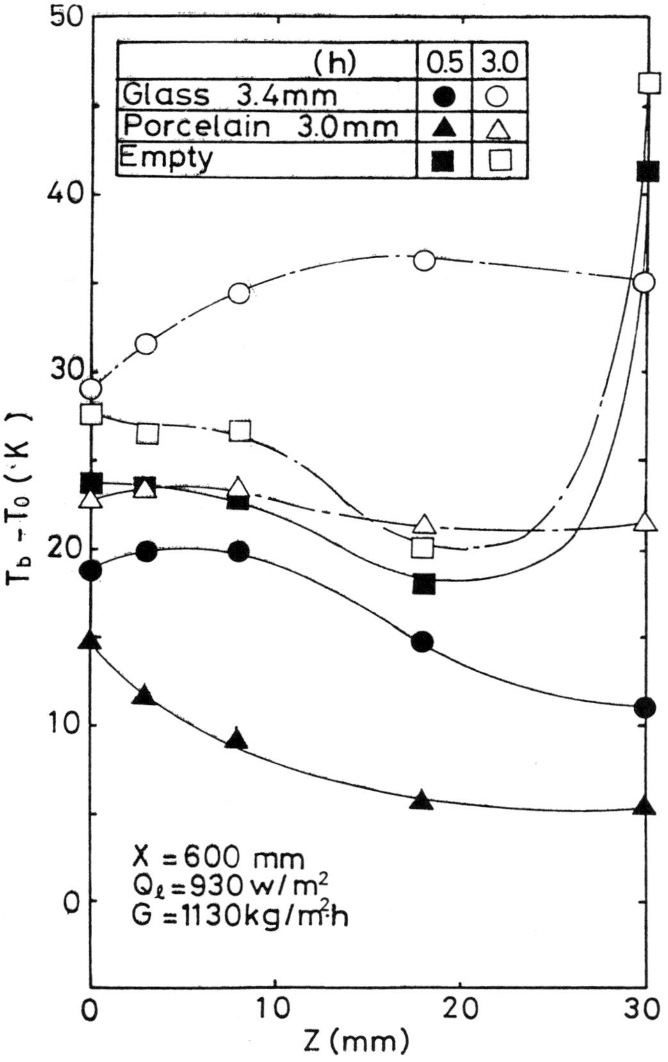

Fig.3 Effect of optical properties of packed material on temperature distribution

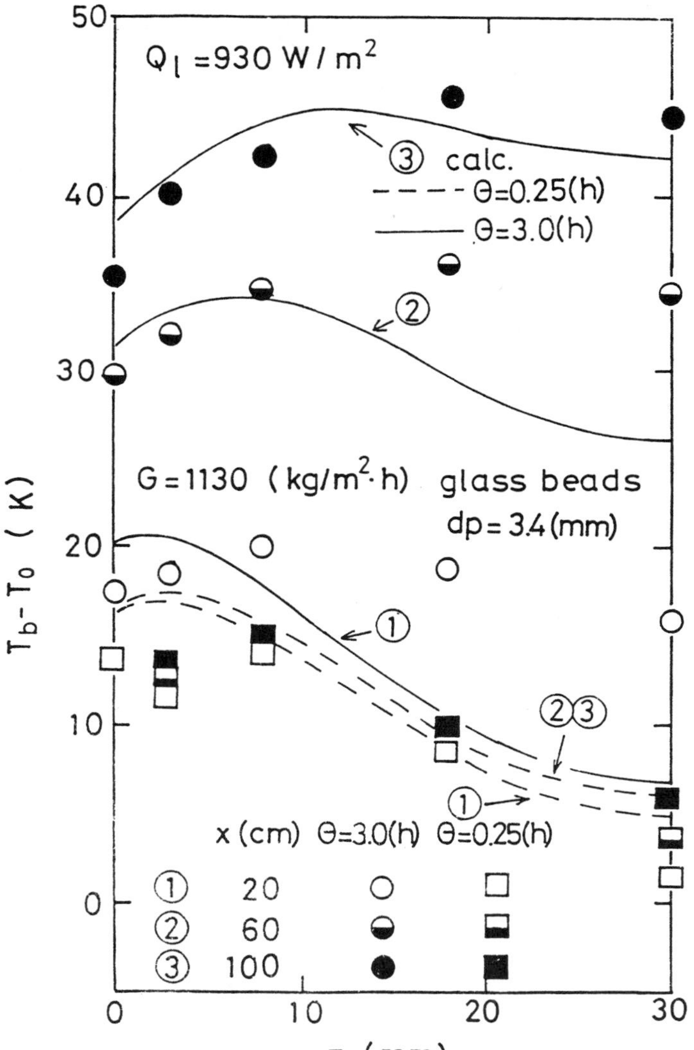

Fig.4(a) Comparison of the experimental data with the calculated results for unsteady temperature distribution within the packed bed

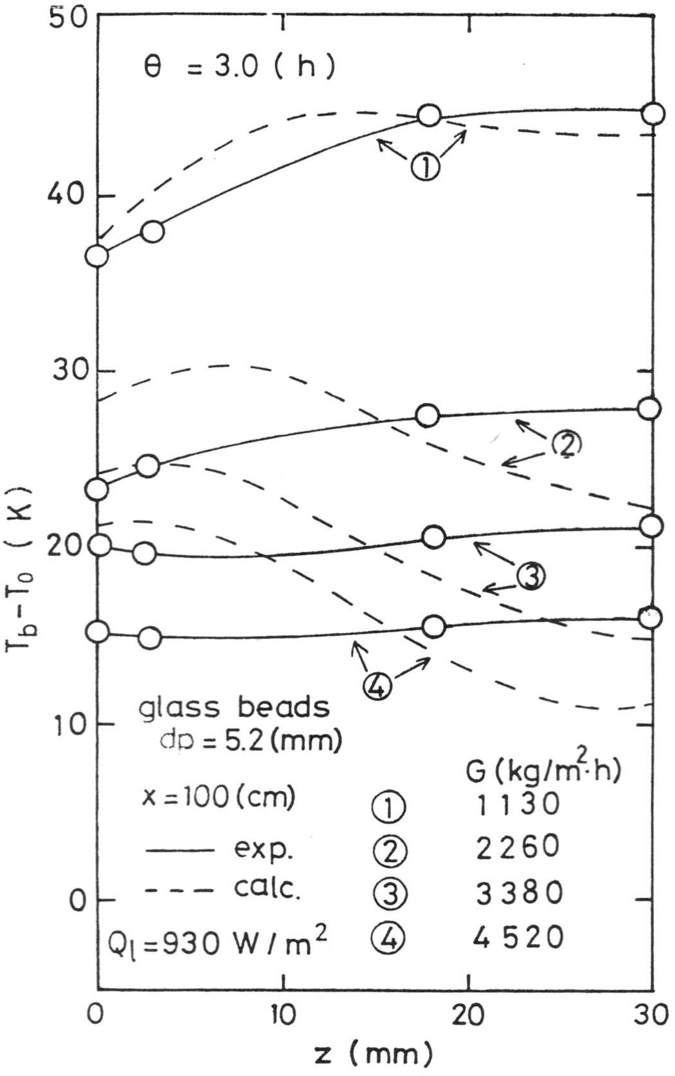

Fig.4 (b) Effect of air mass flow rate on temperature distribution

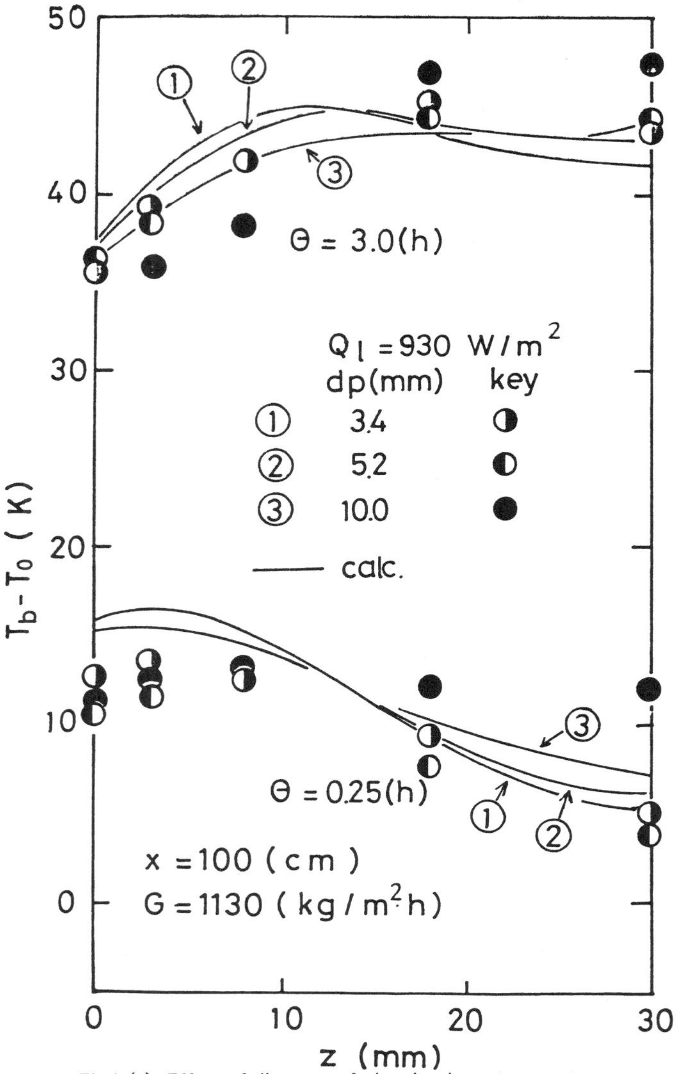

Fig.4 (c) Effect of diameter of glass beads on temperature distribution

$Q_{lamp} = 930 (w/m^2)$

---- calc.

KEY		particles	dp(mm)
A	◐		3.4
B	◐	glass bead	5.2
C	●		10.0
△		porcelain	3.0
□		empty	—
◎		glass tube	5.0

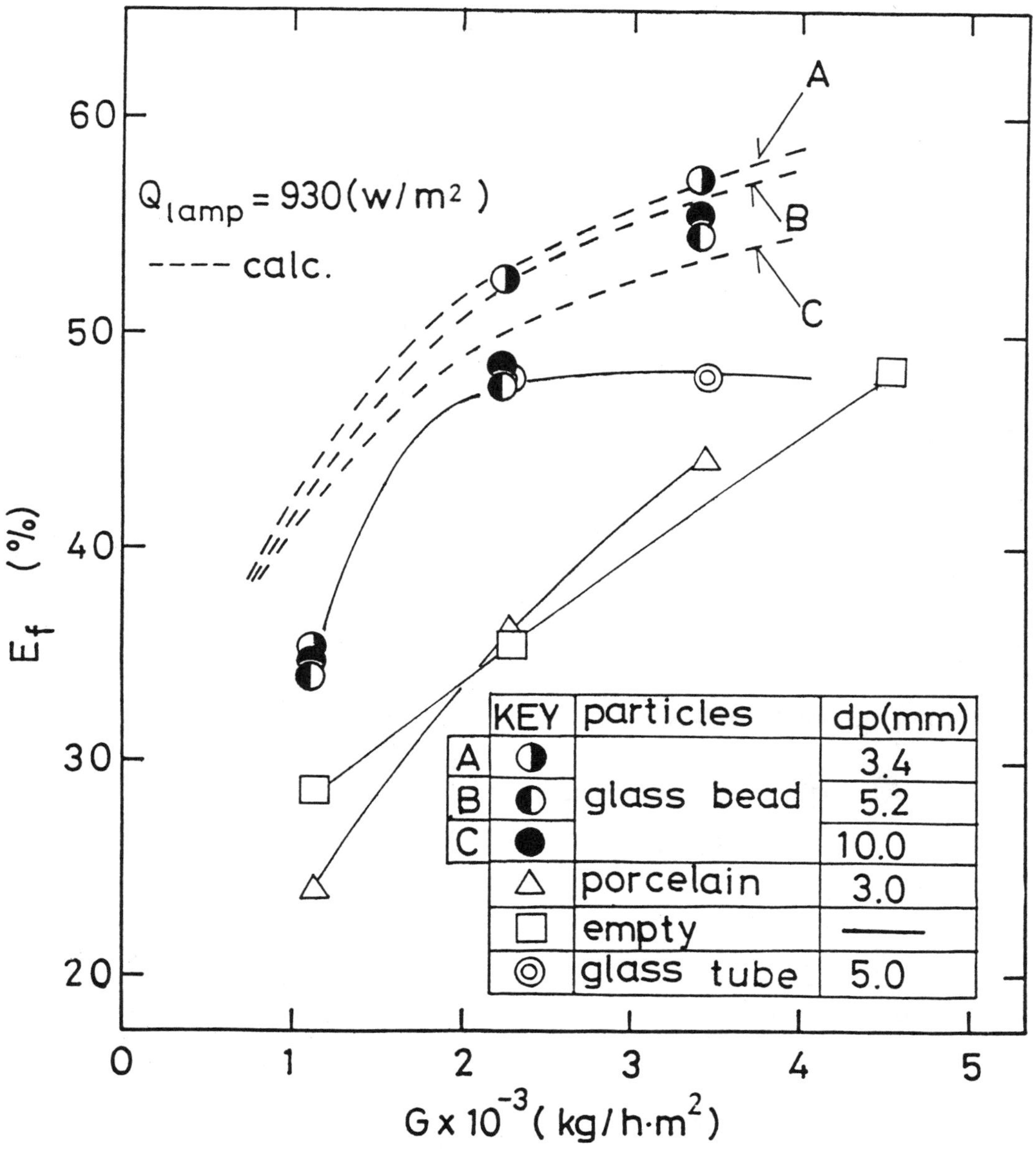

Fig. 5. Comparison of effciency of heat collection in various experimental conditions

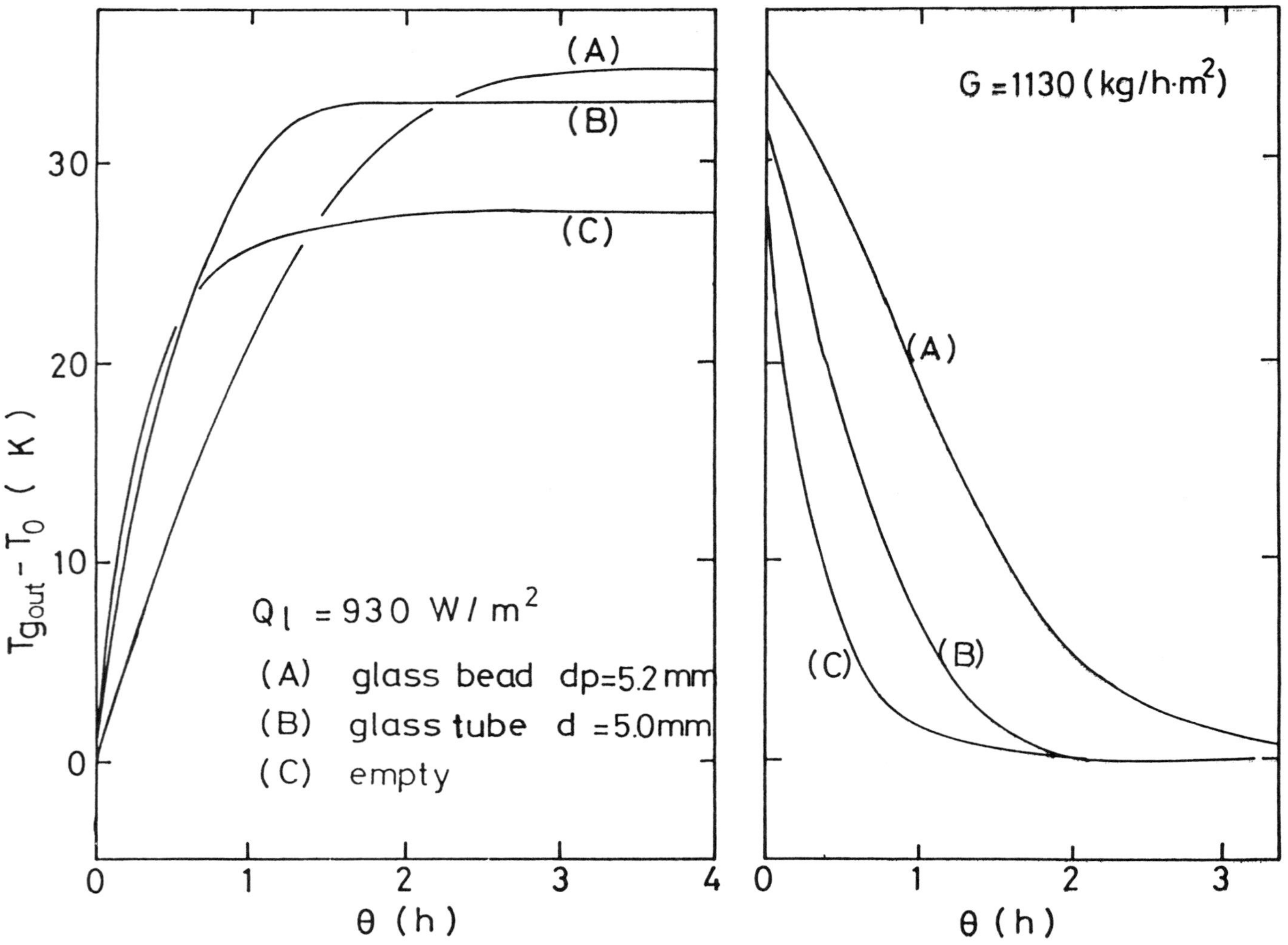

Fig.6 Time-change of outlet air temperature

Laminar Combined Convection from a Rotating Cone to a Thermally Stratified Medium

K. HIMASEKHAR and P. K. SARMA
Department of Mechanical Engineering
Andhra University
Visakhapatnam 530 003, India

The problem of laminar combined convection from a rotating cone to a thermally stratified medium is formulated and numerically solved employing finite-difference techniques. The results indicate that the velocity and temperature fields are profoundly influenced by the values of stratification parameter, mixed convection parameter and Prandtl number. It is observed that with the increase in stratification parameter, the local heat transfer coefficients increase whereas local moment coefficients decrease. The effects of mixed convection parameter on the flow and temperature fields for different levels of stratification, are essentially similar to those for an unstratified case except for magnitudes.

1. INTRODUCTION

Natural convection flows in thermally stratified media have been studied both analytically and experimentally in view of the inherent practical importance arising in several heat rejection and storage processes. The class of problems solved, in this area, pertains to laminar and turbulent flows in plumes and jets both two-dimensional and axisymmetric, laminar two-dimensional flows over a vertical surface [1-10]. The mathematical tools employed are similarity, local non-similarity, finite-difference and integral techniques. An excellent review of the work done so far on these problems has been presented by Jaluria [11]. In a very recent paper, the problem of heat transfer from a vertical surface to a stratified medium, with simultaneous suction applied at the solid boundary has been studied by employing a finite-difference numerical scheme [12].

Studies related to laminar combined convective flows in stratified media are not many, except for the recent investigations of Himasekhar and Jaluria [9,10]. The combined effects of free and forced convection are indicated by the mixed convection parameter (Gr/Re^2).

The present paper is intended to solve another problem related to combined convection from a different geometry, namely rotating vertical cone in a stratified medium. Heat transfer from rotating surfaces, is thoroughly reviewed by Frank Kreith [13] and the importance of these studies is amply indicated in this article. Invariably the investigations, both experimental and theoretical, deal with heat transfer to isothermal ambient medium.

The theoretical investigations are facilitated by employing either similar or non-similar solutions. The variation of surface temperature of the cone should be linear for similarity solution to exist.

The aim of the present paper is to obtain a similarity solution for the problem of rotating cone in a linearly stratified medium which is most frequently encountered in practice. An arbitrarily stratified environment tends to become linearly stratified as time elapses due to the underlying conductive mechanisms [8]. The mixed convective flows are studied at various stratification levels and mixed convection parameters. Prandtl number values of 6.7 and 0.7 corresponding to water and air at normal temperature are considered.

2. ANALYSIS

The configuration of the rotating cone with the coordinate system is as shown in Fig.1 and it is assumed that the cone is rotated under steady state conditions in non-dissipative constant property medium except for the density variations in the buoyancy force term. Further, the nature of variation of stratification of the medium is linear as shown in Fig.1.

Thus, for the coordinate system chosen with usual boundary layer approximations, the system of equations can be written as follows:

$$\frac{\partial u}{\partial x} + \frac{\partial w}{\partial z} + \frac{u}{x} = 0 \tag{1}$$

$$u\frac{\partial u}{\partial x} + w\frac{\partial v}{\partial z} - \frac{v^2}{x} = \nu\frac{\partial^2 u}{\partial z^2} + g\beta(T - T_{\infty,x})\cos\alpha \tag{2}$$

$$u\frac{\partial v}{\partial x} + w\frac{\partial v}{\partial z} + \frac{uv}{x} = \nu\frac{\partial^2 v}{\partial z^2} \tag{3}$$

$$u\frac{\partial T}{\partial x} + w\frac{\partial T}{\partial z} = \frac{k}{\rho c_p}\frac{\partial^2 T}{\partial z^2} \tag{4}$$

The following velocity and temperature functions would admit similarity transformation:

$$u = x\omega\sin\alpha\, F(\eta) \tag{5}$$

$$v = x\omega\sin\alpha\, G(\eta) \tag{6}$$

$$w = (\nu\omega\sin\alpha)^{1/2} H(\eta) \tag{7}$$

$$(T - T_{\infty,x}) = (T_{w,x} - T_{\infty,x})\,\Theta(\eta) \tag{8}$$

$$\eta = (\omega\sin\alpha/\nu)^{1/2} z \tag{9}$$

Further, it is found that in the present investigation, the variation of the surface temperature should be linear for similarity solution to exist. In addition, the wall and ambient temperatures should be equal at the vertex of the cone as already pointed out by Cheesewright [1] and Yang et al. [2] for the case of vertical surface.

Substitution of the variables defined by equations (5) to (9) into equations (1) to (4) leads to the following equations:

$$2F + H' = 0 \quad (10)$$

$$F'' - F^2 + G^2 - HF' + [Gr/Re^2]\ \Theta = 0 \quad (11)$$

$$G'' - HG' - 2FG = 0 \quad (12)$$

$$\Theta'' - Pr\,[F\,S + H\Theta' + F\,\Theta] = 0 \quad (13)$$

where Pr is the Prandtl number, $Gr = g \cos\alpha\ \beta\ (T_{w,x} - T_{\infty,x})\ x^3/\nu^2$, $Re = \omega \sin\alpha\ x^2/\nu$, stratification parameter, S

$$S = \frac{(T_{\infty,L} - T_{\infty,o}) \cos\alpha}{[(T_{w,L} - T_{\infty,o}) - (T_{\infty,L} - T_{\infty,o}) \cos\alpha\ ]} \quad (14)$$

The above equations are to be solved subjected to the following boundary conditions:

$$\text{At } z = 0 \quad u = w = (T - T_{w,x}) = (v - x\omega \cos\alpha) = 0 \quad (15a)$$

$$\text{At } z = 0 \quad u = o,\ v = o,\ T = T_{\infty,x} \quad (15b)$$

In terms of the variables defined by the equations (5) to (9) these conditions become:

$$\text{at } \eta = 0;\ F = (G - 1.0) = H = (\Theta - 1.0) = 0 \quad (16a)$$

$$\text{at } \eta \to \infty;\ F = 0,\ G = 0,\ \Theta = 0 \quad (16b)$$

The local heat transfer rates are calculated from the Fourier equation:

$$q = -k \left.\frac{\partial T}{\partial z}\right|_{z=0} = h\,(T_{w,x} - T_{\infty,x})$$

or in dimensionless form

$$Nu = \Theta'(0)\ Re^{1/2} \quad (17)$$

The local friction factor $C_{fy} = \tau_y / 1/2\ \rho\ (x \cos\alpha\ \omega)^2$ is given by

$$C_{fy}\ Re^{1/2} = 2\ G'(0) \quad (18)$$

The dimensionless moment coefficient is obtained by

$$C_m = M/1/2\ \rho\ (Lw \cos\alpha)^2\ \gamma_0^3 \quad (19)$$

where M is the shaft torque required to overcome the shear of the rotating cone and is given by

$$M = -\int_0^L r\tau_y \cdot 2\pi\, r dx \quad (20)$$

and r_o is the cone radius at the base where x = L. Equation (19) in terms of the non-dimensional variable becomes:

$$C_m\ Re^{1/2} = \frac{-\pi G'(0)}{\sin\alpha} \quad (21)$$

3. NUMERICAL SCHEME

Equations (10) to (13) along with the boundary conditions (16a) and (16b) are solved by a finite-difference technique. In the analysis of Hering and Grosh [14] for an unstratified case, a fourth-order Runge-Kutta forward integration scheme was used. This method required that at the staring point of integration of an nth order differential equation, the function and (n-1) derivatives be specified. For the present case, it means that seven conditions must be known at the starting point of integration since the system consists of three second order equations and a first-order equation. As seen from the boundary conditions (16a) and (16b), only four conditions are available. Thus Runge-Kutta method boils down to a systematic and laborious search for the remaining initial values which upon integration of the equations have to satisfy the three conditions at large value of η still unknown. To circumvent such difficulties, a finite-difference scheme is used.

The equations are expressed in finite-difference form. A central difference form for the second-order derivatives and upwind differencing [15] for the convection terms were used. These finite-difference equations lead to a system of simultaneous equations which are solved by Gauss-Siedel iterative technique [16]. The grid spacing was varied to ensure negligible dependence of the results obtained upon the spacing employed.

The specification of initial conditions for interior grids is arbitrary for the final converged solution and it was confirmed by varying them. The computer time required to obtain the converged solution was found to vary with the initial conditions. However, the converged solution was found essentially, to be independent of the specified conditions. The convergence criterion used for attaining the final converged solution is of the form $|G_i^{n+1} - G_i^n|_{max} = \varepsilon$, where the superscript refers to the number of iterations and subscript refers to the location. The value of ε was varied to ensure a very negligible dependence of the solution obtained of its value and was finally specified as 10^{-4}. A comparison of the results obtained for Pr = 0.7 with stratification parameter equal to zero with these of Hering and Grosh [14] showed an excellent agreement as discussed later. This indicates the consistency of the scheme employed. Numerical computations were performed on an IBM - 1130 computer and the solutions are obtained for a wide range of controlling parameters. Some of the typical results obtained are discussed below:

4. RESULTS AND DISCUSSION

Fig. 2 shows the temperature profiles for various values of S ranging from 0, the unstratified case, to 8.0. The results are shown for Pr = 0.7 and $(Gr/Re^2) = 100$ and the local temperature excess is non-dimensionalised with the surface temperature excess over the ambient temperature. The dimensionless temperature varies from 1.0 at the surface to 0.0 away from the surface. It is found that as S increases, the temperature profile steepens near the surface and attains increasingly negative values at large η. Negative values of temperature in the outer region indicates a temperature defect in the flow.

In a stratified medium, the ambient temperature increases with height. At a given vertical location in the outer region of the boundary layer, the fluid coming from below being colder than the local ambient temperature would create temperature defect when the rise in the ambient temperature is sufficiently large, or in other words, for large values of S. In addition, the temperature defect is profoundly influenced by Prandtl number and the mixed convection parameter (Gr/Re^2). The observed trends are similar to those in the investigations [2,5,6] related to the heat transfer studies from a vertical plate to a stratified medium. Comparision of the results obtained with those of Hering and Grosh

[14] for the unstratified case as shown in Fig.2 indicates a very close agreement between the two.

In Figs. 3,4 and 5 the influence of stratification parameter S on the tangential, normal and circumferential velocities is shown for $(Gr/Re^2) = 100$ and Pr=0.7 The profiles obtained for an unstratified case are found to be identical to those obtained by Hering and Grosh [14]. In Fig.3, the tangential velocity is seen to decrease, with increasing S, indicating an adverse effect on the flow due to the existence of stable stratification. As S increases, the buoyancy forces decrease and consequently, the magnitude of the maximum velocity also decreases. For high values of S, a slight flow reversal is observed away from the solid boundary.

The normal component of velocity, shown in Fig.4, is found to decrease with increase in the value of stratification parameter S and it is mainly due to the decrease in the magnitude of buoyancy force. Fig.5 shows the effect of stratification on the circumferential velocity level. It indicates that with an increase in the level of stratification, the circumferential component increases or in other words the rotation of the cone is felt even by the fluid particles away from the surface resulting in a substantial decrease in the mass flow of ambient fluid into the boundary layer. The effect of stratification on the flow and on the temperature field is found to be very marginal for high Prandtl numbers.

Figs.6,7 and 8 show the effect of mixed convection parameter on the temperature and flow fields for S=2.0 and Pr = 0.7. The boundary layer thickness dimnishes with increasing value of (Gr/Re^2). Fig.6 shows the variation of normal component of velocity for various values of (Gr/Re^2). It is seen that a rise in the value of (Gr/Re^2) leads to an increase in the normal component of velocity. This is due to the fact that an increase in the value of (Gr/Re^2) results in an increase in the level of buoyancy which causes an increase in the mass flow of ambient fluid into the neighbourhood of the cone. In Fig.7 the tangential velocity variation is plotted for various values of (Gr/Re^2). It is obvious that as (Gr/Re^2) increases, the mass flow increases and the boundary layer thickness decreases. These effects cause a rapid rise in the maximum value of the tangential velocity with (Gr/Re^2) Fig.8 shows the effect of (Gr/Re^2) on the variation of circumferential velocity and it is observed that as the value of (Gr/Re^2) increases, the rotational component of velocity decreases which can be attributed due to the increased mass flow of the ambient medium into the boundary layer.

In Fig.9, temperature distribution for various values of (Gr/Re^2) is shown for S=2.0 and Pr=0.7. With (Gr/Re^2) increasing, the influence of buoyancy forces become dominant and therefore, the magnitude of temperature defect decreases. The steep variation in the temperature profiles can be explained by a substantial increase in the mass flow rate which in turn is due to the increase in buoyancy forces. The trends observed here regarding the influence of (Gr/Re^2) are very similar to those obtained by Hering and Grosh [14] for an unstratified case S=0 except for the magnitudes.

The effect of stratification of the medium for the cases viz., forced, mixed and free convective conditions have been shown plotted in Figs. (10) and (11) for Pr = 0.7 and 6.7. The salient observations are that for a given Prandtl number and for the stipulated conditions of the flow field $(Gr/Re^2) = 0$, increase in S would lead to enhanced values of the heat transfer coefficients. From Fig.10, it is evident that the influence of stratification on the heat transfer coefficient for free convection is more pronounced and substantial than in the case of forced convection. The rise in heat transfer coefficient for $(Gr/Re^2)=100$ with respect to S for increasing magnitudes of S is steeper in relation to results obtained for the forced convection. Similar are the trends noted in Fig.11 for Pr=6.7 save for the magnitude of the gradients. However, the influence of stratification on the net heat transfer rate at a given location is to decrease its magnitude in view of the substantial decrease in the temperature potential existing between the wall and the local ambient medium. Similar observations are made in [3,10].

In Fig.(12) the variation of the moment coefficient as a function of stratification parameter for Pr = 0.7 is shown plotted for different values of the mixed convection parameter. A similar plot for Pr = 6.7 can be made. It is obvious from the figure that for a given value of S, the moment coefficient increases with the increase in the value of (Gr/Re^2). In addition, an increase in the stratification level monotonically decreases the moment coefficient which can be attributed to a reduction in the level of buoyancy force that leads to an increase in the value of circumferential velocity. This can be clearly seen from Fig.5. Fig.12 can be employed in the estimation of the torque for any stipulated conditions of rotation and stratification.

5. CONCLUSIONS

The following conclusions can be arrived at from the afore presented theoretical analysis:

1) The present problem happens to be a general case in which isothermal conditions of the medium happened to be a particular case.

2) Similar solutions are viable when the surface and ambient medium temperatures vary linearly with the existence of identical thermal conditions at the vertex of the cone.

3) Stratification of the medium is found to profoundly influence the temperature and velocity fields. The temperature defect can be attributed to the mixing of the fluid particles from the downstream side which are at a temperature lower than the local ambient temperature.

4) The temperature defect and flow reversal for high degrees of stratification are observed only in the outer region of the boundary layer.

5) The mixed convection parameter (Gr/Re^2) affects the velocity and temperature fields in a manner similar to that for an unstratified case S=0.

6) The local heat transfer coefficients are found to increase with an increase in S.

7) The local moment coefficients are found to decrease with the stratification, its implication being that the power requirements for the rotation of the cone in a stratified medium is less than that for the case of an unstratified medium.

ACKNOWLEDGEMENTS

The authors acknowledge the help rendered by the University Grants Commission, New Delhi under Departmental Research Programme.

NOMENCLATURE

x,y,z	coordinate directions
u,v,w	velocity components
V	local cone surface velocity
V_o	cone surface velocity at the base
r	cone radius
r_o	cone radius at the base
L	cone slant height
T	temperature of the fluid
$T_{\infty,x}$	ambient temperature at any x
$T_{w,x}$	surface temperature of the cone at any x
g	gravitational acceleration
k	thermal conductivity
c_p	specific heat of the fluid
C_{fy}	friction coefficient, $\tau y/(\rho v^2/2)$
C_m	moment coefficient $M/(\rho v_o^2 r_o^3/2)$
M	torque for rotation
Pr	Prandtl number
Gr	local Grashof number $g\beta(T_{w,x} - T_{\infty,x}) x^3 \cos\alpha/\nu^2$
Re	local Reynolds number Vx/ν
S	stratification parameter [E9.14]
q	local heat flux
h	local heat transfer coefficient, $q/(T_{w,x} - T_{\infty,x})$
Nu	local Nusselt number, hx/k
y	circumferential shear stress
ω	angular velocity
α	cone apex half-angle
β	coefficient of thermal expansion
ρ	density of the fluid
η	dimensionless coordinate $(\omega \sin\alpha / \nu)$ 1/2 Z
Θ	dimensionless temperature, $(T - T_{\infty,x}) / (T_{w,x} - T_{\infty,x})$
μ	absolute viscosity
ν	kinematic viscosity
ε	convergence criterion

REFERENCES

1. Cheesewright R., Natural convection from a plane vertical surface in non-isothermal surroundings, Int.J.Heat Mass transfer 10, (1967) 1847.

2. Yang K.T., Novotny J.L., and Cheng Y.S., Laminar convection from a nonisothermal plate immersed in a temperature stratified medium, Int.J.Heat and Mass Transfer, 15, (1972) 1097.

3. Eichhorn R., Natural convection in a thermally stratified fluid, Prog. Heat Mass Transfer 2, (1969) 41.

4. Chen C.C. and Eichhorn R., Natural convection from simple bodies immersed in thermally stratified fluids, Univ. of Kentucky Tech. Report No.105-ME (1977) 14-77.

5. Jaluria Y., and Himasekhar K., Buoyancy-induced two-dimensional vertical flows in a thermally stratified environment, J.Computers and Fluids 11, (1983) 39.

6. Jaluria Y., and Gebhart B., Stability and transition of buoyancy induced flows in a stratified medium, J. Fluid Mech 66 (1974) 593.

7. Turner J.S., Buoyancy effects in fluids, Cambridge Univ. Press, England (1973).

8. Himasekhar K., An analystical and Experimental study of laminar free boundary flows in a stratified medium, M.Tech thesis, I.I.T. Kanpur, India (1980).

9. Himasekhar K., and Jaluria Y., Laminar buoyancy induced axisymmetric free boundary flows in a thermally stratified medium, Int. J. Heat Mass Transfer 25 (1982) 213.

10. Himasekhar, K., and Jaluria, Y., Two-dimensional heated vertical jet in a thermally stratified medium submitted for presentation to the seventh National Conference on Heat and Mass Transfer to be held in I.I.T., Kharagpur in Dec.(1983).

11. Jaluria Y., Natural convection Heat and Mass Transfer, Pergamon Press, Oxford (1980).

12. Venkatachalam B.T., and Nath G., Non-similar laminar natural convection to a thermally stratified medium, Int. J. Heat and Mass Transfer 24 (1981) 1848.

13. Kreith F., Convective heat transfer in rotating systems, Advances in Heat Transfer, Vol. 5, (1968) 129.

14. Hering R.G. and Grosh R.J., Laminar combined convection from a rotating cone, J, of Heat Transfer, Trans. of ASME, series C 85, (1963) 29.

15. Roache P.J., Computational Fluid Dynamics, Hermosa, Albuquerque (1972).

16. Carnahan B., Luther H.A., and Wilkes J.O., Applied Numerical Methods, John Wiley, New York (1969).

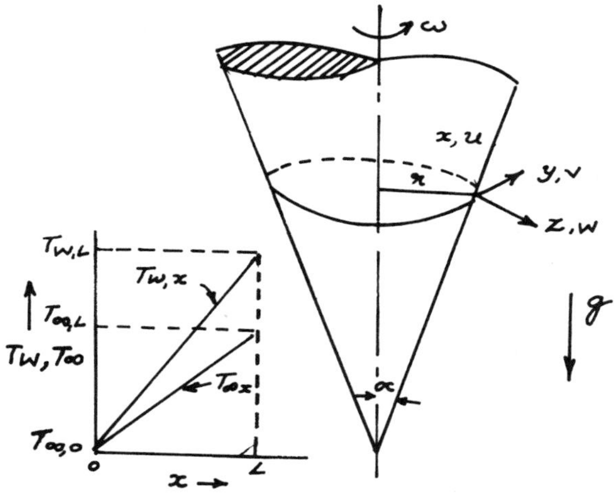

Fig. 1. Physical model

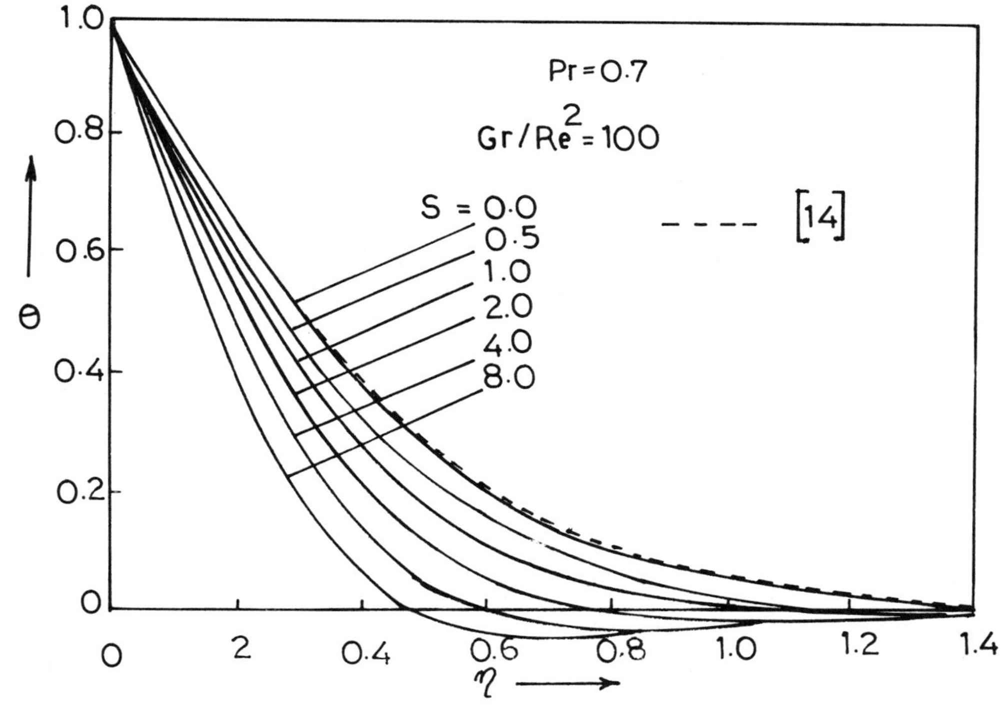

Fig.2. Variation of temperature with η for various values of S and for Pr = 0.7 and Gr/Re^2 = 100

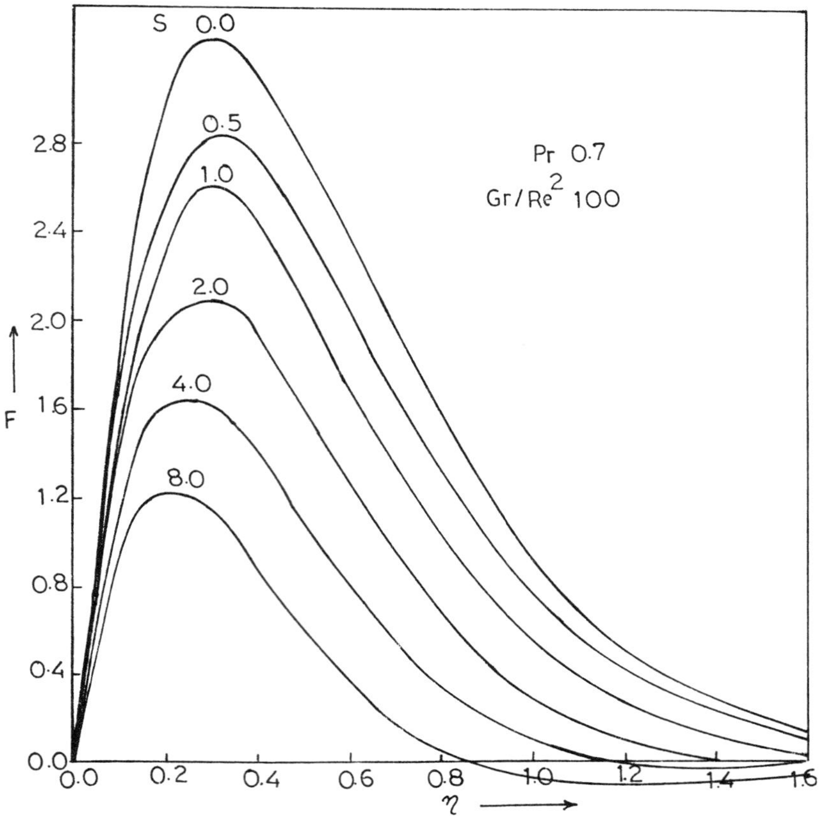

Fig.3. Tangential velocity distribution for various values of S and for Pr = 0.7 and Gr/Re^2 = 100

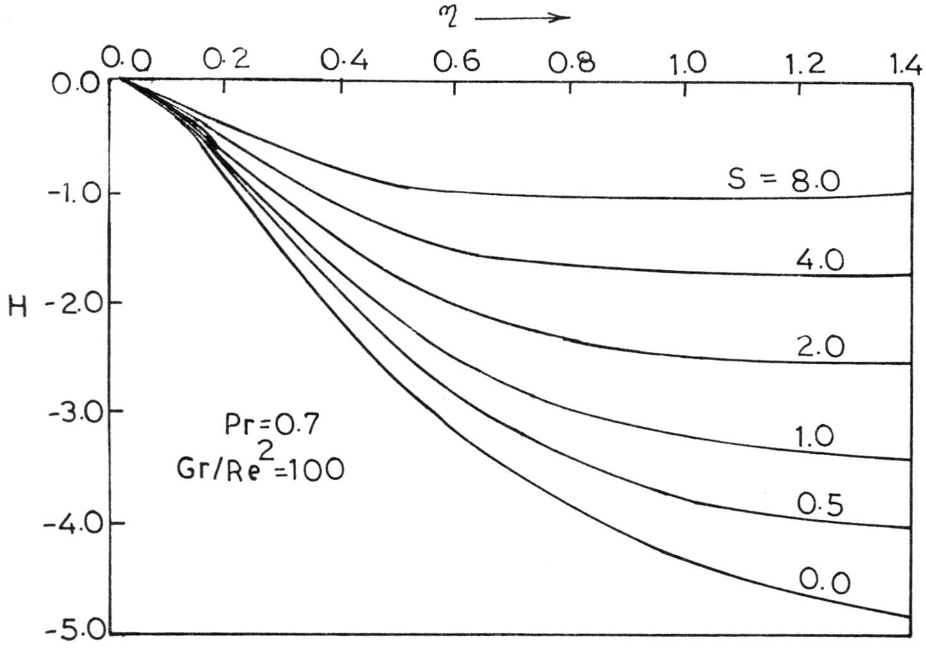

Fig.4. Normal Velocity distribution for various values of S and for Pr = 0.7 and Gr/Re^2 = 100

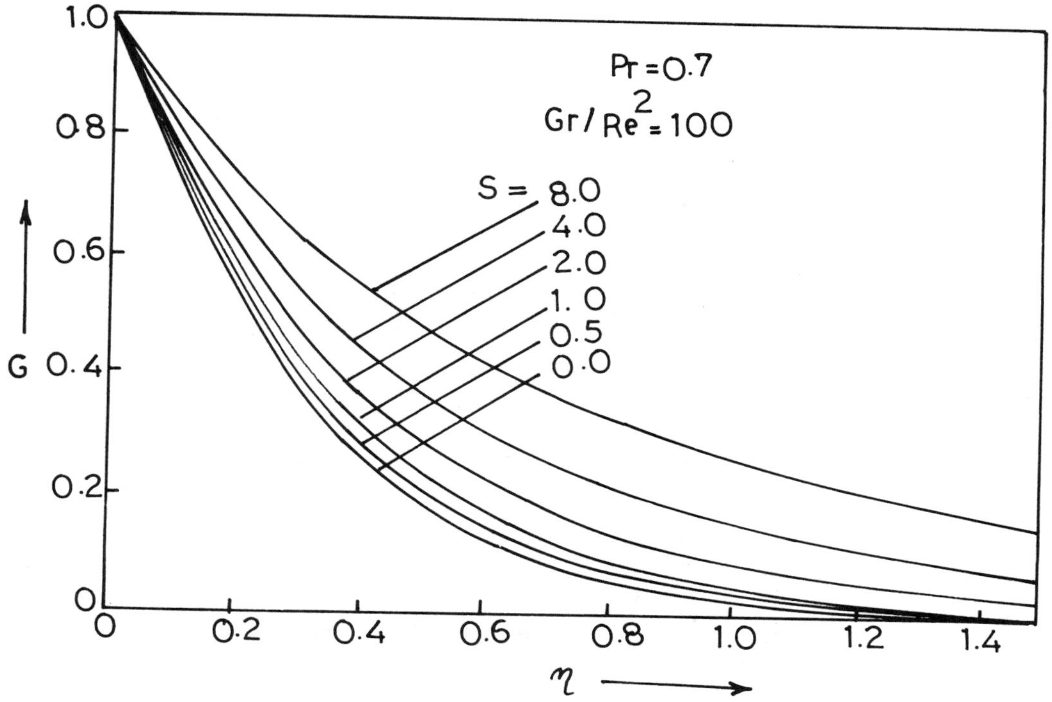

Fig.5. Circumferential velocity distribution for various values of S and for $Pr = 0.7$ and $Gr/Re^2 = 100$

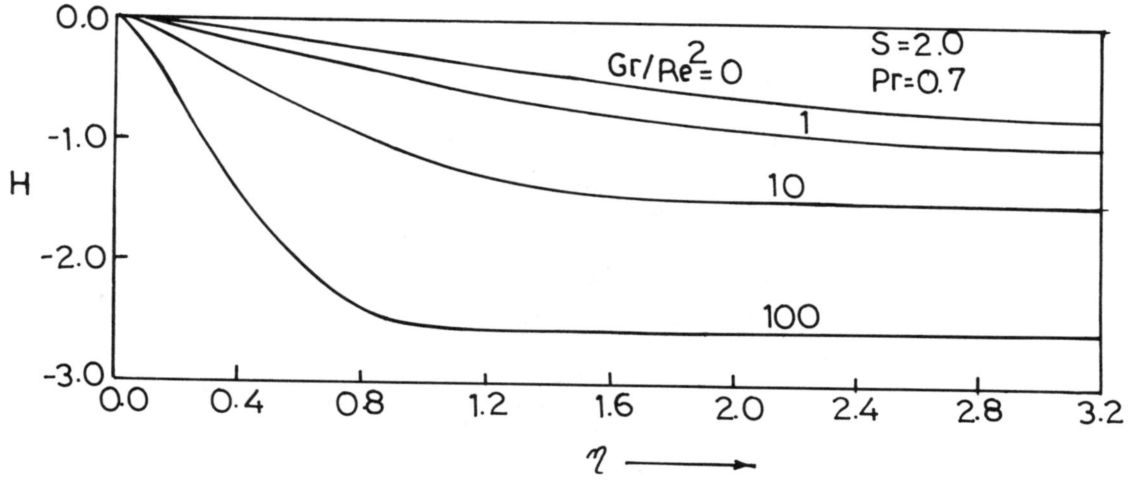

Fig.6. Effect of Gr/Re^2 on the tangential velocity profile for $Pr = 0.7$ and $S = 2.0$

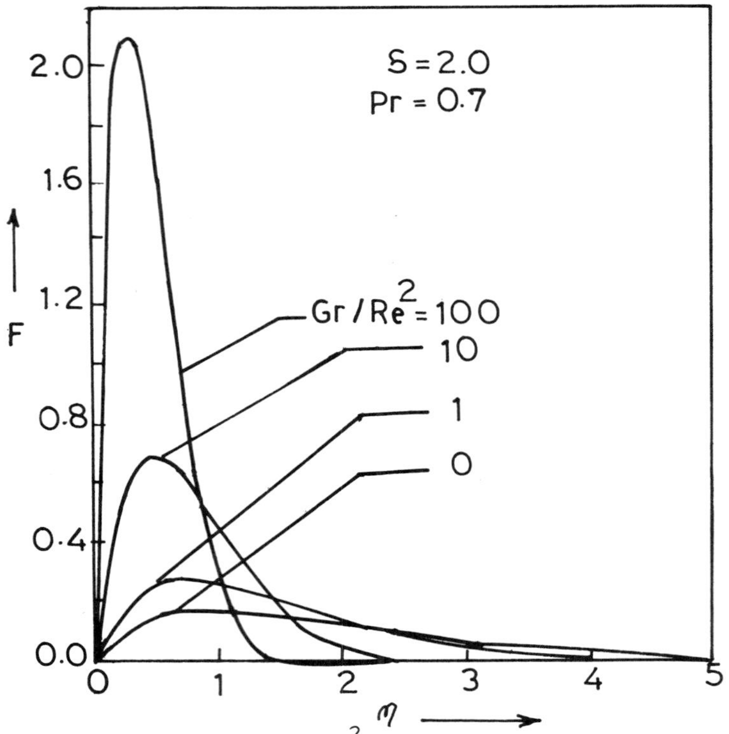

Fig.7. Effect of Gr/Re^2 on the tangential velocity profile for Pr = 0.7 and S = 2.0.

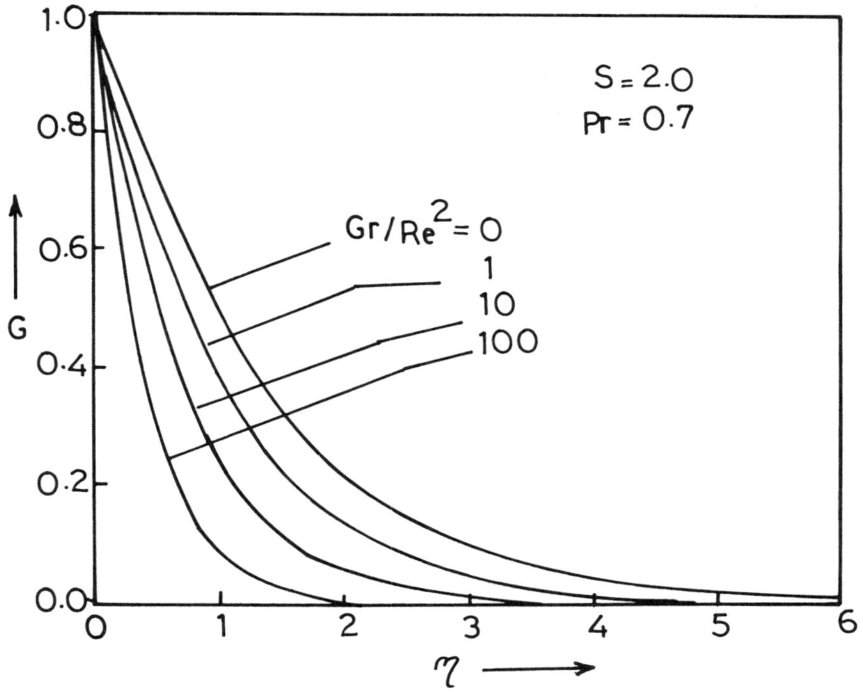

Fig. 8. Effect of Gr/Re^2 on the circumferential velocity profile for Pr = 0.7 and S = 2.0.

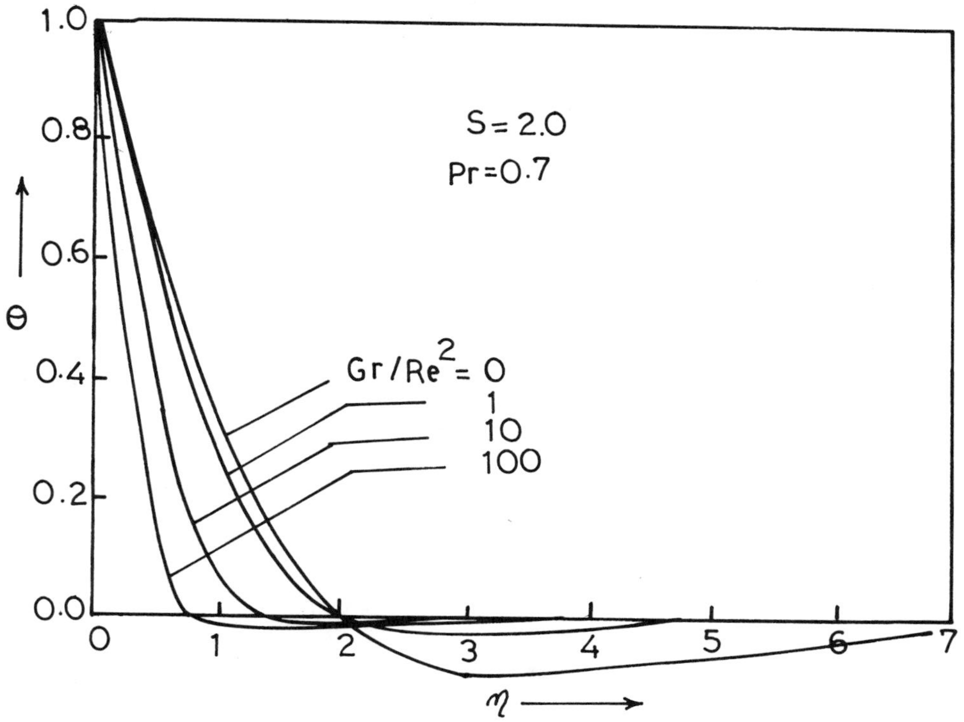

Fig.9. Variation of temperature with η for various values of Gr/Re^2 and for $Pr = 0.7$ and $S = 2.0$

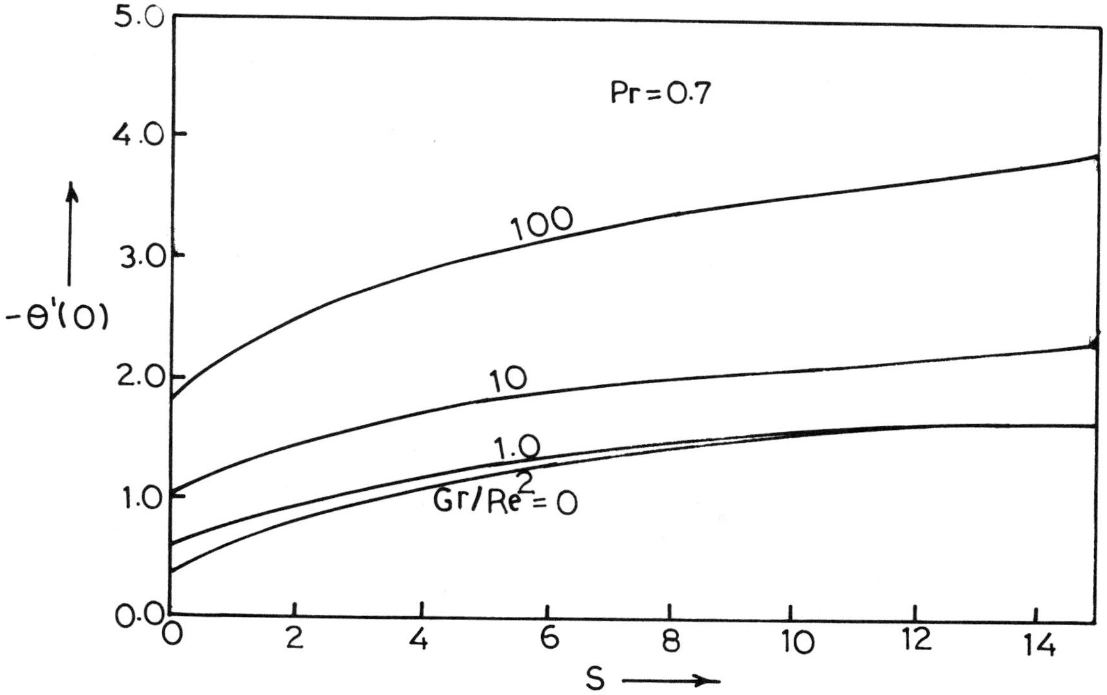

Fig.10. Effect of stratification parameter on heat transfer rate for $Pr = 0.7$ and for various values of Gr/Re^2.

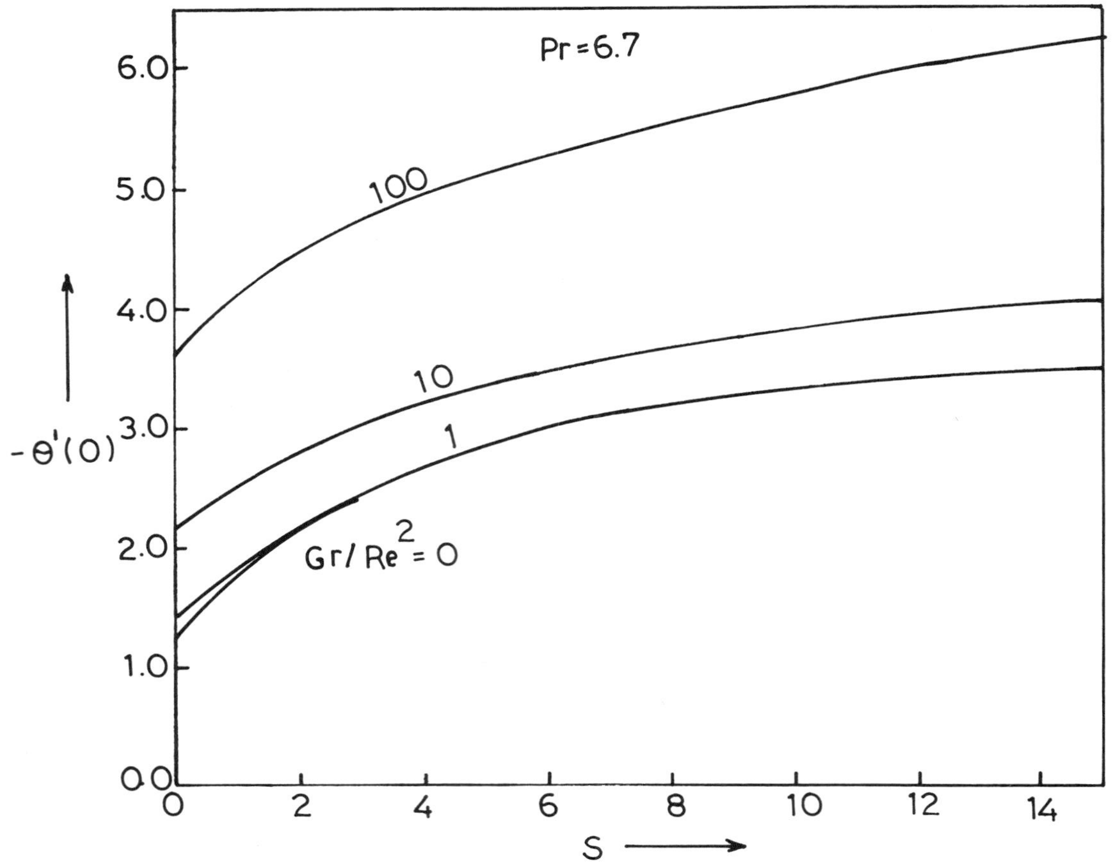

Fig.11. Effect of stratification parameter on heat transfer rate for Pr = 6.7 and various values of Gr/Re^2.

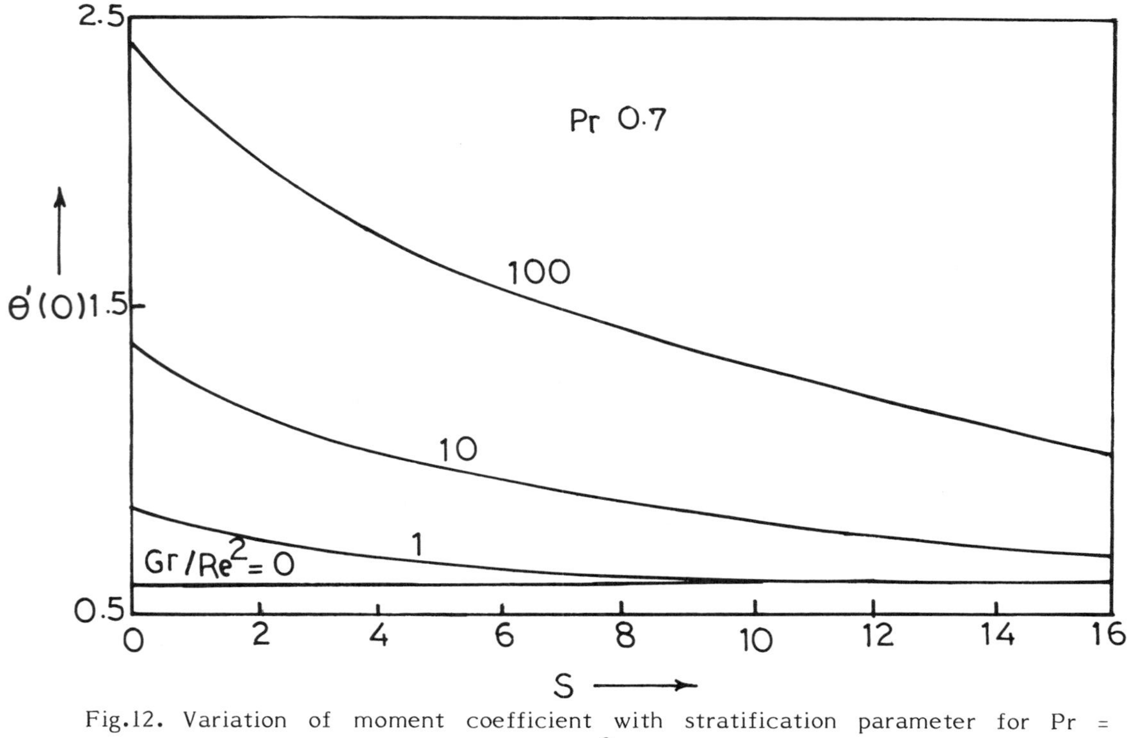

Fig.12. Variation of moment coefficient with stratification parameter for Pr = 0.7 and for various values of Gr/Re^2.

Mechanism for Enhancement of Heat Transfer in Turbulent Impinging Jets

K. KATAOKA, T. HARADA, and R. SAHARA
Department of Chemical Engineering
Kobe University
Kobe 657, Japan

The time-dependent convective heat transfer was measured by using a Gardon heat-flux sensor when a turbulent air jet impinged normal to a flat plate. The turbulent structure was also observed by using hot-wire anemometers. It has been confirmed that the stagnation-point heat transfer parameter (Froessling number) based on the arrival velocity and jet width is in proportion to the turbulent intensity of the approaching jet. The spectral analysis of velocity and wall heat-flux fluctuations suggests that two kinds of low-frequency turbulent motion play an important role in the stagnation-point heat transfer. A simple model is proposed to explain the mechanism for the heat transfer enhancement : large-scale turbulent eddy motion enhances the impingement heat transfer but sway motion of jet axis makes blunt the heat transfer distribution in the stagnation-point region.

1. INTRODUCTION

In the technology of convective heat transfer, there are two methods available for heat transfer augmentation: (1) effective increase in heat transfer area and (2) effective increase in heat transfer coefficient. The latter usually implies to increase gradients of time-averaged temperature near a heat transfer surface. This can be attained by inducing disturbances in fluid velocity near the heat transfer surface. This heat transfer augmentation is usually accompanied by an increase in friction loss that results from non-linear interaction of velocity fluctuations (i.e. the Reynolds stress). Most of the augmentation techniques proposed are effective only for laminar - and free - convective heat transfer [1].

One of the most interesting flows is a turbulent jet impinging on a solid body. As distinct from channel flows, a free jet can easily attain very high intensities of turbulence by virtue of intensive large-scale eddy and vortex motions.

This paper deals with convective heat transfer in a turbulent air jet impinging normally on a large flat plate. For large nozzle-to-plate spacings, flow in submerged impinging jets consists of four separate regions (shown in Fig.1) : (1) potential core region, (2) fully-developed turbulent jet region, (3) impingement region and (4) wall jet region.

It is very important question where a flat plate should be placed apart from a jet exit. The jet centerline velocity remains almost unchanged from its initial value over the potential core region whereas the intensity of turbulence increases with jet development and reaches a maximum at a certain position a little bit downstream of the apex of the potential core. A flat plate should, therefore, be placed at the end of the potential core region in order to attain maximal heat-fluxes at the stagnation point [2]. For the case of isothermal free jets issuing from a well-shaped convergent nozzle, the optimal nozzle-to-plate spacing usually lies between six and eight nozzle diameters (i.e $6 \le H/D \le 8$) [3 - 7].

For small nozzle-to-plate spacings, the radial distribution of wall heat-flux shows the existence of two secondary peaks around the stagnation point [3, 8, 9]. Pamadi and Belov [10] have confirmed by using a semi-empirical turbulence model that the inner peak arises as a result of nonuniform turbulence in the developing jet. However their model cannot be extended to the case of large nozzle-to-plate spacings ($H/D > 4$).

The Reynolds analogy between momentum and heat transfer does not hold in the small region including the stagnation point. This region is called the "stagnation-point region" in what follows. It has been revealed by many previous investigations [e.g 5, 6, 9] that convective heat and mass transfer is greatly enhanced without substantial increase in skin friction. The mechanism for the selective enhancement, however, remains a mysterious question owing to the experimental difficulties.

The effect of turbulence on heat transfer at a stagnation point has been studied for axisymmetric impinging jets by many investigators [3, 5, 6, 8, 9, 11, 12]. In general, to account for the effect of turbulence a correction factor is defined as

$$\frac{Nu}{Re^{1/2}} = (1 + \varepsilon)\left(\frac{Nu}{Re^{1/2}}\right)_{TF} \qquad (1)$$

It is very important to elucidate the turbulence correction term ε . From theoretical and empirical studies of heat transfer from circular cylinders, ε can be regarded as a function of $Tu\sqrt{Re}$ for the case of jet impingement heat transfer [5, 12].

The present work measures local, time-dependent fluxes of heat transfer in the stagnation-point region and proposes a simple model explaining the enhanced heat transfer.

2. EXPERIMENT

Figure 2 shows the experimental set-up. An isothermal air jet, issuing vertically upward from a smoothly-curved convergent orifice (40 mm in diameter), impinges normally on a horizontal circular flat plate. The jet exit has uniform velocity distribution with low initial turbulence (< 0.5%).

The detail of the circular flat plate is shown in Fig.3. The circular flat plate, made of copper, is 420 mm in diameter and 10 mm thick. A hot water jacket is installed on the back side of the flat plate so as to make heat flow from the back to the front surface. Steady-state convective heat transfer is attained between the flat plate and air jet. The flat plate is kept at a constant uniform temperature (usually at 55°C).

Local, time-dependent wall heat-flux on the surface of the flat plate is measured by means of a heat-flux sensor orignallly devised by Gardon[13]. The heat-flux sensor, shown in Fig.4, consists of a copper hollow male screw (as a heat sink, 14 mm OD) and a circular constantan foil (9 mm in diameter and 20 μm thick). One thermocouple junction is formed by spot-welding a 100 μm diameter copper wire on the backside center of the constantan foil. The other thermocouple junction is formed between the periphery of the constantan foil and the top flat plane of the copper male screw. The heat-flux sensor was embedded 50 mm apart from the center of the flat plate. The discontinuity between the surface of the heat-flux sensor and the surrounding surface of the flat plate was carefully eliminated by precise finish.

The hot junction (i.e. the periphery of the constantan foil) is kept at the same temperature as the flat plate. The **cold** junction (i.e. the center of the constantan foil) is cooled by air jet. The temperature difference between these two junctions is proportional to local coefficient of heat transfer

$$h = K_1 \quad \Delta T \qquad (2)$$

Another type of copper plug heat-flux gauge [14], embedded in the wall of the same flat plate, can directly measure time-averaged wall temperature T_w and wall heat-flux q_w for calibration of the Gardon heat-flux sensor. The axis of the convergent orifice was fixed 50 mm away from the rotation axis of the flat plate. The heat-flux sensor can, therefore, be placed at an arbitrary radial position r from the geometrical stagnation point (r = 0) by turning the flat plate. The nozzle-to-plate spacing can be changed by means of the bearing-screw assembly (shown in Fig.3). Turbulent velocities of the air jet are measured by constant-temperature hot-wire anemometers.

3. HEAT-FLUX SENSOR

Local wall heat-fluxes at the periphery and center of the constantan foil (Fig.5) are respectively expressible as

$$q_w = h\,(T_w - T_\infty) \qquad (3)$$

$$q_c = h\,(T_c - T_\infty) \qquad (4)$$

Heat conduction occurs in the foil from the periphery to the center at a rate of

$$Q_c = \kappa_c \frac{\Delta T}{\Delta r} S_c \qquad (5)$$

Assuming that the heat loss q_c is very rapidly compensated by the heat conduction Q_c

$$\kappa_c \frac{\Delta T}{\Delta r} S_c = h\, S(T_c - T_\infty) \qquad (6)$$

In the operating condition, $\Delta T = T_w - T_c$ is small compared to $T_w - T_\infty$. Hence, time-dependent wall heat-flux and heat transfer coefficient have a linear relationship with Δ T:

$$h = K_1 \quad \Delta T \qquad (2)$$

$$q_w = K_2 \quad \Delta T \qquad (7)$$

where $K_1 = \dfrac{\kappa S_c}{(T_w - T_\infty) \quad \Delta r\, S}$

$$K_2 = \frac{\kappa S_c}{\Delta r\, S}$$

The two constants are determined by experiment. Actually the response of the heat-flux sensor depends on the heat capacity of the constantan foil. The time constant of the heat-flux sensor used has been found to be 0.122 s.

4. THEORETICAL BACKGROUND

The stagnation-point heat transfer coefficient for the turbulence-free condition can be calculated theoretically using the boundary layer equation. Although Giralt et al.[5] tried to characterize the impingement flow field, this work adopts the free jet flow conditions for scaling the heat transfer data in a manner similar to that of Donaldson et al.[11]. It is convenient to adopt as the characteristic velocity and length scales the centerline velocity (arrival velocity) U_s and jet diameter D_j that free jet would have at the plane of impingement if a flat plate were removed. According to [2], $U_s = U_o$ and $D_j = D$ in the potential core region and $U_s = U_m$ and $D_j = 2\ r_{1/2U}$ in the fully-developed free jet region.

The flow in the stagnation-point region ($r/D << 1$) can be regarded approximately as an axisymmetric stagnant flow with uniform approach velocity U_s, i.e. axisymmetric Hiemenz flow. According to the boundary layer analysis[2], the stagnation-point heat transfer coefficient is given by the following form of dimensionless group :

$$\left[\frac{Nu_s}{Re_s^{1/2}\ Pr^{1/2}}\right]_{TF} = 0.88 \qquad (8)$$

This is a theoretical prediction for constant-property and turbulence-free conditions. Donaldson et al. [11] also derived a similar laminar solution given by

$$\left[\frac{Nu_s}{Re_s^{1/2}\ Pr^{1/2}}\right]_{TF} = 1.06 \qquad (9)$$

This is 20% higher than our result.

Because the characteristic velocity and length scales are defined at the plane of impingement, the preceding dimensionless group (Froessling number) should not be a function of the jet Reynolds number Re_o and the nozzle-to-plate spacing H/D. However it can be expected that the dimensionless stagnation-point heat transfer coefficient is a function of the so-called free-stream turbulence:

$$\frac{Nu_s}{Re_s^{1/2}\ Pr^{1/2}} = f(Tu) \qquad (10)$$

The intensity of turbulence is defined as the rms value of axial velocity fluctuation u (measured at 0.3D upstream of the geometrical stagnation point) divided by the time-averaged axial velocity on the jet axis U_m within the impingement region. According to Giralt et al.[5], the entrance to the impingement region lies at 1.2D upstream of the flat plate.

Hoogendoorn [12] correlated his stagnation-point heat transfer data in a functional form similar to that found for cylinders in a turbulent flow:

$$\frac{Nu}{Re^{1/2}} = 0.65 + 2.03\left(\frac{Tu\sqrt{Re}}{100}\right) - 2.46\left(\frac{Tu\sqrt{Re}}{100}\right)^2 \quad (11)$$

He adopted U_s and D as the characteristic velocity and length scales, respectively.

5. DISCUSSION OF EXPERIMENTAL RESULTS

The following basic flow properties have been confirmed :
(1) uniform initial distribution of axial velocity, (2) low initial velocity turbulence ($<0.5\%$), (3) constant centerline velocity over the potential core region (up to $Z/D = 4.75$), and (4) axial decay of the centerline velocity in inverse proportion to Z in the fully-developed free jet region. Figures 6 and 7 show the axial variation of the characteristic velocity and length scales.

5.1 Free-stream Turbulence of Approaching Jets

In order to observe the development of free-stream turbulence, time-dependent Reynolds stresses have been mesured at $Re_o = 20{,}000$ by using an X-probe located on the jet axis. Figure 8 shows some traces of the Reynolds stress.

Just downstream of the jet exit (at $Z/D = 1$), the Reynolds stress remains almost zero. It becomes considerably large with axial distance from the jet exit. As described by Yule [15], the development of a free jet is accompanied by the transition of axi-symmetrically coherent vortical structure to less-coherent large eddy structure. The potential core fluctuation is induced initially by trains of ring-shaped vortices produced in the shear layer. Large-scale eddy motion occurs due to the entanglement of vortex cores downstream of the potential core region. As will be described in the model consideration, the enhancement factor becomes very large when $H/D = 6$. It is very important to consider the effect of large-scale eddy motion on the stagnation-point heat transfer when a flat plate is placed at the end of the potential core region, i.e $H/D = 6$.

5.2 Heat Transfer Coefficient

Figure 9 indicates the effect of nozzle-to-plate spacing on the stagnation-point heat transfer. Within the experimental condition, the stagnation-point heat transfer coefficient becomes maximal around $H/D = 6$. The optimal nozzle-to-plate spacing tends to increase slightly with increasing Re_o. This can be attributed to the fact that the potential core is lengthened slightly with increasing Re_o.

Figure 10 indicates that the radial distribution of the normalized heat transfer coefficient becomes similar for $H/D \geq 6$ if the radial coordinate is made dimensionless with respect to the half-radius of wall heat-flux [14].

5.3 Effect of Free-stream Turbulence

Figure 11 shows the enhancement effect of free-stream turbulence on the stagnation-point heat transfer expressed in the form of Eq.(10).

As would be expected, it is found that the dimensionless stagnation-point heat transfer coefficient is a function of free-stream turbulence only. As distinct from the functional form of Hoogendoorn [12], there is no discernible effect of Reynolds number. This is consistent with the result of Donaldson et al.[11]. It has been confirmed that the dimensionless heat transfer parameter is linearly proportional to the intensity of turbulence. The best-fit equation is given by

$$\frac{Nu_s}{Re_s^{1/2}\ Pr^{1/2}} = 0.88 + 13.3\left(\frac{Tu}{100}\right) \quad (12)$$

As Tu goes to zero, the heat transfer parameter approaches the theoretical value for the turbulence-free condition. The heat transfer data for $Re_o = 7{,}000$ deviate a little bit from the linear relationship because the jet is unstable owing to very small arrival velocity. The Donaldson et al. heat transfer data were recalculated from Fig.27 with the aid of Table 1 of their paper[11]. Their heat transfer data points were also plotted in Fig.11 assuming their turbulence intensity of free jets $\sqrt{\tilde{\kappa}'}/w_c$ equal to our turbulence intensity of impinging jets Tu. It seems that their data have a linear relationship with the intensity of turbulence.

The Hoogendoorn model, given by Eq.(11), was derived by making use of the functional form [16, 17] of heated cylinders placed normal to a turbulent stream. A nozzle diameter D was taken as the length scale. Nevertheless the effect of scale of turbulence was not studied. Still it can be conjectured from Fig.5 of his paper that the Hoogendoorn heat transfer data have the same tendency against the intensity of turbulence as found in this work. The integral scale of turbulence for free jets is usually proportional to the width of mixing layer, i.e. to the jet half-radius. Therefore it can be considered that the ratio of the scale of turbulence to the local jet diameter remains constant. As long as the jet diameter is adopted as the length scale, the dimensionless heat transfer parameter can be interpreted to include implicitly the effect of the scale of turbulence.

5.4 Velocity Fluctuation Measurements of Impingning Jets

Figure 12 shows the power spectrum of axial velocity fluctuation measured at the jet axis 0.3D (i.e. 12 mm) upstream of the geometrical stagnation point for $H/D = 2$.

The Nyquist frequency is 1 Hz in the condition of our spectral analysis. Although the intensity of turbulence is very small, low frequency components already exist between 3 and 4 Hz in both the u- and v- spectra. The potential core fluctuation is induced due to trains of vortex rings growing in the shear layer.

Figure 13 shows the power spectra of velocity fluctuations measured at the jet axis 0.3D upstream of the geometrical stagnation point for $H/D = 6$. The measuring point can be considered from [5] to be located within the impingement region. The potential core has disappeared before the jet enters the impingement region. Consequently the intensity of turbulence becomes considerably large owing to large-scale turbulent eddies. According to the u-spectrum, there appear very low frequency components around 3 Hz and somewhat higher-frequency spectrally-continuous components in the range 10 - 100 Hz. It can be considered that the turbulent structure consists of a sway motion of jet axis, a large-scale eddy motion, and the background small-scale turbulent motion. According to the v-spectrum (Fig.13b), there exists low frequency components around 5 Hz. These low frequency components (< 10 Hz) come from the sway motion of the jet axis. The X-probe is, however, insensitive to one velocity component normal to both the two hot-wire elements. Therefore both the u- and v-spectra are apt to underestimate the passing frequency of the jet axis.

Figure 14 shows the power spectrum of v-fluctuation measured by fixing an I-probe at a position (y = 1 mm, r/D = 0.25). This spectrum can be compared to the u-spectrum for H/D = 6 (Fig.13a). It is found that the large-scale motion of the approaching jet is still preserved after the deflection. The plateau in the range 2 - 5 Hz comes from the sway motion. Since the instantaneous stagnation point moves two-dimensionally around the axis of symmetry (r = 0), the passing frequency the measuring point (r/D = 0.25) can detect is lower than the actual passing frequency measured at r = 0. A few large peaks appearing in the range 7 - 20 Hz can be considered as the large-scale eddy motion.

5.5 Heat-flux Fluctuation Measurements

Figure 15 shows time-traces of the wall heat-flux measured at the geometrical stagnation point and the spectrum of the wall heat-flux fluctuation for H/D = 6.

The q-spectrum indicates a single sharp large peak at 8 Hz. This dominant oscillation is discernible in the original time-traces. The predominant fluctuation of the wall heat-flux comes form the sway motion. As mentioned, the v-spectrum (Fig.13b) indicated a sharp peak of the sway motion at a little bit lower frequency, i.e. about 5 Hz. The discrepancy in the characteristic frequency of the sway motion may be attributable to the fact that the heat-flux sensor can detect clearly the sway motion even if the stagnation point moves in any directions.

Figure 16 is the q-spectrum obtained at r/D = 0.25. Any large peaks do not exist in the range 2 to 15 Hz. As mentioned, this implies that the instantaneous stagnation point did not frequently pass the measuring point during the sampling time. It can be considered that since the stagnation-point heat transfer becomes maximal when H/D = 6, the convective heat transfer is enhanced mainly by the large-scale turbulent eddy motion. It is necessary to take into account the effect of the sway motion in model consideration of the enhancement of heat transfer in the stagnation-point region.

6. MODEL CONSIDERATION

According to the flow-visulization experiment for H/D $\simeq$ 6 [18], an instantaneous stagnation point fluctuates at random around the geometrical stagnation point (r = 0). This sway motion of the jet axis is caused by large-scale turbulent eddies. Uchida et al. [19] observed a similar sway motion occuring at the re-attaching point of separated flows.

A simple model will be considered using a two-dimensional impinging jet flow so as to explain momentum and heat transfer in the stagnation-point region. For simplicity, the numercial solution of the laminar boundary layer equation [e.g.20] can be approximated near the stagnation point by the following equations

$$\tau_o = \alpha_o x \tag{13}$$

$$q_o = \beta_o (1 - \beta_1 x^2) \tag{14}$$

where x designates the lateral distance from the geometrical stagnation point. These equations imply that the wall shear stress increases linearly from zero while the wall heat-flux decreases from the stagnation-point value β_o with increasing x. In this sense, the Reynolds analogy does not exist in the stagnation-point region.

It is assumed for model consideration that the turbulent structure of the impingement region consists of a low-frequency sway motion and a spectrally-continuous turbulent motion including the large-scale eddy motion. If the sway motion were not in the approaching stream, therefore, if the instantaneous stagnation point were stationary at x = 0, the wall shear stress and wall heat-flux enhanced by the turbulent eddy motion could be expressed by

$$\tau_w^* = \alpha_o x [1 + \gamma_M(x)] \tag{15}$$

$$q_w^* = \beta_o (1 - \beta_1 x^2) [1 + \gamma_H(x)] \tag{16}$$

where γ_M and γ_H are the enhancement effect factors for momentum and heat transfer, respectively. Owing to the selective enhancement of heat transfer, γ_H is usually much larger than γ_M.

It has also been reported [e.g.21] that the turbulence in the approaching jet is amplified strongly in the stagnation-point region owing to the stretching of vortex filaments. Figure 17 is a schematic diagram illustrating a model of stagnation-point heat transfer enhancement.

The probability density of existence of the instantaneous stagnation point is given by the following normal distribution :

$$\Phi(\xi) = \frac{1}{\sqrt{2\pi}\,\sigma} \exp(-\xi^2/2\sigma^2) \tag{17}$$

Here ξ is a lateral position where the instantaneous stagnation point exists and σ the standard deviation from x = 0.

In Fig.18, the probability density distribution measured for submerged water jet by using a hydrogen bubble method is shown in the form of histogram.

As shown in Fig.17d, the shear force at x is directed to negative x when the stagnation point stays at the right of x, i.e. $\tau_w(x) < 0$. This suggests that the probability at which the instantaneous wall shear stress becomes negative is 50% at $x = 0$ owing to symmetry. At any instant, the instantaneous wall shear stress and wall heat-flux at x become

$$\tau_w'(x) = \alpha_o(x - \xi)[1 + \gamma_M(x)] \quad (18)$$

$$q_w'(x) = \beta_o[1 - \beta_1(x - \xi)^2][1 + \gamma_H(x)] \quad (19)$$

Taking into account the probability of existence of the instantaneous stagnation point, the time-averaged values of τ_w and q_w are obtained, respectively, as

$$\tau_w(x) = \alpha_o[x - \int_{-\infty}^{\infty} \xi\, \Phi(\xi)\, d\xi][1 + \gamma_M(x)] =$$

$$\alpha_o x[1 + \gamma_M(x)] \quad (20)$$

$$q_w(x) = \beta_o[1 - \beta_1 x^2 + 2\beta_1 x \int_{-\infty}^{\infty} \xi\, \Phi(\xi)\, d\xi -$$

$$\beta_1 \int_{-\infty}^{\infty} \xi^2\, \Phi(\xi)\, d\xi][1 + \gamma_H(x)]$$

$$= \beta_o[1 - \beta_1(x^2 + \sigma^2)][1 + \gamma_H(x)] \quad (21)$$

It is found that if the stagnation point were stationary, the stagnation-point heat-flux would become larger than the time-averaged value given by Eq.(21). As $x \to 0$, the preceding equations give the time-averaged values as the geometrical stagnation point :

$$\tau_w(0) = 0 \quad (22)$$

$$q_w(0) = \beta_o(1 - \beta_1 \sigma^2)[1 + \gamma_H(0)] \quad (23)$$

The wall momentum-flux (shear strss) does not share the benefit of the enhancement effect owing to symmetry in the neighbourhood of the geometrical stagnation point. The wall heat-flux is very much enhanced by γ_H but depressed somewhat by σ (i.e.by the sway motion) The two enhancement effect parameters γ_M, γ_H remain to be determined by experiment.

Most of the shear stress measurements have non-zero definite values at the stagnation point. For example, Popiel and Trass [22] and Baines and Keffer [23] measured radial or lateral distribution of the wall shear stress by means of hot-film surface probes. They pointed out that we should take into consideration the size effect of the sensing element on the local shear stress measurements in the stagnation-point region. It should also be considered that the hot-film surface probes used could not detect a change in sign of the wall shear stress at the stagnation point in spite of the sway motion.

The wall shear stress and mass transfer coefficient distributions [6] measured by an electrochemical technique are appropriate to determine γ_M and γ_H. Figure 19 shows some distributions of the mass transfer enhancement effect factor calculated with the aid of Fig.8 of the previous paper [6]. It is found that γ_H remains constant in the stagnation-point region. This is consistent with the turbulence effect reported by Donaldson et al. [11]. The mass transfer coefficient distribution [6] had a small dip at the geometrical stagnation point. At the present stage, this is attributable to an experimental error due to the thickness of the epoxy resin which was used for insulating the test electrodes embedded in the wall of flat plate. The stagnation-point values, therefore, have been obtained by interpolation of the adjacent data points because this error is large only at the geometrical stagnation point. On the contrary, the shear stress enhancement effect factor γ_M is always approximately zero over the whole impingement region according to Fig.9 of the previous paper [6]. It is still unknown why the heat and mass transfer is enhanced without substantial increase in wall shear stress in the stagnation-point region of impinging jets. At the present stage, the mechanism for the selective enhancement of heat/mass transfer continues to be an abstruse question. Still much remains to be done for it.

ACKNOWLEDGEMENTS

This project has been partially supported by The Asahi Glass Foundation for Industrial Technology.

NOMENCLATURE

- D, exit diameter of convergent nozzle or orifice;
- D_j, local jet diameter;
- f, frequency;
- H, nozzle-to-plate spacing;
- h, heat transfer coefficient;
- K_1, K_2, constants determined by experiment;
- Nu, Nussselt number, $h\,D/\kappa$;
- Nu_s, stagnation-point Nusselt number, $h_s D_j/\kappa$;
- Pr, Prandtl number;
- Pq, power spectrum of heat-flux fluctuation;
- Pu, power specturm of axial velocity fluctuation;
- Pv, power spectrum of radial velocity flucutation;
- q, local heat-flux;
- Re, Reynolds number;
- Re_o, jet Reynolds number evaluated at exit, $U_o D/\nu$;
- Re_s, stagnation-point Reynolds number, $U_s D_j/\nu$;
- r, radial distance from geometrical stagnation point or jet axis;
- $r_{1/2}$, jet half-radius;
- Δr, distance between center and periphery of constantan foil;
- S, heat transfer area;
- T, temperature;
- Tu, turbulence intensity, $\sqrt{\overline{u^2}}/U_m$ (%);
- ΔT, temperature difference between center and periphery of heat-flux sensor, $T_w - T_c$;
- t, time;
- U_m, centerline velocity;
- U_o, jet velocity at exit;
- U_s, arrival velocity defined by free jet condition;
- u, axial velocity fluctuation;
- v, radial velocity fluctuation;
- x, lateral distance from stagnation line or point;
- y, distance from and normal to flat plate surface;
- Z, axial distance from jet exit;

Greek symbols

α_o, constant given by Eq.(13);
β_o, β_1, constants given by Eq.(14);
γ, enhancement effect factor;
ε, turbulence correction factor;
κ, thermal conductivity;
ν, kinematic viscosity;
ξ, instantaneous position of stagnation point;
σ, standard deviation of position of stagnation point;
Φ, probability density function of existence of instantaneous stagnation point;

Subscripts

c, constantan;
H, heat;
M, momentum;
o, laminar flow or nozzle-exit value;
q, heat-flux;
s, stagnation point;
TF, turbulence-free condition;
U, velocity;
w, wall;
∞, bulk fluid;

Superscripts

*, without sway motion;
', instantaneous quantity;

REFERENCES

1. Kataoka, K., Special flow techniques in heat transfer augmentation, Rev. Paper (in Japanese), Chem.Engg.,Japan 49 (1985) 269-275.
2. Kataoka, K., Optimal nozzle-to-plate spacing for convective heat transfer in nonisothermal, variable-density impinging jets, DRYING TECH. 3, (1985) 235-254.
3. Gardon R., and Cobonpue,J., Heat transfer between a flat plate and jets of air impinging on it, Int.Dev. in Heat Transfer,(1963)454-460.
4. Nakatogawa, T., Nishiwaki, N., Hirata M.,and Torii K., Heat transfer of round turbulent jet impinging normally on flat plate, Proc. 4th Int. Heat Transfer Conf., FC5.2(1970).
5. Giralt, F., Chia, C.J., and Trass,O., Characterization of the impingement region in an axisymmetric turbulent jet, Ind.Eng.Chem.Fundam 16 (1977) 21-28.
6. Kataoka, K., Kamiyama, Y., Hashimoto, S., and T. Komai, Mass Transfer between a plane surface and an impinging turbulent jet; the influence of surface-pressure fluctuations. J.Fluid Mech., 119 (1982) 91-105.
7. Hrycak, P., Heat transfer from round impinging jets to a flat plate, Int.J.Heat Mass Transfer, 26 (1983) 1857-1865.
8. Gardon, R., and Akfirat, J.C., The role of turbulence in determining the heat-transfer characteristics of impinging jets, Int.J.Heat Mass Transfer, 8 (1965) 1261-1272.
9. Chia C-J., Giralt, F., and Trass, O., Mass transfer in axisymmetric turbulent impinging jets, Ind.Eng.Chem., Fundam., 16 (1977) 28-35.
10. Pamadi, B.N., and Belov, I.A., A note on the heat transfer characteristics of circular impinging jet, Int. J.Heat Mass Transfer, 23 (1980) 783-787.
11. Donaldson, C.D., Snedeker, R.S., and Margolis, D.P., A study of free jet impingement, Part 2. Free jet turbulent structure and impingement heat transfer, J.Fluid Mech. 45 (1971) 477-512.
12. Hoogendoorn, C.J., The effect of turbulence on heat transfer at a stagnation point, Int.J.Heat Mass Transfer, 20 (1977) 1333-1338.
13. Gardon, R., A transducer for the measurement of heat flow rate, J. Heat Transfer, C82, (1960) 396-398.
14. Kataoka, K., Shundoh, H., Matsuo, H., and Kawachi Y., Characteristics of convective heat transfer in nonisothermal, variable-density impinging jets, Chem. Engg. Commun., 34, (1985) 267-275.
15. Yule, A.J., Large-scale structure in the mixing layer of a round jet, J.Fluid Mech., 89, (1978) 413-432.
16. Kestin J., and Wood, R., The influence of turbulence on mass transfer from cylinders, J.Heat Transfer, C93 (1971) 321-327.
17. Lowery, G.W., and Vachon, R.I., The effect of turbulence on heat transfer from heated cylinders, Int.J.Heat Mass Transfer, 18, (1975) 1229-1242.
18. Kataoka, K., Mihata, I., Maruo, K., and Sunguro, M., Study on coherent structure of turbulent impinging jets, Preprint of 21st National Heat Transfer Symposium of Japan (in Japanese), B211, (1984) 88-90.
19. Uchida, Y., Mori, Y., and Sakai, K., Study of time and space microstructure for heat transfer performance near the reattaching point of separated flows (2nd Report), Preprint of 22nd National Heat Transfer Symposium of Japan (in Japanese), B111, (1985) 181-183.
20. Miyazaki, H.,and Silberman, E., Flow and heat transfer on a flat plate normal to a two-dimensional laminar jet issuing from a nozzle of finite height, Int.J.Heat Mass Transfer, 15, (1972) 2097-2107.
21. Yokobori, S., Kasagi, N., and Hirata, M., Characteristic behaviour of turbulence in the stagnation region on a two dimensional submerged jet impinging normally on a flat plate, Proc. Symp. on Turbulent Shear Flows, (1977) 3-17.
22. Popiel, C.O., and Trass, O., The effect of ordered structure of turbulence on momentum, heat and mass transfer of impinging round jets, Proc. 7th Int.Heat Transfer Conf., CP25,Vol.6, (1982) 141-146.
23. Baines, W.D., and Keffer, J.F., Shear stress and heat transfer at a stagnation point, Int.J.Heat Mass Transfer, 19, (1976) 21-26.

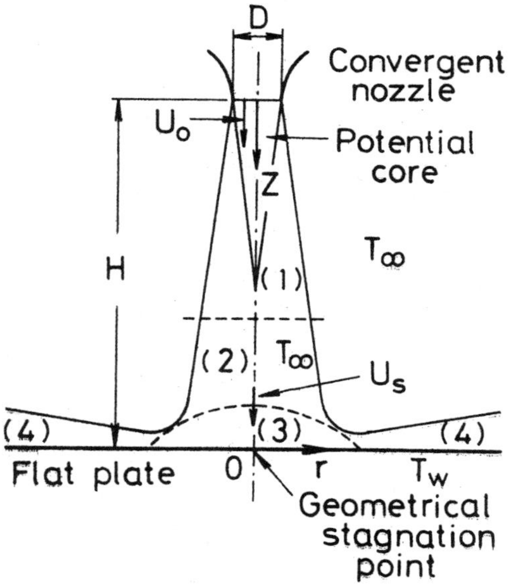

Fig.1 Schematic diagram of an impinging jet, 1: potential core region, 2: fully-developed turbulent jet region, 3: impingement region, 4: wall jet region.

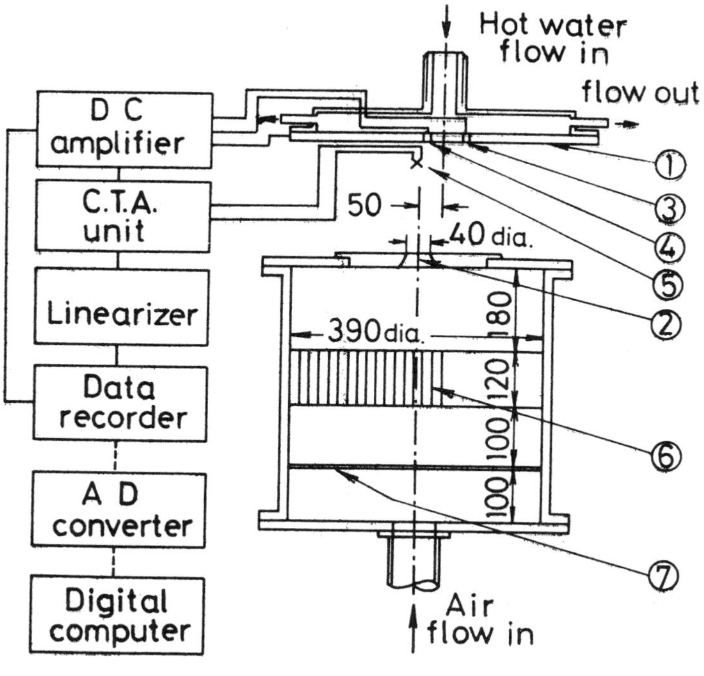

Fig.2 Schematic of experimental set-up. Dimensions given are in mm, 1: circular flat plate, 2: convergent bell-mouth orifice, 3: Gardon heat-flux sensor, 4: copper plug heat-flux gauge, 5: X-probe, 6: straw-tube calming section, 7: perforated plate.

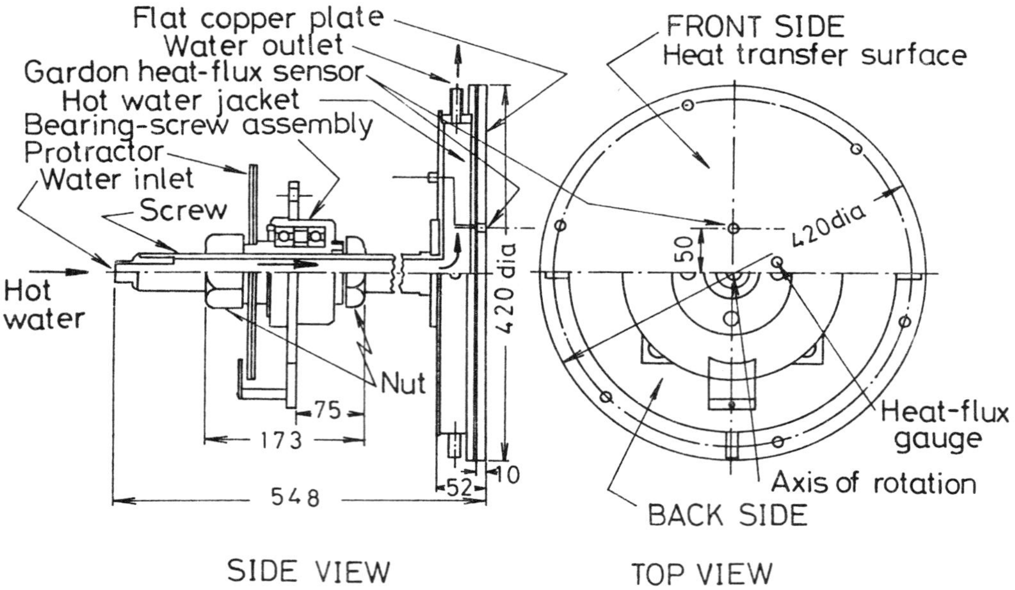

Fig.3 Circular flat plate. Dimensions given are in mm.

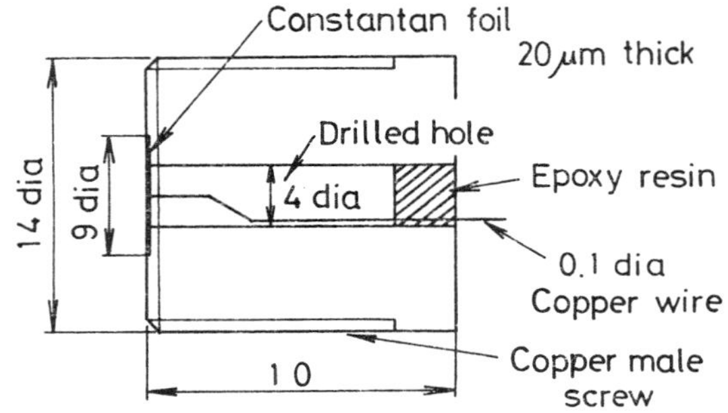

Fig.4 Gardon heat-flux sensor. Dimensions given are in mm.

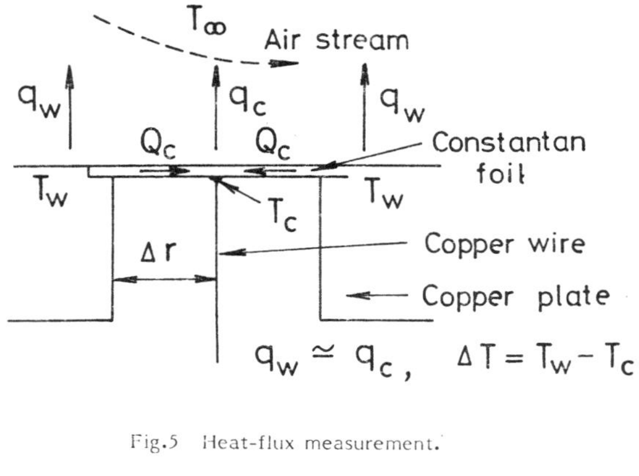

Fig.5 Heat-flux measurement.

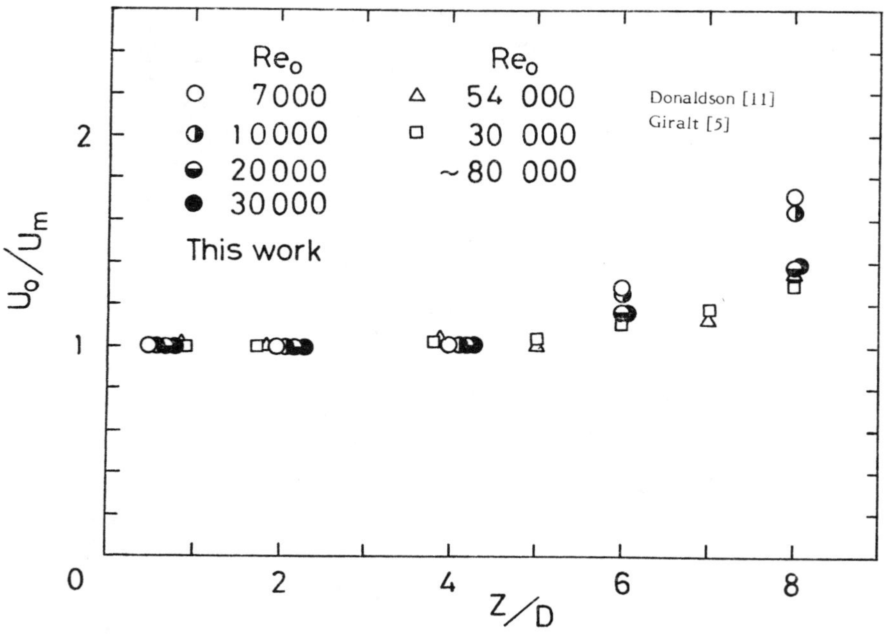

Fig.6 Axial variation of centerline velocity for free jet.

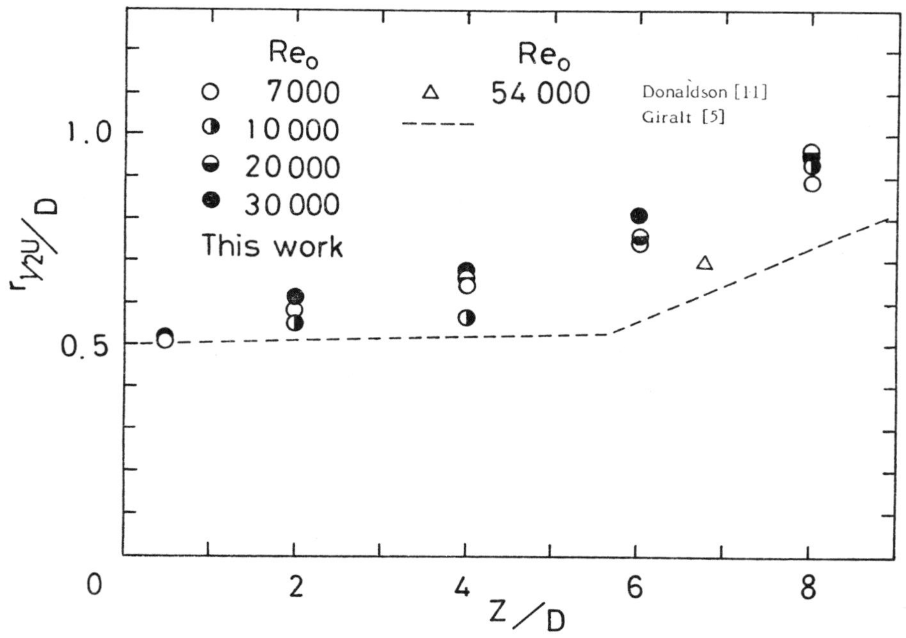

Fig.7 Axial variation of jet half-radius for free jet.

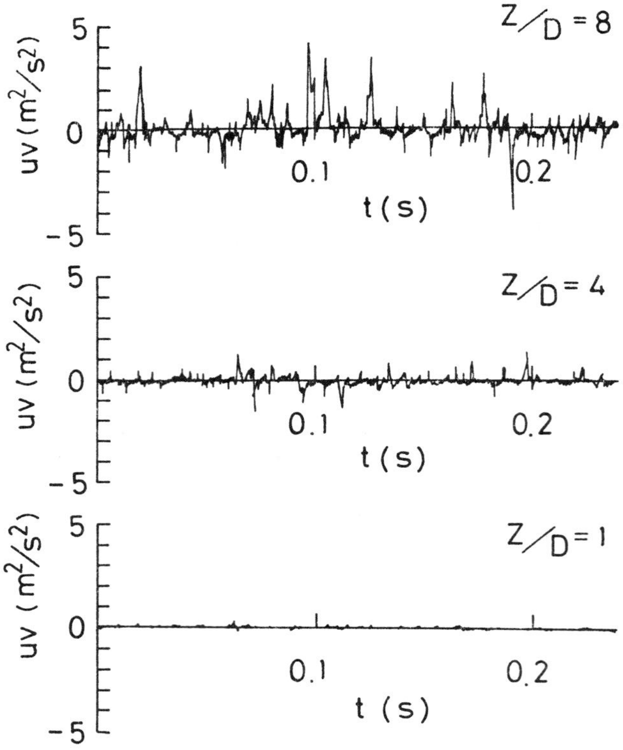

Fig.8. Time-traces of Reynolds stress on jet axis.

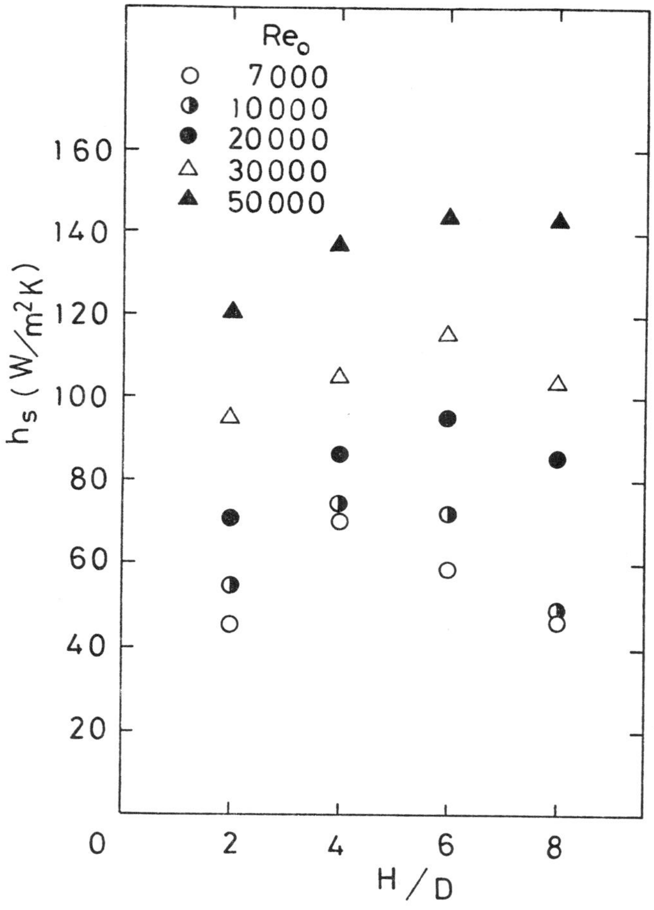

Fig.9 Effect of nozzle-to-plate spacing on stagnation-point heat transfer.

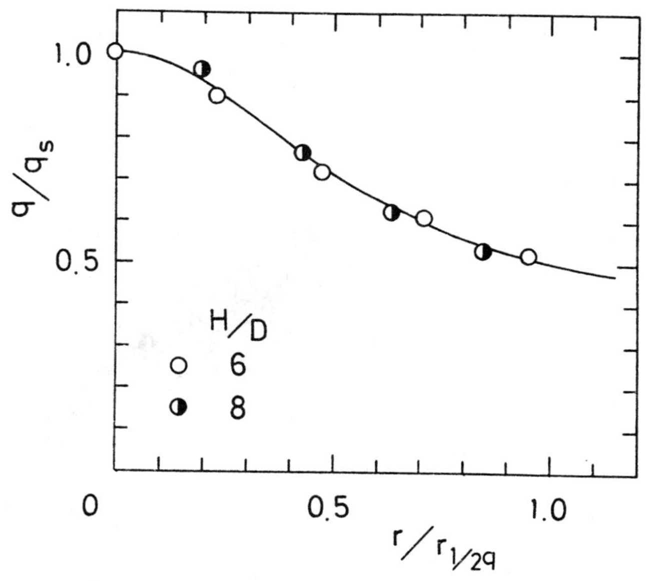

Fig.10 Radial distribution of normalized wall heat-fluxes for H/D ≥ 6,.

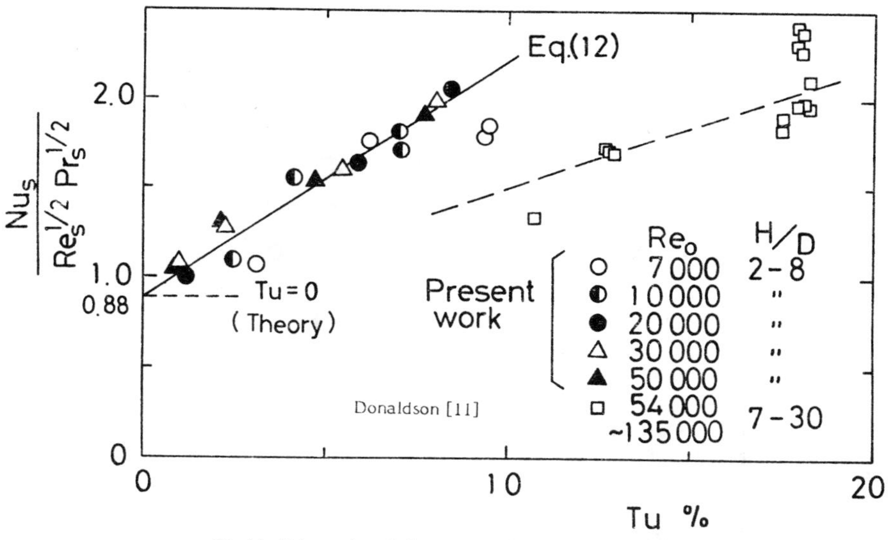

Fig.11 Effect of turbulence on enhancement of stagnation-point heat transfer.

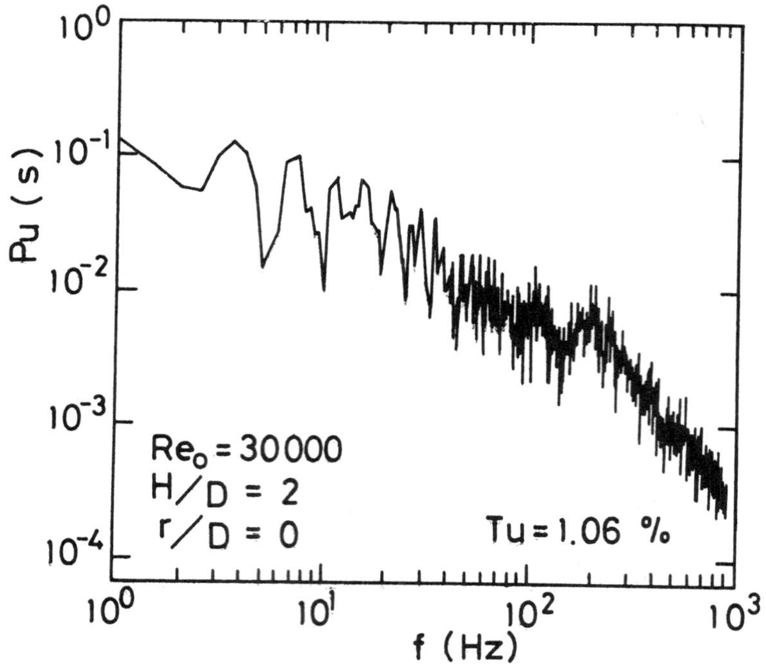

Fig.12 Power spectrum of axial velocity fluctuation for H/D = 2.

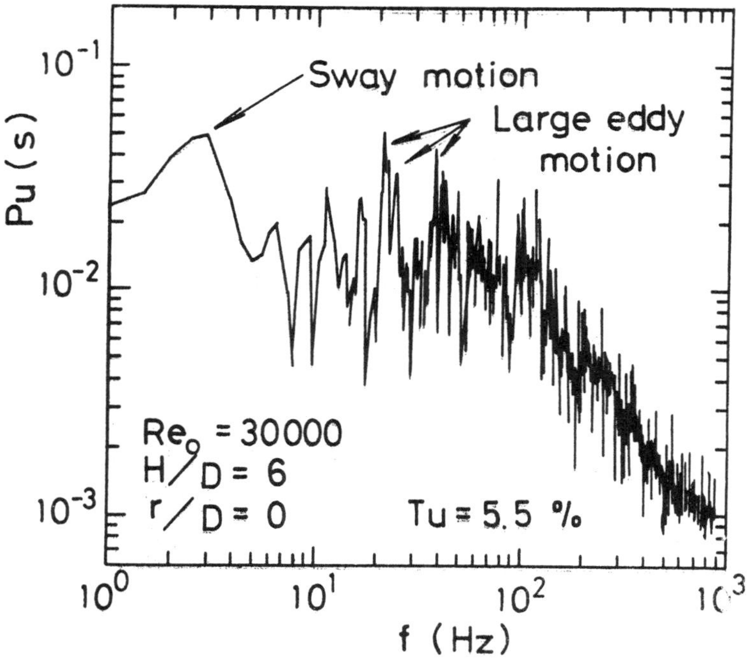

Fig.13(a) Power spectrum of axial velocity fluctuation for H/D = 6.

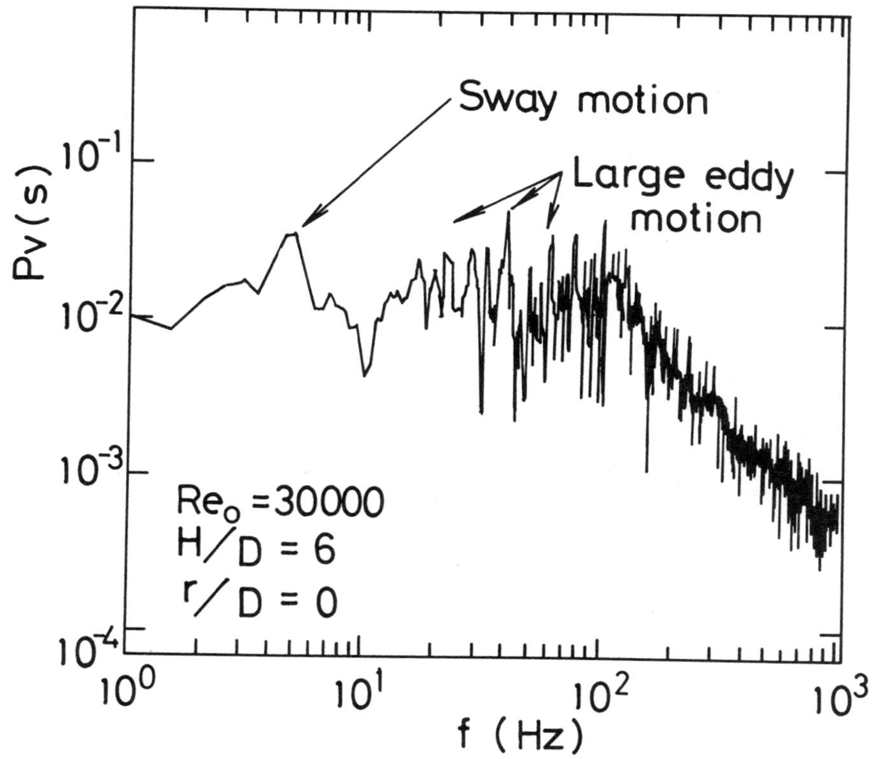

Fig.13(b) Power spectrum of radial velocity fluctuation for H/D=6.

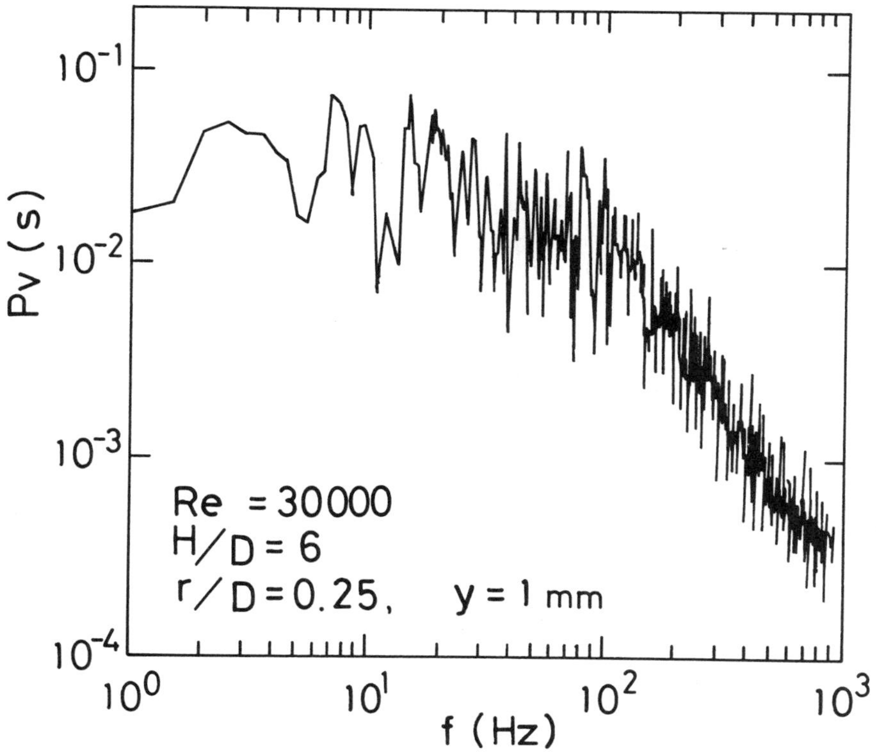

Fig.14 Power spectrum of radial velocity fluctuation at r/D = 0.25

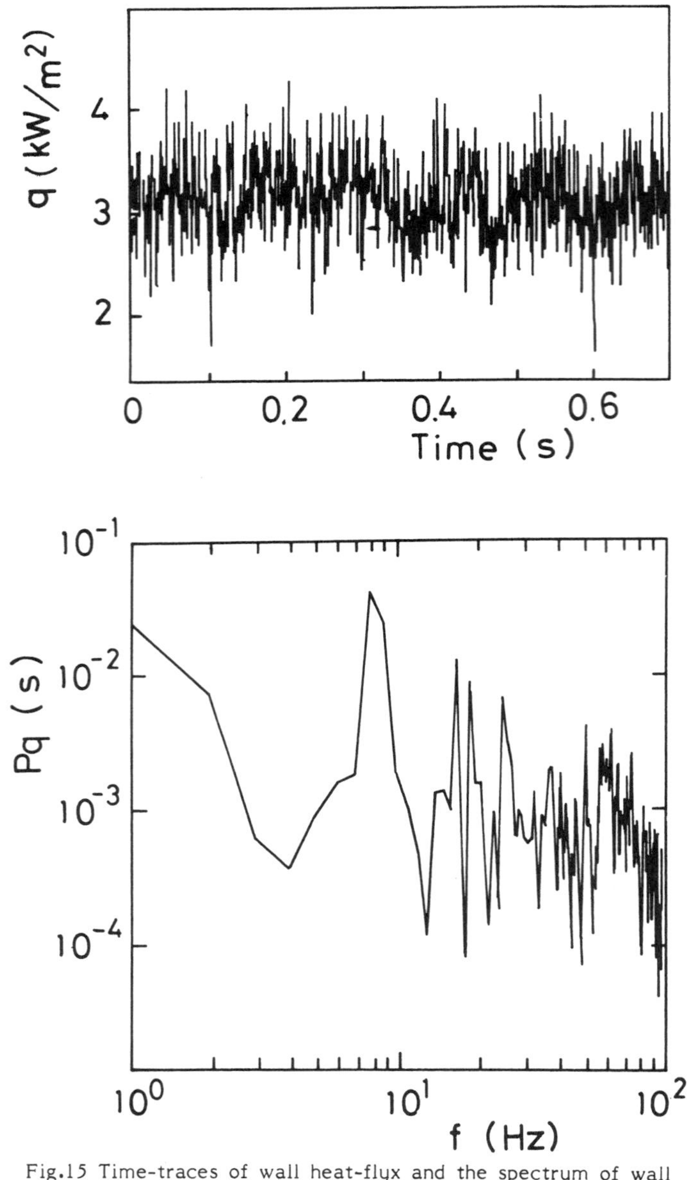

Fig.15 Time-traces of wall heat-flux and the spectrum of wall heat-flux fluctuation, H/D = 6, r/D = 0, Re_o = 30,000

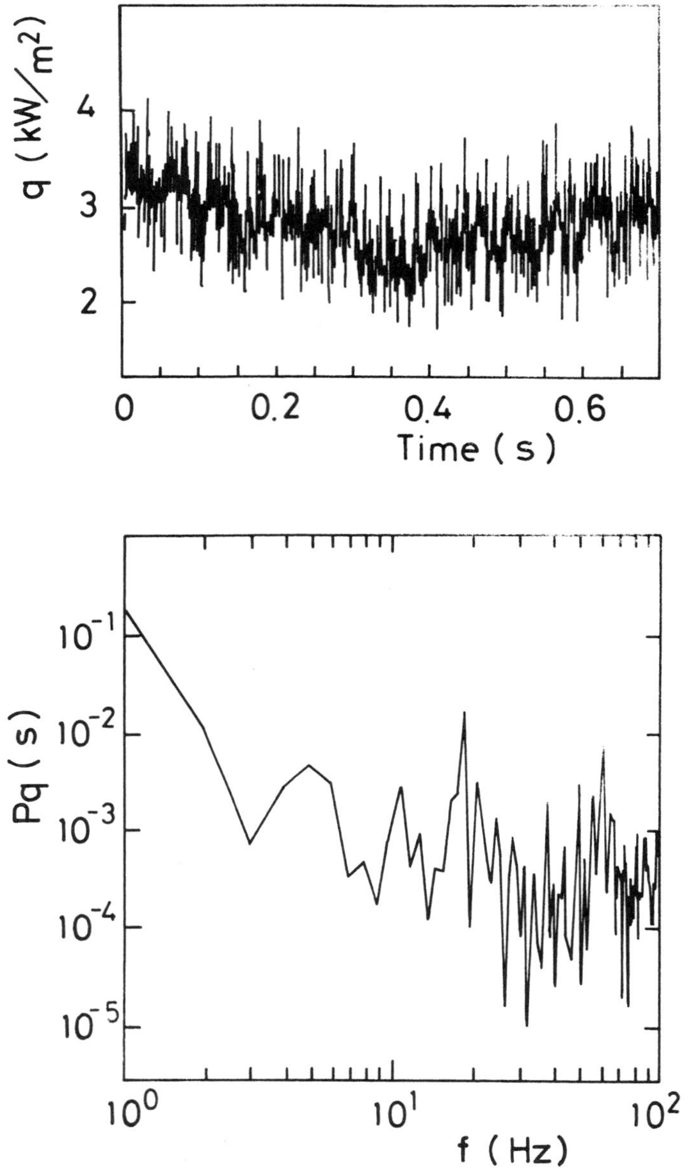

Fig.16. Time-traces of wall heat-flux and the spectrum of wall heat-flux fluctuation, H/D=6, r/D=0.25, Re_o=30,000.

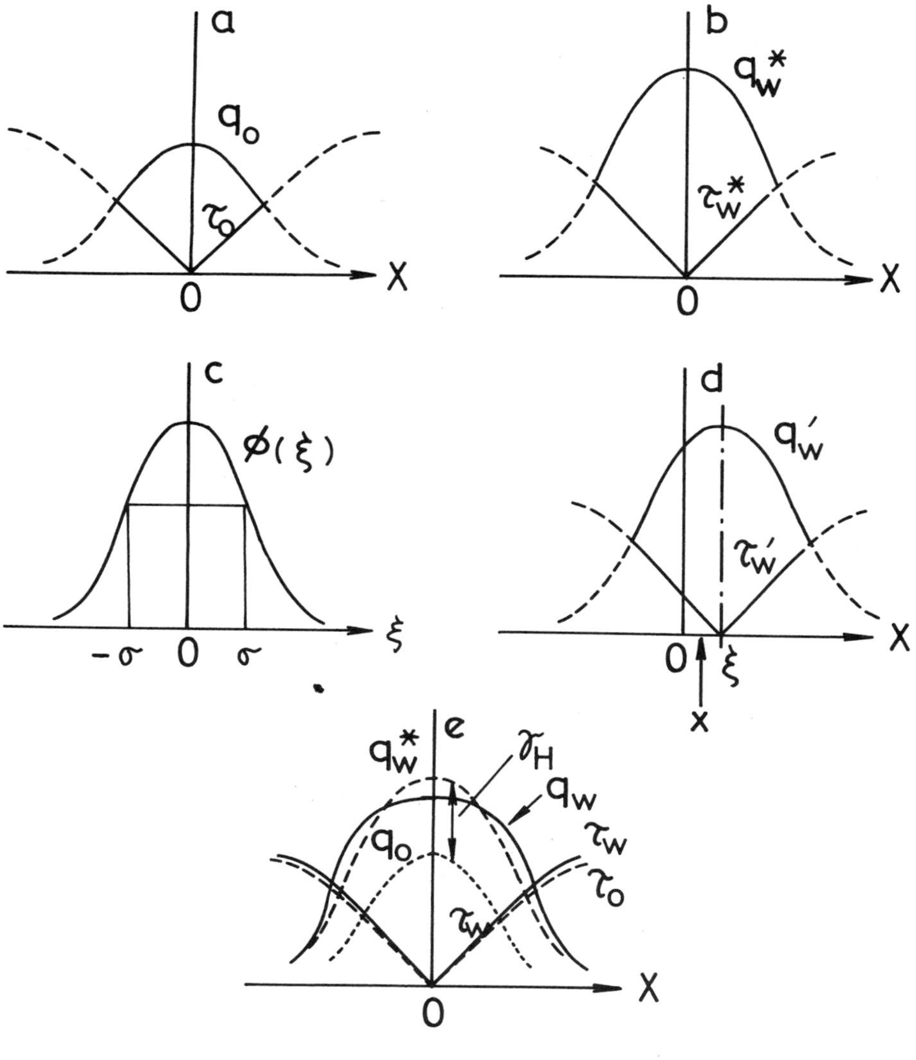

Fig.17. Model for heat transfer enhancement, a: laminar impinging jet, b: turbulent impinging jet without sway motion, c: probability density distribution of existence of instantaneous stagnation point, d: instantaneous distribution of wall shear stress and wall heat-flux, e: time-averaged distribution, x = 0; geometrical stagnation point.

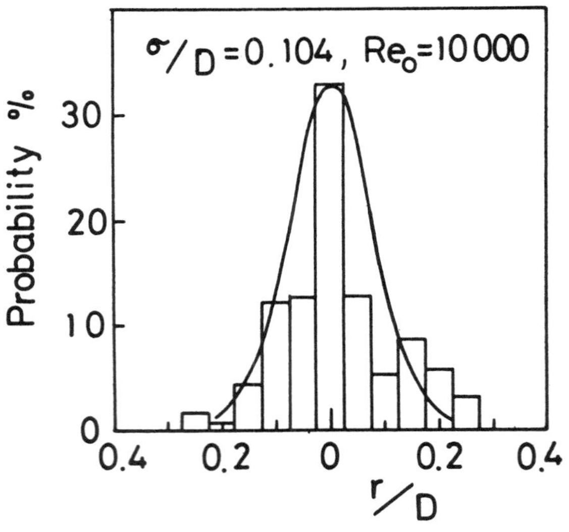

Fig.18 Probability density distribution

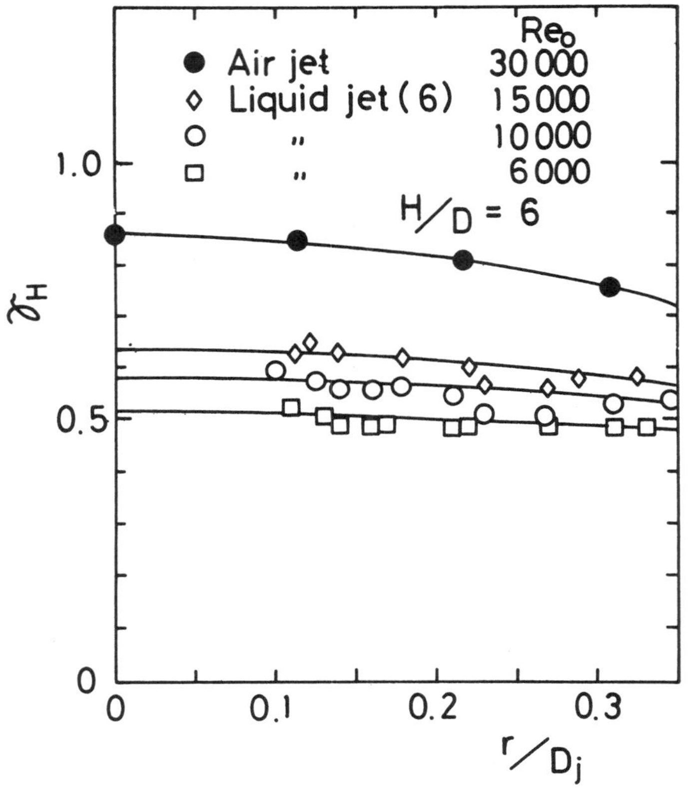

Fig.19 Radial distribution of enhancement effect factors.

Effect of Secondary Flow on Heat Transfer in Ducted Flows

M. V. KRISHNA MURTHY and B. SWAMY*
Department of Mechanical Engineering
Indian Institute of Technology
Madras 600 036, India

A transient two-dimensional model is developed for studying the effect of secondary flow on heat transfer in horizontal square ducts subjected to non-uniform thermal boundary conditions. Computations have been performed upto $Gr^+ = 10^5$ showing the velocity and temperature profiles in the horizontal and vertical planes and time-wise development of streamlines and isotherms. Onset of significant effects of free convection occurs at abour $Gr^+ = 1000$. For $Gr^+ > 10^4$, Nusselt number approaches an asymptotic limit.

Experimental investigation has been carried out on air flowing in a square duct with the top side heated uniformly while the other three sides are insulated. Velocity and temperature of air in the horizontal and vertical planes are measured at different axial positions. The experimental results are compared with the theoretical predictions.

1. INTRODUCTION

Buoyancy influenced heat transfer in channel flows with non-uniform heating is a common occurrence in practice. For example in a solar air heating collector, the flow passages are heated on only one side while the other three sides are insulated. Usually, in these devices, the flow rates are small and/or the temperature differences between the flow field and the boundaries are large. In such situations the fluid flow and heat transfer phenomena can be significantly affected by the imposed secondary fluid motion due to buoyancy effects.

There are some investigations reported in literature on combined convection in rectangular channels with uniform thermal boundary conditions(1, 2,3). However studies on transient analysis of laminar mixed convection in a channel with non-uniform thermal boundary conditions are rather meagre in literature. Zeiberg and Mueller [4] have analysed transient combined free and forced convection in a duct subjected to specified temporal and axial variations of wall temperature on all the four sides. This limited analysis considered the effect of free convection only on the axial velocity without considering the secondary velocities.

Sparrow et al [5] and Tan and Charters [6,7] have reported experimental investigation for forced convective heat transfer in turbulent flow in a rectangular duct with asymmetric heating.

In this paper a numerical study of the initial transients and final steady state heat transfer and fluid flow, and an experimental investigation of the steady heat transfer and primary flow in a fully developed laminar combined convection process occurring in a square duct subjected to non-uniform thermal boundary conditions are presented.

2. ANALYSIS

Consider a hydrodynamically fully developed laminar flow of a Newtonian fluid (with constant properties except for the variation of density in the buoyancy term, negligible viscous dissipation and constant axial pressure gradient) in a horizontal square duct (side D) as shown in Fig.1. Two types of thermal boundary conditions can be considered:

(i) One side of the duct is subjected to uniform heat flux (axial and peripheral) while the other three sides are insulated (case A).

(ii) all the four walls are subjected to constant axial heat flux with constant peripheral wall temperature (case B).

With the scale factors defined in the nomenclature, the govering equations in dimensionless form for two-dimensional transient flow in the orthogonal coordinates X,Y and Z are:

$$\frac{\partial U}{\partial X} + \frac{\partial V}{\partial Y} = 0 \qquad (1)$$

$$\frac{\partial U}{\partial t} + U\frac{\partial U}{\partial X} + V\frac{\partial U}{\partial Y} = -\frac{\partial P}{\partial X} + \nabla^2 U \qquad (2)$$

$$\frac{\partial V}{\partial t} + U\frac{\partial V}{\partial X} + V\frac{\partial V}{\partial Y} = -\frac{\partial P}{\partial Y} + \nabla^2 V + Gr^+\,\theta \qquad (3)$$

$$\frac{\partial W}{\partial t} + U\frac{\partial W}{\partial X} + V\frac{\partial W}{\partial Y} = 1.0 + \nabla^2 W \qquad (4)$$

$$\frac{\partial \theta}{\partial t} + U\frac{\partial \theta}{\partial X} + V\frac{\partial \theta}{\partial Y} = \frac{1}{Pr}\nabla^2\theta - \frac{1}{4\,Pr}\left(\frac{W}{\overline{W}}\right) \qquad (5)$$

Pressure term in equations (2) and (3) can be eliminated by cross-differentiation and introducing a dimensionless vorticity.

$$\xi_1 = -\nabla^2\psi \qquad (6)$$

Where the dimensionless stream function ψ is such that

$$U = \frac{\partial\psi}{\partial Y} \text{ and } V = -\frac{\partial\psi}{\partial X} \qquad (7)$$

* Present address :Department of Mechanical Engineering S.V.U. College of Engineering, Tirupati - 517 502.

The resulting vorticity transport equation is

$$\frac{\partial \xi_1}{\partial t} + U \frac{\partial \xi_1}{\partial X} + V \frac{\partial \xi_1}{\partial Y} = \nabla^2 \xi_1 + Gr^+ \frac{\partial \Theta}{\partial X} \quad (8)$$

The initial conditions (at t = 0) are:

$$U = V = T = \xi_1 = \psi = 0 \quad (9a)$$

and W = two - dimensional fully developed axial velocity distribution for a square duct obtained as a solution for the following equation

$$\nabla^2 W + 1.0 = 0$$

at X = 0 and for $0 \leq Y \leq 2$, W = 0

at X = 1 and for $0 \leq Y \leq 2$, $\frac{\partial W}{\partial X} = 0$ (9b)

and for $0 < X < 1$ and at Y = 0 and 2, W = 0

The boundary conditions ($t > 0$) are:

$\psi = \xi_1 = \frac{\partial W}{\partial n} = \frac{\partial \Theta}{\partial n} = 0$ along the vertical centre line. (10a)

$U = V = W = \psi = \frac{\partial \psi}{\partial n} = 0$ at the duct walls (10b)

For case A,

$$\left.\begin{aligned} \frac{\partial \Theta}{\partial n} &= \frac{1}{2} \text{ for heated wall and} \\ \frac{\partial \Theta}{\partial n} &= 0 \text{ for abiabatic wall} \end{aligned}\right\} \quad (10c)$$

or for case B $\quad \Theta = \frac{1}{4\,\overline{Nu}}$ on the duct walls (10d)

Where $\overline{Nu}$ is the circumferentially averaged Nusselt number defined as

$$\overline{Nu} = \frac{\overline{q}}{(T_w - T_b)} \frac{D_h}{\lambda} \quad (10e)$$

The vorticity transport equation (8), stream function equation (6), axial momentum equation (4) and the energy equation (5) with the initial and boundary conditions (9) and (10) complete the statement of the problem under investigation. The secondary velocities U and V are to be evaluated using equation (7).

The solution in addition to satisfying the governing equations, initial and boundary conditions must be compatible with the definition of bulk temperature, i.e.,

$$\int\int \Theta . W . dX . dY = 0 \quad (11)$$

It is to be noted from the dimensionless governing equations that the magnitude of the modified Grashof number Gr^+ and Pr govern the heat transfer phenomena. Reynolds number does not explicitly influence the process.

The grid for the finite difference approximation is shown in Figure 2. Because of symmetry, one half of the duct is only considered for the solution domain. The finite difference analogue (FDA) of the equations (4), (5) and (8) with forward time and centred space differences for all terms [8] are solved by alternating direction implicit (ADI) method. FDA of equation (6) is solved by point SOR. U and V are evaluated from equation (7) using fourth order formulae.

During the initial stages of numerical computations it was observed that Gr^+ in the source term of the vorticity transport equation (8) has been causing divergence of the whole system of equations due to very large temperature gradients adjacent to heated boundaries especially at high Gr^+. To circumvent this difficulty, vorticity is modified and so defined that Gr^+ appears only in the stream function equation. Thus

$$Gr^+ \xi = \xi_1 \quad (12)$$

This, on substitution, eliminates Gr^+ from the vorticity transport equation. This shifting of computational difficulty from vorticity to stream function equation is found to give reasonable stability for the numerical scheme.

For the numerical solution, equations (5), (8), (7) and (4) are solved in sequence with a relative accuracy of 1 percent to attain steady state condition. Further details can be found in [9].

The maximum non-dimensional time step, Δt, required for stability has been found ;to be ranging from 0.001 for lower Gr^+ to 0.0001 for higher Gr^+. A satisfactory approximation for the steady state is generally attained before t reaches 1.0. A non-uniform grid with a coarse grid of (20x18) and finer grid (20x10) is employed with step sizes of $\Delta X = 0.05$, $\Delta Y_1 = 0.1$ (coarse) and $\Delta Y_2 = 0.02$ (fine). The computing time markedly increased with Gr^+. The CPU time on the IBM 370/155 computer is about 4 seconds for each time step. The computations are limited to $Gr^+ \leq 10^5$ as high Gr^+ would require excessive computer time.

The local Nusselt number Nu is evaluated from

$$Nu = 1 / (\Theta_W)h \text{ for case A} \quad (13)$$

where the suffix h denotes the heated wall. Average Nusselt number, will then be given by

$$\overline{Nu} = \int_0^1 Nu \; dx \quad (14a)$$

Similarly, for case B, considering the average temperature gradient,

$$\overline{Nu} = \frac{2(\partial\theta/\partial n)\overline{W}}{\overline{W(\theta - \theta_W)}} \quad (14b)$$

By considering overall energy balance, the average Nusselt number can also be written as

$$\overline{Nu} = \frac{\overline{W}}{4[\overline{W(\theta_W - \theta)}]}$$

3. EXPERIMENTAL INVESTIGATION

3.1 Test set-up

The schematic layout of the setup is shown in Figure 3a. The set-up consists of an intake chamber, hydrodynamic development section, test section, after-calming length, mixing chamber, flow nozzle and outlet section with suction type flow arrangement.

The test section is made up of three components- a 10 mm - thick fibre reinforced plastic (FRP) channel (Fig.3b), SS strip and a FRP top cover over the FRP channel (Figs.3c and 3d). The assembly of the three components gives the square duct passage of 30 mm side and a length of 2.006 m. The required thermal boundary condition (case A, equation 10(c)) is obtained by electrically heating the SS strip which forms the top side of the duct passage. The FRP channel forms the other three sides. The channel is made up of an electrically insulating and a high heat - resistant grade resin and performs the function of three adiabatic walls. The FRP top cover, made of identically same resin as that of channel, rigidly keeps the SS strip in position over the channel.

At eight selected positions, extra thickness of about 10 mm for a width of 10 mm has been provided peripherally on the outside of the channel to fix hollow SS tubes as guides, for the insertion of transverse probes to measure velocity and temperature of the fluid. Holes, 3 mm diameter to a depth of 10 mm from the outside surface and later 1 mm dia for the rest of the thickness are drilled over the extra thickness portion along the centre lines in the bottom side and on the one vertical side. Hollow thin 3 mm dia., 70 mm length SS tubes are rigidly fixed in these holes with araldite to act as guides for measuring probes.

An AISI 316 SS strip 2 mm thickness, 2 m length, and width cut as shown in Figure 3c is placed in the groove of the channel. The increase in the area due to the small projections is only 0.06% which is negligible. But these projections would facilitate good seating for the SS strip in the groove of the channel. After keeping the SS strip in position, 30 mm central portion of the strip fits in exactly over the channel width making the passage as 30 mm side square. Two copper plates, 3 mm thickness and same width as that of the SS strip are silver-brazed to the SS strip at the two ends and the extra material is filed off. The copper plates are meant for connecting electrical lead wires for heating.

The top cover, made of the same FRP material and with the same length as that of the channel, has a matching recess on the underside and a mating flange like projection on each side as shown in Figure 3d. The out-side face of the end flanges has a recess of 3 mm depth and same width as that of the copper plate so that the copper plate nicely fits into the recess without leaving any projection.

For the measurement of surface temperature on the heated surface, 32 copper-constantan thermocouples (30 gauge) are fixed at 8 selected axial positions (4 Nos. at each axial location) by pressing the beads into already prepared very fine dents on the outside surface of the SS strip and applying a thin layer of copper cement (a blend of technical 'G' copper cement powder and technical 'G' copper cement liquid). After the copper cement becomes dry, a thin coating of araldite is applied over the copper cement and this is found to have good bonding of thermocouples to the surface.

The axial locations for thermocouples measuring surface temperature and for traversing probes are given in Table 1.

Table 1 : Axial locations for measuring surface temperatures and for inserting transverse probes.

No.	1	2	3	4	5	6	7	8
Z	2.8	17.2	38.1	55.4	76.3	93.4	114.8	130.5

At the same axial positions as given in Table 1. 1, 8 numbers 30 gauge copper-constantan thermocouples are fixed on the underside of the top FRP cover one at each axial location. About 13 thermocouples of the same type are fixed at the same axial locations on the horizontal and vertical walls of the FRP channel. For this purpose a fine hole is drilled at the desired location and after inserting the thermocouple bead, the hole is filled with copper cement.

The three components are assembled to form the test section as follows: The SS strip is placed over the channel carefully. The top cover is then placed over the channel after taking out the 40 thermocouple wires (32 on the SS strip surface and 8 on the underside of the FRP cover) through holes, drilled for that purpose. In order to make the SS strip press firmly against the channel groove, 4 mm thick (36 mm width and 10 mm length) hylam distance pieces are kept in the 4 mm gap between the SS strip and the top cover at regular intervals all along the length of the SS strip. The remaining space between the SS strip and top cover is packed with Armaflex (foam rubber insulation) to suppress any convection and radiation heat loss. By providing 6 mm thick thermorex (expanded polystyrene, flexible) as a gasket between the two mating flanges, the channel and the top cover are rigidly fixed by 3 mm bolts and nuts all along its length including the two end flanges. A thin layer of araldite has been applied over all the joints of the test section assembly to ensure leakproofness.

The entire test section is insulated with 70 mm thick glass wool. Three 30-gauge copper-constantan thermocouples measure the mixed mean temperature of air after the mixing chamber. The air flow rate is measured by a long radius flow nozzle placed after mixing chamber. A very smooth and highly steady suction type air flow could be achieved by forcing compressed air at high velocity through a convergent - divergent nozzle which forms the outlet section of the set-up.

3.2 Instrumentation

The flow rate is obtained by measurement of pressure drop across the long radius flow nozzle with the help of a standard single tube inclined manometer having 1/25 inclination. The local velocity of air inside test section is measured by a Pitot-static tube which can be inserted through SS guides with the help of a specially made traversing mechanism. Using the adjacent SS guide as a wall pressure tap, the output of the Pitot-static tube can be measured with a highly sensitive digital micromanometer having an accuracy of 0.01 mm.

The microthermocouple (chromel-alumel, bead size 1mm) can be inserted through the SS guides in the same manner as that of Pitot - static tube for measurement of air temperature inside the test section. A Fluke digital voltmeter having an accuracy of 0.001 mV measures the emf of the traversing microthermocouple of the same type kept at the ice point. All the 32 thermocouples fixed over the tube wall, two thermocouples at the test section inlet, three thermocouples at the outlet of the mixing chamber and a thermocouple to measure the ambient temperature are all connected to the Fluke digital voltmeter through an ice point by using selector switches. The electrical heating current through the test section is measured by a calibrated A.C. ammeter (0-100 A range).

3.3 Experimental Procedure

The heat loss from the test section to the ambient is determined by imposing a suitable electric potential when there is no flow through the test section. After attaining steady state, heating current is switched off and wall temperatures are noted at suitable time intervals. The heat loss coefficient is then calculated from the above data. All final readings are taken after ascertaining that steady state is reached.

As this paper is concerned with fully developed velocity and temperature fields, two conditions that are to be satisfied are constant axial temperature gradient of the SS strip and fully developed velocity and temperature profiles. From the measured data, it was observed that the wall temperature rapidly increases with length near the entry to the test section (upto Z = 40) and becomes practically linear in the downstream part. There is excellent agreement between the vertical and horizontal velocity and temperature profiles at Z = 93.4 to Z = 114.8 for different values Gr^+ and it is taken that the flow and temperature fields are fully developed beyond Z = 93.4. The test section is divided into axial segments for data analysis purposes and the position of wall thermocouples dictated the size of each segment. For each axial segment, electrical power generation, heat loss, heat transferred to the fluid, local bulk temperature, Gr^+ and finally the Nu and $\overline{Nu}$ are calculated.

4. RESULTS AND DISCUSSION

Computations have been performed for Pr = 0.72 and Gr^+ varying from 10^3 to 10^5. In a flow field which has an appreciable temperature difference between the heated boundary and the fluid, considerable variations in density result because of the temperature difference. Therefore natural convection currents occur inside the flow field. Since the heated boundary is the hottest, these convection currents go up along the heated walls and continuity requires that the fluid in the core must come down. Thus secondary currents are generated in a direction perpendicular to the main flow in horizontal confined flows. The intensity of these currents depends naturally upon the intensity of heating and on other factors. Figures 4 and 5 showing the streamlines and isotherms ($\Theta_t = \Theta - \Theta_c$) for cases A and B for different Gr^+ confirm this behaviour. The effects of free convection which appear in the form of secondary flow are now superimposed on forced convection.

Comparing Figures 4 and 5, it may be seen that for the same values of the parameters (Gr^+ and Pr) peripherally uniform wall temperature case (case B) requires a larger time to achieve steady state condition than case A. The intensity of the secondary flow is also more in this case. For case A, thermal stratification is observed whereas for case B the fluid temperature is the highest near the heated boundaries. In Figure 5 (for case B), it can be seen that the isotherms are clustering near the bottom and lower portions of the side walls whereas they are spread out in the top layers of the fluid. The centre of circulation is more or less at the horizontal central plane while for case A it is below the centre line.

It should be noted that the thermal configuration in case A is relatively more stable than the configuration in case B. The isotherms and streamlines pattern for case A very clearly indicates that the heat transfer process in this case is more conduction-dominated and less buoyancy-dependent than in the case of case B.

Figures 6 and 7 show the steady state streamlines and isotherms for different Gr^+. It may be noted that the steady state condition is the asymptotic limit of the transient solution.

It is seen from these figures that as Gr^+ increases, the time required to attain steady state increases. This has significant implications in solar energy collection systems. This would be clear if we rephrase the modified Grashof number Gr^+ in terms of other dimensionless parameters. It can be shown that

$$Gr^+ = Gr_e \, Re \, Pr \, (d\theta_e/dz) \tag{15}$$

$$\text{Where } Gr_e = g \gamma D_h^3 T_{be}/(8\gamma^2) \tag{16}$$

Gr_e is Grashof number based on T_{be} (suffix e represents entrance or initial condition) and

$$\theta_e = (T_b - T_{be}) / T_{be} \tag{17}$$

A large Gr^+ will either mean a large value of Gr_e, or Re or $(d\theta_e / dz)$ for a given fluid. A higher Gr_e for a given fluid and flow passage implies a higher implies a higher heat input per unit length and this is directly related to the incident solar radiation. Thus when the fluid is entering the collection device at a higher temperature or at a high flow rate it will require large time to attain steady state. This is also true when the incident solar radiation is high. This fact should be taken note of while designing the control strategies for the collector fields.

As can be seen from the figures 6 and 7, in general, the onset of significant free convection effects occurs at about $Gr^+ = 1000$. At this value of Gr^+, the stream lines are elliptical in shape and the isotherms follow the contour of the heated boundary. Further the centre of circulation is on the central plane.

As mentioned earlier, the centre of circulation for the secondary flow shifts below the horizontal central line with increasing Gr^+ for case A. However, for case B, it is observed that the eye of circulation moves towards the heated boundary. As can be seen from Figures 6 and 7 the magnitude of the contour levels for streamlines increase considerably with Gr^+ for case B and is moderate for case A.

Figures 8 and 9 show the computed axial velocity profiles for $Gr^+ = 10^3$, 10^4 and 10^5. The non-dimensional axial velocity for the isothermal flow through a square duct is also shown in these figures. It is seen that the horizontal velocity profile is different for different Gr^+ and deviates from the isothermal case only for case B. The deviation increases with Gr^+. However, the vertical velocity profiles tend to become asymmetric with increasing Gr^+ for both the cases A and B. It is also seen that this asymmetry is caused as a result of increased axial velocities in the lower half of the duct while in the upper half the axial velocity remains more or less the same as that of isothermal flow. This was also evident from the streamline plot of Figures 6 and 7 where the centre of circulation was found to be lying below the horizontal central plane. It is also seen that for $Gr^+ = 1000$, the velocity profiles are not much different from the isothermal flow case confirming the earlier observation that the onset of significant free convection effects occurs at about $Gr^+ = 1000$.

For the purpose of comparison of the computed temperature profiles (case A) with those obtained experimentally the fluid temperature is redefined as

$$\theta_1 = \frac{\theta_{wh} - \theta}{\theta_{wh} - \theta_c} \tag{18}$$

where θ, θ_{wh} and θ_c are the dimensionless temperatures of the fluid, of the fluid at the heated wall, and of the fluid at the centre of the duct. Figs. 10 and 11 show such normalized temperature profiles in both the horizontal and the vertical central planes for case A and B respectively. From these profiles, it can be seen that for case A, the variation of temperature along the horizontal central plane is quite small compared with that of case B. This means that for case A, the temperature of the fluid is of the same order of magnitude all along the horizontal central plane. It can also be observed that the buoyancy has a more significant effect on heat transfer, (as G_r^+ increases) in case B than in case A.

Fig. 12a shows the circumferentially averaged Nusselt numbers ($\overline{Nu}$) for different Gr^+ for case A. It is evident from this figure that there is a substantial increase in $\overline{Nu}$ upto $Gr^+ = 10^4$ and beyond this value $\overline{Nu}$ tends to attain an asymptotic value. This fact is also evident from the pattern of isotherms which showed a strong thermal stratification of the fluid below the heated boundary, thus indicating a more diffusion-dominated rather than a buoyancy-dependent heat transfer process.

Variation of $\overline{Nu}$ with G_r^+ for the case B is shown in Fig. 12 b.

To facilitate easy evaluation of Nusselt number, the heat transfer results of each case have been correlated with an equation of the type

$$\overline{Nu} = \left[\overline{Nu}_o^{\,2} + \left\{ a \, (Gr^+)^b \right\}^2 \right]^{1/2} \tag{19}$$

The values of the coefficients a and b are tabulated below. The value of $\overline{Nu}_o$ is taken as 2.7 for case A and 48/11 for case B, corresponding to pure forced convection situation.

Table - 2

Case	Coefficients	
	a	b
A	0.5651	0.1333
B	0.0196	0.4997

Fig. 13 shows the experimentally measured temperature profile for different Gr^+ but nearly the same Reynolds number where as Fig. 14 presents the temperature profiles for nearly the same Gr^+ but different Re. As could be expected, Re does not have any effect on the temperature profiles. The horizontal temperature profile is little affected by Gr^+ whereas the vertical temperature profile is influenced to some extent by the value of Gr^+.

Fig. 15 shows the comparison between the theory and experiment; while there is good agreement between the two in the central core of the fluid, there is considerable deviation in the regions near the boundary. Such deviations are not totally unexpected in view of the restrictive assumptions made for the analysis and the difficulties one has to encounter in experimental measurements of this kind.

The variation of wall temperature along the length of the test section is shown in Fig. 16. It is seen that the wall temperature increases rapidly near the entry to the test section and becomes practically linear in the downstream part. The bulk temperature of the fluid is also shown plotted in the same figure. It is seen that the variation of both these temperatures is linear upto a certain point. However, the rate of increase is not the same for both the cases. This results in a variation of heat flux along the axial length.

Fig. 17 shows a comparison of the experimentally obtained Nusselt numbers with the analytically predicted values. It is seen that the experimental values are somewhat higher because of the reasons already stated.

5. CONCLUSIONS

A numerical model for the transient laminar combined convection process in horizontal ducts subjected to non-uniform thermal boundary conditions around the periphery and an experimental investigation of the same under steady conditions have been presented.

From the numerical and experimental results the following conclusions are drawn.

(i) The velocity and temperature profiles in the vertical direction are affected by buoyancy to a greater extent than the horizontal velocity and temperature profiles.

(ii) Buoyancy effects increase as Gr^+ increases.

(iii) For values of $Gr^+ > 10^4$, the Nusselt number tends to attain an asymptotic value.

(iv) The onset of significant free convection effects occurs at about $Gr^+ = 1000$.

(v) There is good agreement between experimentally measured velocity temperature profiles and the numerically predicted values.

(vi) The numerically computed heat transfer results can be correlated by equations of the type

$$\overline{Nu} = \left[Nu_o^2 + \left\{ a\,(Gr^+)^b \right\}^2 \right]^{1/2}$$

NOMENCLATURE

a.	a constant	
b.	a constant	
D	side of square duct	m
Dh	hydraulic diameter	m
G_r^+	modified Grashof number, $g\gamma Q D_h^3 / 8\lambda\nu^2$	
g	acceleration due to gravity	m/s^2
I_h	electrical heating current	amps
Nu	local Nusselt number	$\alpha D_h / \lambda$
$\overline{Nu}$	peripherally averaged Nusselt number	$\bar{q} D_h / [(T_w - T_b)\lambda]$
P	dimensionless pressure	$P'(D_h^2 / 4\rho\nu^2)$
P'	Pressure	N/m^2
Q	heat transfer rate per unit length	$D\bar{q}$, W/m
$\bar{q}$	peripherally averaged heat flux	W/m^2
Re	Reynolds number	$\bar{W} D_h / \nu$
T	temperature	C
t	dimensionless time	$4\tau\nu / D_h^2$
U	dimensionless velocity	$U D_h / 2\nu$
u	secondary velocity in the X-direction	m/s
V	dimensionless velocity	$v D_h / 2\nu$
v	Secondary velocity in the Y- direction	m/s
W	dimensionless axial velocity	$w/(-dp'/dz\ D_h^2/4\eta)$
w	axial velocity (Z - direction)	m/s
X	dimensionless coordinate	$2x/D_h$
x	cartesian coordinate	m
Y	dimensionless coordinate	$2y/D_h$
y	cartesian coordinate	m
Z	dimensionless coordinate	$2z/D_h$
z	cartesian coordinate	m
α	Local heat transfer coefficient,	W/Km^2
$\bar{\alpha}$	averaged heat transfer coefficient	W/Km^2
γ	thermal coefficient of expansion	K^{-1}
η	dynamic viscosity	kg/ms
Θ	dimensionless temperature, $\frac{T - T_b}{(Q/\lambda)}$	
θ_1	normalised temperature,	$(\theta_{wh} - \theta) / (\theta_{wh} - \theta_c)$
θ_t	non-dimensional temperature excess	$(\theta - \theta_c)$
λ	thermal conductivity	W/mK
ν	Kinematic viscosity	m^2/s
ξ	modified dimensionless vorticity	ξ_1 / Gr^+
ξ_1	dimensionless vorticity	$\frac{\partial V}{\partial X} - \frac{\partial U}{\partial Y}$
ρ	mass density	kg/m^3
τ	time	s
Ψ	dimensionless stream function	

∇^2 Laplacian operator $\frac{\partial^2}{\partial X^2} + \frac{\partial^2}{\partial Y^2}$

Subscripts

b Bulk condition
c centre
h heated wall
o pure forced convection condition
W wall

REFERENCES

1. Chen, K.C. and Hwang, G.J.," Numerical solution for combined free and forced convection in horizontal rectangular channels", Trans. ASME, J. of Heat Transfer, 91 (1969) 59-66.

2. Nakamura, H., Matsunra, A., Kiwaki, J., Hiraoka, S. and Yamada, J., "Numerical solutions for combined free and forced laminar convection in horizontal rectangular ducts by conjugate gradient method", J. of Chemical Engg., Japan, 11 (1978) 354-360.

3. Ou, J.W., Cheng, K.C. and Lin, R.C., "Combined free and forced laminar convection in rectangular channels,", Int. J. Heat Mass Transfer, 19 (1976) 277-283.

4. Zeiberg, S.L., and Mueller, W.K., "Transient laminar combined free and forced convection in a duct," Trans. ASME, J. of Heat Transfer, 84 (1962) 141-148.

5. Sparrow, E.M., Lloyd, J.R. and Hixon, C.W., "Experiments on turbulent heat transfer in an asymmetrically heated rectangular duct", Trans ASME, J. of Heat Transfer, 88 (1966) 170-174.

6. Tan, H.M. and Charters, W.W.S., "Effect of thermal entrace region on turbulent forced convective heat transfer for an asymmetrically heated rectangular duct with uniform heat flux", Solar Energy, 12 (1969) 513-516.

7. Tan, H.M. and Charters, W.W.S., "An experimental investigation of forced convective heat transfer for fully developed turbulent flow in a rectangular duct with asymmetric heating," Solar Energy, 13, (1970) 121-125.

8. Roache, P.J., Computational Fluid Dynamics, Hermosa publishers, Albuquerque, New Mexico, (1972) 91-95.

9. Swamy, B., "Buoyancy effects on heat transfer in confined flows with non-uniform thermal boundary conditions," Ph.D. Thesis, Indian Institute of Technology, Madras. (1982).

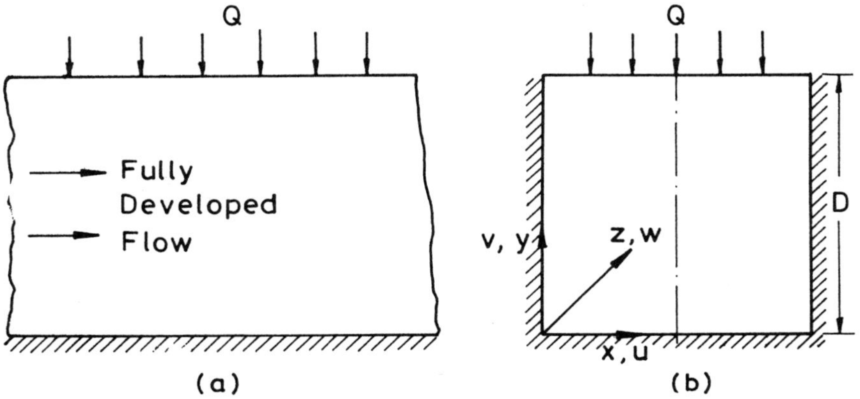

Fig.1 Physical Model and Coordinate System

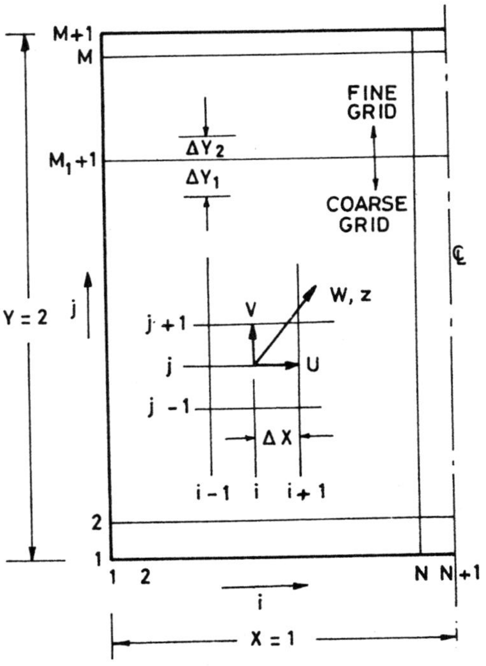

Fig.2 Numerical Grid

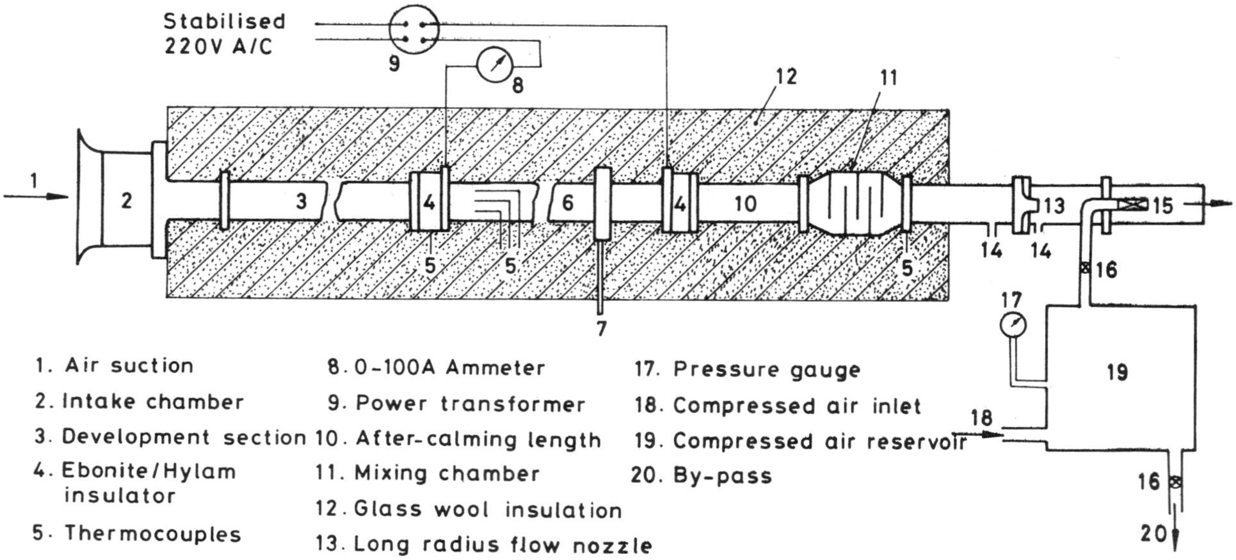

1. Air suction
2. Intake chamber
3. Development section
4. Ebonite/Hylam insulator
5. Thermocouples
6. Test section
7. Stainless steel guide
8. 0-100A Ammeter
9. Power transformer
10. After-calming length
11. Mixing chamber
12. Glass wool insulation
13. Long radius flow nozzle
14. Pressure taps
15. C-D Nozzle
16. Needle valve
17. Pressure gauge
18. Compressed air inlet
19. Compressed air reservoir
20. By-pass

Fig.3(a) Experimental Set-up (Schematic)

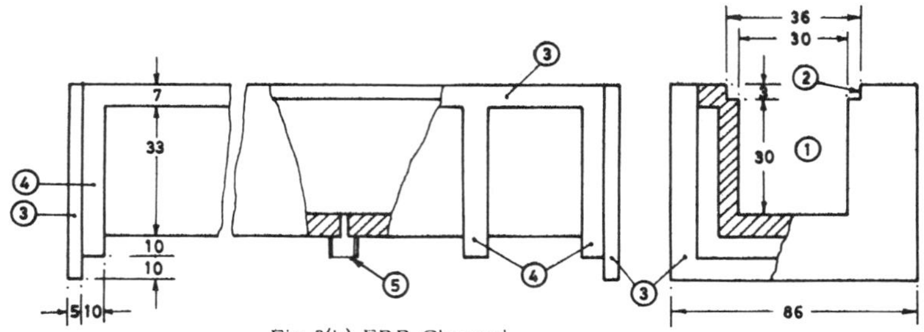

Fig.3(b) FRP Channel

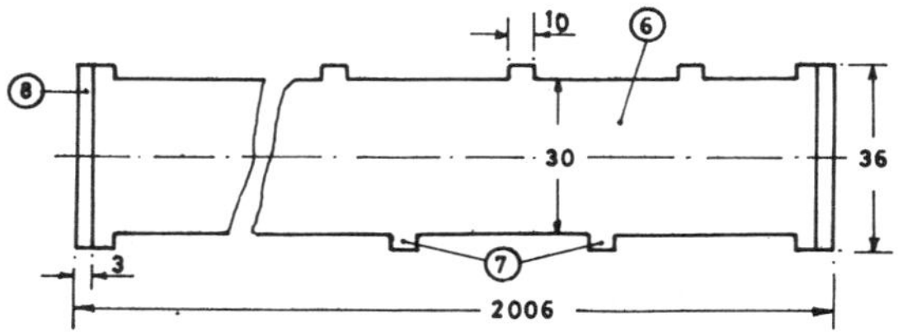

Fig.3(c) Stainless Steel Strip

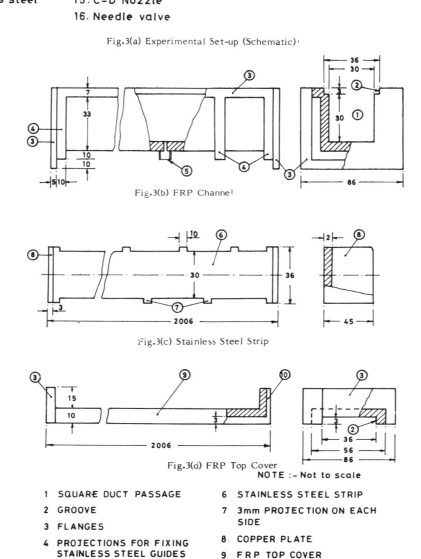

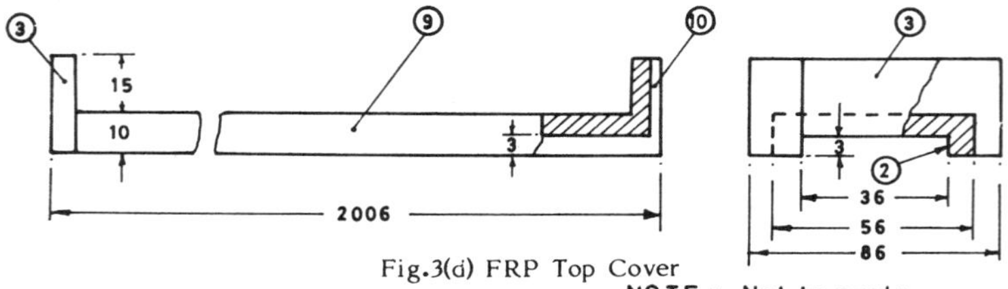

Fig.3(d) FRP Top Cover

NOTE :- Not to scale

1 SQUARE DUCT PASSAGE
2 GROOVE
3 FLANGES
4 PROJECTIONS FOR FIXING STAINLESS STEEL GUIDES
5 HOLES FOR FIXING S.S. GUIDES
6 STAINLESS STEEL STRIP
7 3mm PROJECTION ON EACH SIDE
8 COPPER PLATE
9 FRP TOP COVER
10 RECESS

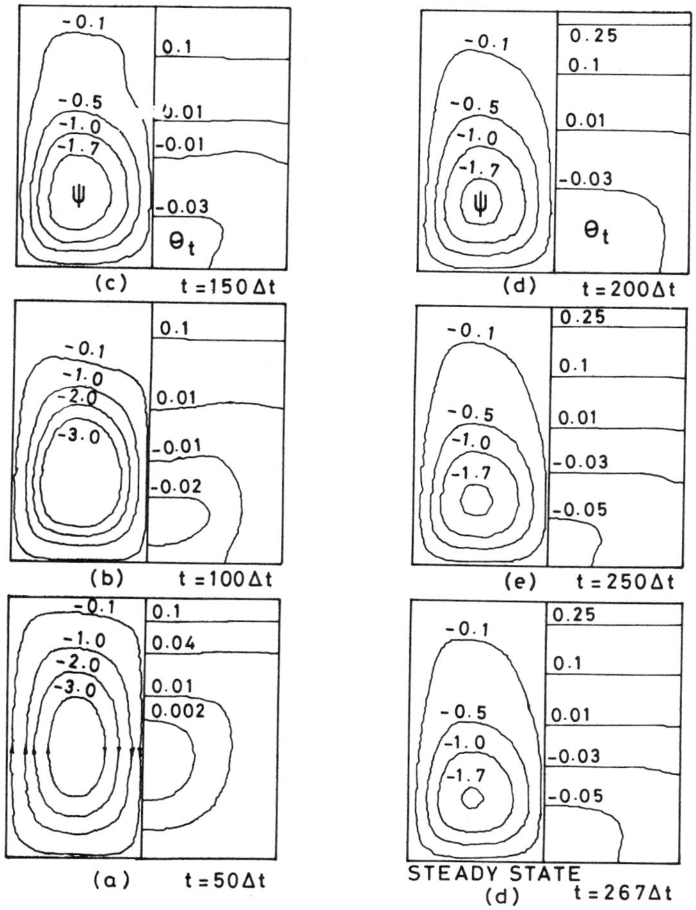

Fig.4 Time-wise Development of Streamlines ψ and Isotherms θ_t. Case ⁚ $Gr^+ = 10^5$, $Pr = 0.72$, $\Delta t = 0.001$.

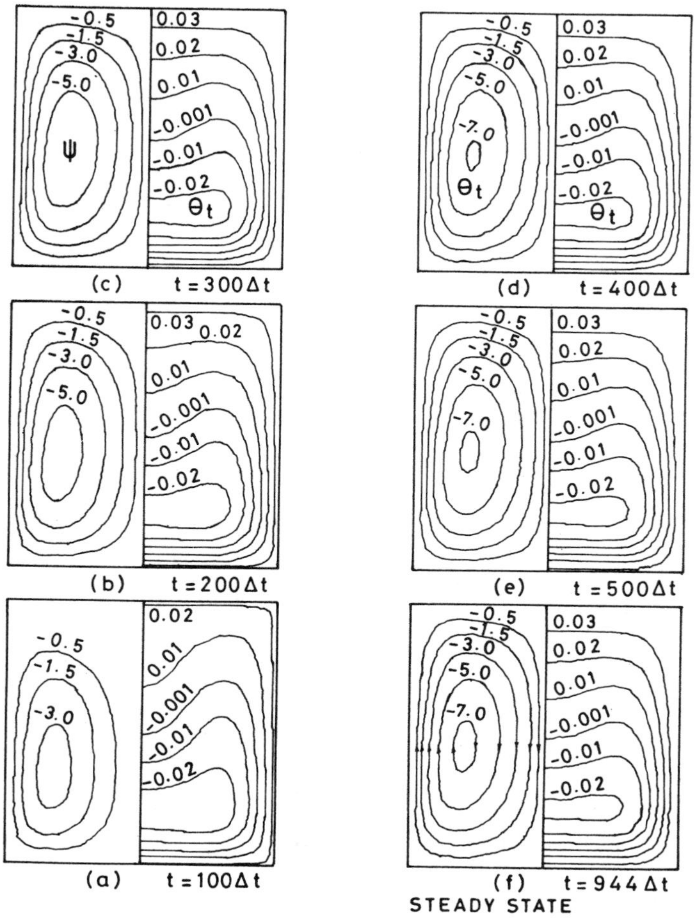

Fig.5 Time-wise Development of Streamlines ψ and Isotherms θ_t. Case A : $Gr^+ = 10^5$, $Pr = 0.72$, $\Delta t = 0.001$

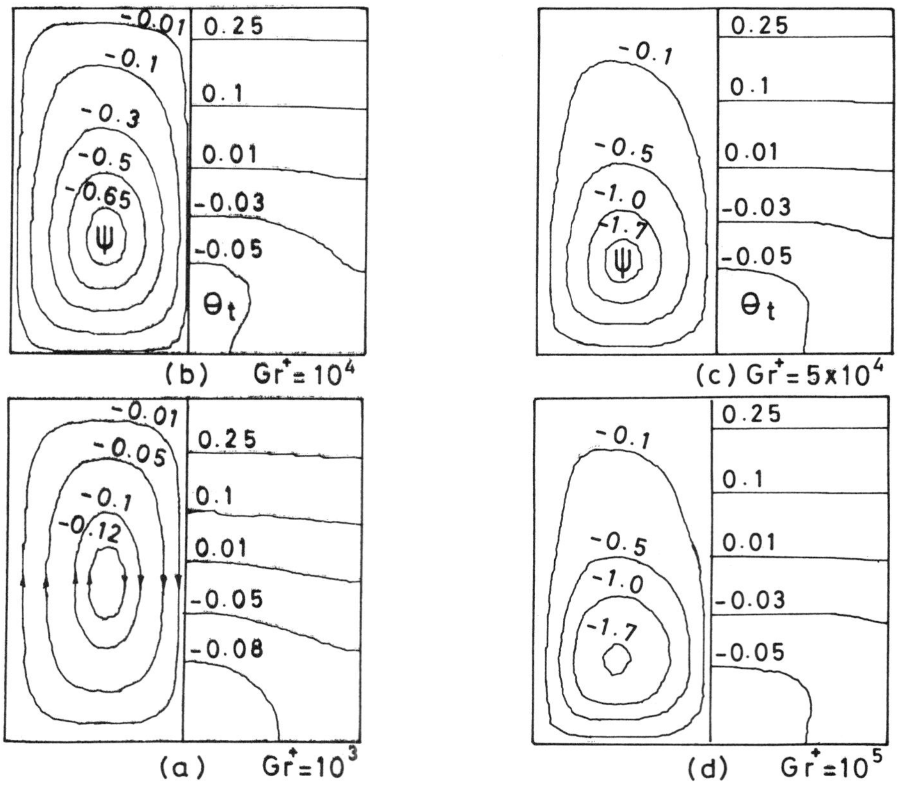

Fig.6 Steady State Streamlines ψ and Isotherms θ_t Case A: : Pr = 0.72

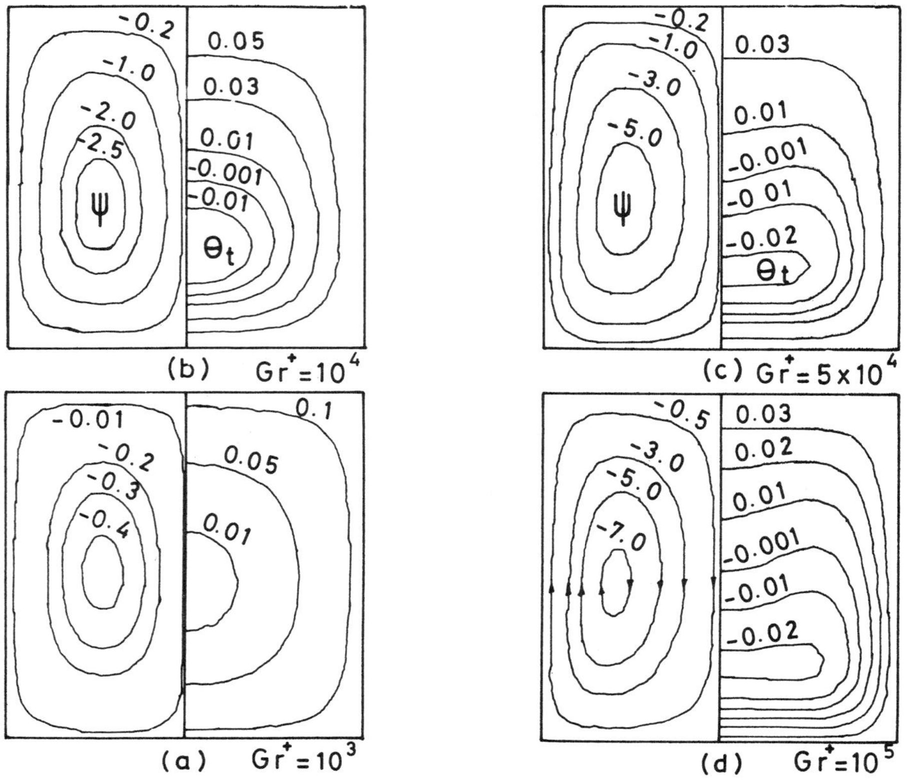

Fig.7 Steady State Streamlines ψ and Isotherms θ_t Case B: : Pr = 0.72

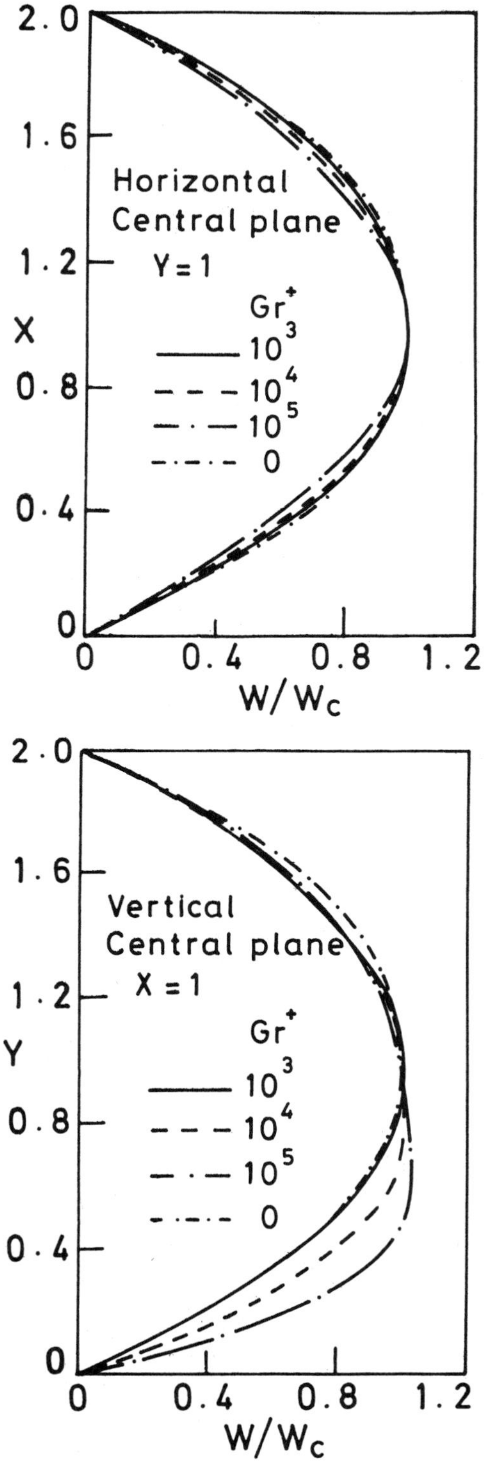

Fig. 8 Normalised Axial Velocity Profiles Along the Horizontal and Vertical Central Planes for Case A

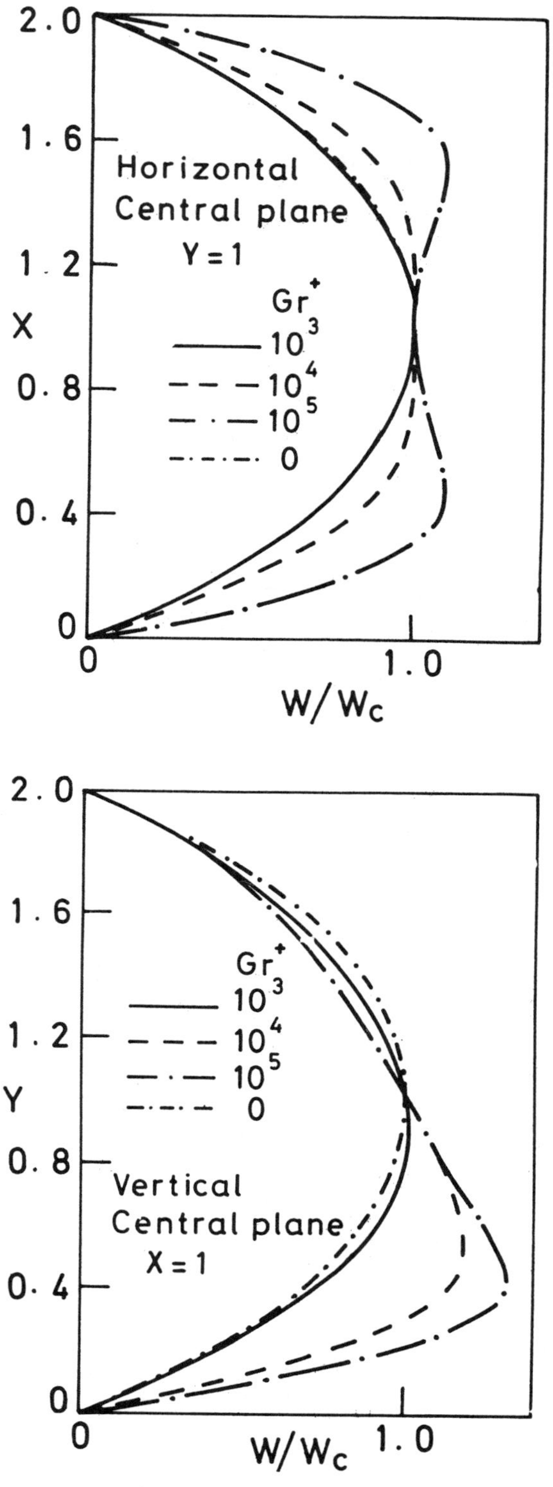

Fig.9 Normalised Axial Velocity Profiles Along the Horizontal and Vertical Central Plane for Case B

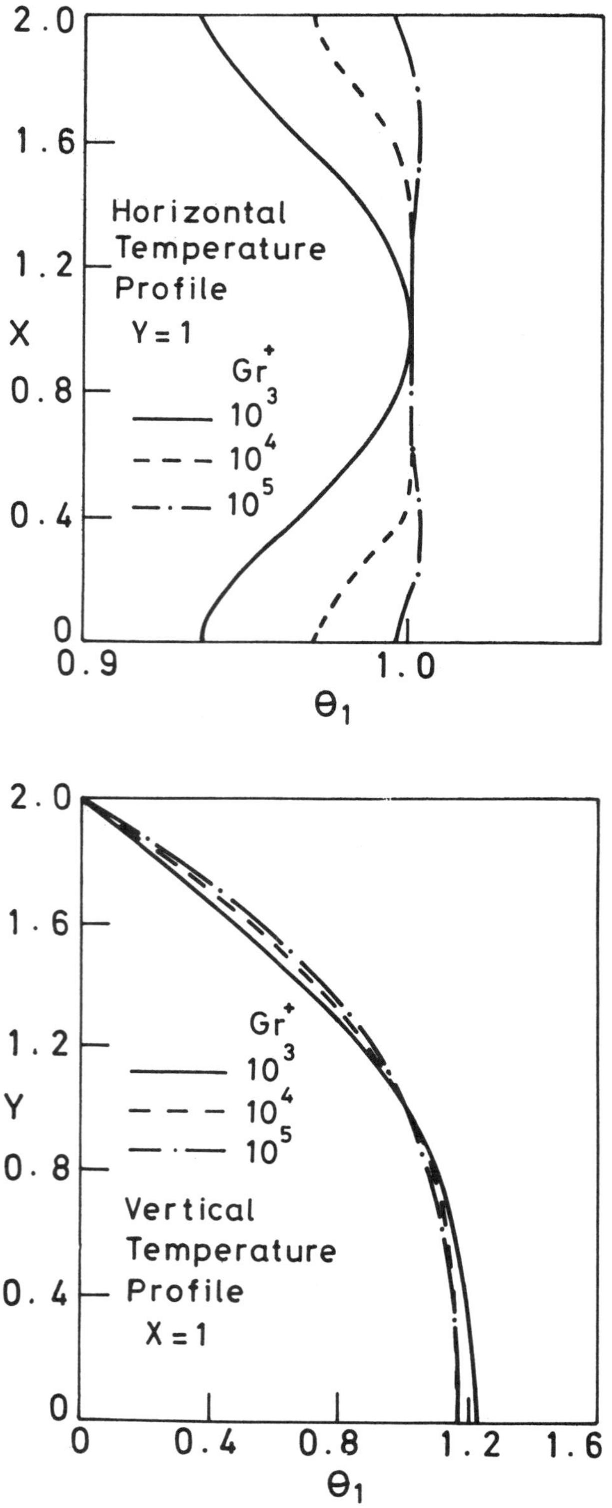

Fig. 10 Normalised Fluid Temperature Distribution $\Theta_1 = (\Theta_{wh} - \Theta)$ $(\Theta_{wh} - \Theta_c)$ Along the Horizontal and Vertical Planes for Case A

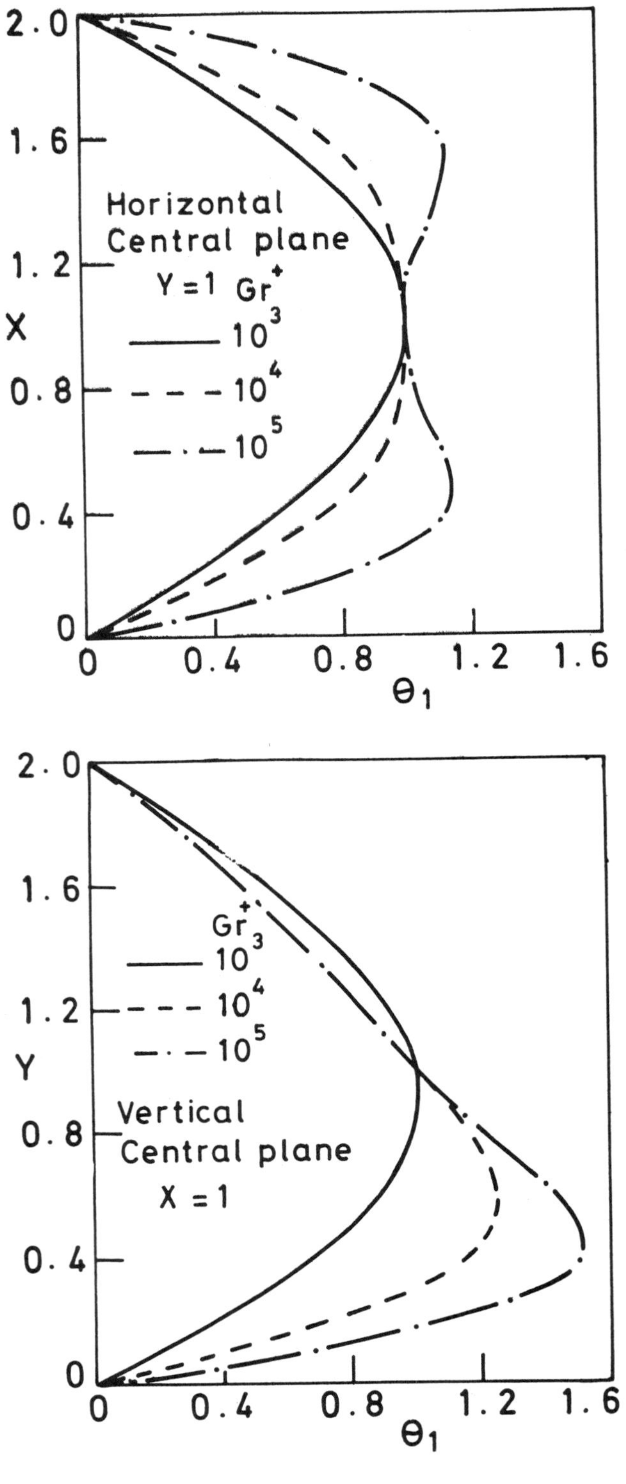

Fig.11 Normalised Fluid Temperature Distribution Θ_1 $(\Theta_w - \Theta)/$ $(\Theta_w - \Theta_c)$ Along the Horizontal and Vertical Central Planes for Case B

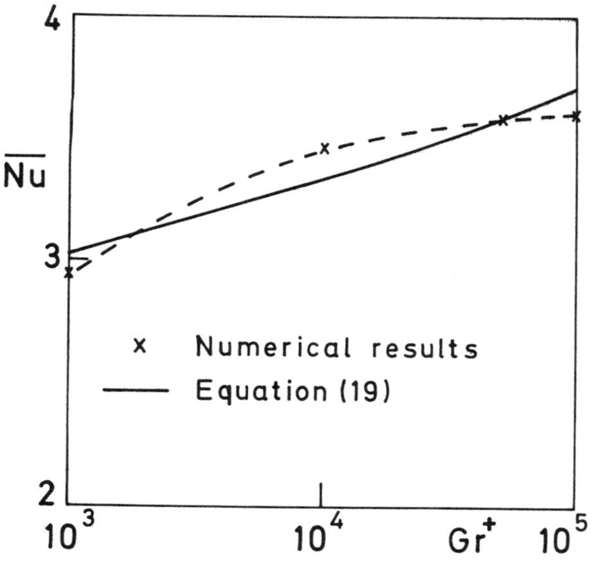

Fig

Fig.12a Circumferentially Averaged Nusselt Number for Case A

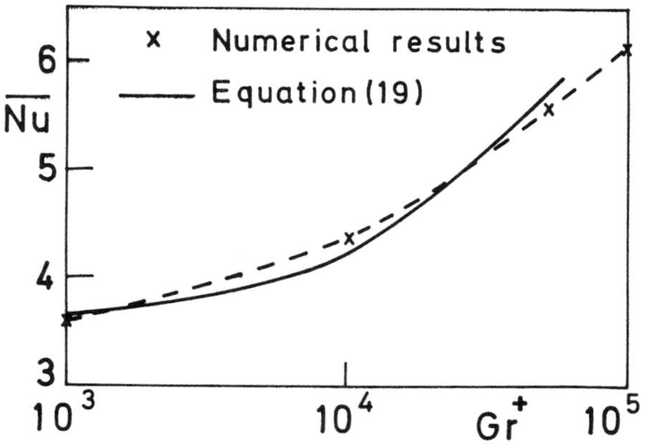

Fig

Fig.12 b Circumferentially Averaged Nusselt Number for Case B

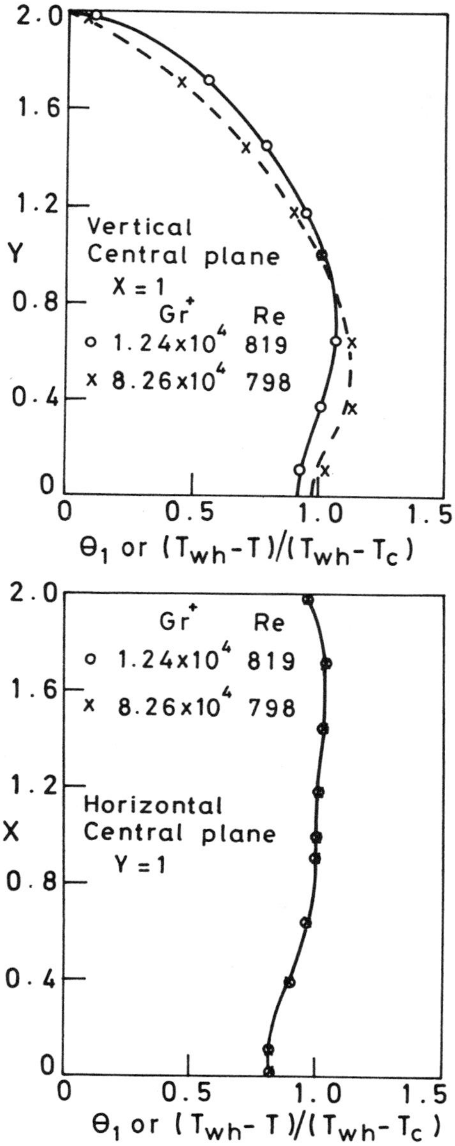

F
f

Fig. 13 Temperature Profiles for Nearly the Same Re but Different Gr^+

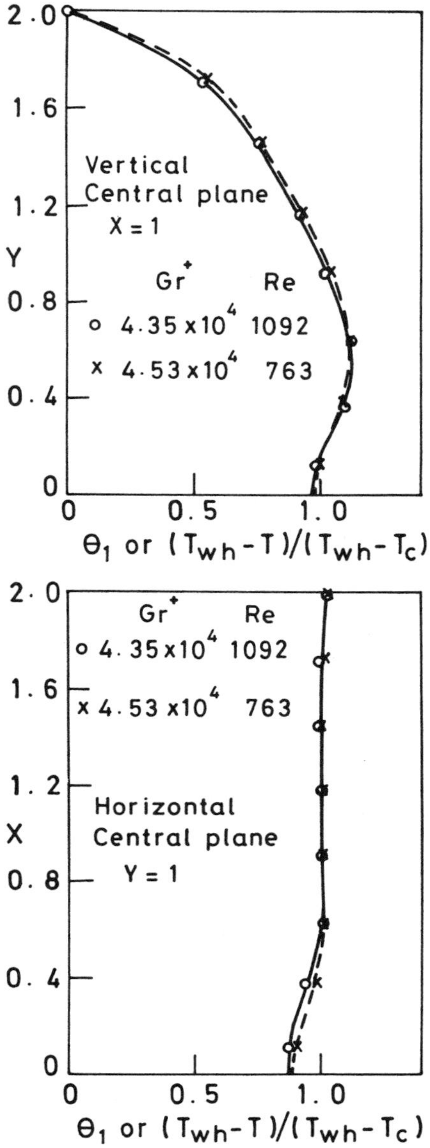

F

Fig. 14 Measured Temperature Profiles for Nearly Same Gr^+ but Different Re

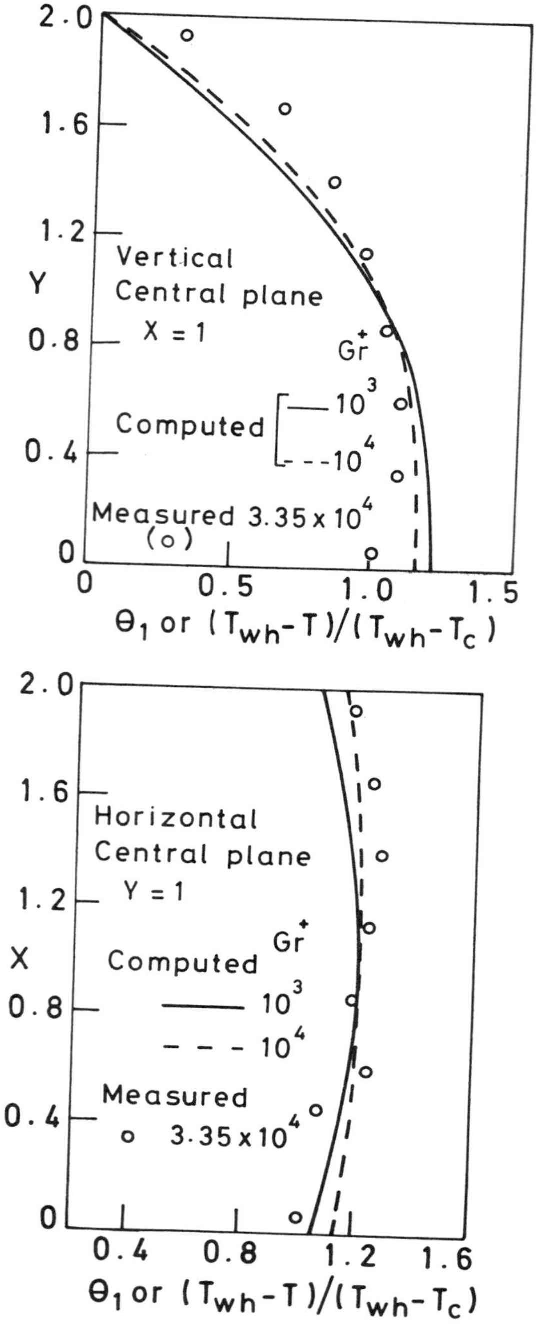

Fig. 15 Computed and Measured u- Temperature Profiles

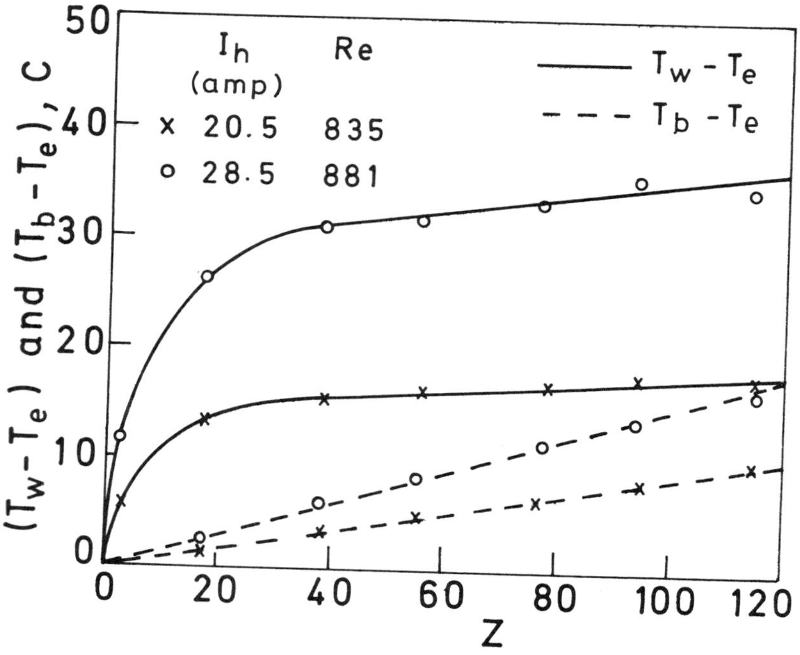

Fig.16 Bulk Mean and Wall Temperature Variation in Axial Dirction

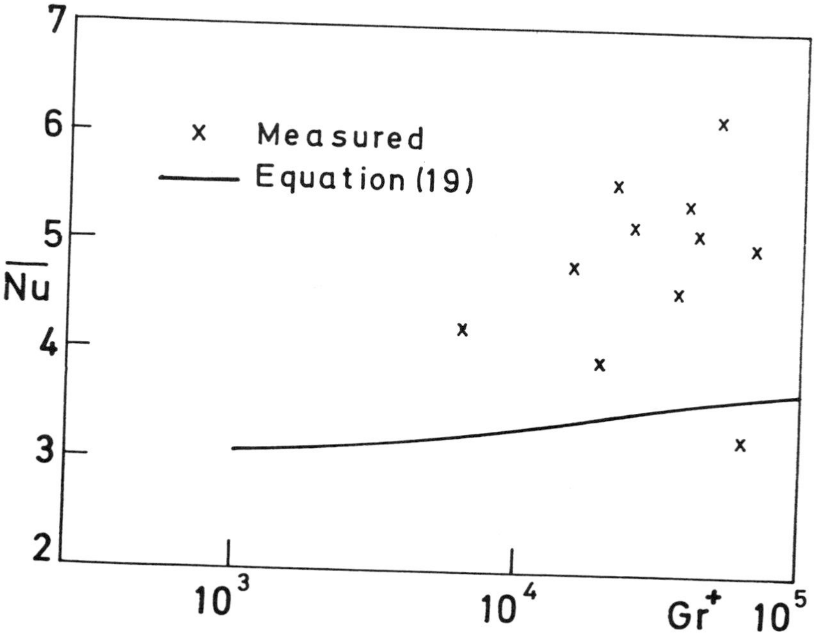

Fig.17 Comparison of Circumferentially Averaged Nusselt Numbers

On Some Problems of Up-to-Date Thermo-Hydrodynamics

S. S. KUTATELADZE and YE. M. KHABAKHPASHEVA
Institute of Thermophysics of the Siberian Branch
of the USSR Academy of Sciences
1 Lavrentyev Avenue
Novosibirsk-90 630090, USSR

Some fundamental problems of the convective heat transfer theory are considered which have not so far been solved. The absence of a comparitively full physico-mathematical model for developed nucleate boiling is pointed out and an analysis of the empirical formulae available is offered. The models for the boiling crisis (displacement effect in bubbling) are discussed. The authors consider the relation between the intensities of the turbulent momentum and heat transfer and the possibilities for experimental determination of the turbulent Prandtl number in the vicinity of the wall. The point of view is expressed which concerns the necessity to select canonical problems of the convective heat transfer theory for inclusion in courses of studies and handbooks.

The heat conductivity theory is the product of the 19th century. The thermal radiation theory is the product of the end of the 19th and of the beginning of the 20th century. The convective heat transfer theory is the product of the 20th century. By its nature, convective energy transfer is connected with the motion of its material carrier and therefore is inseparable from fluid mechanics. The revolutional changes in the latter (appearance of the hydrodynamics of non-Newtonian, electric current-conducting and magnetic media, the super - and hypersonic gas dynamics, dynamics of plasma, free molecular and heterogeneous flows, the hydro- and gasdynamic effects during physical and chemical transformations) have affected the theory of heat and mass transfer in moving media to the full extent. In this connection there arose a number of fundamental and methodical problems which are far from having been solved even by the end of the century. The sporadic understanding of unsolved problems is an important element in the development of each science. Below we will try to present some considerations based on our personal research experience and understanding of the general development of the convective heat transfer theory. Of course, they do not pretend to the full interpretation of the problem in general.

Turbulence is the basic form of macromotion of fluids. Its general importance has been realised just in this century. This is an interesting example of how an extremely deep and widespread phenomenon was first recognised and developed by engineers and only after that was realised in its global importance by professional investigators. The closure problem for a chain of Reynolds-Keller-Friedman's equations still remains unsolved in the theoretical aspect. Nevertheless, the semi-empirical closure methods available (in terms of Reynolds' equations proper, i.e. a Taylor-Prandtl type model, and in terms of the cutting-off at the level of one or another senior momentum, which go back to Kolmogorov-Richardson-Rotta's models) allow us to create practically realizable algorithms, effective in sufficiently extensive and clearly defined situations. Here partial closure models become euristically important.

The relation between the intensities of turbulent momentum and heat transfer processes is one of the subtle problems of the convective heat transfer theory. Formally it is reduced to the determination of the "turbulent" Prandtl number. For a two dimensional averaged flow.

$$Pr_T = \frac{\langle (\rho v)' u' \rangle}{\langle (\rho v)' T' \rangle} - \frac{\partial \langle T \rangle / \partial y}{\partial \langle u \rangle / \partial y} \qquad (1)$$

Beyond the viscous sublayer the value of Pr_T is of the order of unity (0.75 in a submerged jet and 0.50 in a wake behind a poorly flowed body). However, in the immediate vicinity of a solid body, at the depth of the viscous sublayer, $Pr_T > 1$, so that $\partial Pr_T / \partial \eta < 0$. In this case the nonisothermality of the flow results in that temperature fluctuations penetrate into the surface layer of the flowed body, which, in its turn, affects the value of Pr_T as $\eta \to 0$. The general qualitative dependence $Pr_T(pr; \eta; \Lambda)$ is shown in Fig.1 (see ref.[1]). This dependence is extremely important during the transfer of an inert additive in a medium with the diffusion Prandtl number $Pr_D > 100$. Within the framework of a rather natural model, for large physical Prandtl numbers and a thermally inertialess substrate the following relations are valid (see ref. [2])

$$Pr \sim \infty; \; \mu_T / \mu \sim \eta^3; \; \lambda_T / \lambda \sim Pr\, \eta^4;$$
$$Pr_T \sim \eta^{-1}; \; Nu \sim Pr^{1/4}\, Re \sqrt{C_f} \qquad (2)$$

At Pr > 100 the turbulent thermal boundary layer is submerged in the viscous sublayer of the turbulent hydrodynamic boundary layer and so formula (2) is also valid for the gas-liquid flow when the liquid which wets the wall forms a wall film. Fig.2 shows experimental data corroborating this fact (see ref. [3]). The coefficients in these relations can be determined from heat transfer experiments. Approximately the proportionality factor in the expression for λ_T / λ is of the order of 10^{-4}:

Unfortunately, to directly measure the values, necessary for the determination of the turbulent Prandtl number in the viscous sublayer, is extremely hard. At our Institute such measurements are based on the determination of the actual velocity field by pulse visualization of a liquid flow with a micro-suspension, and of the actual temperature field by low-inertial thermocouples ($\phi \simeq 10\,\mu$). Single-point velocity and temperature fluctuations in water flows were determined by synchronizing the reading of a temperature probe and the actual velocity field in its vicinity. Data files were processed in accordance with an algorithm which makes it possible to find the correlations of $\langle u'v' \rangle$ and $\langle v' t' \rangle$ and to analyze the structure of turbulent momentum and heat fluxes (see refs. [4-6]).

The values of the correlation coefficients are given in Fig.3. At $y^+ > 10$ the correlation coefficient for velocity fluctuations practically attains a value which is characteristic for the developed-turbulence region ($\langle u' v' \rangle / \sigma_u \sigma_v \simeq 0.4$). The correlation coefficient $\langle u' t' \rangle / \sigma_u \sigma_t$ exceeds this value somewhat and the coefficient $\langle v' t' \rangle / \sigma_v \sigma_t$ turns out to be considerably lower. (Here t' is a wall-flow temperature difference fluctuation). With approaching the wall the values of the correlation coefficients decrease.

The detailed analysis of the structure of Reynolds stresses carried out in ref. [7] has shown that events with fluctuations of different signs ($u' > 0; v' < 0$ and $u' < 0; v' > 0$) contribute to the value of Reynolds stresses more appreciably than events with fluctuations of the same signs. In the magnitude of the turbulent heat flux in the viscous sublayer, the role of instantaneous combinations of the fluctuations $v' > 0$ and $t' > 0$ increases. This fact which leads to a more noticeable decrease in the correlation $\langle v' t' \rangle$ possibly reflects the difference in the momentum and heat transfer rate when a liquid with the number $Pr > 1$ moves near the wall.

Fig.3 gives statistical errors for the correlation coefficients with the 95% confidence interval. One can see that with decreasing the value of this correlation coefficients, the accuracy of measurement noticeably decreases. It is quite evident that the error in determining Pr_T on the basis of these data will be great and its reliable estimation turns out to be very hard. Therefore only the trend of variation in the value of Pr_T with approaching the wall is shown in Fig.4 by the dotted line. Thus, despite the careful measurements of velocity and temperature profiles we cannot provide reliable qualitative values for the turbulent Prandtl number in the vicinity of the wall. Nevertheless, the trend of the value Pr_T to increase with approaching the wall is observed to be quite reliable.

The further improvement of these measurements may proceed by way of considerably increasing experimental data files. For visual measurements of velocity fluctuations, the processing of the material becomes too labour-consuming. At the same time, hot-wire anemometers are not suitable for measuring the correlations in the immediate proximity of the wall. The situation may improve with the progress in laser-doppler anemometers whose measurement volumes in the vicinity of the wall are still significant.

Heat transfer in heterogeneous systems is a much more complicated and versatile problem than turbulent convection in single-phase media. Here we can find all the difficulties in the description of the turbulent motion within the ranges of existence for each phase of the system, as well as the turbulent interface momentum and heat transfer processes and various structural reconstructions which are connected with them. The wave dynamics in such systems changes substantially. It is characterised by a large dispersion, powerful attenuation of perturbations, existence of specific forms for propagation of oscillations. As a result, the gas dynamic parameters (e.g., sound velocity in the gas phase) affect the heat transport even with small linear velocities of convection. Despite the appreciable progress in the dynamics of gas-liquid systems, which has been observed during the last ten or fifteen years, it (i.e the dynamics) still includes extensive areas for which no physico-mathematical models have been developed that would be considered to a certain extent satisfactory.

Boiling dynamics is in this respect the most complicated problem for theoretical analysis. Here a peculiar situation exists when a number of problems (integral outflow characteristics, propagation of disturbances in a bubble medium, kinetics of evaporation centers) have good mathematical models which either do not contain semi-empirical elements, or include a small quantity of the latter. The other problems, which are extremely important in applications, remain practically on a pure empirical basis. Most distinctly this situation is reflected in the problem of heat transfer in developed nucleate bubbling and boiling.

At present in the literature of the English-speaking countries, which is devoted to boiling in the presence of free convection, the authors most frequently use Rosenow's formula (see ref.[8]):

$$\frac{C' \Delta T}{\gamma} = C\, Re_*^{0.33}\, Pr^S \tag{3}$$

According to Rosenow for water S = 1, for other liquids S = 1.7, which points to the insufficient universality of the structure of this formula. The Russian-speaking authors most frequently use Kruzhilin-Labuntsov's formula (see refs.[9,10]) :

$$\frac{\alpha\, l_*}{\lambda} = c\, Pr'^{1/3} \left(\frac{U''\, l_*}{\nu} \right)^r \tag{4}$$

(at $\frac{U''\, l_*}{\nu} < 10^{-2}$ C = 0.062 and n = 0.50; at $\frac{U''\, l_*}{\nu} > 10^{-2}$ C = 0.125 and n = 0. 65) and Kutateladze-Malenkov's formula (see ref. [11]):

$$Nu_* = c K_*^n \tag{5}$$

where at n = 2/3 C = $1.5 \cdot 10^{-3}$ and at n = 0.7 C = 0.85×10^{-3}. Rosenow's and Kutateladze's formulae satisfy the condition of the existence of an analogy between bubbling and boiling after the simple substitution $U'' = \frac{q}{r\rho''}$. Labuntsove-Kruzhilin's formula does not satisfy this condition, because it includes such a purely thermodynamical characteristic, as the saturation temperature. Since at present the ranges of existence for the above analogy have been established with sufficient confidence, the structure of Kruzhilin-Labuntsov's formula cannot therefore be considered as reasonable. This does not mean that the similarity numbers, which include the value of the minimum radius of the vapour - phase nucleus, have no sense and know their own application areas in the generalisation of experimental data according to the thermodynamics of boiling.

Kruzhilin-Labuntsov's formula does not satisfy the fact that the value α is self-similar with respect to viscosity in the case of developed nucleate boiling. Rosenow's and Kruzhilin-Labuntsov's formulae do not also reflect the experimentally established fact concerning the influence of capillary - acoustic interactions on the heat transfer in nucleate bubbling (this fact has been established by direct experiments) and boiling (this fact has been established indirectly).

All this has not been said only to criticize Rosenow's and Kruzhilin-Labuntsov's formulae and to establish Kutateladze - Malenkov's formula. The case is that the latter is also an empirical generalisation of experimental data in a system of similarity numbers selected also on the basis of experimental facts. Thus there is still no sufficiently complete physico-mathematical model describing the process of developed nucleate boiling. Such a situation, unique for the up-to date heat transfer theory, exists despite the fact that the first efforts of generalisation go back to the papers of Jacob which appeared in the thirties of our century.

The problem of boiling crisis (displacement effect with bubbling) has no completed physico-mathematical models too. The hydrodynamic crisis manifest themselves most appreciably in pool boiling and in channel liquid flows with an averaged entropy which is smaller than the saturation entropy. Nevertheless, this problem has been developed to a much great extent than the heat transfer problem of nucleate boiling and bubbling. The additive model for the stability of the appearing vapour (gas) layer (see ref.[12]). is satisfactory for pool boiling both in the qualitative aspect and as a first approximation:

$$c_1 g \Delta \rho \delta + c_2 \rho' U'^2 = \rho'' U''^2 + c_3 \frac{U' \mu'}{\delta} + c_4 \rho' U' a''_* + c_5 \, wlU'' \sqrt{\rho' \rho''} \, \rho' U'^2 \sim \rho'' U''^2 \quad (6)$$

Here the first term is the gravitation-effect measure and the second term is the measure of the turbulent fluctuation effect in a liquid flow. Both forces destabilize the vapour layer. The right-hand side contains the forces which stabilize the vapour film (and prevent the liquid from breaking through to the heating surface). The third term is the measure of the dynamic effect by the vapour flow. The fourth term is the measure of the viscous drag of liquid films. The fifth term is the acoustic - effect measure. The sixth term is the measure of Coriolis drift in boiling near the wall layer.

In terms of similarity numbers this model has the following form :

$$K^2(1-c_2)+K(c_3/\sqrt{A}_{2*}+c_4/M^* \sqrt{\Delta \rho} + c_5 c_0) - c_1 = 0 \quad (7)$$

Fig.5 compares calculated results obtained using this model with a number of experimental data. It well elucidates the results obtained by Kirichenko in experiments (see ref.[13]) which have shown that with considerable overloads in the field of centrifugal forces, the critical heat flux passes through a certain maximum in the dimensional coordinates (see fig.6).

When a vapour-liquid mixture flows in a heated tube, there arises the so-called second-type thermal crisis which is due to the destabilisation and drying of the wall liquid layer. The hydrodynamic model also provides here a good description, as a first approximation in the form of a connection between two dimensionless values-the boundary (critical) void fraction and the Weber-number constructed on the basis of the mass velocity of the mixture.

However, in toto the problem of generalising data on critical heat fluxes with boiling in tubes and channels of complicated form remains open. We should note that here a vast amount of experimental material has been accumulated which has been obtained practically in all the leading energetic research centers of the world.

The above examples refer to the fundamental problems of the convective heat transfer theory and the fact that they remain unsolved testifies to the necessity to estimate the state of the canonical presentation of the heat transfer theory course in general. The state of the art of applied physics is such that it has to solve a plentitude of complicated (but partial in general) problems which are conditioned by the development of technology. The more important becomes the selection of canonical problems for courses of studies and handbooks. It also refers to the full extent to such a field of applied physics as the fundamentals of the heat transfer theory.

In this sense the theories of heat conductivity, radiation in transparent media, heat transfer in laminar flows may be assumed as being completed, since these theories include a set of mathematical models for canonical geometrical objects, boundary conditions, methods of solutions. To a considerable extent these fields have become the objects of computational physics. The physics proper, being an experimental science which extracts new facts from the examination of the Nature, now concentrates its attention, as regards the heat transfer theory, on turbulence, heterogeneity, interaction of radiation with absorbing and dissipating semi-transparent media. Here it is also important to create a canonical set of objects and their mathematical models.

NOMENCLATURE

u : mean local velocity

T : mean local temperature

u'N v' : velocity fluctuation components

t' : temperature fluctuation

σ_u, σ_v N σ_T : RMS velocity and temperature fluctuations

$\eta = \frac{y v_*}{\nu}$: dimensionless distance from the wall

u_* : dynamic velocity on the wall

μ, ν : dynamic and kinematic viscosities, $\bar{\mu} = \mu_T / \mu$

c : specific heat

D : diffusion

ρ : density, $\Delta\rho' = \rho' - \rho''$, $\Delta\rho = (\rho' - \rho'') / \rho'$

λ : thermal conductivity, $\bar{\lambda} = \lambda_T / \lambda$

Λ $(c \rho \lambda)_w / (c \rho \lambda)'$

$P_r = \nu / a$ - : Prandtl number, $Pr_D = \frac{\nu}{D}$ - diffusion Prandtl number

Pr_T : turbulent Prandt number

α : heat transfer coefficient

$Nu = \frac{\alpha \ell}{\lambda}$ - Nusselt number, $Nu_* = \frac{\alpha \delta_\sigma}{\lambda'}$

δ_σ : Laplace's constant

R_e : Reynolds number

c_f : drag coefficient

Re_* : $\frac{q \delta_\sigma}{\mu' r}$

q : heat flux density

r : heat of evaporation

U'' : $q / \rho'' r$ - rate of evaporation (bubbling)

$l_* = \frac{\sigma T_s C_p}{\rho'' r^2}$ - characteristic linear scale in formula (4)

K_* : $\frac{P U'' c_p}{g \lambda'}$ - heat transfer criterion for nucleate bubbling and boiling

P : pressure

g : gravitational acceleration

$\bar{g} = \frac{g}{g_0}$: overload, ($g_0 = 9.81$ m/sec^2)

$Ar_* = \rho' \sigma^{1.5} / (\mu')^2 \, (g \Delta \rho)^{0.5}$

$M_* = \sqrt{\frac{\rho''}{\rho}} \sqrt[4]{\frac{g \sigma}{\Delta \rho}}$: capillary-acoustic interaction criterion

$C_0 = \frac{\omega l \sqrt{\rho'}}{(g \sigma \Delta\rho)^{0.25}}$ Coriolis number

$X = \frac{1}{2(1 - C_2)} \left(C_3 \sqrt{Ar_*} + C_4 / M_* \sqrt{\Delta \rho} + C_5 C_0 \right)$

- see Fig.5.

Sub - and superscripts:

w : wall

' and " - : liquid and vapour, respectively

T : Turbulent

REFERENCES

1. Geshev, P.I., The effect of wall heat conductivity on the value of turbulent Prandtl number in a viscous sublayer. IFZh, No.2, 35 (1978) 292-296.

2. Kutateladze, S.S., Wall turbulence, Novosibirsk, Nauka, (1973) 227.

3. Pecherkin, N.I., Chekhovich, V.Yu., Heat and mass transfer in a vertical gas-liquid flow. In: Thermophysics and hydrodynamics in boiling and condensation, Novosibirsk, (1985) 24-26.

4. Perepelitsa, B.V., Khabakhpasheva, Ye.M., Correlations between temperature and velocity fluctuations in the near-wall region of a turbulent flow. In: Heat and Mass Transfer-YI, Minsk, pt.1, 1 (1980) 148-159.

5. Kutateladze, S.S., Volchkov, E.P., Mironov, B.P., Rubtsov,N.A.,Khabakhpasheva,Ye.M., Turbulent heat and mass transfer under interaction of various disturbing factors. In: Heat Transfer - 1982, Proc. of the 7-th Intern. Heat Transfer Conference, Washington, 3 (1982) 269-281.

6. Khabakhpasheva, Ye.M., Perepelitsa, B.V., Pshenichnikov, Yu.M., Mikhailova, Ye.S., The use of flow visualization to study the turbulent transfer in a viscous sublayer. In: The structure of forced and thermogravitational flows, Novosibirsk, Institute of Thermophysics, Siberian Branch of the USSR Academy of Sciences, (1983) 87-96.

7. Khabakhpasheva, Ye.M., Perepelitsa, B.V., Pshenichnikov, Yu.M., Mikhailova, Ye.S.,, Gruzdeva, I.M., Investigation into the mechanism of turbulent transfer in the immediate vicinity of the wall. In: Wall turbulent flows, Novosibirsk, Institute of Thermophysics, Siberian Branch of the USSR Academy of Sciences, (1984) 126-138.

8. Rosenow, W.M., A methed of correlation heat transfer data for surface boiling of liquids. Trans. ASME, (1952) 969-974.

9. Kruzhilin, G.N., Generalization of experimental data on heat transfer with boiling of liquids under free convection. Izv. AN SSSR, OTN, No.5,(1949) 701-710.

10. Labuntsov, D.A., Approximate theory of heat transfer in developed nucleate boiling. Izv. AN SSSR, Energetika i transport, No.1, (1963) 58-71.

11. Kutateladze, S.S.,Boiling and bubbling heat transfer under free convection of liquid. Int. J. Heat and Mass Transfer, No.2, 22 (1979) 289-299.

12. Kutateladze, S.S., Fundamental problems of heat transfer theory and fluid dynamics stimulated by nuclear technology. Heat and Technology (Calore e Tecnologia), No.1, '12 (1984) 1-28.

13. Kirichenko, Yu.A., Kozlov, S.M., Levchenko, N.M., The study of heat transfer crisis in boiling of He-I in the field of centrifugal forces. IFZh, No.2, 42 (1982) 207-213.

14. Ogata, H., Nakayama, W., Heat transfer to subcritical and supercritical helium in centrifugal acceleration fields. 1. Free convection regime and boiling regime. Cryogenics, No.8, 17 (1977) 461-470.

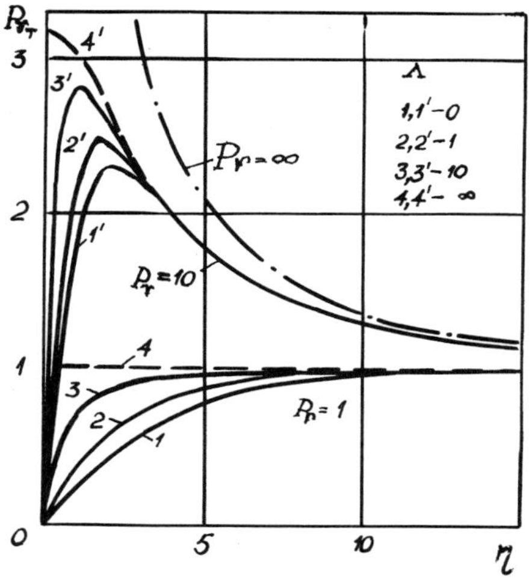

Fig.1 Dependence $Pr_T(Pr,\eta,\Lambda)$ according to the data of ref. [1].

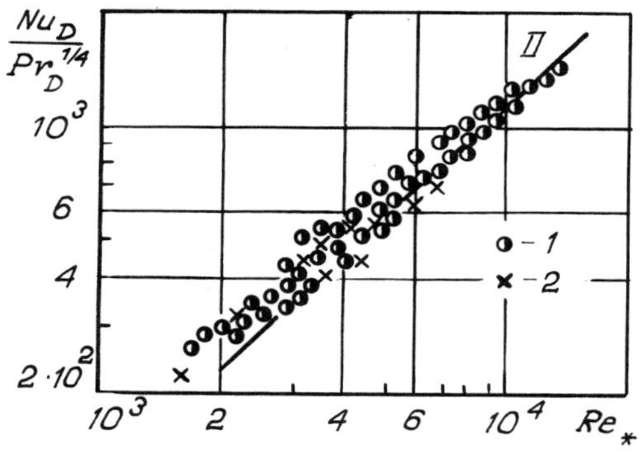

Fig.2 Mass transfer on the wall in a two-phase flow in in a vertical pipe (see ref. [3]); $Re_* = \frac{v^* d}{\gamma}$
1 - gass liquid flow; 2 - vapour-liquid flow; II - calculation according to formula (2)

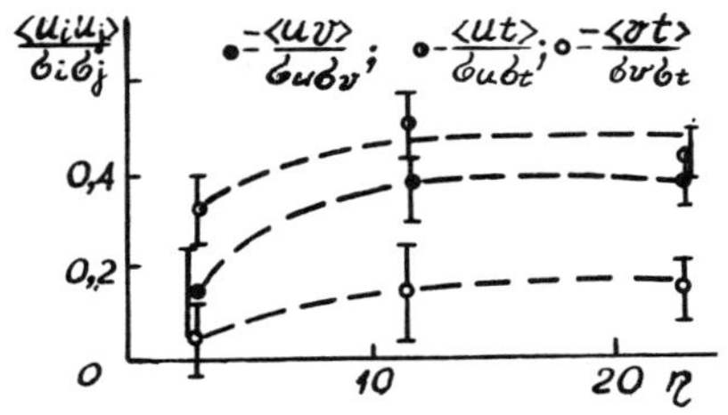

Fig.3 Correlations of velocity and temperature fluctuations.

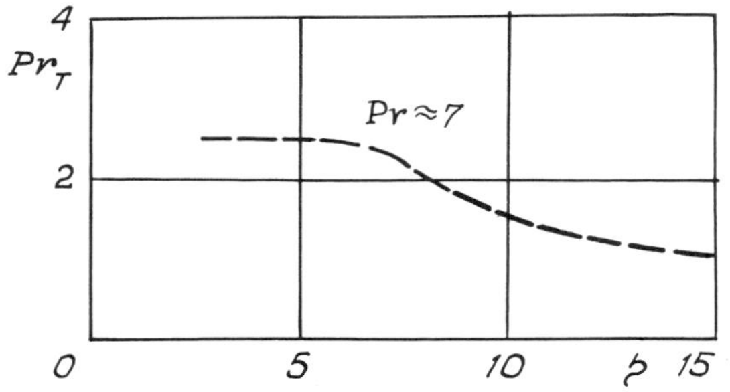

Fig.4 Turbulent Prandtl number in the vicinity of the wall.

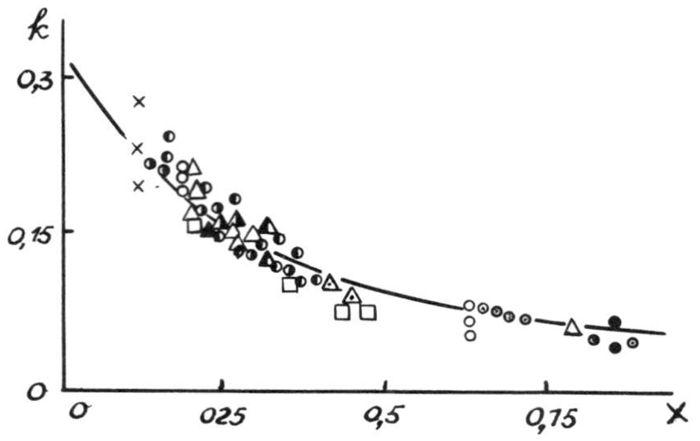

Fig.5 Bubbling of 1) water by: X - xenon: Δ - argon; ▲- nitrogen; 0 - helium; ● - hydrogen; 2) ethanol by: ▲ - argon; ▲ - nitrogen; 3) water-glicerin by: ⊖ - helium; ▲ - nitrogen; Boiling of: ,●● - helium (different pressures); - nitrogen;

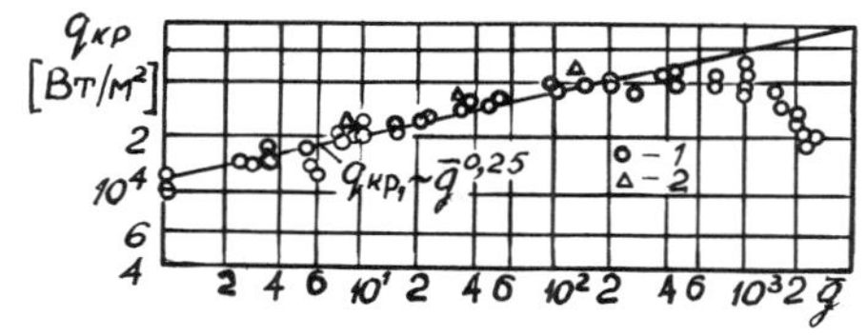

Fig.6 Experimental dependence q_{cr} on g.
1 - date of ref. [13] ; 2 - date of ref. [14].

The Properties of Superheated Liquids and the Limits of Liquid Superheat

JOHN H. LIENHARD
Heat Transfer/Phase-Change Laboratory
Mechanical Engineering Department
University of Houston
Houston, Texas 77004, USA

Liquids must be superheated before nucleation -- either boiling or flashing -- can occur in them. Thus an extremely important part of understanding boiling is that of understanding two related matters: the mechanics of nucleation, and the physical properties of superheated liquids. The paper reviews what we know about homogeneous nucleation, and heterogeneous nucleation as it occurs in highly superheated liquids. It also deals with the prediction, estimation, and correlation of the thermodynamic properties of metastable fluids -- and particularly of superheated liquids.

1. SCOPE

The process of boiling never occurs before a liquid has first been brought to a temperature beyond its local boiling point -- before it has been "superheated." Sometimes physical processes cause liquids to be heated very rapidly. When this happens, they can reach temperatures so far beyond their boiling points that when boiling occurs it does so with great violence. This kind of boiling is called "flashing".

Two highly inter-related issues are thus placed directly in our path. We must learn all we can about the circumstances that give rise to nucleation in a liquid -- the rates and extent of superheating or depressurization, the cleanliness of the system, the limitations of the fluid, and so forth. We shall find that what is, or even can be, known about nucleation is limited since it is often tied to such nuisance variables as cleanliness and surface roughness.

We must also find out how to evaluate or estimate the physical properties of superheated liquids: the thermodynamic and transport properties, the intrinsic limits to superheating and depressurization, the nature of thermodynamic paths, and the potential for doing damage when nucleation finally occurs. While knowledge of these matters is also incomplete; such behaviour is not dictated by nuisance variables. Research in this area is therefore more promising than it is in the case of nucleation. Our efforts have the potential for producing clean, useable, and badly-needed results.

The practical situations that require a knowledge of the limits of liquid superheat, and the physical properties of superheated liquids, are seemingly limitless. A few of them include :

- The creation of aerosols by flashing.
- Thermohydraulic explosions as might occur in nuclear coolant line breaks, liquid metal release into water following a core meltdown, liquified light-hydrocarbon spills, or Kraft paper process boiler leaks.
- Quenching, as occurs in heat treating metals, rewetting nuclear cores, the cooling of shot, or the diagnosis of boiling heat transfer.
- Predicting the behaviour of liquids heated beyond their boiling points in nucleate and transition boiling, and in the momentary surface contact that might occur in film boiling.
- Estimating how much damage a thermohydraulic explosion can do.
- Improving combustion sprays by emulsifying small amounts of water in liquid fuels and heating them until flashing occurs.
- Predicting the depressurization that follows the blowing off of excess vapour from hot liquid storage vessels.

2. THE METASTABLE FLUID STATES AND THE SPINODAL LIMIT

The p-v-T Equation of State. Figure 1 shows isotherms in the p-v plane of the p-v-T surface[1] of a real fluid. All states along an isotherm are equilibrium states. When the slope of an isotherm is positive, that equilibrium is unstable. When the slope is negative, the equilibrium is either stable or metastable. The two spinodal lines connect the points where the isotherms have zero slope. Then, for example, to locate the liquid spinodal line is to specify the absolute limit beyond which a liquid can never be superheated.

The Gibbs potentials, g_f and g_g of real isotherms must be equal in the saturated liquid and vapor states. Thus:

$$g_g - g_f = 0 = \int_f^g (vdp - sdT)_T \tag{1}$$

The last term vanishes giving the "Gibbs-Maxwell" relation, which requires that Area, A, in Fig. 1 equals Area, B :

$$\int_f^g vdp = 0 \tag{2}$$

For a long time, van der Waals' equation

$$p = \frac{RT}{v - b} - \frac{a}{v^2} \tag{3}$$

Symbols not defined in context are conventional ones. They are defined in the Nomenclature Section.

provided the only theoretical knowledge of real fluid isotherms. Van der Waals argued, on the basis of molecular behaviour, that there is an inherent continuity from the liquid to the vapor states. An important feature of his equation is that it can be nondimensionalized using critical data. Thus :

$$P_r = \frac{8T_r}{3v_r - 1} - \frac{3}{v_r^2} \qquad (4)$$

where $P_r = p/P_C$, etc. The dimensionless van der Waals equation suggests the Law of Corresponding States -- that one equation of state, written in reduced coordinates, should describe all fluids.

We have subsequently learned that the Law of Corresponding States should be amended in the following way :

$$P_r = f(T_r, v_r, \text{primary molecular parameter, other molecular parameters}) \qquad (5)$$

Where :

- The strongest influences in equation (5) are those of T_r and V_r. Indeed the need for any further parameters was not clarified until the mid-1950's.
- The primary molecular parameter is usually taken to be Z_c; the Riedel factor, α_R, (see[1]); or the Pitzer acentric factor, ω (see, e.g.,[2]):

  $$\omega = -1 - \log_{10}[P_r, sat(T_r = 0.7)]$$

  Figure 2 is a recent Corresponding States correlation [3] of the saturation temperature as a function of pressure. In it data for all kinds of fluids including liquid metals are correlated perfectly with the help of the Pitzer factor. It is also important to note that the van der Waals fluid itself takes its place in the family of real fluids at the appropriate value of Pitzer factor.
- Little has been done with secondary molecular parameters. Figure 3 shows how data for molecules with high dipole moments deviate slightly from an otherwise successful correlation of burnout heat fluxes, based on the Law of Corresponding States [4]. Thus the Law sometimes needs correction when it is applied to polar molecules.

The most important corollary of the preceding results is that van der Waals' equation should accurately describe any real fluid with Z_c, α_R, or ω equal to the van der Waals' value. This notion, was originally demonstrated by Lienhard [5] based on the molecular parameter, Z_c. Peak [6], subsequently showed that the notion was even better borne out using the Pitzer factor.

The importance of the membership of van der Waals' equation in the family of real fluids is that we are now presented with enormous leverage in fitting Corresponding States correlations. This fact is made evident in Fig.2 in which the simple function:

$$\ln p_r, sat = 5.37270 \, (1 - 1/T_r) + \omega(7.49408 - 11.18177T_r^3 + 3.68769T_r^6 + 17.92998 \ln Tr) \qquad (6)$$

fit all the data within ± 0.42 percent. This provided significant improvement and simplification, in the range of fluids similar to the van der Waals fluid, over an earlier correlation by Lee and Kesler [7].

A second point is that van der Waals' equation defines a pair of spinodal lines. But since the van der Waals equation is the basis for the Law of Corresponding States, we can also expect real-fluid spinodal lines to obey it.

The van der Waals spinodal lines are obtained by setting $(\partial P_r / \partial V_r)_T = 0$ and eliminating T_r between this result and the original equation. The result:

$$P_r = \frac{3}{V_r^2} - \frac{2}{V_r^3} \qquad (7)$$

describes the liquid spinodal for $V_r < 1$ and the vapor spinodal for $V_r > 1$.

3. HOMOGENEOUS NUCLEATION

3.1. The Homogeneous Nucleation Limit and its Relation to the Spinodal Line.

We would like to use measurements of the limiting liquid superheat to obtain the location of the spinodal line. But can that be done? Are the homogeneous nucleation limit and the spinodal line related? To bring a real liquid all the way up to the spinodal limit, one would have to do so without any disturbances or imperfections in the system. However, real liquids are made of molecules that constantly move. As the liquid temperature rises these motions provide the disturbances needed to upset liquid stability at a temperature less than the spinodal temperature.

Frenkel (see e.g. [8]) first calculated the least disturbance needed to create a minimum stable vapor bubble or the "potential barrier" to nucleation. He calculated the difference in Gibbs function of the liquid with and without an unstable vapor bubble in it and obtained the critical work, Wk_{crit}, needed to create the bubble:

$$Wk_{crit} = \frac{4 \pi R_0^2 \sigma}{3} \qquad (8)$$

The radius, R_o, is the well-known unstable equilibrium radius:

$$R_o = \frac{2}{\text{Psat at Tsup} - P} \qquad (9)$$

where T_{sup} is the local temperature of the superheated liquid, and P is the pressure in the surrounding liquid.

Wk_{crit} must now be compared with a characteristic energy of the superheated liquid. Two energies are appropriate to this purpose:

- The average level of molecular vibrational energy is of the order of kT, where k is Boltzmann's constant and T is T_{sup} -- the temperature of the superheated liquid. Conventional theories of homogeneous nucleation are based on this energy.

- The energy required to separate two molecules from one another. This is the depth of the "potential well" and it is well known to be of the order of kT_c. The use of kT_c was introduced in [9], and is discussed below.

The ratio of Wk_{crit} to kT (or kT_c) is the Gibbs number:

$$Gb \equiv Wk_{crit}/kT \qquad (10)$$

What is the minimum Gb for which nucleation absolutely must occur? The least possible value of R_o will give the minimum Wk_{crit} and Gb. To find it, we first define:

j ≡ probability of nucleating a bubble in a given molecular collision (11)

Now j must be 1 for Gb = 0, and we expect that:

$$\frac{dj}{j} = -\frac{dWk_{crit}}{kT \text{ or } kT_c} = -dGb \qquad (12)$$

so :

$$j \frac{\text{nucleation events}}{\text{molecule collisions}} = e^{-Gb} \qquad (13)$$

The problem thus reduces to establishing j. To do this, consider a spherical region of undisturbed liquid in which the smallest critical bubble, with radius, R_o, might nucleate, and then count the rate at which collisions can occur in this region. Nucleation must occur if just one of these collisions triggers a critical nucleus within one relaxation time (or about 10 collisions.) This gives.

$$\text{Probability of nucleation} = j \leq 10^{-5}$$

It was thus shown in [9] that the limiting value of j is 10^{-5}. This limit, of course, is only an order of magnitude estimate. But, fortunately, nucleation is only sensitive to large variations in j. (Later we suggest the value should probably be a little higher than 10^{-5}.)

We now have a prediction for the limit of homogeneous nucleation. To find how close to the spinodal line it lies, we use the thermodynamic availability of spinodal liquid, with respect to liquid at the limit of homogeneous nucleation :

$$\Delta a\Big|_{h.n.}^{sp} = Gb_{min}\,(kT \text{ or } kT_c) \qquad (14)$$

Where Gb_{min} is of the order of $-\ln(10^{-5})$ or about 11.5, and the isobaric difference in availabilities is :

$$\Delta a\Big|_{h.n.}^{sp} = [\Delta h - T_{h.n.}\,\Delta s]_{h.n.}^{sp} \qquad (15)$$

It is shown in [9] that equation (15) implies a temperature difference of the order of 1°C between the homogeneous nucleation and spinodal lines on the liquid, but not the vapor, side. Therefore the liquid spinodal line can be located accurately using a homogeneous nucleation prediction. We make this prediction by combining equations (8) and (9) in equation (13). The final result, which includes an additional correction for the effect of curvature on surface tension (see [8]), $(1 - v_f/v_g)^2$, is :

$$-\ln j = \frac{16\pi\sigma^3}{3(kT)\,[P_{sat}(T_{sp}) - p]^2\,(1 - v_f/v_g)^2} \qquad (16)$$

The terms T_{sp}, v_f, and v_g are defined in Fig. 1. The term kT is based on the local temperature in the conventional theory; but we show subsequently that it probably should be kT_c .

When this argument is applied to the nucleation of liquid drops in a vapor -- which is less dense -- far fewer nucleation events are needed in the limiting case of complete nucleation. Thus, j must be far smaller and Gb far larger. This leads to large temperature differences between homogeneous nucleation and the vapor spinodal.

3.2 Experimental Data for Homogeneous Nucleation at High Superheat, or Large Values of j.

V. Skripov and his coworkers at the Ural Institute at Sverdlovsk have pushed the limit of liquid superheat much further in the laboratory than anyone else. Much of this work is summarized in two books [8,10]. Avedision [11] recently provided an extensive compendium of measured liquid superheat and related j values[3].

Many techniques exist for creating high liquid superheats. The most effective has been Skripov's method of pulse heating a fine wire filament. When the wire is subjected to a known electrical pulse, its temperature rises rapidly and predictably within a few microseconds. As the temperature rises, a few isolated instances of nucleation occur; but then -- at a certain temperature -- a complete blanket of vapor appears on the wire. It is at this point that no further temperature increase is possible in the liquid.

Skripov reached different limiting temperatures depending upon the rate at which he heated the liquid. However, as the heating rate rose, the temperature approached an asymptote. The value of j at that limit was about 10^{-5}. (Skripov reports some values of j slightly above 10^{-5}, but only at heating rates for which a different and less reliable experimental technique was used.)

Figure 4 shows those data of Skripov et al., that are available in 1976 [12], for j values of the order of 10^{-13}. The difference between his homogeneous nucleation temperature, and the saturation

[3] Actually, Skripov (and Avedisian as well) use J instead of j. J is equal to j multiplied by the rate of molecular collisions per cm^3. J is 10^{39} times j for water at 1 atm., in these units.

temperature at the same pressure, is plotted against T_r. The notational distinction between fsp (liquid spinodal) and gsp (vapor spinodal) is introduced temporarily in Fig. 4 because it includes the best data available for nucleation of droplets in subcooled vapor. The liquid and vapor spinodal limits calculated from van der Waals' equation are also included.

Figure 4 makes two things clear: One is that the homogeneous nucleation limits for the 12 liquids do conform to Corresponding States correlation. Furthermore, the shape of the dashed correlating line through them is very similar in form and placement to the van der Waals prediction. This corroborates the demonstration [9] that these data should be almost the same as the spinodal line which is a thermodynamic variable. The second point made clear in Fig. 4 is that -- as we anticipated -- the vapor nucleation data do not in any way conform to Corresponding States correlation.

In [12] the following equation was fitted to the Skripov data :

$$\Delta T_r = 0.905 - T_{r,\ sat} + 0.095\ T^8_{r,\ sat} \qquad (17)$$

This gives values slightly less than the true generalized spinodal value, because j for this set of Skripov's data is a factor of 10^{-8} smaller than j for the spinodal line[4].

All the variables in equation (16) are subject to corresponding states correlation. Such a correlation for the limit of pressure undershoot was formed in [13]. Substituting KT_c for KT, gives:

$$\Delta P_r = P_{sp} - P_{sat} = \frac{112.82 + 224.42\,\omega}{\sqrt{-\ln j}}\ (1 - T_{r,\ sp})^{1.83} \qquad (18)$$

See Fig.1 for notation and note that, while j is of the order of $(10)^{-5}$ gives the spinodal line, any smaller j gives a homogeneous nucleation limit that does not reach the spinodal line.

Real systems always suffer some heterogeneous imperfections, and real depressurization processes are normally slow enough that an improbable high-energy molecule can trigger nucleation at a higher pressure than we would expect. Either situation will yield a smaller value of j. For example (see e.g.[11]):

- The pulse heating technique of Skripov et al. (described earlier) typically yielded $j \simeq 10^{-13}$
- Experiments in which one fills a very thin and clean glass capillary tube with fluid, and suddenly immerses it in a bath of heated liquid, are inherently slower that the pulse heating experiments and give j of the order of 10^{-28}
- Experiments, in which a droplet of volatile liquid floats upward in a thermally stratified nonvolatile liquid until it explodes, are also slow, and give j's of the order of 10^{-28}.
- If depressurization times reach the order of seconds or minutes, and if systems are not extremely clean, j can drop almost without limit below the values given above.

The Corresponding States correlation can also be written in terms of temperature. The result for the spinodal line can also be approximated with high accuracy in the form:

$$\Delta T_r = 0.923 - T_{r,\ sat} + 0.077\ T^9_{r,sat} \qquad (19)$$

This lies about 2 percent above the correlation given by equation (17), and reflects the increase of j from the value of 10^{-13} (which was characteristic of Skripov's experiments) to $(10)^{-5}$.

4. THE DEVELOPMENT OF EQUATIONS OF STATE AND FUNDAMENTAL EQUATIONS TO DESCRIBE THE METASTABLE REGIONS.

We can now place the spinodal limit with reasonable accuracy. However thermodynamic properties are generally unavailable in the metastable and unstable ranges. Such data are constantly needed in boiling and two-phase heat transfer work, but they are extremely hard to come by, and little analysis has been done.

It has been customary in boiling work to estimate thermodynamic properties in these regimes by extrapolating them linearly in temperature. This works in slightly superheated liquids, but at higher superheats such a simple strategy becomes impossible. (The specific heat at constant pressure approaches infinity on spinodal lines, liquids become highly compressible at high superheat, etc.)

4.1 The Formulation of Fundamental Equations.

The "fundamental" or "canonical" thermodynamic equation is one from which all equations of state can be derived and in which all thermodynamic properties are embodied. For a pure substance, four common ones are: s=s(u,v), h=f(s,p) $\psi=\psi(T,v)$ and g=g(T,p). The several equations of state are obtained from the derivatives of these equations with respect to their independent variables. For example, the p-V-T and energy equations of state are obtained from:

$$P/T = (\partial s/\partial v)_u \quad \text{and}$$

$$1/T = (\partial s/\partial u)_v \qquad (20)$$

One tends to associate the p-V-T equation with the term "equation of state"; but it gives incomplete thermodynamic information and must be combined with a "caloric", or specific heat, equation of state to give -- say -- enthalpies or entropies. A fundamental equation, when it can be written, is more convenient than a set of equations of state. It provides all thermodynamic information through straightforward calculations. The Helmholtz function form of the fundamental equation, for example, can be obtained from the p-v-T and c_p^o equations:

---

4 People unfamiliar with these calculations are understandably startled at the notion that a factor of 10^{-8} makes so little difference in this range.

$$\psi(v,T) = \psi_{ref} + \int_{v}^{RT/p_{ref}} p dv' + \int_{T_{ref}}^{T} c_p^o (T') \left[\frac{T}{T'} - 1 \right] dT' - R(T - T_{ref}) \qquad (21)$$

where (P_{ref}, T_{ref}) is a reference point where the ideal gas equation is valid and the entropy is taken as zero.

The Steam Tables (see, e.g., [14] or [15] are normally based on empirical fundamental equations in the Helmholtz form with scores of constants in them. Such equations permit one to evaluate any thermodynamic property within four decimal places. But we seldom have anything nearing such complete property information for fluids other than water. Furthermore, such equations do not normally provide valid information in the metastable regimes.

4.2 Surface Tension

The property, surface tension, σ, is important in its own right for predicting almost any aspect of boiling heat transfer. And unfortunately, it has not been measured over wide ranges of temperature for many fluids.

Furthermore, σ is intimately related to the p-v-T equation of state. Van der Waals -- well known for his remarkably simple and successful equation of state -- also developed this remarkable and completely precise equation for predicting the surface tension, σ, from a knowledge of the p-v-T equation of state [16] :

$$\sigma = \sigma_o \int_{v_{r_f}}^{v_{r_g}} \frac{1}{v_r^{5/2}} \left[p_{r,sat}(v_r - v_{r,f}) - \int_{v_{r_f}}^{v_r} p_r \, dv_r \right]^{1/2} dv_r \qquad (22)$$

The integration of equation (22) requires complete p-v-T values in the metastable and unstable regimes. Thus surface tension can be used to extend metastable property information and vice versa.

4.3 Karimi's Fundamental Equation for Water

Karimi made the first attempt to creat an accurate fundamental equation for water in the metastable regimes. He and Lienhard first [17] made a relatively crude modification of the Keenan-Keyes-Hill-Moore (KKHM) equation of state for water [14]. Karimi then refined the curve-fit and produced a greatly improved equation [18].

Some of the important results predicted by Karimi's equation are the modified p-v-T and T-s-p surfaces shown in Figs. 5 and 6. Notice that all properties exhibit continuous change as they pass into the "two-phase" regime. The stable, metastable, and unstable regimes are all shown, and the conventional straight-line representations of mixtures are not included.

Although Karimi's results reveal the potential for accurately predicting metastable and unstable behaviour of real substances, he had to use different coefficients on either side of T = 150°C in his final equation. Consequently the equation displayed inconsistencies and discontinuities in the neighbourhood of 150°C.

Karimi based his work on the KKHM equation,

$$p = \rho R T \left[1 + \rho \frac{\partial}{\partial \rho} (\rho Q) \right] \qquad (23)$$

Where Q is a function of ρ and T, obtained by fitting the data to a suitable expression with many terms of the type $(T - a)^m (\rho - b)^n e^{-c\rho}$. KKHM only used stable data in their fitting. Karimi used additional theoretical constraints and the meager existing data for metastable fluids to improve the coefficients.

The KKHM equation is the sum of an ideal gas term and a correction. No such features as phase-changes, stable and metastable regions, etc., can be modelled by the ideal gas "reference function". Many of Karimi's difficulties arose because he had to absorb all of the complexity of water in the correction term, Q.

4.4 An Improved Strategy for Creating a Fundamental Equation.

A new and highly accurate equation of state was developed at the National Bureau of Standards (NBS) by Haar, Gallagher, and Kell in 1979 [19]. It, too, gives pressure as the sum of a reference function and a residual correction, but the reference function has a theoretical basis. It displays all real fluid features qualitatively, and yields errors less than 1 percent in wide ranges of T and p Thus, its residual correction is much smaller than in the KKHM equation. It must only improve the accuracy - it need not provide all the features of real fluid behaviour by itself. Thus it can be fitted accurately with fewer terms.

We are currently modifying the NBS equation in much the same way as Karimi modified the older KKHM equation. A key factor to doing this is the use of cubic equations to extrapolate p-v-T data. We turn to this matter next.

4.5 Cubic Equations of State

Cubic equations of state, of which the van der Waals equation is the prototype, have the smallest number of free parameters that can represent real data accurately. They are therefore much safer to use in estimating data than more complicated ones. They display major features of physical behaviour correctly in relatively large interpolations, with no need for smoothing. When such equations have enough inherent flexibility and are carefully fitted, they are remarkably accurate at a very low cost of calculation.

To this end we turn to a new and far more general form of cubic equation developed by Shamsundar and Murali [20,21] :

$$\frac{P}{P_{sat}} = 1 - \frac{(v - v_f)(v - v_m)(v - v_g)}{(v + b)(v + c)(v + d)} \qquad (24)$$

The quantities P_{sat}, v_f, v_m, v_g, b, c, and d all vary with temperature. The advantage of this form is that it automatically satisfies critical point criteria, but it need not be tied to them. Another nice feature is that it is prefactored to do away with the need for finding roots of the cubic in fitting the constants.

Three of the coefficients in equation (24) are the known temperature-dependent properties: P_{sat}, v_f, and v_g. Thus the most straightforward use of the equation is one in which isotherms are fitted one at a time. Two of these constants can be fixed with the ideal gas law limit at low pressure and the Gibbs-Maxwell equal-areas condition, equation (2). Then just two pieces of data will fit an equation that should be quite accurate along a given isotherm.

The isothermal compressibility of saturated liquid, and one compressed liquid point, have proven to give the best results. It turns out that if the two missing pieces of data are reduced to just one by the seemingly arbitrary assumption that c = d, the resulting equation is still extremely accurate. For a given temperature, the coefficients for the equation can be found using very few data, and making a direct, not a statistical calculation.

The remaining problem associated with this kind of use of equation (24) -- that of generalizing the temperature dependence of the coefficients -- remains to be solved.

The temperature-by-temperature application of equation (24) to water has yielded accuracies that are within the tolerance of the IFC [15] Skeleton Tables in the subcritical range of pressures. A comparison of IFC data with our equation is given over a wider range of pressures for water in Fig. 7. The only other cubic equation (actually a modified cubic that was not intended for use with water) close enough to the data to appear on this plot is that of Redlich [22]. Fuller's [23] cubic equation -- intended for water -- predicts values outside the figure.

Another such comparison is presented for ethylene in Fig. 8. The present cubic again outclasses the existing cubic equations and other "simple" equations. The key to this success is, of course, the fact that the coefficients are free from having to obey any predetermined dependence upon temperature.

We emphasize liquid properties because they are so hard to predict; however Murali [20] has shown that equation (24) is extremely accurate in predicting superheated vapor properties.

But a most important use of the new equation of state is that of predicting properties in the metastable regimes. Figure 9 compares the liquid spinodal of the cubic equation with the one predicted in [9]. The agreement is very good except at such low temperatures and high liquid tensile stresses that both theories are being pushed to their limits of applicability.

The nucleation limits shown in Fig. 9 are both obtained from equation (16), using both kT and kT_c : the use kT_c is far superior. The choice makes little difference in the positive pressure range, which is the only range in which experiments have ever been made for large j's. At lower pressures the two diverge very strongly. Our recommendation to use kT_c is largely based on the compelling evidence derived from the cubic equation. The value of j used in Fig 9 is $3(10)^{-5}$ instead of 10^{-5} (as recommended in [9]) — a minor alteration. The replacement of kT with kT_c is the revolutionary issue here.

4.6 The Prediction of Surface Tension

The acid test of any p-v-T equation that purports to predict metastable and unstable properties is that it must correctly predict the temperature dependence of surface tension when it is used in van der Waals' surface tension equation (22). The cubic equations for each isotherm of water have been subjected to this test, and the results are shown in Fig.10 (this result was developed in [24].) Figures 10 makes it quite clear that, except at the very lowest temperatures, the cubic equation passes this test with flying colours. In Fig.11 we [24] repeat this test for several other fluids with equal success.

While we use equation (22) to verify our p-v-T representations here, it is also clear that the present methods can also be used with good accuracy to predict surface tension if we do not already know it -- at least within an as-yet-unknown lead constant, σo . However, for each case shown in Figs. 10 and 11, we had to calculate [24] the average value of σo from the data.

All of these σo 's are plotted in Fig.12. These data fall into two categories: those based on property data in which we had complete confidence (the solid symbols); and those for which we entertained some shadow of doubt (the open symbols). We have nondimensionalised surface tension as recommended by Brock and Bird [25] and looked for its dependence on the Pitzer factor.

The result, based on the solid symbols, takes the form of a perfect straight-line dependence within a correlation coefficient of 0.995.

$$\sigma o / P_c^{2/3} (kT_c)^{1/3} = 1.08 - 0.65\,\omega \qquad (25)$$

The remaining points do not alter the correlation -- they merely lower the correlation coefficient slightly.

One can thus predict surface tension with acceptable accuracy for most applications, using nothing but a modest number of p-v-T data, and the methods we have developed here.

4.7 On the Formulation of Corresponding States Correlations for p-v-T Data in the Metastable and Unstable Regimes.

Figures 13 and 14 show vapor pressure data and spinodal line results (including van der Waals values) on generalized compressibility (or z = pV/RT coordinates, and Fig.15 shows several stable-metastable-unstable isotherms developed from the cubic equations. (These results are taken from [3].)

These graphs show that the spread of data in ω is generally modest in these coordinates -- particularly for non-polar molecules with relatively low positive ω's. The scatter is predictably larger on the liquid side than on the gaseous side, and it increases strongly toward the critical point.

The reader should note that we extrapolate the vapor spinodal line to Z = 0.5 at low pressure.

This result, noted by Karimi and Lienhard (1979), stems from the fact that one may write any equation of state in virial form :

$$P = RT/V + B/V^2 + C/V^3 + \quad (26)$$

One may then differentiate it and set the result equal to zero on the spinodal line. At low pressures in the vapor region we may further drop terms of degree beyond v^{-2}. We solve the result :

$$(\partial p/\partial v)_T = -RT/V^2 - 2B/V^3 = 0 \quad (27)$$

for $B = -RTV/2$ which, substituted back into the low pressure virial, gives

$$P = RT/V - RTV/2V^2 = RT/2V \text{ or } Z = 1/2 \quad (28)$$

Figures 13, 14 and 15 thus suggest that it is feasible to create a Z-chart in the metastable and unstable regimes, particularly for a restrictive range of ω's-just as is conventional for the stable ranges. Figure 16 is such a Z-chart.

Figure 16 is centered on pentane (with $\omega = 0.258$) and it represents data very well in the ranges ;

$0.10 < \omega < 0.35$, P_r, liquid < 0.8

P_r, vapor < 0.9

Figure 16 can still be used outside these ranges, but the accuracy can be expected to fall off further away from them.

5. DEPRESSURIZATION OF HOT LIQUIDS

5.1 On Predicting the Limiting Undershoot

Consider a liquid in a container, suddenly depressurized from an initial point, (P_i, T_i), as shown in Fig. 17. Such a system was studied both analytically and experimentally by Alamgir, Lienhard, and Trela [26,27,28].

When the liquid is initially not too hot, or when the depressurization ends in nucleation well before the spinodal limit, the isentropic depressurization lies close to an isothermal path. We note in Fig. 18 (a typical pressure-time history for such a process [26]) the abruptness with which depressurization ends in nucleation. Experiments [27] show that, while some nucleation occurs on the pipe wall before depressurization abruptly reverses, it is not enough to affect depressurization. Alamgir and Lienhard [26] therefore concluded that a real homogeneous nucleation limit was reached on the two-dimensional pipe wall.

Thus the homogeneous nucleation theory had to be reconstructed in a sort of "Flatland" form. Nucleation was taken to occur everywhere on the wall when the two-dimensional J was equal to 10^{18} nucleation events/ m^2-s. This translates to $j = 0.5661(10)^{-12}$ which is very close to j for Skripov's pulse heating results.

The pressure undershoot, in this case, depends on the rate of depressurization, ξ^1 atm/s (see Fig. 18.) The reason is that a two-dimensional array of bubbles will have more time to grow and completely fill the wall if the rate is slower. Alamgir and Lienhard defined a "heterogeneity factor,"

$$\phi \equiv \frac{\text{actual potential barrier}}{\text{potential barrier for homogeneous nucleation}}$$

and an $\eta \equiv Gb/\phi$. Then they demonstrated that :

$$e^{-\eta\phi} - \sqrt{\pi\eta\phi}\ \text{erfc}\ \sqrt{\eta\phi}$$

$$= \frac{\Sigma'\ (1 - v_f/v_g)}{(N_A/v_f)^{2/3}\ (\text{collision rate})\ \sqrt{36 k T\eta\sigma/\pi}} \quad (26)$$

and used equation (16) in the form applicable to real (not homogeneous) nucleation :

$$\eta = \frac{16\pi\sigma^3}{3kT(1 - v_f/v_g)^2 (P_{sat} - P_n)^2} \quad (28)$$

to eliminate η from equation (27). They then used equation (27) to calculate ϕ for observed nucleation events.

The resulting values of Φ are correlated as a function of ξ^1 and temperature in Fig.19. This correlation brings together the data of many investigators and it only breaks down when ξ^1, becomes so low that single isolated nucleation events can stop the depressurization. This occurs when ξ^1 is less than 4000 atm/s.

The correlation in Fig. 19 can be summarized in the form :

$$\Phi(T_r, \Sigma') = 0.1058 T_r^{\,28.46} (1 + 14\Sigma'^{\,0.8}) \quad (29)$$

for

$0.62 \leq T_r \leq 0.935$,

$0.004 \leq \Sigma' \leq 1.8$ Matm/s,

which correlates all available blowdown data with a 10.4 % rms error.

The data in Fig. 19 imply that j is $0.5661(10)^{-12}$. This gives $Gb = 28.2$. By substituting Gb/ϕ for η in equation (28) and using equation (29), we recast the undershoot prediction in dimensional form :

$$P_{sat}(T_i) - P_{nucleation} = 0.252\frac{\sigma^{3/2} T_r^{13.73}(1 + 14\Sigma'^{0.8})^{0.5}}{\sqrt{kT_c}(1 - v_f/v_g)} \quad (30)$$

For the reader's convenience, this pressure undershoot is displayed in dimensional form for water, as a function of Σ' and the initial saturation temperature, T_i, in Fig. 20.

5.2 The Damage-Doing Potential of a Sudden Depressurization

A most important practical problem is that of answering the question: "A large pipe carrying nearly-saturated water at high pressure is suddenly reptured. How much damage can each pound of water do the system in the resulting thermohydraulic explosion?". Two thermodynamic issues lurk in this important safety-related question. The first -- that of saying how far below its saturation pressure the water must fall before an explosion is initiated -- is answered by equation (30).

The second issue is that of determining the thermodynamic availability of the liquid at this limit. The availability of a liquid with respect to its surroundings specifies the maximum "useful" work -- actually damage in this case -- that it can deliver to these surroundings. The latter issue requires a knowledge of the thermodynamic properties of metastable liquids.

The isobaric availability of a slightly superheated liquid with respect to its saturation state is (recall equation (15)) :

$$\Delta a = \int_{T_{sat}}^{T_{sup}} c_p \, dT - T_{sat} \int_{T_{sat}}^{T_{sup}} (c_p/T) \, dT \qquad (31)$$

By assuming c_p is constant -- which is certainly untrue at high superheats -- we obtain as a limit for low superheat (see [29].)

$$\Delta a = c_p \; \Delta T^2 / 2 \, T_{sat} \qquad (32)$$

A description of the exact calculation, and the evaluation of Δa for van der Waals fluids at their spinodal lines are given in [30]. We have also computed availabilities of superheated water for EPRI using Karimi's fundamental equation. One of these calculations involved challenging the often-used assumption that isentropic depressurizations can be treated as isothermal. Figure 21 shows what the left-hand region of Fig. 1 or 17 would look like plotted to scale. Notice that at $T_r = 0.8855$ (or T = 300°C) the assumption is not bad as long as the pressure stays positive; but at $T_r = 0.9627$ (350°C) it breaks down much more quickly.

This information is shown in another way in Fig. 22 which shows how the relation between P and T changes along isentropic paths, starting from various initial saturation temperatures.

Figure 23 shows how the availability of water above the normal boiling point varies with the temperature at which it first crosses the saturated liquid line. As P decreases, the availability diminishes, but the reduction is too little to improve safety noticeably. A very rapid depressurization of 1.5 Matm/s brings water to nucleation at a point whose availability almost matches the saturation point from which depressurization began. The point where nucleation occurs in Fig. 23 is obtained with the help of equation (28). It is sobering that Δa is of the order of RT_c, or about $10^5 \text{ft-lb}_f/\text{lb}_m$ for water. Thermohydraulic explosions can do a lot of damage.

6. SUMMERY

1.) There is a real need to be able to predict spinodal lines, homogeneous nucleation, and properties of superheated liquids.

2.) These issues present an inherent inter-relation of purposes :

- A knowledge of metastable and unstable p-V-T behaviour facilitates the extrapolation of single measurements of σ over the full range of temperatures. By the same token, a knowledge of $\sigma(T)$ provides an important check on any equation of state.
- The ability to predict homogeneous nucleation and the ability to predict the liquid spinodal are equivalent.
- Thus knowledge of the p-V-T surface includes knowledge of the liquid spinodal and the limit of homogeneous nucleation as well.

3.) Equation (16) predicts homogeneous nucleation in a variety of practical situations :

- It predicts the absolute homogeneous nucleation limit directly.
- By replacing the left side with the -inj appropriate to a real situation, one can use it to handle real situations.
- If v_f, v_g, and σ are not available for equation (16), correlations (18) and (19) can be used with good accuracy.

4.) The vapor spinodal is unrelated to the limit of homogeneous nucleation of liquid droplets.

5.) Strong evidence suggests that homogeneous nucleation theories should be based on the characteristic energy, kT_c, instead of kT.

6.) Metastable p-V-T data can be accurately predicted by fitting a general cubic equation to the known saturation points, a known saturated liquid compressibility, a third stable equalibrium point ; and the Gibbs-Maxwell condition.

7.) When more than just p-v-T data are needed, Karimi's method can be used to develop a fundamental equation, but it should be altered to use a more realistic reference function.

8.) Karimi's fundamental equation provides the best information for metastable water, to date.

9.) Equation (30) predicts the pressure undershoot in the rapid depressurization of hot water. It reflects the fact that sufficiently rapid depressurizations reach a true (two-dimensional) homogeneous nucleation limit.

NOMENCLATURE

a = availability function, or

a, b = van der Waal constants (eqn.(3)). (also used in other equations as undetermined constants.)

B,C = second and third virial coefficients, respectively

c, d = undetermined constants (which could be complex conjugates)

C_p, c_p^o = specific heat at constant pressure, low pressure c_p

g = specific Gibbs function, or function of ω defined in Fig.3

Gb, Gb_{max} = Gibbs number (see eqn.(10)); smallest attainable Gb

h = specific enthalpy

j, J = nucleation probability (see eqn. (11)); j expressed as a rate per unit volume

k = Boltzmann's constant

m, n = undetermined constants

p = pressure

Q = residual correction function for a fundamental equation

R = ideal gas constant

R_o = critical radius of nucleation (see eqn. (9))

s = specific entropy

T = temperature

u = specific internal thermal energy

v = specific volume, $1/\rho$

v_f,v_g = saturated liquid and vapor volume. (at the saturated spinodal condition in equation (16))

v_m = v at the root of a cubic p-V-T equation between v_f, and v_g

Wk_{crit} = potential barrier to nucleation (see eqn.(8))

Z, Z_c = compressibility, pv/RT; critical compressibility, $p_c v_c / RT_c$

α_R = the Riedel factor

ΔT = $T_{h.n.} - T_{sat}$

η = Gb/ϕ

λ = nondimensionalizing factor, defined in [4].

ρ = fluid density, $1/v$

σ, σ_0 = surface tension, an undetermined reference value of σ

ϕ = ratio of actual to ideal potential barrier to nucleation

ψ = Helmholtz function

ω = the Pitzer factor, $-1-\log_{10}[p_{r,sat}(T_r=0.7)]$

General Superscripts and Subscripts

= denotes a dummy variable of integration

c = a property at the critical point

f, g = saturated liquid or vapor properties

fsp,gsp = identify liquid and vapor spinodals in Fig. 4

h.n. = a homogeneous nucleation limit

i = an initial state

r = a "reduced" property (e.g. $X_r = X/X_c$)

ref = a reference state

sat = a property at a saturation condition

sp = a property at a spinodal point

sup = a property at some surpeheated liquid state

REFERENCES

1. Riedel, L., "Eine Neue Universelle Dampfdruckformel", Chemie-Ing. Technik, 26 (1954) 83.
2. Pitzer, K.S., Lippman, D.Z., Curl, R.F., Huggins, C.M., Peterson, D.E., J. Am. Chem. Soc., 77(1955) 3433.
3. Dong, W.G., and Lienhard, J.H., "Corresponding States Correlation of Saturated and Metastable Properties," Can. J. Chem. Engr. (in press.)
4. Sharan, A., Kaul, R., and Lienhard J.H., "A Corresponding States Correlation for the Peak Pool Boiling Heat Transfer, No.2, 107 (1985) 392-397.
5. Lienhard, J.H., "Relation Between van der Waals' Fluid and Real Substances," Ir.J.Sci. and Tech., 5, (1976) 111-116.
6. Peck, R.E., "The Assimilation of van der Waals' Equation in the Correspondence States Family," Can.J.Chem.Engr., 60 (1982) 446-449.
7. Lee, B.I., and Kesler, M.G., "A Generalized Thermodynamic Correlation Based on Three-Parameter Corresponding States," AiChE Jour.No.3, 21 (1975) 510-527.
8. Skripov, V.P., Metastable Liquids, (1970). English transl. John Wiley and Sons, New York, (1974).

9. Karimi, A., and Lienhard, J.H., "Homogeneous Nucleation and the Spinodal Line," J.Heat Transfer, No.3, 102 (1980) 457-460.

10. Skripov, V.P., Sinitsin, F.N., Pavlov, P.A., Ermakov G.V., Muratov, G.V., Bulanov, N.V., and Baidakov, V.G., Thermophysical Properties of Liquids in the Metastable State, Atomizdat, USSR, (1980).

11. Avedisian, C.T., "The Homogeneous Nucleation Limits of Liquids," J.Phys. Chem. Ref. Data, (in press).

12. Lienhard, J.H., "Correlation for the Limiting Liquid Superheat," Chem. Engr. Sci., 31 (1976) 847-849.

13. Lienhard, J.H., "Corresponding States Correlations of the Spinodal and Homogenous Nucleation Limits," J.Heat Transfer, No.2, 104 (1982) 379-381.

14. Keenan, J.H., Keyes, Hill P.G., and Mooree, J.G., Steam Tables, John Wiley and Sons, New York, (1969).

15. Steam Tables, 4th ed., American Society of Mechanical Engineers, New York, (1979).

16. Van der Waals, J.D., "Thermodynamische Theorie der Kapillaritat under Voraussetzung stetiger Dichteanderung," Zeit. Physc. Chemie. 13 (1984) 657-725.

17. Karimi, A., and Lienhard J.H., "Toward a Fundamental Equation for Water in the Metastable States," High Temperatures - High Pressures, 11 (1979) 511-517.

18. Karimi, A., and Lienhard, J.H., "A Fundamental Equation Representing water in the Stable, Metastable, and Unstable States," EPRI Report NP-3328, Dec.1983.

19. Haar, L., Gallagher, J.S., and Kell, G.S., "Thermodynamic Properies for Fluid Water," Water and Steam, (J.Straub and K.Scheffler, eds.) Pergamon Press, (1980) 69-82.

20. Murali, C.S., Improved Cublic Equations of State for Polar and Non-polar Fluids," MSME thesis, Dept. of Mech. Engr., Univ. of Houston, (1983).

21. Lienhard, J.H., Shamsundar,N., and Biney P.O., "Spinodal Lines and Equations of State," -- A Review, Nuclear Sci. and Engr., (in press).

22. Redlich, O., "On the Three-Parameter Representation of Equation of State," I.E.C. Fundtls, 14 (1975) 257.

23. Fuller, G.G., "A Modified Redlich-Kwong-Soave Equation Capable of Representing the Liquid State," I.E.C. Fundamentals, 15 (1976) 254.

24. Bin , P.O., Dong, W.G., and Lienhard J.H., "Use of Cublic Equation to Predict Surface Tension and Spinodal Limits," AiChE/ASME Heat Transfer Conference, Denver, Aug.4-7 (1985).

25. Brock J.R., and Bird R.B., "Surface Tension and the Principle of Corresponding States," AiChE Jour., Vol.1 (1955) 3433-3440.

26. Lienhard, J.H., Md.Alamgir, and Trela, M., "Early Response of Hot Water to Sudden Release from High Pressure," J. Heat Transfer, No.3, 100 (1978) 473-479.

27. Alamgir, Md., and Lienhard J.H., "Correlation of Pressure Undershoot During Hot Water Depressurization," J. Heat Transfer, No.1, 103 (1981) 52-55.

28. Alamgir, Md., Kan, C.Y., Lienhard, J.H., "An Experimental Study of the Rapid Depressurization of Hot Water," J. Heat Transfer, No.3, 102 (1980) 433-438.

29. Lienhard, J.H., "Some Generalizations on the Stability of Liquid-Gas-Vapor Systems," Int. J. Heat Mass Transfer, 7 (1964) 813.

30. Shamsundar, N., and Lienhard, J.H., "Properties of the Saturated and Metastable van der Waals Fluid," Can. J. Chem. Engr., 61 (1983) 876-880.

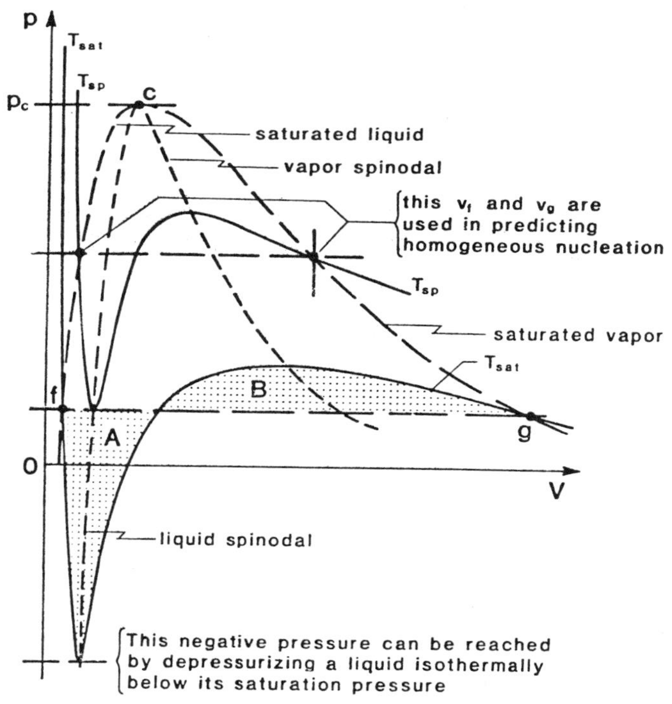

Fig.1 Typical real-gas isotherms

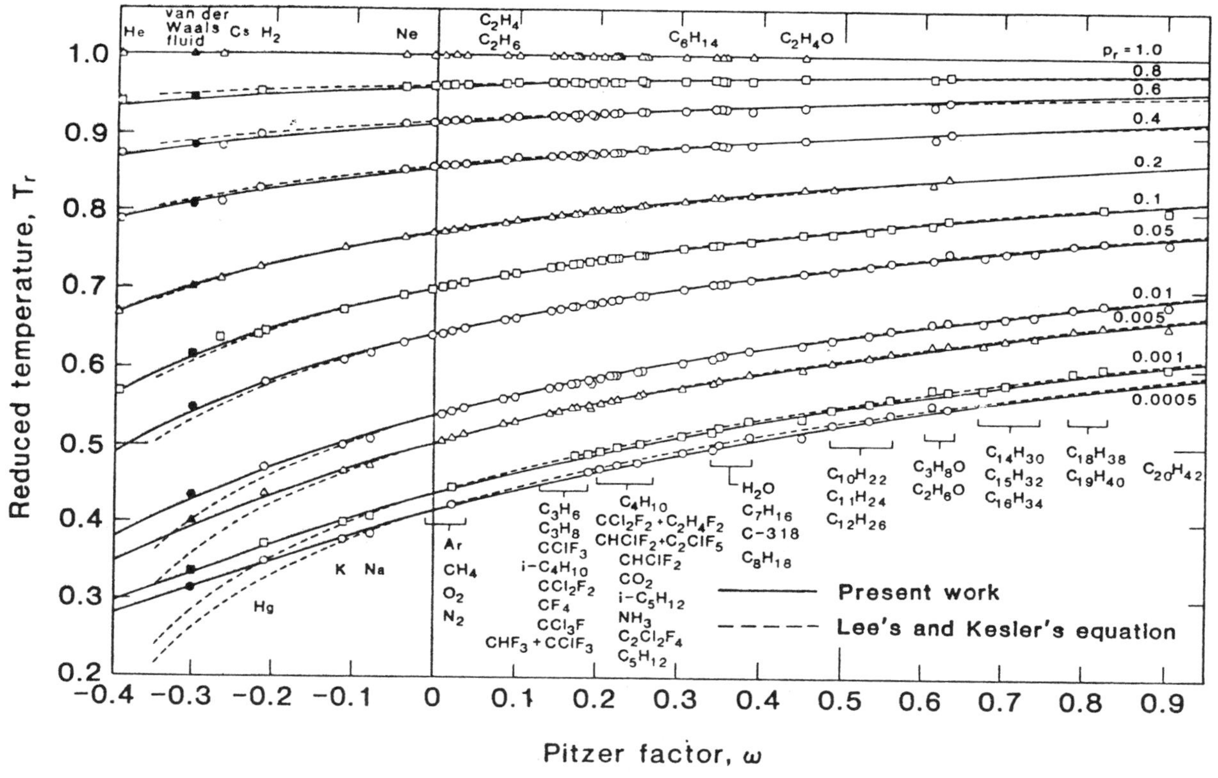

Fig.2 The use of the Pitzer factor to complete a Corresponding States correlation of the saturation temperature of fluids (from [3]).

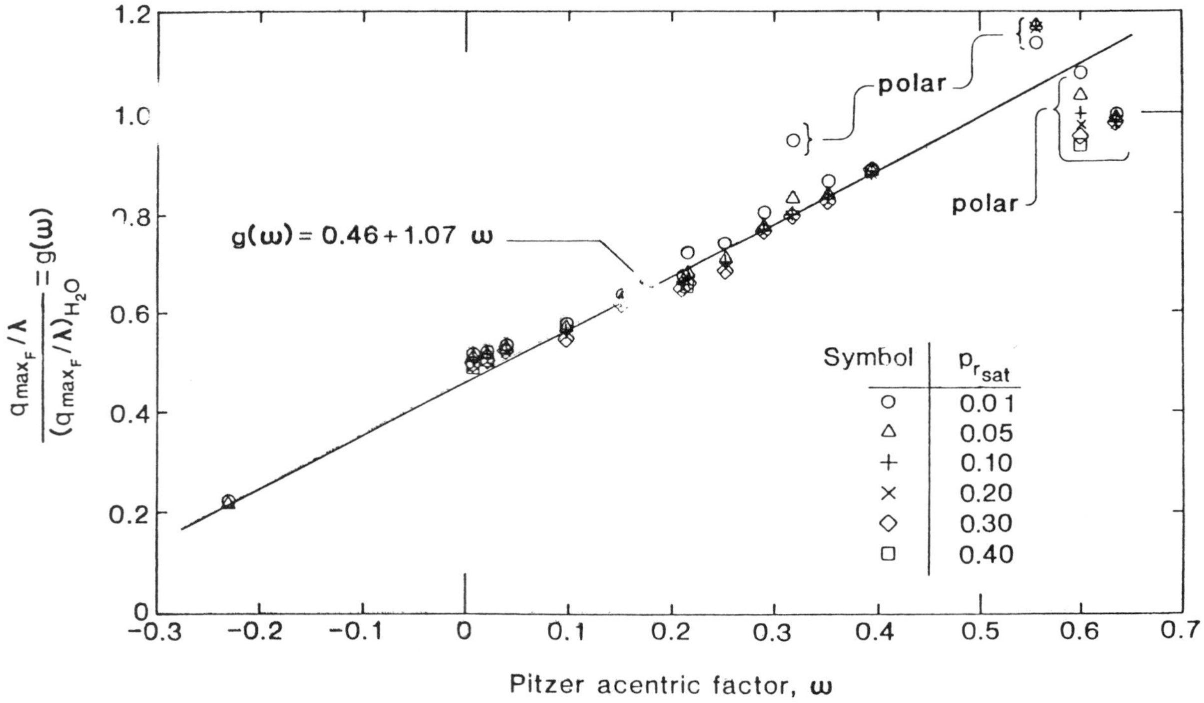

Fig.3 Illustration of the failure of polar fluids to conform to a Corresponding states correlation (from [4]).

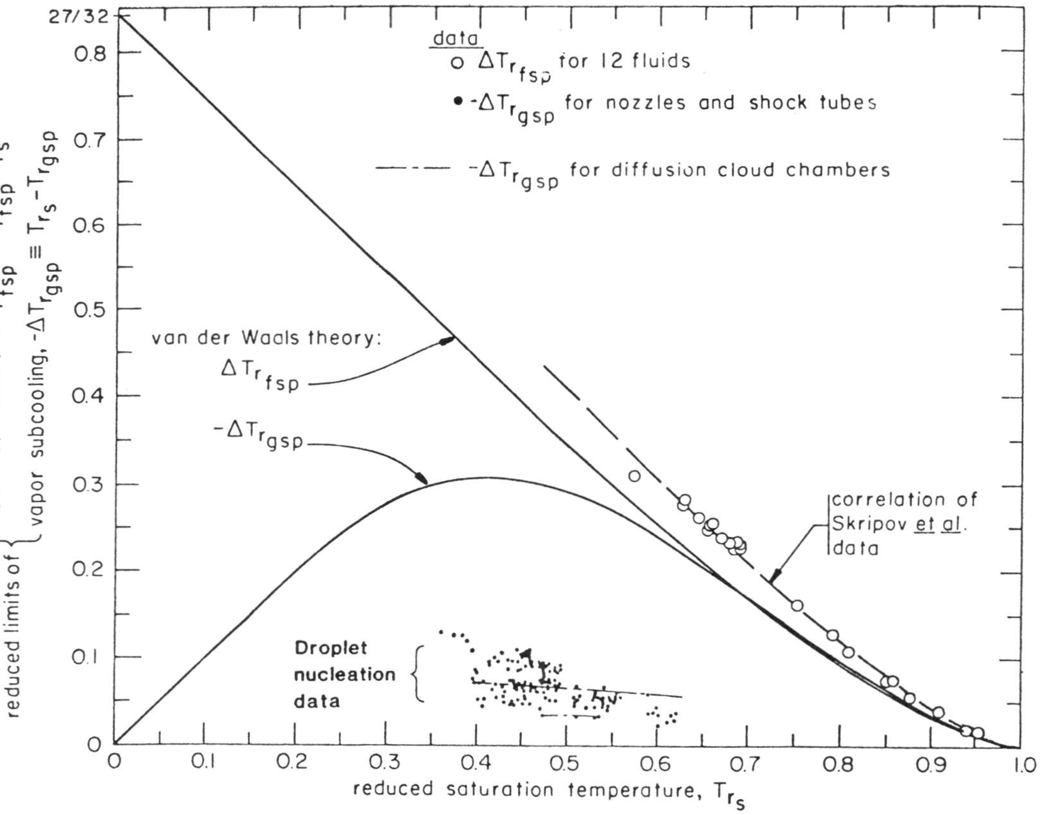

Fig.4 Corresponding States correlations of the limiting liquid superheat and vapor subcooling.

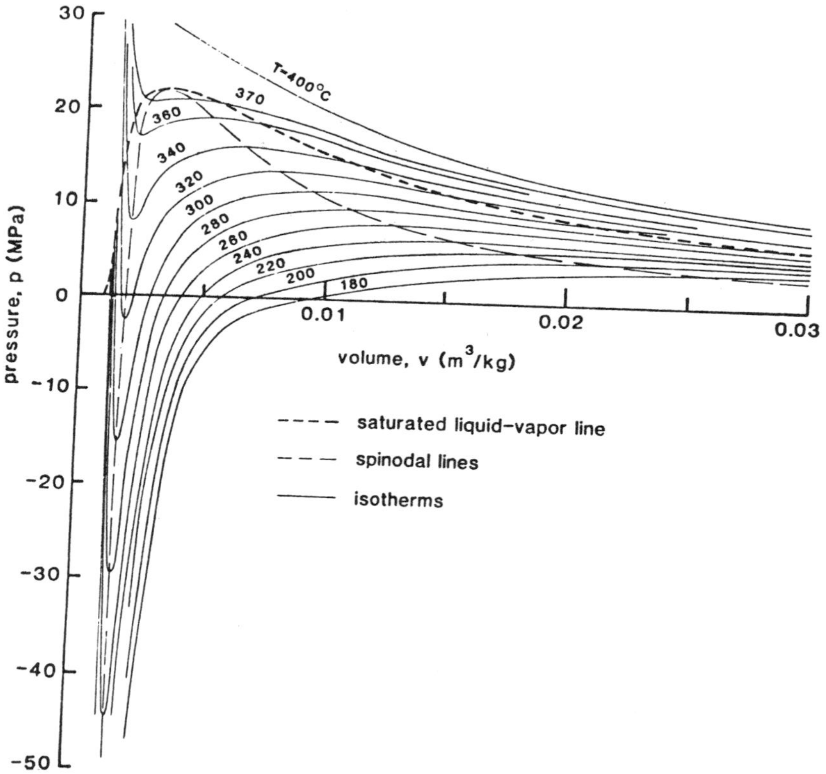

Fig.5 The p-v-T surface for water based on Karimi's [18] fundamental equation.

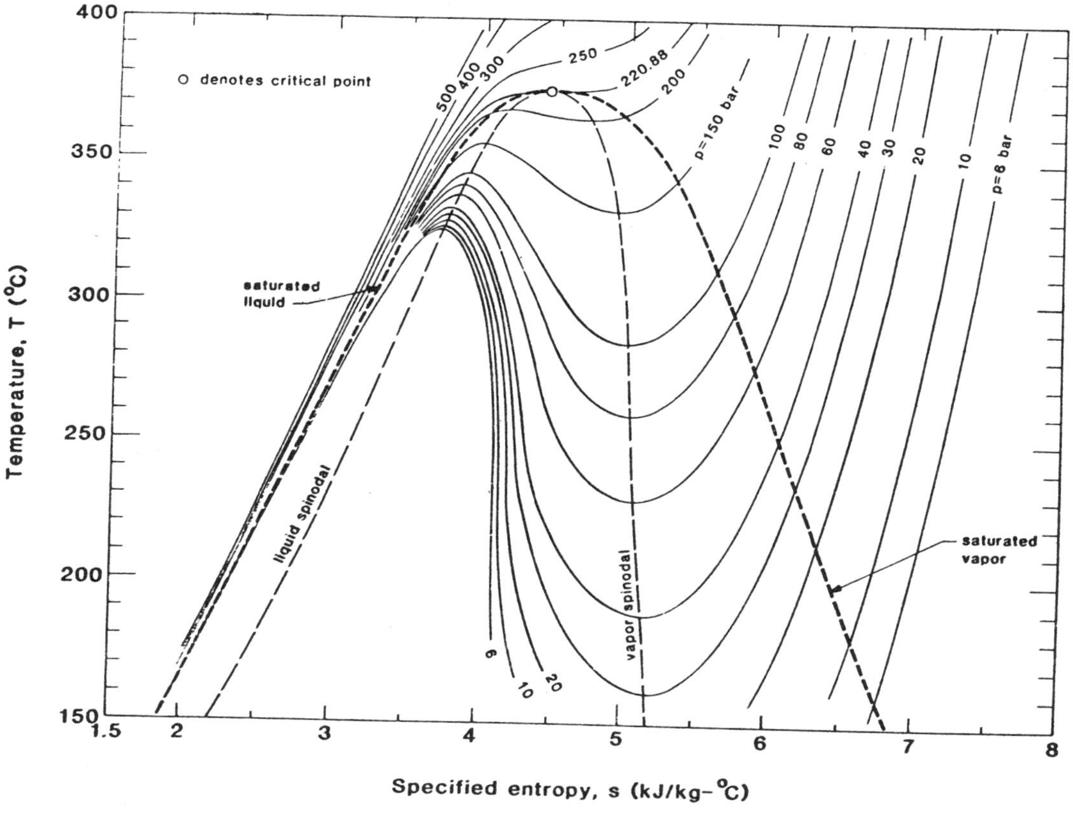

Fig.6 The T-s-p surface for water based on Karimi's [18] fundamental equation.

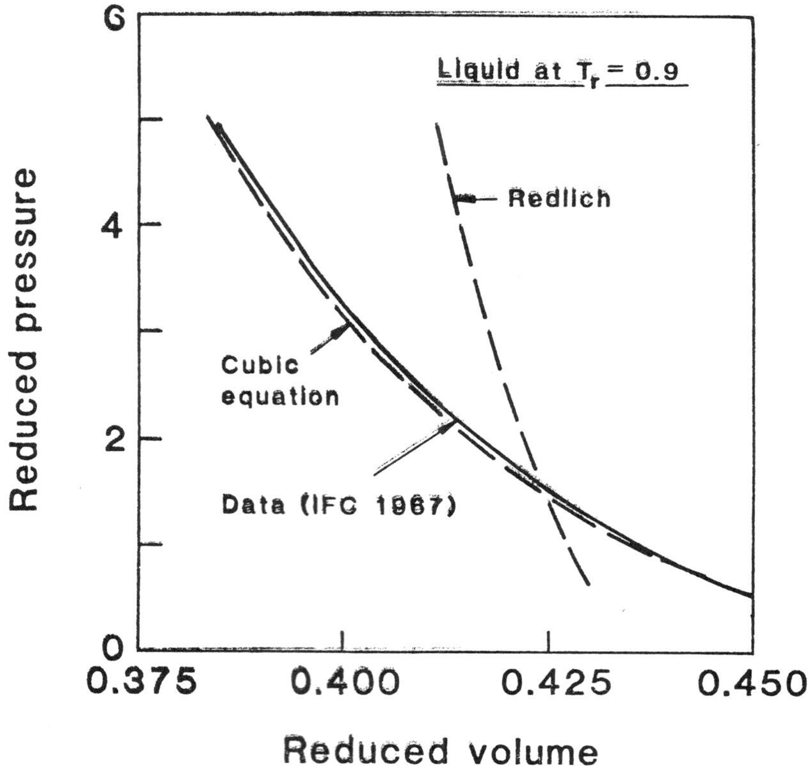

Fig.7 Comparison of cubic equation, equation (24), with IFC data [15] for liquid at high pressures.

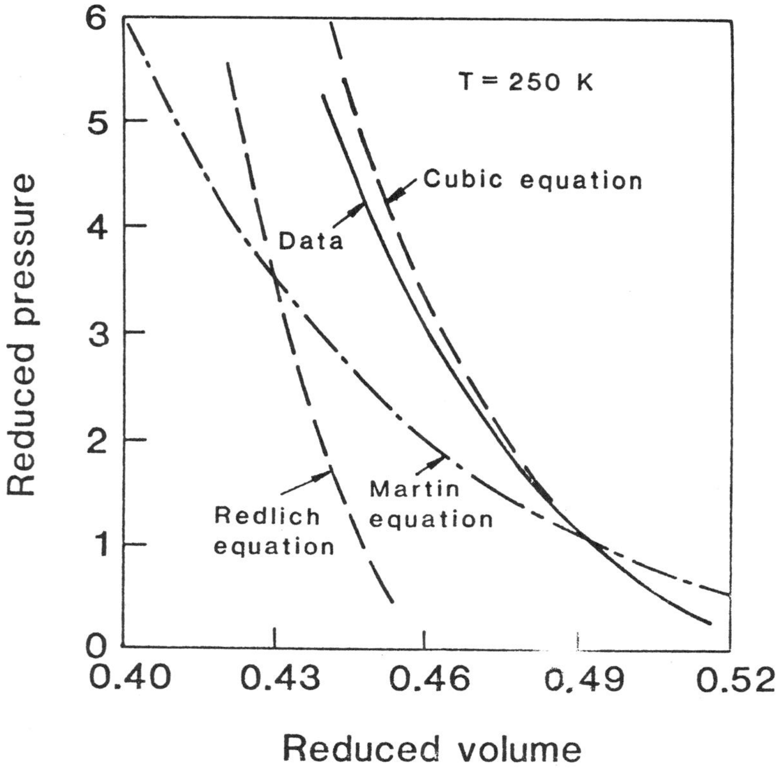

Fig.8 Comparison of cubic equation, equation (24), to data for liquid ethylene at high pressures.

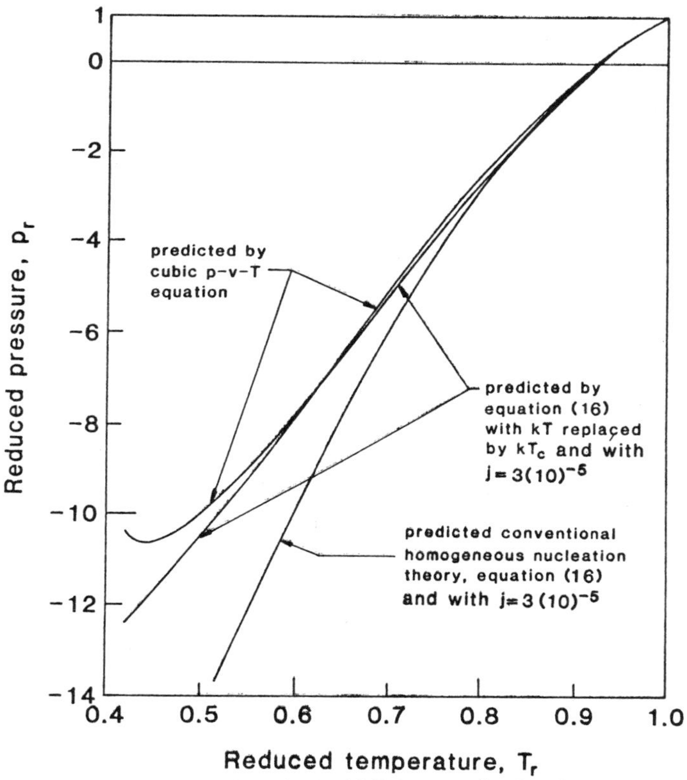

Fig.9 Comparison of spinodal limit as predicted by homogeneous nucleation theory and by the general cubic equation

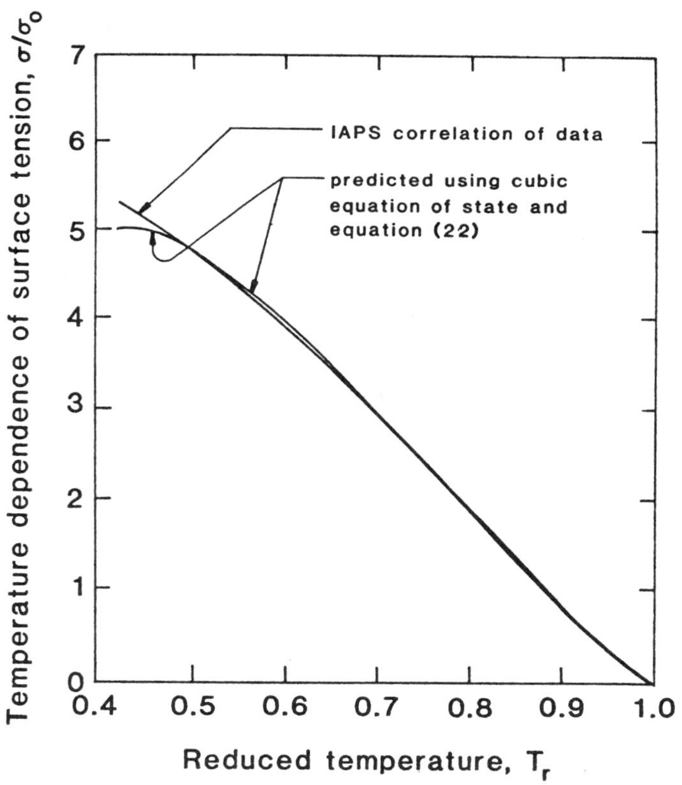

Fig.10 Predicted and measured temperature dependence of surface tension

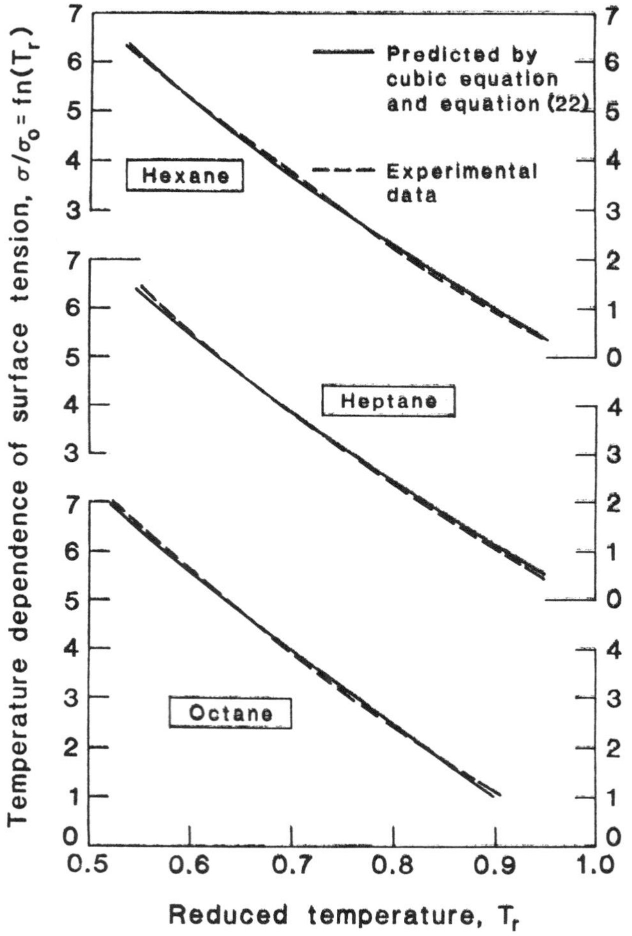

Fig.11 Predicted and measured temperature dependence of surface tension of several organic fluids.

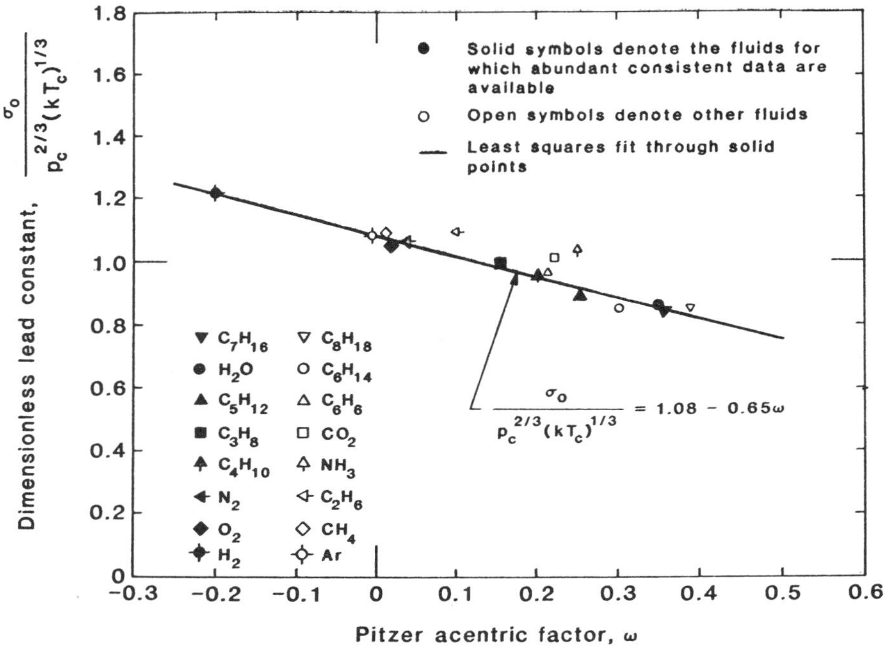

Fig.12 Correlation of surface tension lead constants

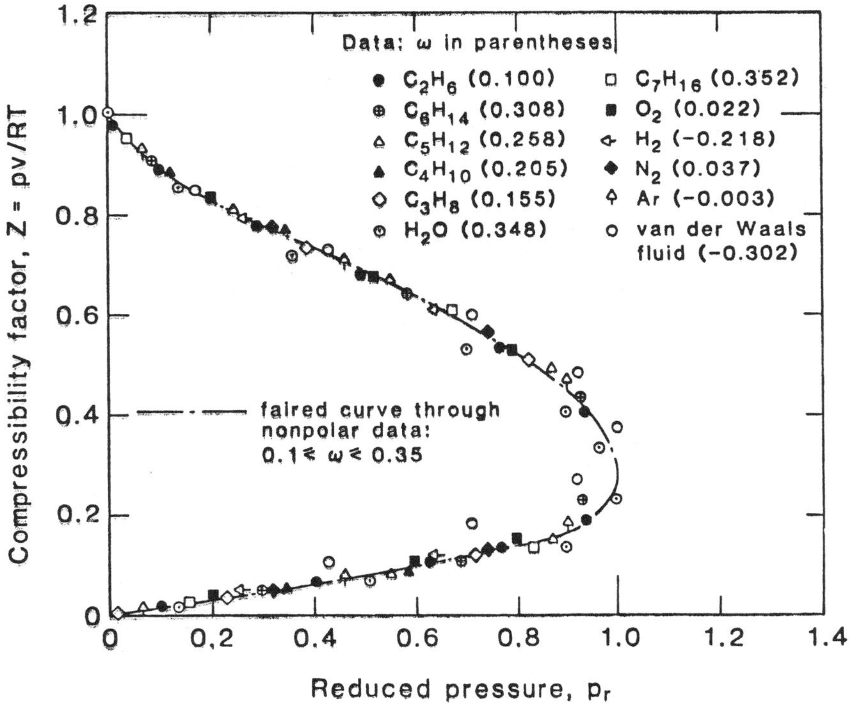

Fig.13 Z-chart correlations of the vapor-pressure data.

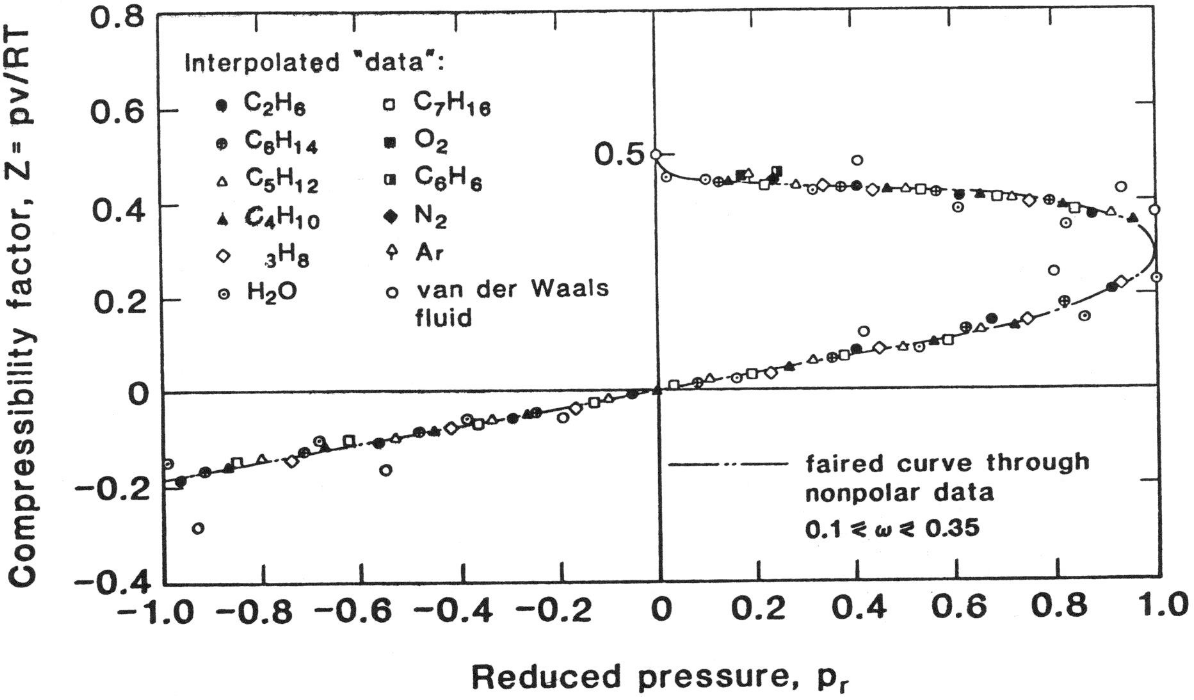

Fig.14 Z-chart correlation of the vapor and liquid spinodal lines for various substances.

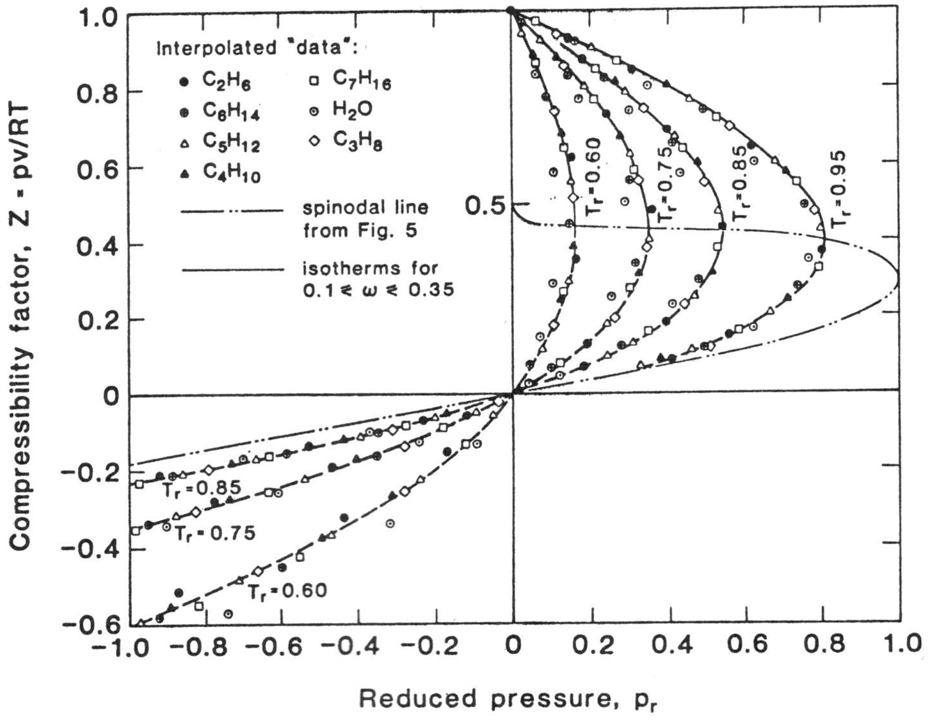

Fig.15 Z-chart correlation of metastable/unstable isotherms for various substances.

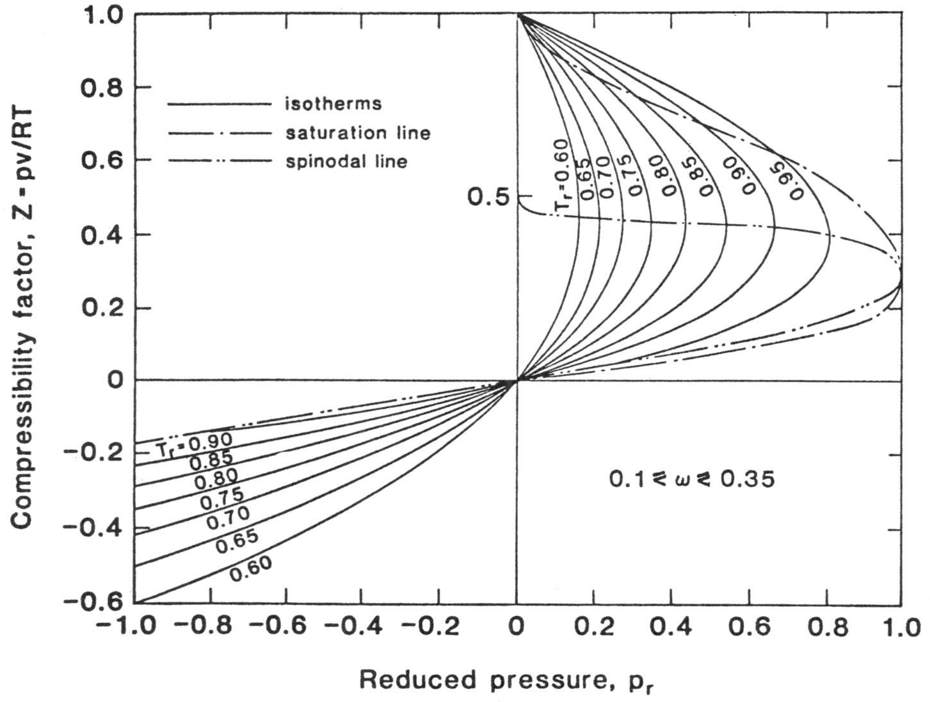

Fig.16 The Z-chart for metastable, unstable, and saturated fluids. Accurate for non-polar fluids in the range $0.10 < \omega < 0.35$.

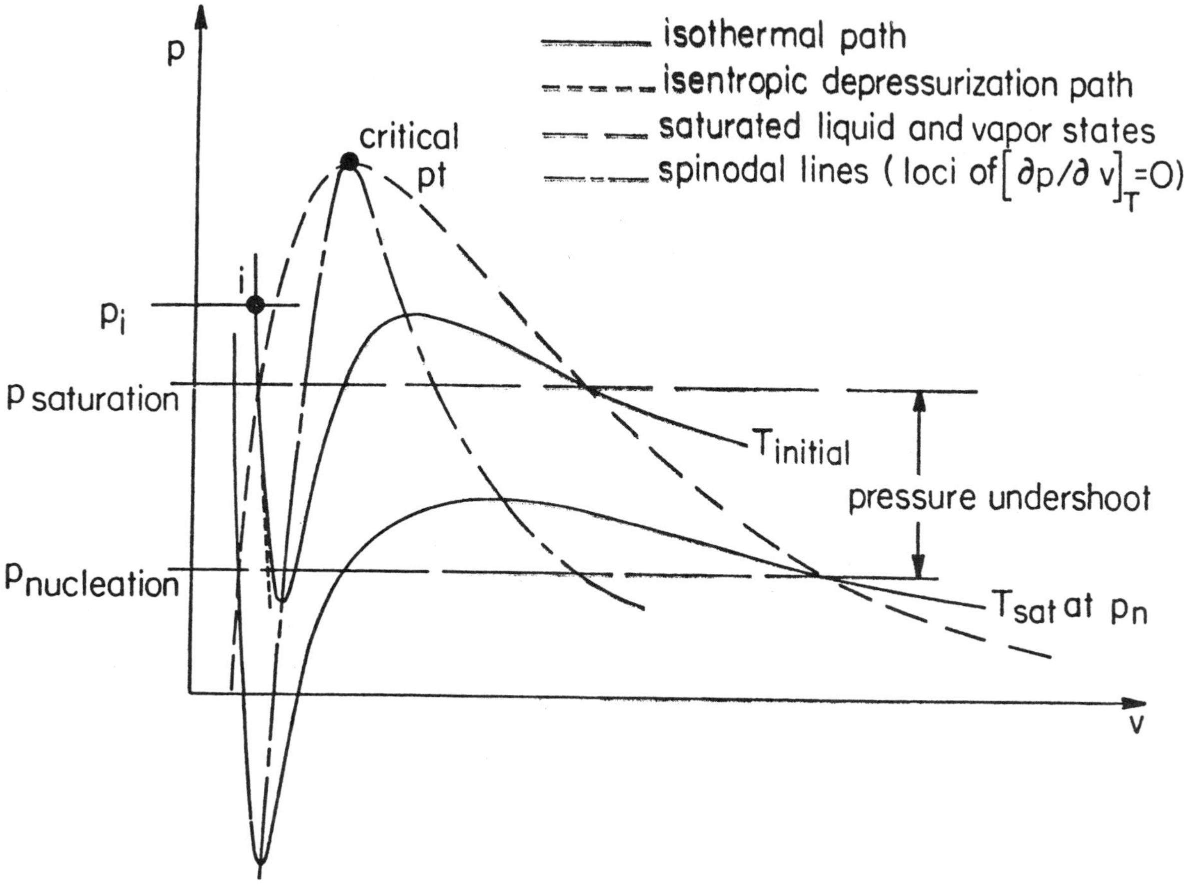

Fig.17 Depressurization of a hot liquid.

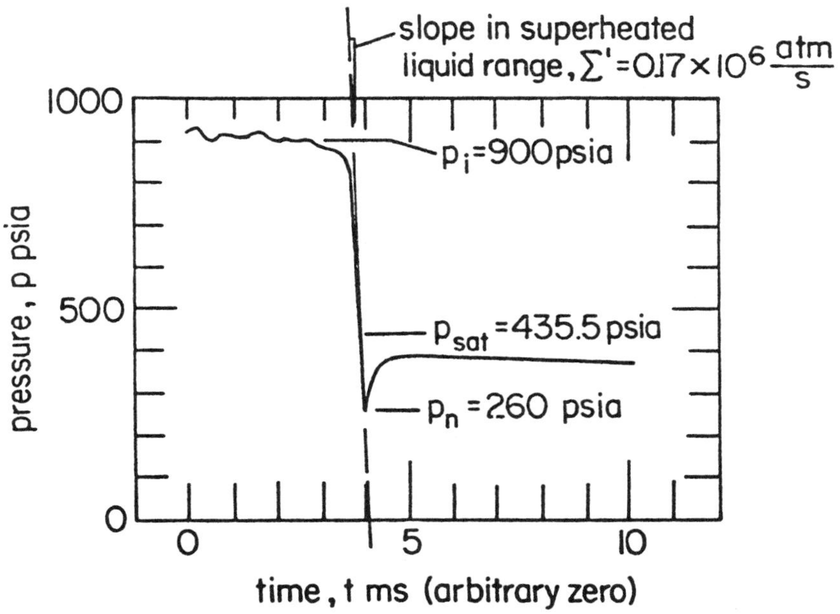

Fig.18 A typical pressure-time history during depressurization in 453°F (233.9°C) liquid.

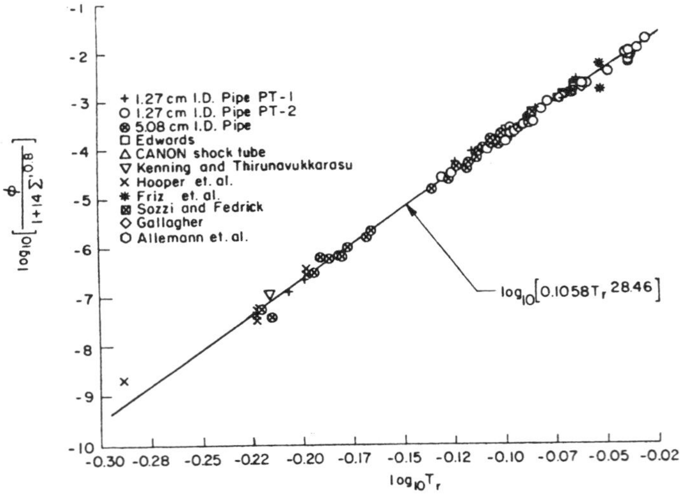

a) The function, $f_1(T_r) = \phi / f_2(\Sigma')$

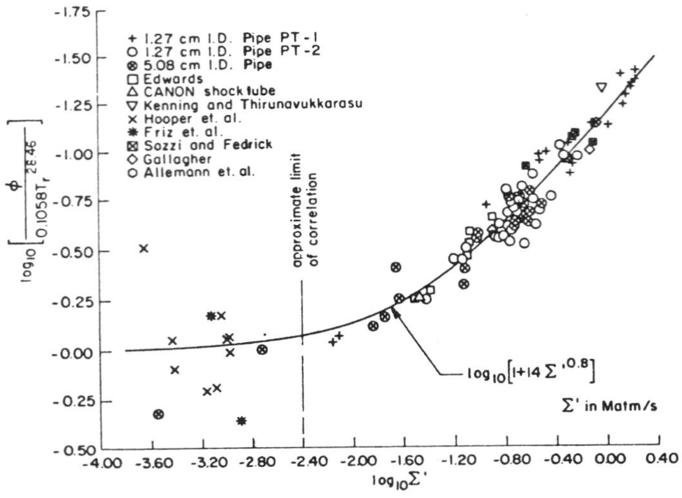

b) The function, $f_2(\Sigma') = \phi / f_1(T_r)$

Fig.19 Correlation of ϕ as a function of liquid temperature and the rate of depressurization.

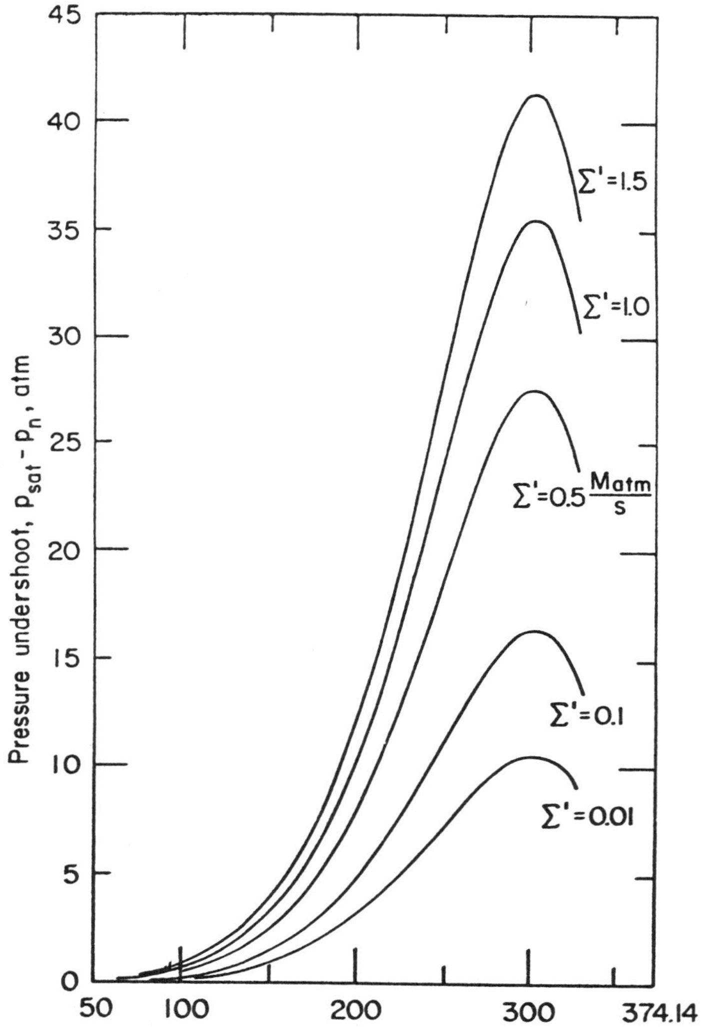

Fig.20 Prediction of pressure undershoot applied to water

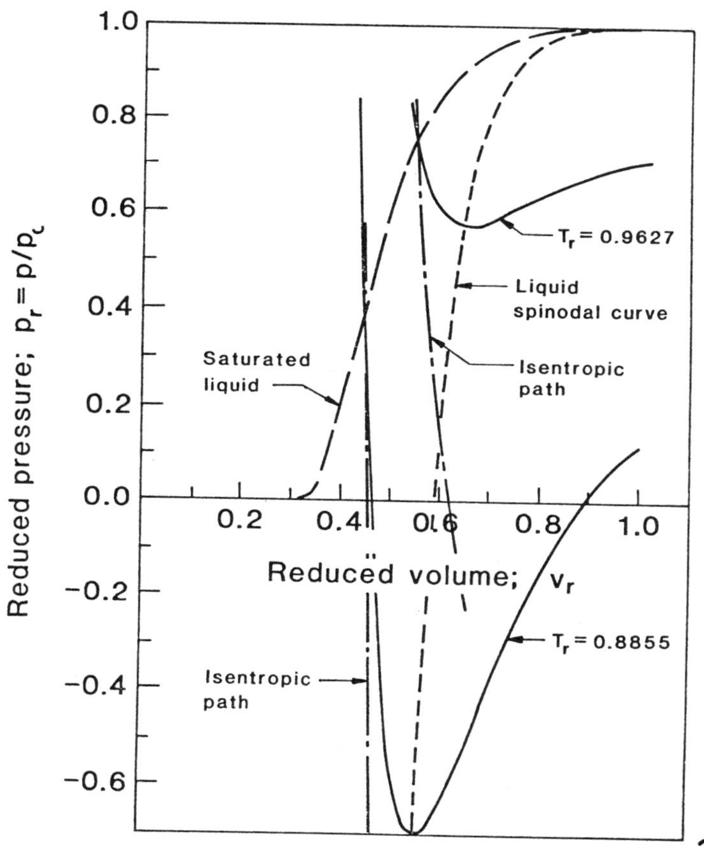

Fig.21 Comparison between isentropic and isothermal depressurizations in water

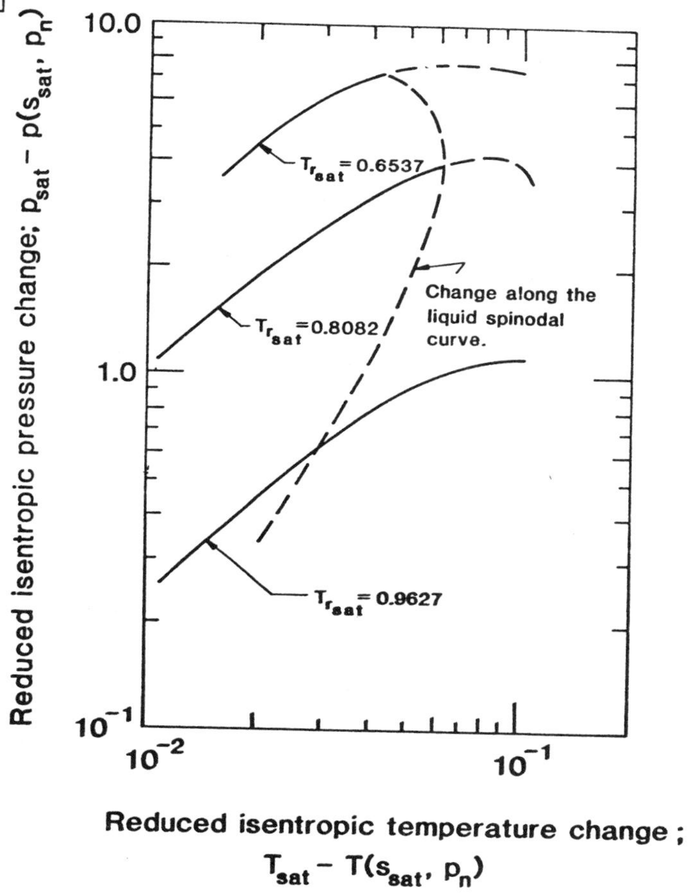

Fig.22 The relation between pressure and temperature during three isentropic processes in water

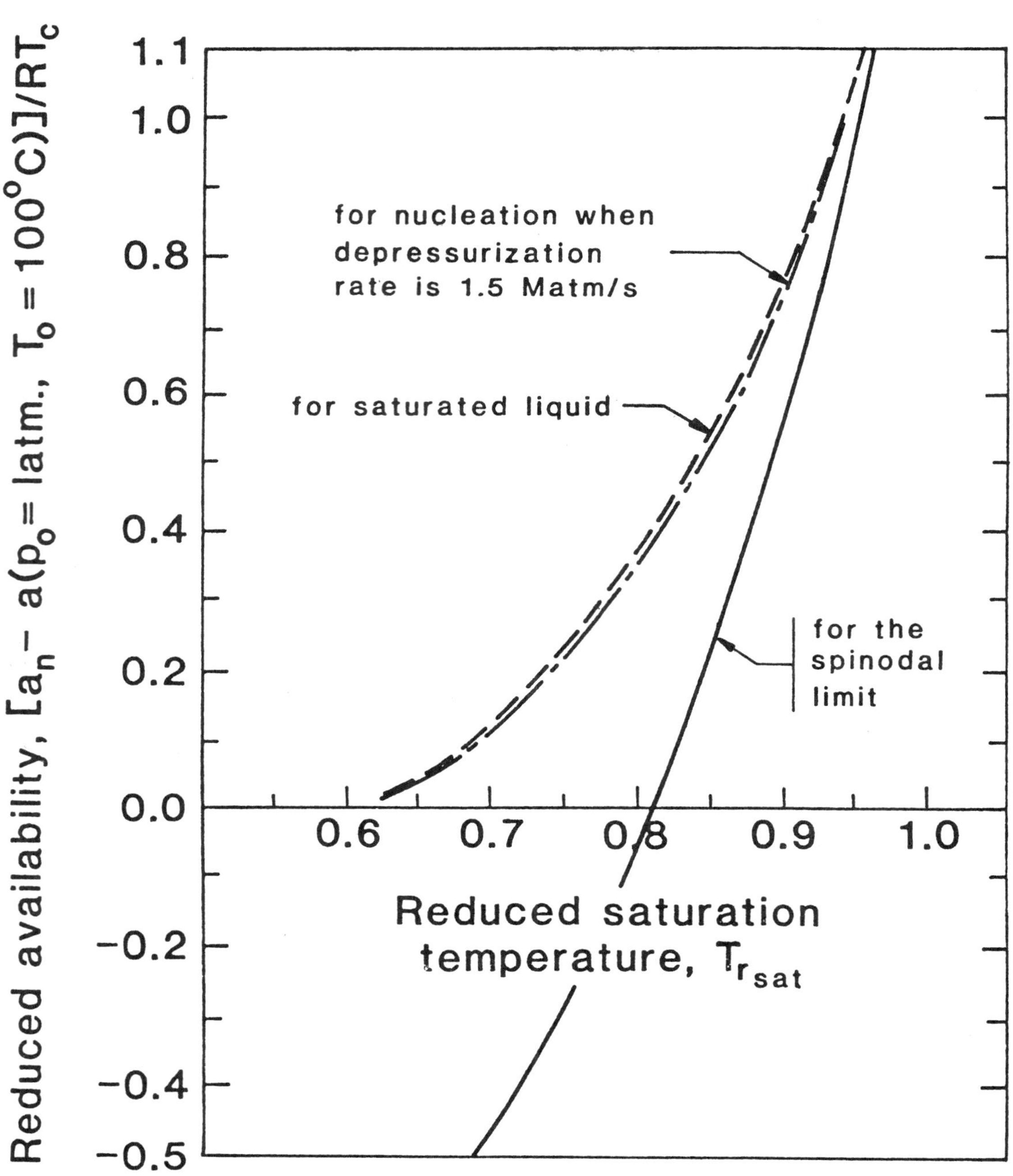

Fig.23 Availability of water, when it nucleates, in comparison with saturated water at one atmosphere. The independent variable is the reduced temperature at which the liquid becomes saturated during depressurization.

A Swirling Vertical Turbulent Buoyant Jet

O. G. MARTYNENKO, V. N. KOROVKIN, and YU. A. SOKOVISHIN
The Luikov Heat and Mass Transfer Institute
Byelorussian Academy of Sciences
Minsk, 220728, USSR

The results of the study of averaged characteristics of turbulent weakly swirling jets issuing vertically into a uniform stationary medium are presented. The semi-empirical turbulence model is used to close the set of initial equations of a turbulent boundary layer. Exact and approximate analytical solutions are constructed and the results of calculation are compared with the experimental and numerical data of other authors.

1. INTRODUCTION

The solution of the problem on viscous liquid motion in laminar and turbulent jets with the density (temperature) different from that in the surrounding medium has a wide spectrum of technical applications. This explains the appearance of a large number of publications (see, for example, the bibliography to [1 - 3]) dealing with the study of jet flows involving buoyancy effects. However, even for the case of laminar motion, only few problems are known, for which exact and approximate analytical solutions to the boundary-layer equations [4-11] have been obtained. The reason is that the problems considered are mathematically rather complex. The difficulties arise, first of all, due to the fact that the basic initial equations are non-linear (and nonclosed for turbulent motion) and that they are interrelated through the volumetric force. The search for the solution of such problems has been the object of a large number of works whose detailed survey can be found elsewhere [1,3].

The present work deals with the theoretical study of the development of turbulent weakly swirling vertical buoyant jets. It should be noted that this type of jet flows has been yet very little investigated [12-15].

2. BASIC EQUATIONS

In the turbulent boundary layer theory approximation, the averaged equations for the axysymmetric vertical motion with swirling in the gravity field have the form [12]:

$$u\frac{\partial u}{\partial x} + v\frac{\partial u}{\partial y} = -\frac{1}{\rho}\frac{\partial \Delta p}{\partial x} + \frac{1}{y}\frac{\partial}{\partial y}\left(\nu_t y\frac{\partial u}{\partial y}\right) + g\beta_s \Delta T;$$

$$\frac{\partial}{\partial x}(yu) + \frac{\partial}{\partial y}(yv) = 0, \quad \frac{w^2}{y} = \frac{1}{\rho}\frac{\partial \Delta p}{\partial y}; \qquad (1)$$

$$u\frac{\partial w}{\partial x} + v\frac{\partial w}{\partial y} + \frac{vw}{y} = \frac{1}{y^2}\frac{\partial}{\partial y}\left[\nu_t y^2\left(\frac{\partial w}{\partial y} - \frac{w}{y}\right)\right];$$

$$u\frac{\partial \Delta T}{\partial x} + v\frac{\partial \Delta T}{\partial y} = \frac{1}{\sigma_t}\frac{1}{y}\frac{\partial}{\partial y}\left(\nu_t y\frac{\partial \Delta T}{\partial y}\right);$$

boundary conditions:

$$y = 0: \ v = \frac{\partial u}{\partial y} = \frac{\partial \Delta T}{\partial y} = w = 0;$$

$$y \to \infty: \ u \to 0, \ \Delta T \to 0, \ \Delta p \to 0, \ w \to 0. \qquad (2)$$

To couple the correlational moment with the averaged flow parameters, the Boussinesq hypothesis is used [16];

$$\langle u^1 v^1 \rangle = -\nu_t\frac{\partial u}{\partial y}, \quad \langle v^1 w^1 \rangle = -\nu_t\left(\frac{\partial w}{\partial y} - \frac{w}{y}\right),$$

$$\langle v^1 T^1 \rangle = -\frac{\nu_t}{\sigma_t}\frac{\partial T}{\partial y}.$$

In the adopted system of coordinates (Fig.1), the axis x is directed vertically upward. The conditions for the existence of the nontrivial solution to the problem can be obtained by prescribing the flux of the excess heat content Q_o and moment of momentum L_o

$$2\pi\rho C_p\int_0^\infty u\,\Delta T y\,dy = Q_o, \quad 2\pi\rho\int_0^\infty uwy^2\,dy = L_o. \qquad (3)$$

It should be noted that in contrast to a swirling jet, where [17]

$$2\pi\int_0^\infty (\rho u^2 + \Delta p)y\,dy = K_o \qquad (4)$$

for the case of development of a swirling buoyant jet

$$\frac{d}{dx}\int_0^\infty \left(u^2 + \frac{\Delta p}{\rho}\right)y\,dy = g\beta_s\int_0^\infty \Delta T y\,dy \qquad (5)$$

This is explained by the continuous influence of buoyancy forces on the flow. However, the set of equations (1)-(3) is not closed and, naturally, cannot be solved without any additional assumptions.

The majority of the earlier theoretical models of vertical buoyant submerged swirling jets [12,13,15] are based on the similarity properties of transverse velocity, temperature, etc. profiles within the jet over all heights and on the use of the basic integral relations being derived by investigating the basic initial equations (1)-(2) across the jet. Moreover, the coefficient of suction [1] which characterizes the change in the volumetric flow rate along the jet, or the hypothesis for determination of turbulent viscosity ν_t [18] are employed. This work gives the preference to the latter version with the following relation being used for the turbulent viscosity:

$$\nu_t = \frac{4\chi_o \psi x}{\alpha y^2} \quad (6)$$

suggested in [19,20] and based on the second Prandtl hypothesis

$$\nu_t = \sqrt{\chi}\ b\, u_m . \quad (7)$$

Expression (6) results from (7) after the substitution of u_m by u^* which is the flow cross section-mean value of the vertical velocity

$$u^* \sim \frac{1}{y^2} \int_0^y uy dy$$

Note, that according to (7), ν_t remains constant in this cross-section of the flow (varying only from section to section), whereas according to (6), ν_t decreases smoothly from the flow axis to the external boundary. The latter lowers the discontinuity of the coefficient ν_t at the boundary between the ascending flow and the surrounding medium, which is in the state of rest.

3. FORCED CONVECTION

First, let us investigate the regime of purely forced convection ($\beta_s = 0$). In this case the governing equations (1), boundary (2) and integral (3),(4) conditions admit the similarity solution of the form

$$\psi(x,\eta) = \chi_o f_o(\eta)\, x, \quad \Delta T(x,\eta) = h_o(\eta) x^{-1},$$
$$w(x,\eta) = b_o(\eta) x^{-2}, \quad \frac{\Delta p}{\rho}(x,\eta) = a_o(\eta) x^{-4}; \quad (8)$$
$$x = x, \quad \eta = \frac{y^2}{4\chi_o x^2}$$

Substitution of (8) into (1) yields

$$(f_o f_o'')' + \frac{\alpha}{2}(f_o f_o')' = 0 ;$$

$$\frac{1}{\sigma_t}(f_o h_o')' + \frac{\alpha}{2}(f_o h_o)' = 0;$$

$$(f_o b_o')' + \frac{\alpha}{2}(f_o b_o)' - \frac{1}{2}\frac{f_o' b_o}{\eta} + \frac{1}{4\eta^2} f_o b_o (1+\alpha\eta) = 0; \quad 2 a_o' \eta = b_o^2 , \quad (9)$$

where functions f_o, h_o, b_o, a_o satisfy the following boundary conditions

$$f_o(0) = \sqrt{\eta}\, f_o''(0) = \sqrt{\eta}\, h_o'(0) = b_o(0) = 0;$$
$$f_o'(\infty) = h_o(\infty) = b_o(\infty) = a_o(\infty) = 0. \quad (10)$$

Integration of (9) with account for (10) gives

$$f_o = 2(1 - e^{-\alpha\eta/2}), \quad h_o = c_1 e^{-\frac{\sigma_t}{2}\alpha\eta}$$

$$b_o = \beta\sqrt{\alpha\eta}\, e^{-\alpha\eta/2}$$

$$a_o = -\frac{\beta^2}{2} e^{-\alpha\eta}, \quad \alpha = \frac{K_o}{\pi\rho\chi_o}, \quad c_1 = \frac{(1+\sigma_t)Q_o}{4\pi\rho\chi_o C_p}$$

$$= \frac{L_o\sqrt{\alpha}}{4\pi\rho\chi_o\sqrt{\chi_o}} \quad (11)$$

with the integration constants α, c_1, β being determined from (3), (4). Thus, in the framework of the model [19,20]

$$\nu_t = \chi_o \frac{2(1 - e^{-\alpha\eta/2})}{\alpha\eta} \quad (12)$$

we obtain

$$u = \frac{K_o}{2\pi\rho\chi_o} e^{-\alpha\eta/2} x^{-1}, \quad \Delta T = \frac{(1+\sigma_t)Q_o}{4\pi\rho\chi_o C_p} \cdot e^{-\frac{\sigma_t}{2}\alpha\eta} x^{-1}$$

$$w = \frac{L_o}{4\pi\rho\chi_o\sqrt{\chi_o}} \left(\frac{K_o}{\pi\rho\chi_o}\right)^{1/2} \sqrt{\alpha\eta}\, e^{-\alpha\eta/2} x^{-2} \quad (13)$$

$$\frac{\Delta p}{\rho} = -\frac{L_o^2 K_o}{32\pi^3\rho^3\chi_o^4} e^{-\alpha\eta} x^{-4}.$$

If we refer to the hypothesis (7), then, owing to the fact that $b \sim x$, $u_m \sim x^{-1}$, we have $\nu_t = \chi = \text{const.}$ Consequently, the expressions for the turbulent flow shear stresses will not formally differ from appropriate expressions for a laminar flow. Therefore, all the results obtained for the laminar swirling jet [17] may be regarded valid for the turbulent flow if the values of velocity, temperature and pressure are assumed to be averaged and the coefficient ν is substituted by ν_t:

$$u = \frac{2\alpha}{(1+\alpha\eta)^2} x^{-1}, \quad \Delta T = \frac{C_1}{(1+\alpha\eta)^{2\sigma_t}} x^{-1},$$

$$w = \beta\frac{\sqrt{\alpha\eta}}{(1+\alpha\eta)^2} x^{-2}, \quad \frac{\Delta p}{\rho} = -\frac{\beta^2}{6}\frac{1}{(1+\alpha\eta)^3} x^{-4}$$

$$\alpha = \frac{3K_o}{16\pi\rho\chi}, \quad \beta = \frac{3L_o\sqrt{\alpha}}{8\pi\rho\chi\sqrt{\chi}}, \quad c_1 = \frac{(2\sigma_t+1)Q_o}{8\pi\rho\chi C_p} \quad (14)$$

Note, that there is a marked difference between (13) and (14). According to (14): $u \sim (\alpha\eta)^{-2}$, $v \sim (\alpha\eta)^{-1/2}$ $w \sim (\alpha\eta)^{-3/2}$ when $\alpha\eta \to \infty$, while (13) gives: $u, v, w \sim \exp(-\alpha\eta/2)$ when $\alpha\eta \to \infty$.

4. FREE CONVECTION

Using the transformations

$$\psi = Af_o(\eta)\, x^{5/3}, \quad \Delta T = Bh_o(\eta)\, x^{-5/3}, \quad W = Db_o(\eta) x^{-8/3},$$

$$\frac{\Delta P}{\rho} = D^2 a_o(\eta)\, x^{-16/3}; \quad \eta = \frac{y^2}{4\chi_o x^2} \; ; \tag{15}$$

$$A = \left[\frac{g\beta_s \chi_o^2 Q_o}{\pi\rho C_p}\right]^{1/3}, \quad B = \frac{1}{2}\left[\frac{Q_o}{\pi\rho C_p \chi_o (g\beta_s)^{1/2}}\right]^{2/3}$$

$$D = \frac{L_o}{4\pi\rho\sqrt{\chi_o}\; A}$$

to the system of equations (1) - (3), we obtain

$$(f_o f_o'')' + \frac{5}{6}\alpha f_o f_o'' + \frac{1}{6}\alpha f_o'^2 + \alpha h_o = 0;$$

$$\frac{1}{\sigma_t}(f_o h_o')' + \frac{5}{6}\alpha (f_o h_o)' = 0;$$

$$(f_o b_o')' + \frac{5}{6}\alpha (f_o b_o)' - \frac{1}{2\eta} f_o' b_o + \frac{5}{12\eta}\alpha f_o b_o + \frac{1}{4\eta^2} f_o b_o = 0; \tag{16}$$

$$2 a_o' \eta = b_o^2; \quad \int_0^\infty f_o' h_o\, d\eta = 1, \quad \int_0^\infty f_o' b_o \eta^{1/2} d\eta = 1.$$

The problem (16) / boundary conditions are of the form of (10) / at σ_t = 0.6 and 2 admits the exact solution [14]. Fot σ_t = 0.6

$$f_o = 2(1 - e^{-\alpha\eta/2}),\; h_o = \frac{1}{3}(\alpha)^2 e^{-\alpha\eta/2}, \quad \alpha = \sqrt{3},$$

$$b_o = \frac{16}{9}(\alpha)^{1/2}(\alpha\eta)^{1/2} e^{-5/6\,\alpha\eta},$$

$$a_o = -\frac{128}{135}\alpha e^{-5/3\,\alpha\eta} \tag{17}$$

For σ_t = 2

$$f_o = \frac{6}{5}(1 - e^{5/6\,\alpha\eta}),\; h_o = \frac{2}{3}(\alpha)^2 e^{-5/3\,\alpha\eta},\; \alpha = \sqrt{\frac{15}{2}}$$

$$b_o = \frac{5}{3}(\alpha)^{1/2}(\alpha\eta)^{1/2} e^{-5/6\,\alpha\eta},\; a_o = -\frac{5}{6}\alpha e^{-5/3\alpha\eta} \tag{18}$$

At other values of σ_t different from 0.6 and 2, the solution is reduced to numerical integration of (16). However, it is possible to find an approximate analytical solution to (16) which qualitatively accurately indicates the limits of exact numerical results [14]. Let us assume that $f_o = c_2 \exp(-C_3\alpha\eta)$, determine h_o from the second equation of (16) and substitute it into the first equation of (16). Then, after some simple transformations, we have

$$f_o = \frac{6(4+5\sigma_t)}{35\sigma_t}\left(1 - e^{\left[-\frac{35\sigma_t}{6(4+5\sigma_t)}\cdot\alpha\eta\right]}\right),$$

$$h_o = \frac{1}{9}\sigma_t \frac{32+5\sigma_t}{4+5\sigma_t}(\alpha)^2 e^{-\frac{5}{6}\sigma_t \alpha\eta}$$

$$\alpha = \left[\frac{15(11+5\sigma_t)}{2(32+5\sigma_t)}\right]^{1/2} \tag{19}$$

Using the results obtained we find the solution to the remainder equations:

$$b_o = \left(\frac{30\sigma_t + 10}{12 + 15\sigma_t}\right)^2 (\alpha)^{1/2}(\alpha\eta)^{1/2} e^{-\frac{5}{6}\alpha\eta}$$

$$a_o = -\frac{3}{10}\left(\frac{30\sigma_t + 10}{12 + 15\sigma_t}\right)^4 \alpha e^{-\frac{5}{6}\alpha\eta} \tag{20}$$

It can be easily seen that at σ_t = 0.6 and relations (19) and (20) pass over into (17) and (18), respectively. Figure 2 presents the relation between the specific layer thicknesses b_T/b_u calculated from (19):

$$\frac{b_T}{b_u} = \frac{7}{4 + 5\sigma_t} \tag{21}$$

The dots in this figure show the results of numerical solutions [14]. It is seen that the agreement between the approximate analytical solution (21) and numerical data [14] within the range of σ_t values from 0.2 to 2 is quite satisfactory. In accordance with (21), at σ_t = 2, the effective dynamic boundary layer is thicker than the thermal one i.e. $b_u > b_T$, while at σ_t= 0.6 there is similarity between the velocity and temperature profiles: $b_T = b_u$. The latter reveals the basic difference between the buoyant and forced jets, where $b_T = b_u$ at σ_t = 1. An explanation is that in forced jets heat propagates as a passive impurity, while in the buoyant ones, it induces motion.

The relation (21) allows one to find the value of the turbulent number σ_t from the measured averaged vertical velocity and temperature profiles in buoyant jets. Thus, e.g., according to [12] b_T/b_u = 1.16 for the similarity segments of a convective swirling jet above the point heat source. To this relationship between the effective thermal and dynamic boundary layers thicknesses there corresponds the value σ_t = 0.41 . The calculation by (19) gives a satisfactory coincidence with the measured [2] averaged buoyant jet parameters at σ_t =0.3 (χ_o = 1.08 $\cdot 10^{-3}$). If we refer to the experimental data of work [22]:

$$h_o/h_o(0) = \exp(-65\frac{y^2}{x^2}),\; f_o^1/f_o^1(0) = \exp(-55\frac{y^2}{x^2}),$$

then, according to (21), this is equivalent to the assignment of σ_t = 0.86 .

As to the determination of the turbulent number σ_t for buoyant jets on the basis of direct measurements of flow fluctuation characteristics, then this problem has not been as yet solved.

Proceeding the analysis, let us consider the solution of equations (1) - (3) with the hypothesis (7) taken into account. Applying again the transformations (15), where

$$A = \left[\frac{2g\beta_s \chi^2 Q_o}{\pi \rho C_p}\right]^{1/3}, \quad B = \frac{1}{4}\left[\frac{2Q_o}{\pi \rho C_p (g\beta_s)^{1/2}\chi}\right]^{2/3}$$

$$D = \frac{L_o}{4\pi\rho\sqrt{\chi}A}, \quad \eta = \frac{y^2}{4\chi x^2},$$

to equations of the system (1)-(3), we obtain

$$(\eta f_o'')' + \frac{5}{3} f_o f_o'' + \frac{1}{3} f_o'^2 + h_o = 0;$$

$$\frac{1}{\sigma_t}(\eta h_o')' + \frac{5}{3}(f_o h_o)' = 0;$$

$$(\eta b_o')' + \frac{5}{3}(f_o b_o)' - \frac{1}{4\eta} b_o\left(1 - \frac{10}{3} f_o\right) = 0; \tag{22}$$

$$2 a_o' \eta = b_o^2;$$

$$\int_0^\infty f_o' h_o \, d\eta = 1, \qquad \int_0^\infty f_o' b_o \eta^{1/2} \, d\eta = 1.$$

The boundary conditions retain the previous form (10). As was shown in [23], the first two equations of system (22) admit the exact solution at σ_t =1.1 and 2. At $\sigma_t = 1.1$:

$$f_o = \frac{18}{11}\frac{\alpha\eta}{1+\alpha\eta}, \quad h_o = \frac{288}{121}(\alpha)^2\frac{1}{(1+\alpha\eta)^3}$$

$$\alpha = \frac{11\sqrt{11}}{36} \tag{23}$$

When $\sigma_t = 2$

$$f_o = \frac{6}{5}\frac{\alpha\eta}{1+\alpha\eta}, \quad h_o = \frac{48}{25}(\alpha)^2\frac{1}{(1+\alpha\eta)^4}$$

$$\alpha = \frac{25\sqrt{2}}{24} \tag{24}$$

Now it remains to find the solution of the rest equations, whose direct integration yields:

$\sigma_t = 1.1$

$$b_o = \frac{205}{33}(\alpha)^{1/2}\frac{(\alpha\eta)^{1/2}}{(1+\alpha\eta)^{30/11}} \tag{25}$$

$$a_o = -\frac{42025}{9702}\alpha\frac{1}{(1+\alpha\eta)^{49/11}}$$

$\sigma_t = 2.0$

$$b_o = 5(\alpha)^{1/2}\frac{(\alpha\eta)^{1/2}}{(1+\alpha\eta)^2}$$

$$a_o = -\frac{25}{6}\frac{1}{(1+\alpha\eta)^3} \tag{26}$$

It should be noted that the above method allows also the construction of the approximate analytical solution to problem (22):

$$f_o = q\frac{\alpha\eta}{1+\alpha\eta}, \quad h_o = \frac{q(6-q)}{3}(\alpha)^2\frac{1}{(1+\alpha\eta)^{5/3\, q\sigma_t}},$$

$$b_o = \frac{5(5q+3)}{9}(\alpha)^{1/2}\frac{(\alpha\eta)^{1/2}}{(1+\alpha\eta)^{5/3\, q}}, \quad \alpha = \left[\frac{3(1+5/3\, q\sigma_t)}{q^2(6-q)}\right]^{1/2}$$

$$a_o = -\frac{25(5q+3)^2}{54(10q-3)}\alpha(1+\alpha\eta)^{-\frac{10-q}{3}} \tag{27}$$

$$q = \frac{36}{7 + 5\sigma_t + \sqrt{49 + 10\sigma_t + 25\sigma_t^2}}$$

From this it follows that at $\sigma_t = 0.6$

$$f_o = 2\frac{\alpha\eta}{1+\alpha\eta}, \quad h_o = \frac{8}{3}(\alpha)^2\frac{1}{(1+\alpha\eta)^2}, \quad \alpha = \frac{3}{4} \tag{28}$$

$$b_o = \frac{65}{9}(\alpha)^{1/2}\frac{(\alpha\eta)^{1/2}}{(1+\alpha\eta)^{10/3}},$$

$$a_o = -\frac{4225}{918}\alpha\frac{1}{(1+\alpha\eta)^{17/3}}$$

just as in the hypothesis (6), there is the similarity between the vertical velocity and temperature. In other words, when $\sigma_t > 0.6$, the thermal boundary layer is thinner than the dynamic one while when $\sigma_t < 0.6$ it is wider.

A most specific feature of the found solutions of the form (15) is their independence of the initial jet pulse K_o. This indicates that expressions (15) describe the distribution of velocity, temperature, etc. in the far field of the jet, in the zone of the prevailing effect of the buoyancy forces (the zone of a free-convection flow). In other words, relations (15) are the limiting similarity solutions for vertical jets with free convection or milarity solutions for vertical buoyant jets abov a point heat source (the so-called plume [24]). Now, consider the problem of the ejection ability of the jet, which can be characterized by the flow rate

$$m(x) = 2\pi \int_0^\infty \rho u y \, dy \sim x^{5/3} \tag{29}$$

From (29) it follows that at $x = 0$, $m = m_o = 0$, i.e. the solution considered corresponds to the zero flow rate of liquid from the source. Let us construct the nonsimilarity solution of the problem, i.e. assume that the jet escapes not from a source but from a nozzle of finite dimension ($\dot{m}_o = 0$). Following [17], we seek the stream function ψ in the series form

$$\psi(x, \eta) = A\{f_o x^{5/3} + f_1 x^{2/3} + f_2 x^{-1/3} + \dots\} \tag{30}$$

The other values, ΔT and ν_t, will also be prescribed by asymptotic expansions:

$$\Delta T(x,\eta) = B\sum_{i=0}^{\infty} h_i x^{-5/3-i},$$

$$\nu_t = \frac{1}{2} A \sum_{i=0}^{\infty} C_i x^{2/3-i} \tag{31}$$

where f_i, h_i are the unknown functions of the variable $\eta = \dfrac{y^2}{4\chi x^2}$. Having estimated the corresponding derivatives and substituted their expressions into (1) at $w = 0$, we obtain an infinite set of equations:

$$(\eta f''_i)' + \frac{5}{3} f_o f''_i + \frac{1}{3}(3i+2) f'_o f'_i + \frac{1}{3}(5-3i) f''_o f_i + h_i$$

$$= \sum_{j=1}^{i-1} \{ \frac{1}{3}(3j-5) f_j f''_{i-j} - \frac{1}{3} [3(i-j)+1] f'_i f'_{i-j} \} -$$

$$\sum_{j=1}^{i} C_j (\eta f''_{i-j})',$$

$$\frac{1}{\sigma_t} (\eta h'_i)' + \frac{5}{3} f_o h'_i + \frac{1}{3}(3i+5) f'_o h_i = \sum_{j=1}^{i}$$

$$\{ \frac{1}{3}(3j-5) f_j h'_{i-j} - \frac{1}{3}[3(i-j)+5] f'_j h_{i-j} - \frac{1}{\sigma_t} C_j (\eta h'_{i-j})' \} \tag{32}$$

which should be integrated under the following boundary and integral conditions:

$f_i(0) = \sqrt{\eta}\, f''_i\ (0) = \sqrt{\eta}\, h'_i\ (0) = 0$; $f'_i \to 0$, $h_i \to 0$ when $\eta \to \infty$

$$\int_o^\infty \{ f'_o h_i + \sum_{j=1}^{i} f'_j h_{i-j} \} d\eta = 0 \tag{33}$$

Moreover, the coefficients C_1, C_2,.... ($C_o = 1$) should satisfy relation (7). The problem (32) - (33) can be solved in a closed form its solution satisfying the condition (33), is of the form:

$$f_{i+1} = - \frac{\gamma}{i+1} (2 f_i \eta - \frac{1}{3}(5-3i) f_i),$$
$$h_{i+1} = - \frac{\gamma}{i+1} (2 h'_i \eta + \frac{1}{3} (5+3i) h_i). \tag{34}$$

Hence,

$$u = \frac{1}{2}\frac{A}{\chi} \{ f'_o x^{-1/3} - \gamma \sum_{i=0}^{\infty} \frac{1}{1+i} (2 f''_i \eta + \frac{1}{3}(1+3i) f'_i) x^{-4/3-i} \tag{35}$$

$$T = B \{ h_o x^{-5/3} - \gamma \sum_{i=0}^{\infty} \frac{1}{1+i} (2 h'_i \eta + \frac{1}{3}(5+3i) h_i) x^{-8/3-i}$$

As was shown in [25], the relations of the form (35) are the infinite series expansion for $/\gamma/x/ < 1$ of the exact solution:

$$u(x,\zeta) = \frac{1}{2} \frac{A}{\chi} f_\zeta x^{-1/3} (1 + \frac{\gamma}{x})^{-1/3} \tag{36}$$

$$\Delta T(x,\zeta) = Bh(\zeta) x^{-5/3} (1 + \frac{\gamma}{x})^{-5/3}$$

Here the functions $f(\zeta)$, $h(\zeta)$ satisfy equations (22) and, consequently, are defined by expressions (27). Note, that solution (36) differs from (15) by the variable

$$\zeta = \frac{y^2}{4\chi x^2} (1 + \frac{\gamma}{x})^{-2} \tag{37}$$

To complete the solution of the nonsimilarity problem, it remains to determine the integration constant γ, which persists through the boundary and integral conditions. This constant is characterized by the mass flow rate per second through the initial cross-section of the jet:

$$\gamma = \left| \frac{\dot{m}_o}{2\pi\rho A q} \right|^{3/5} \tag{38}$$

When $x \to \infty$, equations (36) yield formulae (15). Figures 3-5 contain the results of comparison between the present theoretical calculations and experimental data of work [26] where a buoyant jet above a point heat source is investigated. It is seen from Fig.3 that a satisfactory agreement of the experimental excess temperature and vertical velocity profiles with the theoretical ones is achieved at $\sigma_t = 0.82$. In this case the constant χ for the adopted model of turbulent transfer (7) has turned to be equal to $\chi = 7.74 \cdot 10^{-3}$. The distributions of $\bar{u}_m$ and $\Delta \bar{T}_m$ along the axis are given in Fig.4. Except for the starting length data, where the acceleration of liquid under the action of buoyancy forces is observed, the results [26] are described by a single curve [formulae (36) $\bar{\gamma} = \gamma/d_o = 0.96$] This very figure also contains the similarity solution of equation (15) presented by a dashed line. Figure 5 presents the dependence of b_u/d_o on x/d_o. Here too, one can see a satisfactory agreement between the experiment and theory (37).

Now, let us consider the problem of the search for the nonsimilarity solution for a swirling jet.

Taking into account the fact that $u_m \sim x^{-1/3} \Delta T_m \sim x^{-5/3}$, $w_m \sim x^{-8/3}$, $\Delta p_m \sim x^{-16/3}$ when $x \to \infty$ it is possible, when analyzing the initial set of equations (1), to neglect the term $\frac{\partial \Delta p}{\partial x}$ as compared with the remaining terms of the first equation. Note, that the velocity of this statement is justifiable in the region of the similar flow and less justifiable in the region of nonsimilar flow, where the corresponding turbulent flow will be much more complex. However, as the results of measurements [12] show, such simplified scheme of turbulent motion gives a qualitatively correct pattern of the development of vertical swirling jets. Omitting detailed results (the procedure for the solution construction is given above), we shall write down the finite formulae:

$$w(x,\zeta) = Db(\zeta) x^{-8/3} (1 + \frac{\gamma}{x})^{-8/3},$$

$$\frac{\Delta p}{\rho} (x,\zeta) = D^2 a(\zeta) x^{-16/3} (1 + \frac{\gamma}{x})^{-16/3} \tag{39}$$

with the functions $b(\zeta)$, $a(\zeta)$ being determined by relations (27) and the nonsimilarity variable ζ, by (37).

It is not difficult to see, that the obtained results (36) - (39) can also be considered as similarity solutions (15) but shifted along the axis x by γ. Thus, the search for the nonsimilarity solution to the stated problem is completed by a shifted similarity one, which earlier was successfully used for the analysis of jet flows with account for the buoyancy forces [24].

NOMENCLATURE

u, v, w	axial, radial and tangential components of averaged velocity;
x, ψ, y	vertical, angular and radial coordinates;
T	averaged temperature; $\Delta T = T - T_\infty$;
p	averaged pressure, $\Delta p = p - p_\infty$;
ρ	density;
C_p	specific heat at constant pressure;
ν_t	turbulent viscosity coefficient;
σ_t	turbulent Prandtl number;
β_s	volume expansion coefficient;
g	gravity acceleration;
χ_0 , χ	turbulent transfer constants;

Subscripts

∞	surrounding medium;
m	maximum value.

REFERENCES

1. Turner, J.S., Buoyancy effects of fluids. Cambridge University Press, Cambridge, (1973).
2. Chen, C.J., Rodi W., Vertical turbulent buoyant jets: A review of experimental data. London/ New-York: Pergamon, (1980) 83.
3. List, E.J., Turbulent jets and plumes, Ann.Rev. Fluid Mech., 14 (1982) 189-212.
4. Yih, C.S., Free convection due to a point source of heat. Proc. 1st U.S.Nat.Congr.Appl. Mech., (1951) 941-947.
5. Fujii, T., Theory of steady laminar natural convection above a horizontal line heat source and a point heat source, Int. J. Heat Mass Transfer, 6 (1963) 597-606.
6. Brand, R.S., and Lahey F.J., Heated laminar vertical jet, J. Fluid Mech., 29 (1967) 305-315.
7. Crane, L.J., The round plume due to a falling point source of heat ZAMP, 28 (1977) 599-606.
8. Crane, L.J., Axially symmetric plumes at very small Prandtl numbers, ZAMP, 26 (1975) 427-435.
9. Sullivan, P.J., and Sutherland P.J., Laminar free convection due to a point source of buoyancy, ZAMP, 27 (1976) 671-675.
10. Dzhaugashtin, K.E., and Soldatkin, A.V., Propagation of an axysymmetric jet with interaction of buoyant forces, Izv. SO AN SSSR, Ser. Techn. Nauk, 2(8) (1981) 60-63.
11. Korovkin, V.N., and Sokovishin, Yu.A., A laminar swirling jet with buoyant forces taken into account, Izv. Akad. Nauk SSSR, Mech. Zhid. i Gaza, No.4, (1983) 29-54.
12. Lee Shao-Lin, Axisymmetrical turbulent swirling natural-convection plume, Trans. ASME, J. Appl. Mech., 33, (1966) 647-661.
13. Narain, J.P., Uberoi M.S., The swirling turbulent plume, Trans. ASME, J. Appl. Mech., 41 (1977) 337-342.
14. Gostintsev, Yu.A., Sukhanov L.A., and Solodovnik A.F., A steady-state similarity turbulent jet over the point concentration-thermal source, Izv. Akad. Nauk SSSR, Mech. Zhidk. i Gaza No.2, (1983) 129-135.
15. Sigeti, A.I., and Bruyatsky, E.V., Calculation of the main portion of vertical turbulent weakly swirled buoyant jets by the integral method, Hydrodynamics (Kiev) No.50, (1984) 37-42.
16. Abramovich, G.N., Krasheninnikov S.Yu., Sekundov A.N., and Smirnova I.P. Turbulent displacement of gas jets, Moscow: Nauka, (1974).
17. Loitsyansky, L.G., Propagation of a swirling jet in an infinite space submerged by the same liquid, Prikl. Mat. i Mekh., 17 (1953) 3-16.
18. Szablewski, W., Asymptotische Gesetzmassigkeiten der turbulenten Ausbreitung eines vertikalen Gasstrahls im gleichf formigen Medium under Berucksichtigung des Archimedischen Auftriebs, ZAMM, Bd.46, (1968) 385-392.
19. Vasil'chenko, I.V., An approximate thermodynamic analysis of local ascending flows in the atmosphere, Tudy Gl. Geofiz. Observat., 72 (1957) 3-18.
20. Vasil'chenko, I.V., Concerning the problem on a stationary flow, Trudy Gl. Geofiz. Observat., 93 (1959) 29-36.
21. Rouse, H., Yih, C. S. and Humphreys, H.W., Gravitational convection from a boundary source, Tellus, 4 (1952) 201-210.
22. George, W.K., Alpert, R.L., and Tomanini, F. Turbulence measurements in an axisymmetric buoyant plume, Int. J. Heat Mass Transfer, 20 (1977) 1145-1154.
23. Yih, C. S., Turbulent buoyant plumes, Phys. Fluids, 20 (1977) 1234-1237.
24. Jaluria, Y., Natural convection: Heat and Mass transfer. Moscow: Mir, (1983).
25. Korovkin, V.N., and Sokovishin, Yu.A., Some problems of the viscous jets theory, Prikl. Mech. i Teoret. Fiz., No.2, (1984) 27-34.
26. Hakagome, H., and Hirata, M., The structure of turbulent diffusion in an axisymmetrical thermal plume, In: Heat transfer and turbulent buoyant convection, Washington, 1 (1977) 361-372.

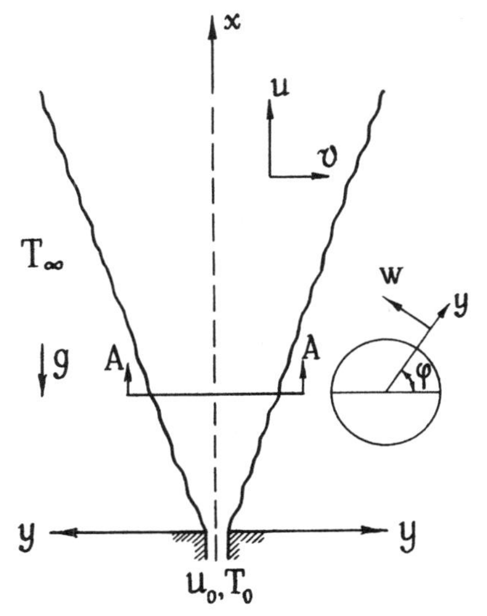

Fig.1

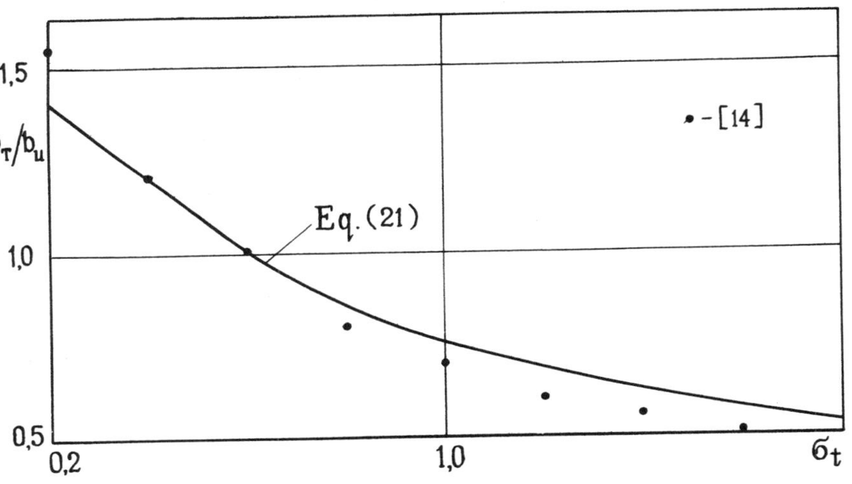

Fig.2

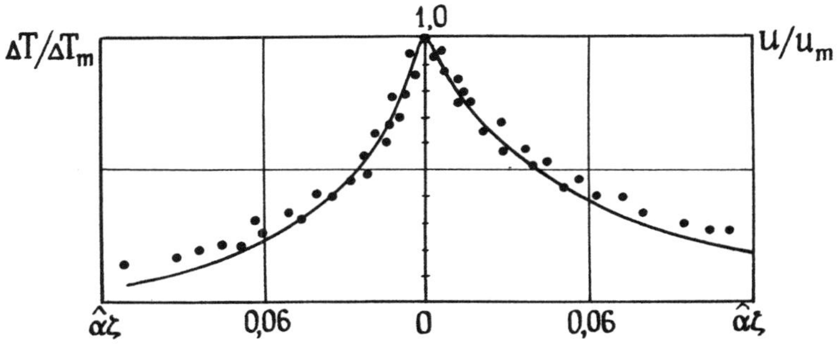

Fig.3

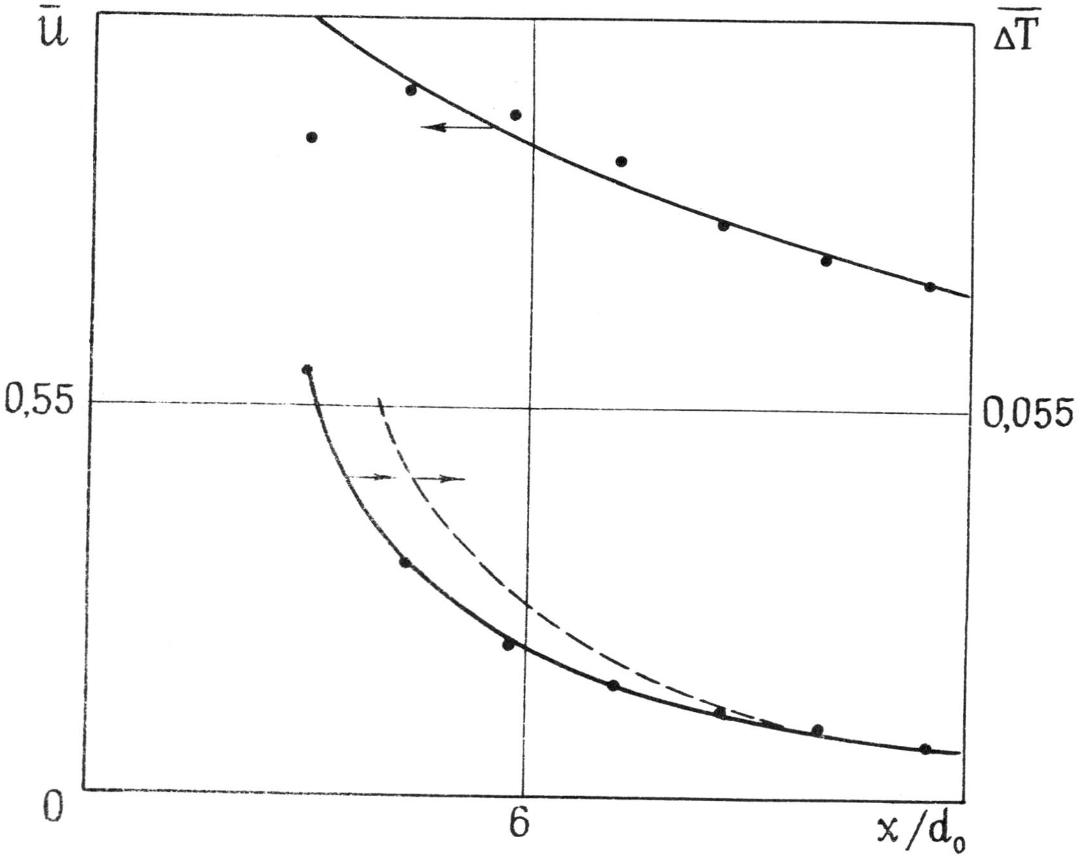

Fig.4

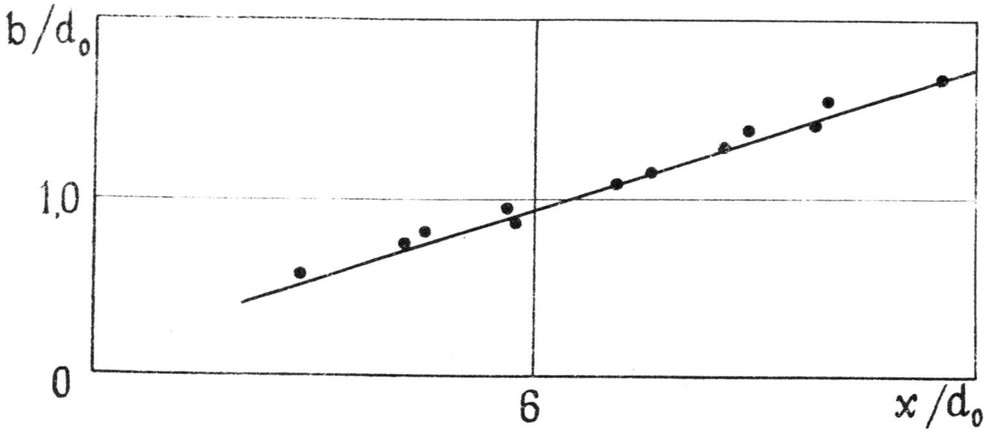

Fig.5

A Correlation for Heat Transfer between Immersed Surfaces and Gas-Fluidized Beds of Large Particles

A. MATHUR and S.C. SAXENA
Department of Chemical Engineering
University of Illinois at Chicago
Box 4348, Chicago, Illinois 60680, USA

Heat transfer between immersed surfaces and fluidized beds of "large" particles is investigated. Large particles here refer to those belonging to groups IIB and III of the powder classification scheme of Saxena and Ganzha (1984), and characterized by Ar > 130,000. Four correlations proposed so far, for large particles are examined on the basis of available experimental data. The data correspond to operating pressures ranging from ambient to 8.1 MPa, and Reynolds numbers ranging up to 5000. A total of three hundred and thirty-six data points have been used in the present study. It is found that none of the correlations are able to adequately predict the heat transfer coefficient over the entire range of experimental conditions, and their ranges of applicability and probable uncertainties are established. A new correlation is developed on the basis of these data and the known mechanistic details of heat transfer by particle and gas convection. The proposed correlation is:

$$Nu = 5.95\,(1 - \varepsilon)^{2/3} + 0.055\,Ar^{0.3}\,Re^{0.2}\,Pr^{1/3},$$

and it is accurate within ± 35% for gas-fluidized bed systems characterized by Ar > 130,000.

1. INTRODUCTION

The high heat transfer rates achieved for boiler tubes immersed in fluidized-bed coal combustors alongwith their in situ capability to remove So_x, and suppress the emission of NOx in high pressure systems, have lead to their exploitation as a promising technology for the generation of electrical power from sulfur-rich coal. However, for adequate design and reliable scale-up, many aspects of their operation need to be carefully understood and mathematically modelled. The present investigation deals with one such aspect concerning the establishment of heat transfer rates from an immersed surface in a gas-fluidized bed at ambient and higher pressure.

Most of the work undertaken towards the understanding of the heat transfer mechanism in gas-fluidized beds has been confined to small particles and these investigations have been reviewed by Zabrodsky (1966), Botterill (1978), Saxena et al. (1978), and Saxena and Gabor (1981). For such systems, bubble dynamics and bubble-induced motion of solid particles play a very significant role in as much as the predominant mechanism of heat transfer is due to particle convection. Suitable modifications of the packet model, Gelperin and Einshtein (1971), have been found adequate to simulate the heat transfer process. The particle residence time at the heat transfer surface and particle concentration in the vicinity of the heat transfer surface are the controlling parameters. In such systems, the interstitial gas velocity is smaller than the bubble velocity, and fluidization behaviour is referred to as the fast-bubble regime.

On the other hand, in beds of large particles the interstial gas velocity is much larger than the bubble velocity, and as the fluidizing velocity is increased this slow-bubble regime changes to rapidly-growing bubble and finally to turbulent regime, Catipovic et al. (1978). It has been shown by Canada and McLaughlin (1978) and Borodulya et al. (1980 b) that in turbulent flow regime, the solids mixing is poor, and further the particle temperature essentially remains constant, Zabrodsky (1966). This causes particle convection to be well-approximated by a steady-state conduction. For such systems, the gas convection contribution is large compared to the particle convection contribution. So far, we have used the words "small" and "large" particles in a qualitative sense and in the next section, we present a criterion which distinguishes between these two types of particles in a quantitative manner.

2. PARTICLE CLASSIFICATION

Geldart (1973) presented a powder classification scheme based on the fluidized behaviour of particles. His groups B and D particles are usally referred to as small and large particles respectively. Group B comprises of particles with $40\,\mu m < d_p < 500\,\mu m$, and $1400\,kg/m^3 < \rho_s < 4000\,kg/m^3$. On the other hand, group D particles satisfy the criterion that $(\rho_s - \rho_g)\,\bar{d}_p^{\,2} \geq 10^{-3}$. Saxena and Ganzha (1984) have shown that this classification scheme does not simulate the hydro dynamical and thermal behaviours simultaneously. To illustrate this point, let us consider a bed of sand particles fluidized by air at ambient temperature and pressure. The above relation would predict that for $d_p > 0.63$ mm, the bed would behave like a large particle bed. However, the heat transfer for such sand particles is still controlled by particle convection so that while from hydrodynamical considerations, a sand bed of 0.63 mm particles will be regarded as a large particle system, from heat transfer point of view, it will be a small particle system.

Saxena and Ganzha (1984) have developed a particle classification scheme that links the hydro dynamical and thermal behaviours of a fluidized bed through the Archimedes number and the Reynolds number at minimum fluidization. They distinguish the different fluidization regimes on the nature of gas boundary layer around individual particles and their mutual interaction. Their group I particles are defined such that:

$$3.35 \leq Ar \leq 21,700 \text{ or}$$

$$1.5 \left[\frac{\mu_g^2}{\rho_g(\rho_s - \rho_g)g}\right]^{1/3} \leq \bar{d}_p \leq 27.9 \left[\frac{\mu_g^2}{\rho_g(\rho_s - \rho_g)g}\right]^{1/3} \qquad (1)$$

These have the characteristics of Geldart group B particles, or those of "small" particles referred to above. The gas boundary layer around such particles is laminar and heat transfer is predominantly controlled by particle convection.

Saxena and Ganzha (1984) defined the so-called "large" particles, or Geldart group D particles into two groups, II B and III. For group II B, the flow field around the particles is turbulent and fully developed so that the active interphase contact surface becomes equal to the total surface of the particles in the bed, and the heat transfer is mainly controlled by the turbulence in the thermal boundary layer and wake. The subgroup II B is defined such that:

$$130,000 \leq Ar \leq 1.6 \times 10^6, \text{ or}$$

$$50.7\left[\frac{\mu_g^2}{\rho_g(\rho_s - \rho_g)g}\right]^{1/3} \leq \bar{d}_p \leq 117\left[\frac{\mu_g^2}{\rho_g(\rho_s - \rho_g)g}\right]^{1/3} \qquad (2)$$

For group III particles, thermal wakes of the particles are disrupted by the following particles leading to overall bed turbulence whose intensity increases with Reynolds number. This group is characterized by:

$$Ar \geq 1.6 \times 10^6, \text{ or}$$

$$\bar{d}_p \geq 117 \left[\frac{\mu_g^2}{\rho_g(\rho_s - \rho_g)g}\right]^{1/3} \qquad (3)$$

For powders of group II B, the two components $h_{w\,pc}$ and $h_{w\,gc}$ of h_w are comparable to each other and both these terms must be computed with good accuracy to obtain a reliable estimate for h_w. For group III particles, the magnitude of h_w is strongly controlled by the $h_{w\,gc}$ part which makes the dominant contribution. The $h_{w\,pc}$ component is a relatively smaller part of h_w.

3. AVAILABLE CORRELATIONS FOR h_w

In the past decade, several correlations have been developed for Nu and Nu_{max} based on semi-theoretical arguments and experimental data. Thus, Maskaev and Baskakov (1972), and Denloye and Botterill (1977) proposed correlations for Nu_{Max}; while Staub (1979), Borodulya et al. (1980 a), Catipovic et al. (1980), Glicksman and Decker (1980), and Zabrodsky et al. (1981) present correlations for Nu. The correlation of Staub (1979) is for tube bundles. The correlations derived for single tubes are briefly reproduced in the following as these will be discussed subsequently for their ability to reproduce all the available experimental data, one of the goals of this work.

Borodulya et al. (1980 a) proposed the following empirical correlation which they derived by expressing their own experimental data on sand particles and glass beads in the pressure range 0.6 - 8.1 MPa, within an accuracy of + 25% and valid for 20 < Re < 5000:

$$Nu = 0.37\, Re^{0.71}\, Pr^{0.31} \qquad (4)$$

Catipovic et al. (1980) evaluated the total heat transfer coefficient as the sum of three different contributions. These are due to solids, bubbles and the emulsion gas. The solids contribution is estimated by treating it as steady-state conduction, the bubble fraction contribution is obtained by regarding it to occur by conduction through a gas film, and the emulsion gas contribution is regarded as constant over the entire range of fluidizing velocities, being equal to that at minimum fluidization. Their final expression is:

$$Nu = 6(1 - \beta) + (0.0175\, Ar^{0.46}\, Pr^{0.33})(1 - \beta) + (\bar{d}_p/D_T)(0.88\, Re_{mf}^{0.5} + 0.0042\, Re_{mf})\, Pr^{0.33}\, \beta \qquad (5)$$

where β, the time fraction a bubble is in contact with the immersed surface, is given by:

$$1 - \beta = 0.45 + \frac{0.061}{(U - U_{mf}) + 0.125} \qquad (6)$$

Glicksman and Decker (1980) employed the concept of steady-state conduction and lateral mixing of the gas to evaluate the particle and gas covective contributions respectively. The gas in excess of that required for minimum fluidization was considered to be transported through the bed as bubbles. Their final expression is:

$$Nu = (1 - \delta)(9.3 + 0.042\, Re\, Pr) \qquad (7)$$

Zabrodsky et al. (1981) estimate the particle convective component by approximating the analytical equation of steady-state conduction given by Zabrodsky (1966). The gas convective component was assumed to be proportional to $U^{0.2} C_{pg} \rho_g \bar{d}_p$. They finally recommend the following semi-theoretical equation based on some of their experimental data at ambient conditions:

$$h_w = 7.2\ (k_g / \bar{d}_p)\ (1 - \varepsilon)^{2/3} + 26.6\ U^{0.2}\ \rho_g C_{pg}\ d_p \quad (8)$$

4. EXPERIMENTAL DATA FOR h_w

We will now examine the experimental data available in the literature and which pertains to large particles as defined by Saxena and Ganzha (1984) powder classification scheme mentioned earlier. Such systems are characterized by Ar > 130,000,and fall in groups II B and III of Saxena and Ganzha (1984) classification scheme. Such data for horizantal tubes and tube bundles are those of Maskaev and Baskakov (1972), Canada and McLaughlin (1978), Catipovic et al. (1980), Chandran et al. (1980), Zabrodsky et al. (1981) and Borodulya et al. (1984). Similar data for vertical tubes and tube bundles are given by Botteril and Denloye (1978) and Borodulya et al. (1980 a, 1983). For our present analysis, the data of Maskaev and Baskakov (1972) could not be employed as their graphical representation of data could not be accurately read. Further, experimental data of only sparsely packed tube bundles is used in the single tube mode based on the information given by the particular investigator. In general, the tube pitch effect disappears whenever $(P/D_T) > 3$. Also, the tube orientation effect is ignored for these large particles. This is substantiated by the experimental data and the large uncertainties which are seen in the data when experimental data of different investigators are synthesized. All the experiments of different workers were conducted with beds at ambient temperature and electrically heated tubes so that temperature is not an important parameter while analyzing these data.

Experimental data sets of eight investigations on five bed materials and comprising of thirteen different beds and seven different heat transfer tubes are considered in the present work. These details alongwith the pressure range of each investigation are given in Table I. It will be noticed that the data ranges between 0.1 and 8.1 MPa, and are reported at fifty-nine pressure levels, in each case at several velocities. In total, there are three hundred and thirty-six data points. One hundred and ninteen of these data points belong to group II B, while the remaining two hundred and seventeen fall in group III. The range of Archimedes number for the ninteen data sets are given in Table 1. In representing these data in the figures discussed later, each investigation is characterized by a particular symbol given in the last column of Table 1 to facilitate evaluation and discussion. The ability of the earlier discussed correlations to represent these data is assessed by graphically displaying the relative percentage deviation of the predicted and experimental Nusselt numbers, $100(Nu - Nu_{exp}) / Nu_{exp}$, as a function of Reynolds number, Re.

In Fig. 1, the ability of the correlation of Borodulya et al. (1980 a), as given by eq. (4), to represent the experimental data is displayed. It will be noticed that the percentage deviations range from -65 to 340. However, most of the deviations beyond about ± 50% are for the data points belonging to investigations at 0.1 MPa. This is not surprising if one examines the form of eq. (4) and the fact that this correlation was obtained on the basis of numerical regression analysis of their own experimental data in the range 1.1. to 8.1 Mpa, Borodulya et al. (1980 a). This study also reveals that the uncertainty of their correlation judged by them as ± 25% is quite an underestimate. We propose on the basis of the present study that majority of the pressure data (1.1 - 82 MPa) can be correlated by eq. (4) within an uncertainty of ± 50%. The correlation of eq. (4) is judged to be inadequate for pressures around ambient.

Catipovic et al. (1980) correlation of eq.(5) in conjunction with the experimental data is examined in Fig.2. The apparent percentage deviations between the predicted and observed Nusselt numbers range between about-130 and 275. However, in majority of the cases, the scatter remains within a somewhat conservative estimate of ± 100%. In certain respects, this is quite a remarkable agreement when one recalls that the entire model development is based on data taken at ambient pressures and on the physical picture of a two-phase bubbling bed model. The latter is particularly controversial for beds of particles of groups II B and III where turbulent heat transfer dominates.

The comparison of experimental data with the predictions of the correlation of Glicksman and Decker (1980) as given by eq. (7) is shown in Fig.3. Majority of the experimental data is overpredicted, the percentage deviations range between - 30 and 105. On the basis of Fig.3, we assign the most probable percentage uncertainty of Eq. (7) as - 20 to 80. The most serious drawback of this correlation appears to be its tendency to overpredict the experimental data. We feel that this has crept in the model development through the assumption that the gas convection contribution is directly proportional to Reynolds number and hence to gas density. This approximation is increasingly at fault as the pressure increases, and experimental investigations of Xavier et al. (1980) have shown that $h_{w\,gc}$ is proportional to $\sqrt{\rho_g}$ The results of Fig.3 also indicate that the first term of eq. (7) poorly simulates the particle convection contribution. As a result, the experimental data over the entire range of Reynolds number are overpredicted.

The predictions based on the correlation of Zabrodsky et al. (1981), eq. (8), are compared with the direct measurements in Fig.4. This figure shows some very systematic trends. Most of the data are overpredicted and the degree of overprediction increases with Reynolds number. At higher Reynolds numbers the percentage disagreement increases to as much as 670. A careful examination revealed that the reproduction of heat transfer data at ambient pressure is much better, the percentage disagreement always being within -50 and 70. The increasing disagreement as the pressure increases can be explained. As explained earlier in connection with the correlation of Glicksman and Decker (1980), the $h_{w\,gc}$ term here also has the same deficiency. The exponent of ρ_g was assumed to be one in model development and the numerical coefficient, 26.6, was adjusted from heat transfer data at ambient pressure. This correlation, therefore, is only satisfactory for heat transfer processes taking place at ambient pressure and its uncertainty for

for such a condition is as stated above.

5. A NEW CORRELATION

The critical examination of all the four correlations on the basis of the available experimental heat transfer data for large particles at ambient and higher pressures such that Ar > 130,000 clearly indicates that none of these correlations are adequate to reliably represent the experimental data. The reasons of their inadequacies differ from correlation to correlation and some of these are elaborated in the previous section. However, it is clear that an urgent need of a correlation consistent with the present theoretical and experimental understanding of the heat transfer mechanism, and capable of representing the experimental data exists. An effort in this direction is presented in the following.

It is well known that in large particle fluidized beds, both at ambient as well as at higher pressures, the heat transfer coefficient, h_w, can be expressed as the sum of the heat transfer coefficient due to particle convection, $h_{w\ pc}$, and the heat transfer coefficient due to gas convection, $h_{w\ gc}$. In dimensionless form, it can be expressed as:

$$Nu = Nu_{pc} + Nu_{gc} \ . \qquad (9)$$

The particle convective heat transfer degenerates to a steady-state heat conduction for large particles due to the relatively small change in their temperatures, and consequently Nu_{pc} can be approximated by a constant. However, as only a fraction of the heat transfer surface is bathed by particles, Nu_{pc} will be a function of the particle concentration, $(1-\varepsilon)$. Following Zabrodsky et al. (1981), and Ganzha et al. (1982), we may express:

$$Nu_{pc} = C_1 \ (1-\varepsilon)^{2/3} \ . \qquad (10)$$

Here C_1 is a numerical constant, and the exponent, (2/3), of $(1-\varepsilon)$ accounts for the fact that heat transfer is dependent on the heat transfer surface area while the particle concentration is a volumetric concept, Zabrodsky (1966).

In evaluating the gas convective heat transfer component of the total heat transfer coefficient, we employ two experimental findings. First, that $h_{w\ gc}$ is proportional to $\sqrt{\rho_g}$ as shown by Xavier et al. (1980), and secondly that $h_{w\ gc}$ is proportional to $U^{0.2}$ as observed by Zabrodsky et al. (1981). Further, as discussed earlier, Archimedes number has proven very successful in characterizing the hydrodynamic and heat transfer behaviour of a two phase gas-solid system. In view of these facts, we express:

$$Nu_{gc} = C_2 \ Ar^{0.3} \ Re^{0.2} \ Pr^{1/3} . \qquad (11)$$

Here, C_2 is a numerical constant.

For group III particles, Nu_{pc} is much smaller than Nu_{gc}, being about 10% of Nu, and therefore it will be a reasonable approximation to treat it as a constant while evaluating the constant C_2 in eq. (11). We thus have for group III particles that

$$Nu = C_3 + C_2 \ Ar^{0.3} \ Re^{0.2} Pr^{1/3} \qquad (12)$$

On the basis of the experimental data belonging to group III, we evaluated the constant C_2 as 0.055 such that the general equation for Nu becomes

$$Nu = C_1 \ (1-\varepsilon)^{2/3} + 0.055 \ Ar^{0.3} \ Re^{0.2} \ Pr^{1/3} . \qquad (13)$$

For establishing the value of the numerical constant C_1, the experimental data of group II B were employed. This yielded a value of 5.95 for C_1. The final form and correlation for Nu is then

$$Nu = 5.95 \ (1-\varepsilon)^{2/3} + 0.055 \ Ar^{0.3} \ Re^{0.2} Pr^{1/3} \qquad (14)$$

The adequacy and appropriateness of the proposed correlation of eq. (14) may be judged from Figs.5 and 6 where data corresponding to groups II B and III respectively, are displayed. In all cases, the deviations are invariably within ± 35%, which is judged to be the probable uncertainty of the proposed correlation, eq. (14). The scatter of data in Fig.6 does present an apparent concern at first sight inasmuch as an unproportionate amount of experimental data is unpredicted. It is interesting to note that the bulk of these data are due to Borodulya et al. (1983, 1984). Further, almost all of their data exhibit this trend, one hundred and thirty-seven data points out of the total of one hundred and forty-three data points. One set of these data shown by filled circles refers to vertical tube and other data set represented by unfilled circles refers to horizontal tube. On the average, the former shows about twice as large a negative deviation as the latter. Whether or not, this is due to the tube orientation is not clear at the present time. The existing understanding of the large particles systems were the predominant mode of heat transfer is by turbulent gas flow, will suggest that tube orientation is likely to have a negligible effect. The stated accuracy of these data is judged as ± 10% and within these limitations, it will be premature to attribute this systematic deviations to any particular source. We, therefore, conclude that till more accurate data become available, the tube orientation effect for large particle .systems may be neglected and that the correlation of eq. (14) may be considered reliable within ± 35% to predict heat transfer coefficient.

ACKNOWLEDGEMENT

We are thankful to the National Science Foundation for partly supporting this work. Computing facilities were provided by the University of Illinois at Chicago Computer Centre.

Table 1

Summary of experimental data employed in the present work

Investigators	D_T (m)	Bed material	$\bar{d}_p$ (mm)	p (MPa)	Ar $\times 10^{-6}$	Symbol
Borodulya et al. (1980 a)	0.018	Sand Glass beads	0.794 3.1	1.1-8.1 0.1-8.1	0.6-3.3 3.3-204	◐
Borodulya et al. (1983)	0.013	Sand Glass beads	0.794 1.225 1.25 3.1	1.1-8.1 0.1-8.1 0.1-8.1 0.1-8.1	0.6-3.4 0.2-11.6 0.2-13.1 3.3-199	●
Borodulya et al. (1984)	0.013	Sand Glass beads	0.794 1.225 1.25 3.1	1.1-8.1 0.1-8.1 0.1-8.1 0.1-8.1	0.6-3.2 0.2-11.6 0.2-12.7 3.9-231	○
Botterill and Denloye (1978)	0.0127	Copper shot Sand	0.62 1.02	0.4-0.9 0.9	0.3-0.8 1.0	□
Chandran et al. (1978)	0.0318	Glass beads	1.58	0.1	0.4	■
Canada and McLaughlin (1979)	0.032	Glass beads	0.65 2.6	0.1 0.1-1.0	0.2 1.9-18.2	◑
Catipovic et al. (1980)	0.0508	Glass beads	1.3 2.6	0.1 0.1	0.2 6.8	●
Zabrodsky et al. (1981)	0.03	Fireclay Millet	2.0 3.0	0.1 0.1	0.26 2.01	●

NOMENCLATURE

Ar = Archimedes number = $g\, d_p^3 (\rho_s - \rho_g) / \mu_g^2$, dimensionless

C_{pg} = specific heat of the fluidizing gas, J/(kg)(K)

C_{ps} = specific heat of the solid particles, J/(kg)(K)

$\bar{d}_p$ = mean particle diameter, m

D_T = diameter of the immersed heat transfer, tube, m

g = acceleration due to gravity, m/s^2

h_w = bed-to-immersed surface heat transfer coefficient, W/(m^2) (K)

$h_{w\,exp}$ = experimentally determined heat transfer coefficient, W/(M^2) (K)

$h_{w\,gc}$ = component of h_w due to gas convection, W/(m)^2 (K)

$h_{w\,max}$ = maximum value of h_w, W/(M^2) (K)

$h_{w\,pc}$ = component of h_w due to particle convection, W/(m)^2 (K)

k_g = thermal conductivity of the fluidizing gas, W/(m)(K)

Nu = Nusselt number = $h_w\, d_p/k_g$, dimensionless

Nu_{exp} = Nusselt number based on experimentally determined heat transfer coefficient = $h_{w\,exp} d_p / k_g$, dimensionless

Nu_{gc} = component of Nusselt number due to gas convection = $h_{w\,gc}\, d_p/k_g$, dimensionless

Nu_{max} = maximum value of Nusselt number = $h_{w\,max} d_p/k_g$, dimensionless

p = tube pitch, center-to-center distance between adjacent tubes, m

pr = Prandtl number = $C_{pg}\, \mu_g / k_g$, dimensionless

Re = Reynolds number = $U\, \bar{d}_p\, \rho_g / \mu_g$, dimensionless

Re_{mf} = Reynolds number at minimum fluidization $U_{mf}\, \bar{d}_p\, \rho_g / \mu_g$, dimensionless

U = fluidizing velocity, m/s

U_{mf} = fluidizing velocity at minimum fluidization, m/s

Greek Symbols

β = time fraction that the immersed surface is in contact with bubbles, dimensionless

δ = bubble fraction in the bed, dimensionless

ε = bed voidage, dimensionless

ρ_g = density of the fluidizing gas, kg/m^3

ρ_s = density of the solid particles, kg/m^3

μ_g = viscosity of the fluidizing gas, kg/(m)(s)

REFERENCES

1. Borodulya, V.A., Ganzha, V.L., Podberezsky, A.I., in "Fluidization" (eds. J.R. Grace and J.M. Matsen), 201, Plenum Press, New York, 1980 a.

2. Borodulya, V.A., Ganzha, V.L., Upadhyay, S.N., Saxena, S.C., Int. J. Heat Mass Transfer, 23 (1980 b) 1602.

3. Borodulya, V.A., Ganzha, V.L. Podberezsky, A.I. Upadhyay, S.N., Saxena S.C., Int. J. Heat Mass Transfer 26 (1983) 1577.

4. Borodulya, V.A., Ganzha, V.L., Poberezsky, A.I., Upadhyay, S.N., Saxena, S.C., Int. J. Heat Mass Transfer, 27 (1984) 1219.

5. Botterill, J.S.M., Fluid Bed Heat Transfer, Academic Press, New York, 1978.

6. Botterill, J.S.M. Denloye, A.O.O., AIChE Symp. Ser. No.176, 74 (1978) 194.

7. Canada, G.S., McLaughlin, M.H., AIChE Symp. Ser. No.176 74 (1978) 27.

8. Catipovic, N.M., Jovanovic, G.N., Fitzgerald, T.J., AICHE J., 24 (3) (1978) 543.

9. Catipovic, N.M., Jovanovic, G.N., Fitzgerald, T.J., Levenspiel, O., in "Fluidization" (eds. J.R. Grace and J.M. Matsen), 225, Plenum Press, New York, (1980).

10. Chandran, R., Chen. J.C., Staub, F.W., ASME J. Heat Transfer, 102 (1980) 152.

11. Denloye, A.O.O., Botterill, J.S.M., Chem. Engng. Sci., 32 (1977) 461.

12. Ganzha, V.L., Upadhyay, S.N., Saxena, S.C., Int. J. Heat Mass Transfer, 25 (10) (1982) 1531.

13. Geldart, D., Powder Technol., 7 (1973) 285.

14. Gelperin, N.I., Einshtein, V.G., in "Fluidization" (eds. J.F. Davidson and D. Harrison), 471, Academic Press, New York, (1971).

15. Glicksman, L.R., Decker, N., Proc. 6th Int. Conf. Fluidized Bed Combustion, 3 (1980) 1152.

16. Maskaev, V.K., Baskakov, A.P., J. Engng. Phys., 24 (1977) 589.

17. Saxena, S.C., Gabor, J.D., Prog. Energy Combust. Sci., 7 (1981) 73.

18. Saxena, S.C., Ganzha, V.L., Powder Technol., 39 (1984) 199.

19. Saxena, S.C., Grewal, N.S., Gabor, J.D., Zabrodsky, S.S., Galershtein, D.M., Adv. Heat Transfer, 14(1978) 149.

20 Staub, F.W., A.S.M.E. J. Heat Transfer 101 (1979) 391.

21. Xavier, A.M., King, D.F., Davidson, J.F., Harrison, D., in "Fluidization" (eds. J.R. Grace and J.M. Matsen), 209, Plenum Press, New York, (1980).

22. Zabrodsky, S.S., Hydrodynamics and Heat Transfer in Fluidized Beds, M.I.T., Press, Cambridge, (1966).

23. Zabrodsky, S.S., Epanov, Yu. G., Galershtein, D.M., Saxena, S.C., Kolar, A.K., Int. J. Heat Mass Transfer, No.4, 24 (1981) 571.

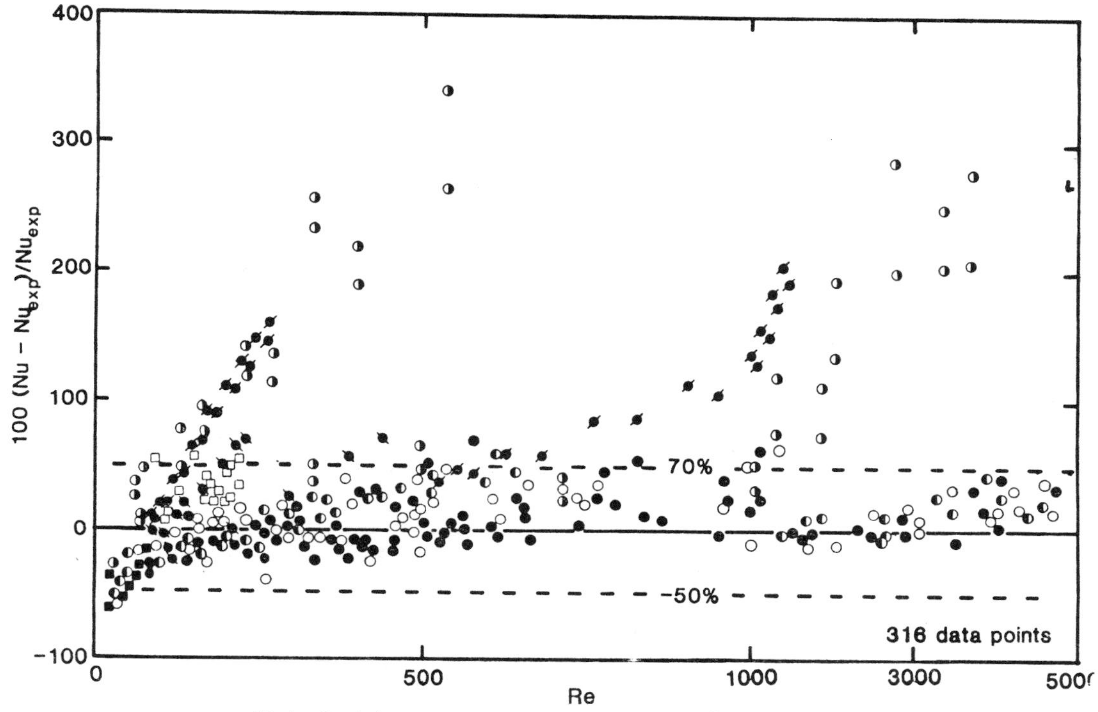

Fig.1 Deviation of Nusselt number predicted by the correlation of Borodulya et al. (1980) from experimental Nusselt number. Note the change of scale at Re = 1000.

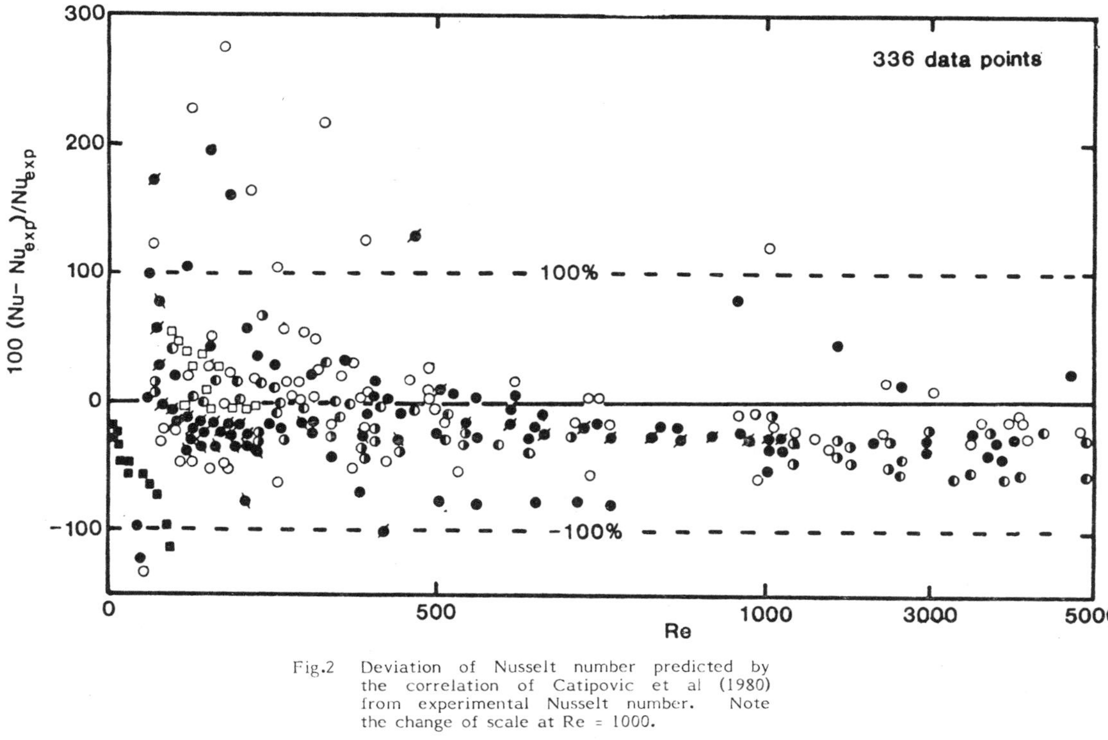

Fig.2 Deviation of Nusselt number predicted by the correlation of Catipovic et al (1980) from experimental Nusselt number. Note the change of scale at Re = 1000.

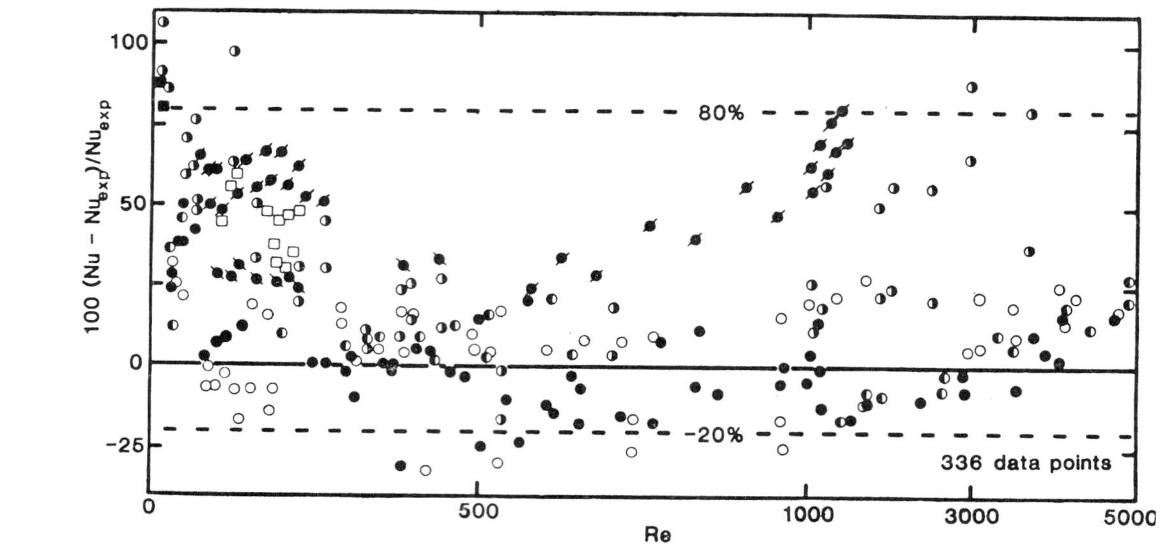

Fig.3 Deviation of Nusselt number predicted by the correlation of Glicksman and Decker (1980) from experimental Nusselt number. Note the change of scale at Re = 1000.

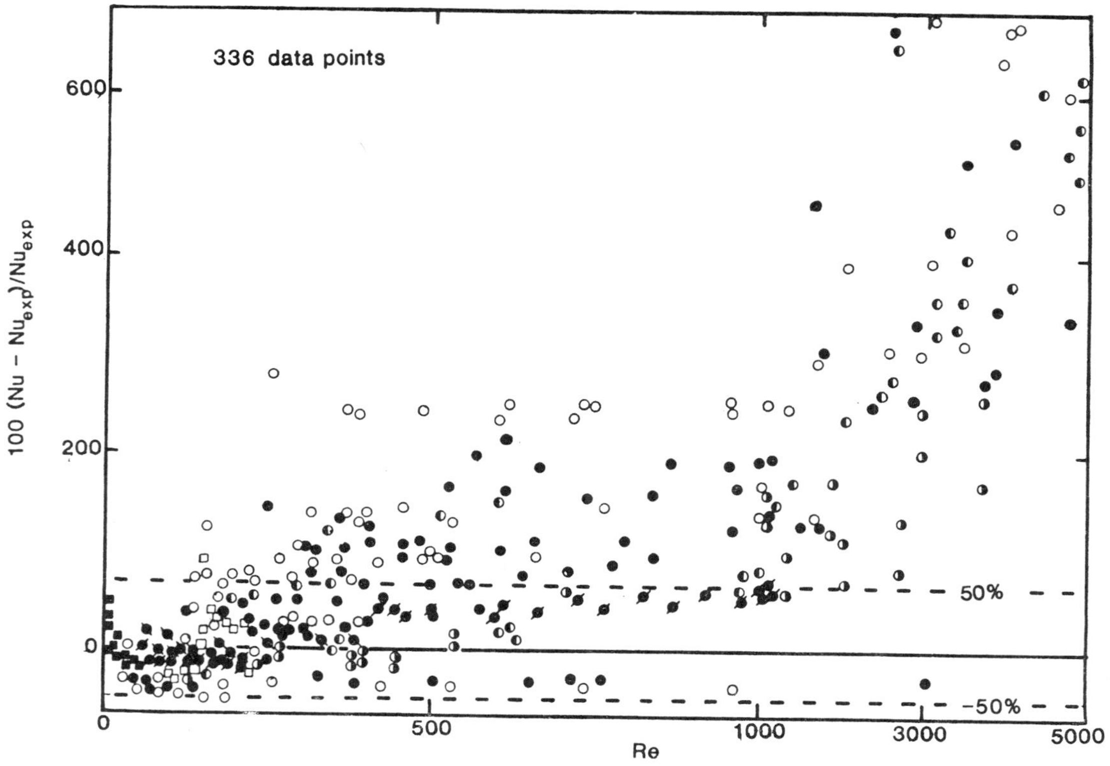

Fig.4 Deviation of Nusselt number predicted by the correlation of Zabrodsky et al. (1981) from experimental Nusselt number. Note the change of scale at Re = 1000.

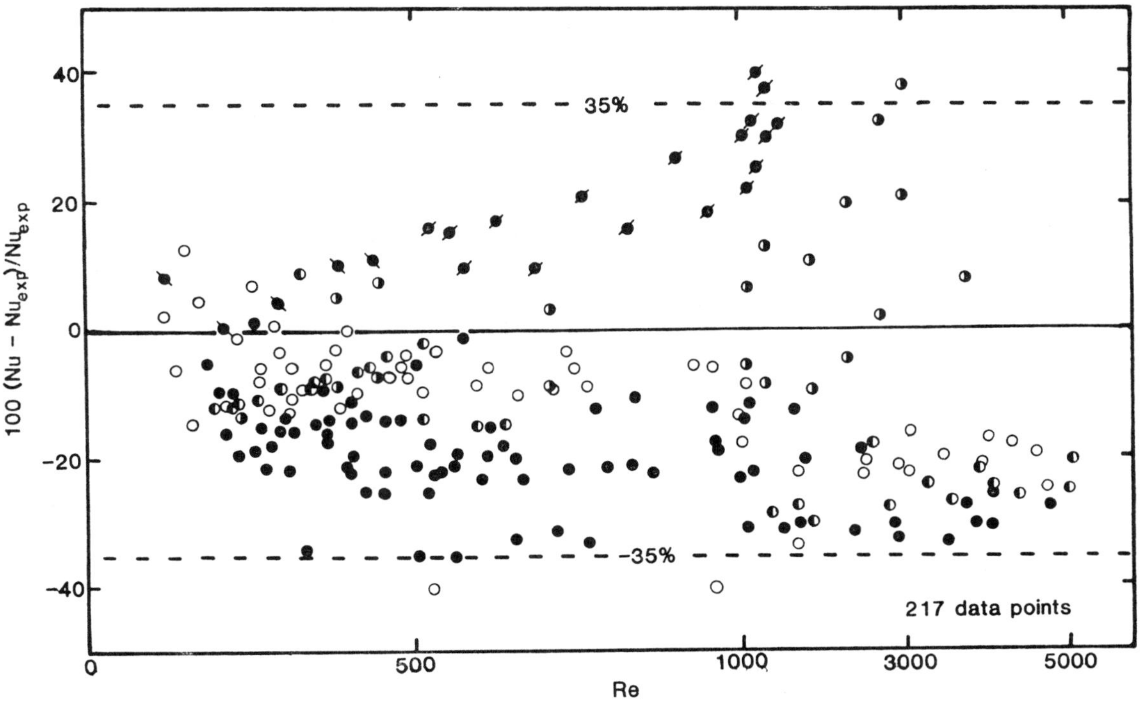

Fig.6 Deviation of Nusselt number predicted by eq. (14) from experimental Nusselt number for beds of group III particles as a function of Reynolds number. Note the change of scale at Re = 1000.

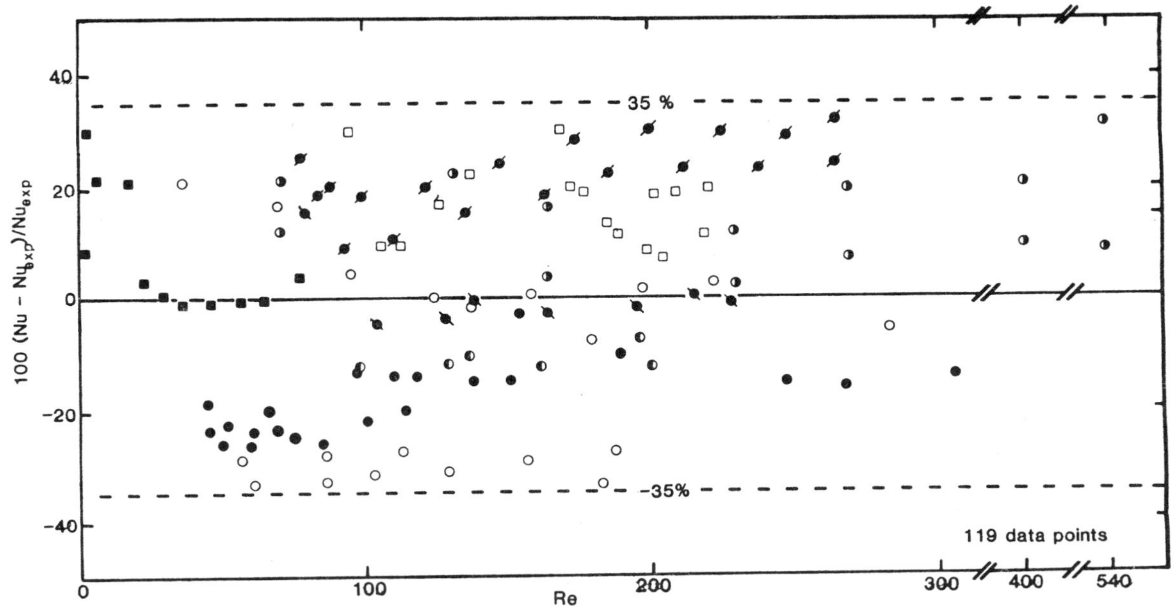

Fig.5 Deviation of Nusselt number predicted by eq. (14) from experimental Nusselt number for beds of group II B particles as a function of Reynolds number.

Post-Dryout Heat Transfer in Tubes with Uniform and Circumferentially Nonuniform Heating

F. MAYINGER and R. SCHNITTGER

Measurements of wall temperatures under post-dryout conditions with uniformly and non-uniformly heated tubes in the refrigerant R12 are presented. Supplementary to these measurements a semi-empirical model was developed which predicts the temperatures of the wall and of the superheated vapour and the real quality downstream of the dryout spot.

Using the model one has to perform a stepwise numerical integration along the flow path of the droplet-vapour mixture, starting from the dryout spot. The basic idea of the model is the assumption that droplets are splitted into smaller particles under the influence of the shear stress in the boundary layer at the wall and during dry collisions with the wall.

Measured and predicted data are compared and give good agreement for uniformly heated tubes, and a satisfactory one for nonuniform heating. A comparison with data from the literature measured in water, proves that the model is also valid for this substance.

1. INTRODUCTION

For designing steam generators and for selecting the material of its tubes the exact knowledge of the wall temperatures to be expected is of high importance. Once-through steam generators - for example of the Benson type - have a heat transfer area where the cooling of the tube walls is produced by a dispersed flow with droplets embedded in vapour and with wall temperatures above wetting conditions. This situation is called post-dryout heat transfer.

The dryout of the wall is caused by two different phenomena: Evaporation reduces the thickness of the liquid film existing before at the wall and also the entrainment of droplets by the momentum forces resulting from the vapour flow decreases the amount of liquid at the wall. Mechanisms of this droplet formation are described for example by Hewitt [1], Langner [2] and Mayinger [3].. A large number of correlations exist to predict the onset of dryout, see for example [1] and [4-9]. In designing the thermohydraulic behaviour of Benson-boilers being operated at subcritical pressures frequently the dryout correlations by Bertoletti [9] or by Konkov [4] are used.

In the post-dryout region the heat transport from the tube wall can be subdivided into the following paths:

a) Convective heat transfer from the tube wall to the vapour,

b) heat transfer from the wall to droplets impinging on to the wall by radial movement in the boundary layer, however, not wetting the wall (dry collisions),

c) heat transfer from the wall to droplets wetting the wall for a short time (wet collisions),

d) heat transfer by radiation from the wall to vapour and droplets.

The contribution of these heat transfer paths were studied in detail by Iloeje [10]. One of his important findings was that the heat transport by wetting droplets becomes negligible small at wall temperatures 80 K above saturation temperature. Also the significance of dry collisions is strongly reduced with increasing vapour quality. So the main mechanism is the heat transfer from the wall to the vapour, which is super-heated, and from which - as a consequence - a thermodynamic non-equilibrium between vapour and droplets originates. A measure for this non-equilibrium is the ratio of the real quality $\dot{x}_{tat}$ and the quality $x_{G,g}$ which would exist if thermodynamic equilibrium would be present, that is, if the vapour would not be superheated or would have given all this superheating to evaporate droplets.

$$\frac{\dot{X}_{tat}}{\dot{X}_{Gg}} = \frac{r}{h_{D,tat} - h_{Fl,s}} \tag{1}$$

In this equation r stands for the latent heat of vaporization and $(h_{D,tat} - h_{Fl,s})$ is the difference between the real enthalpy of the vapour (D) and the enthalpy of the saturated liquid (Fl). Radiation usually can be neglected with the tube diameters used in boilers.

In the literature there is a large number of measurements and correlations presenting the heat transfer in vertical tubes [11-20]. The correlations can be subdivided into the following categories:

1. Empirical correlations assuming thermodynamic equilibrium between the phases.

2. Correlations considering the thermodynamic non-equilibrium by empirical factors.

3. Semi-theoretical, non-equilibrium models calculating the heat transport stepwise.

A correlation belonging to the second category, for example, was presented by Groeneveld and Delorme [21]. Plummer [22] developed a two-step correlation and newer models were presented, for example, by Ganic and Rohsenow [23 and Chen [24].

2. DELIBERATIONS TO PREDICT POST-DRYOUT HEAT TRANSFER

A new model, which upto now is only presented in the Ph.D. - thesis of one of the authors [25], was developed by Schnittger. According to Fig.1 the following energy balance can be established for a steady state, one-dimensional dispersed flow:

$$d\dot{Q} + \dot{M}_D \cdot h_{D.tat} + \dot{M}_{Tr} \cdot h_{Tr} - [\dot{M}_D \cdot h_{D.tat} + d(\dot{M}_d \cdot h_{D.tat})] - [\dot{M}_{Tr} \cdot h_{Tr} + d(\dot{M}_{Tr} \cdot h_{Tr})] = 0 \quad (2)$$

The abbreviations and especially the indices in this equation and in the following ones are explained in the nomenclature.

The heat added to the dispersed system results from the heat flux density $\dot{q}$ at the wall.

$$dQ = \dot{q} \cdot \pi \cdot D_i \cdot dL \quad (3)$$

With the definition $M_D = M_{ges} \cdot x_{tat}$ and by substituting Equ.(3) in Equ.(2) one can write

$$\dot{q}\,\pi\,D_i = \underbrace{\dot{M}_{ges}\,(h_{D,tat} - h_{Fl,s})\,\frac{d\dot{x}_{tat}}{dL}}_{\text{Term 1}} + \underbrace{\dot{M}_{ges}\,\dot{X}_{tat}\,\frac{dh_D}{dL}}_{\text{Term 2}} \quad (4)$$

In this equation the first term on the right side gives the increase of vapour quality due to droplet evaporation and the energy needed to heat up the newly produced vapour from the saturation temperature to the real superheated temperature. In term 2 the enthalpy-rise of the vapour, due to additional superheating from the wall, is presented.

The increase of vapour quality, which would occur if thermodynamic equilibrium would exist, can be calculated from a simple energy balance:

$$dX_{G.g} = \frac{q \cdot \pi \cdot D_i}{\dot{M}_{ges.}\, r}\, dL \quad (5)$$

The change of the real vapour quality is only depending on the heat additions dQ_{Tr} to the droplets

$$d\dot{X}_{tat} = \frac{dQ_{Tr}}{\dot{M}_{ges} \cdot r} \quad (6)$$

Re-arranging the Equs. (4), (5) and (6) we get the differential equation

$$\frac{dh_{D,tat}}{dL} = \left(\frac{d\dot{x}_{G.g}}{dL} - \frac{h_{D,tat} - h_{Fl,s}}{r} \frac{d\dot{Q}_{Tr}}{\dot{M}_{ges} \cdot r \cdot dL} \right) \frac{r}{\dot{X}_{tat}} \quad (7)$$

describing the change of vapour enthalpy $h_{D,tat}$ along the flow path in the tube, starting from the spot where the dryout occurs. At this spot we assume - as starting conditions for the integration - thermodynamic equilibrium between the droplets and the vapour. However, to start this integration we have to know the heat transfer between the vapour and the droplets, and the heat transfer from the wall to the vapour. For the latter one we can use the equation by Groeneveld-Delorme:

$$\alpha_D = \frac{q}{T_w - T_{D,tat}} = \frac{\lambda_f}{D_i}\, 0.008348 \left[\frac{m\ Di}{\eta_f} \left(\dot{X}_{tat} + \frac{g_{D,tat}}{g_{Fl}} (1 - \dot{X}_{tat}) \right) S) \right]^{0.8774} Pr_f^{0.6112} \quad (8)$$

The real vapour quality $\dot{x}_{tat}$ we can express by re-arranging Equ. (1) with the differential superheating $dh_{D,tat}$ of the vapour, starting from the dryout spot

$$\frac{\dot{X}_{tat}}{\dot{X}_{Gg}} = \frac{r}{h_{D,Do} + dh_{D,tat} - h_{Fl,s}} \quad (9)$$

The thermodynamic properties in Equ.(8) have to be selected at the reference temperature T_f

$$T_f = \frac{T_W + T_{D,tat}}{2} \quad (10)$$

We have now to discuss the heat transport dQ_{Tr} before we can integrate Equ.(7):

The heat transport between the superheated vapour and the droplets can be treated similarly to that between a small sphere and a gas flow, however, with a correction for the saturated vapour cushion around the liquid droplet.

Such a correction was proposed by Hoffmann and Ross [26] by using the Spalding-number B

$$B = \frac{h_{D,tat} - h_{D,s}}{r} \quad (11)$$

Doing this, one gets for the heat transfer coefficient at the vapour-droplet interface the equation

$$\alpha_{Tr} = \frac{\lambda_{D,f}}{D_{Tr}} (2 + 0.459 \cdot Re_{Tr}^{0.55} \cdot Pr_f^{0.33}) \cdot (1 + B)^{-0.6} \quad (12)$$

with the definition for the droplet-Reynolds-number Re_{Tr}

$$Re_{Tr} = \frac{D_{Tr} \cdot g_D \cdot w_{rel}}{\eta_D} \quad (13)$$

For small droplets following the vapour flow without slip we can simplify Equ.(12)

$$\alpha_{Tr} = \frac{\lambda_{D,f}}{D_{Tr}} \cdot 2 \cdot \left(\frac{1}{1 + B}\right)^{0.6} \quad (12a)$$

The total heat transported to the droplets in the tube section dL then reads

$$d\dot{Q}_{Tr} = \alpha_{Tr}\ N\ \pi\ D_{Tr}^2\ (T_{D,tat} - T_s) = \frac{2 \cdot \lambda_{D,f} \cdot N \cdot D_{Tr}}{(1 - B)^{0.6}}, (T_{D,tat} - T_s) \quad (14)$$

with N representing the number of droplets existing in this tube section. From the droplet flow $\dot{N}$.

$$\dot{N} = \frac{\dot{M} \cdot (1 - \dot{x})}{\pi/6 \cdot D_{Tr}^3 \cdot g_{F l}} \tag{15}$$

and with the assumption that the droplets have the same velocity as the vapour we get

$$N = \frac{\dot{N}}{w_D} \, dL \tag{16}$$

substituting the vapour velocity in Equ.(16) and introducing it in Equ.(14) we find

$$d\dot{Q}_{Tr} = \frac{3 \cdot \lambda_{D,f} \cdot D_i^2 \cdot (1 - \dot{x}_{tat}) \cdot g_D \cdot dL}{(1 - B)^{0.6} \cdot D_{Tr}^2 \cdot g_{F l}} (T_{D,tat} - T_s) \tag{17}$$

we now need a correlation for the mean droplet diameter D_{Tr} at the dryout spot, because the integration of Equ.(7) has to start there. Measurements of droplet diameters are, for example, presented in [2] and [27]. Using these data, and based on a dimensionless analysis, the following empirical correlation was developed for the mean droplet diameter [25] at the dryout spot:

$$D_{Tr,Do} = 683.5 \cdot \frac{\eta_D}{w_{Do} \cdot \rho_D} \cdot \left(\frac{\sigma}{w_{Do} \cdot \eta_{F l}}\right)^{0.4} \left(\frac{1 - \dot{x}_{Do}}{\dot{x}_{Do}}\right)^{0.25} \cdot \left(\frac{\rho_{F l}}{\rho_D}\right)^{0.325} \cdot \left(\frac{\eta_D}{\eta_{F l}}\right)^{0.1} \tag{18}$$

In this equation the dimensionless group $\sigma / w_{Do} \eta_{F l}$ represents the ratio between the surface tension acting on a droplet departing from the liquid film at the dryout spot and the shear stress, due to vapour flow at the same place. The velocity w_{Do} of the vapour at the dryout spot can be calculated with the mass flow rate density $\dot{m} = \dot{M}/A$ of the mixture:

$$w_{Do} = \frac{\dot{m} \cdot \dot{x}_{Do}}{g_D} \tag{19}$$

Principally, we could now start to integrate Equ.(7) which, however, would comprise the assumption that the number of droplets along the flow path through the tube remains constant, unless they disappear due to total evaporation. Comparing results, calculated this way with measured data, we would realize that the correlation predicts too high wall temperatures or too low heat transfer coefficients, respectively. Therefore, an additional physical phenomenon obviously has to be taken into account. One can think about different possibilities of an incorrect description of the fluid dynamic behaviour in the above mentioned calculating procedure. So the mean droplet diameter may not be chosen correctly, or the assumption that the droplets have the same velocity as the vapour may be too conservative. However, attempts to fit the calculated data to the measured ones by corrections of that kind would bring the theoretical and experimental results into line only within narrow regions, but not over the full length of the tube where post-dryout heat transfer is present, and also for a limited set of parameters only.

So the idea came up that the droplets may undergo an alteration on its way to the tube. Droplets travelling near the wall can have dry collisions with the wall and, due to the steep velocity profile in the boundary layer, are subjected to a higher shear stress than in the center of the flow. Both phenomena may result in a splitting of a droplet into two or several liquid particles. Larger droplets may lose its spherical form.

In addition, a droplet in the boundary layer near the wall is surrounded by a vapour of higher temperature than in the center of the tube. Even without wetting, dry collisions produce a higher momentary evaporation rate than during the transport of the droplets in the core of the superheated vapour.

Schnittger [25] performed detailed deliberations on different effects acting onto the droplets in the boundary layer near the wall and he came out with a semi-empirical correlation which takes into account the increase of the number of droplets and the decrease of the mean droplet diameter, due to droplet splitting by dry collisions and flow-shear stress. He gave the equation for the change in droplet flow rate $d\dot{N}$ versus the differential flow path dL

$$d\dot{N} = K_1 \cdot \dot{N} \left(\frac{\dot{N}}{\dot{N}_{Do}}\right)^{k_4} \cdot \left(\frac{D_{Tr}}{D_i}\right)^{k_2} \cdot Re_D^{k_3} \cdot \frac{1}{D_i} \cdot \left(\frac{\pi/6 \; D_{Tr}^3 \cdot g_{Tr}}{\pi/6 \; D_{Tr,max}^3 \cdot g_{Tr}} - 1\right) \cdot dL \tag{20}$$

The maximum droplet diameter

$$D_{Tr,\,max} = \frac{\sigma \; We_{krit}}{\Delta w^2 \cdot g_D} \tag{21}$$

is calculated with the critical Weber-number

$$We_{krit} = \frac{D_{Tr} \cdot g_D \cdot w_{rel}^2}{\sigma} = 4.5 \div 21 \tag{22}$$

Integrating Equ.(20) we get

$$\int_{\dot{N}_1}^{\dot{N}_2} \frac{d\dot{N}}{\dot{N}^{(1+k_4)}} = \int_0^{\Delta L} k_1 \left(\frac{1}{\dot{N}_{Do}}\right)^{k_4} \cdot \left(\frac{D_{Tr}}{D_i}\right)^{k_2} \cdot Re_D^{k_3} \cdot \frac{1}{D_i} \cdot \left(\frac{\pi/6 \cdot D_{Tr}^3 \cdot g_{Tr}}{\pi/6 \cdot D_{Tr,max}^3 \cdot g_{Tr}} - 1\right) \cdot dL \tag{23}$$

In a stepwise numerical integration, starting from the dryout spot, we now can define $\dot{N}_1$ as the droplet flow rate at the entrance, and $\dot{N}_2$ as the droplet flow rate at the outlet of the tube section dL. We then can re-write Equ.(23) and get

$$N_2 = \left(N_1^{-0.3} - 8.4 \cdot 10^{-7} \cdot \left(\frac{1}{N_{Do}}\right)^{0.3} \left(\frac{D_{Tr}}{D_i}\right) \cdot Re_D^{0.3} \cdot \frac{1}{D_i} \cdot \left(\frac{\pi/6 \cdot D_{Tr}^3 \cdot \rho_{Tr}}{\pi/6 \; D_{Tr,max}^3 \cdot \rho_{Tr}} - 1\right) \cdot \Delta L\right)^{-\frac{1}{0.3}} \tag{24}$$

For the critical Weber-number in Equ.(21) a value of 7.5was derived from measurements. Combining Equs.(7), (12a), (14), (15) and (24), we can now perform a stepwise integration along the flow path through the tube in the post-dryout region and get the real vapour quality and the temperature of the superheated vapour. Returning then to Equ.(8) and by using the well-known interrelation

$$q = \alpha (T_W - T_s) = \frac{Nu\lambda_D}{Di} (T_W - T_s) \qquad (25)$$

we can predict the wall temperature T_w.

3. EXPERIMENTAL TECHNIQUES

Measurements of heat transfer under post-dryout conditions were performed with the refrigerant R12 in 6.12 m long tubes of 12.5 and 24.3 mm inner diameter. The principal arrangement of the experimental apparatus is shown if Fig.2.

The test section itself - the cylindrical tube - was equipped with several thermocouples measuring the wall temperature along the flow path. The real temperature of the superheated vapour in the test section was measured with a special thermocouple arrangement according to Nijhawan [28], which guaranteed that the welding of the thermocouple was only exposed to superheated vapour, but not to liquid droplets. The design of the thermocouple arrangement is shown in Fig.3.

For uniform heating the tube wall served as ohmic resistance. With non-uniform heating over the circumference of the tubes the heat addition was a little more difficult. As shown in Fig.4, an arrangement of indirect heating was chosen. Heating elements were pressed onto the upper half of the outer tube surface by means of an insulating material and a clamp. Each of the heating elements produced the same heat flux so that the heat addition was constant over the half perimeter of the tube within a certain waviness. The tubes had a wall thickness of 2.5 and 4.5 mm respectively, depending on the tube diameter. Therefore, a certain amount of heat can be tangentially transported by conduction in the tube wall, flattening the circumferential power profile. The tangential heat transport is a function of the thermal conductivity of the tube material and of the heat transfer coefficient on the cooled side of the tube.

4. COMPARISON OF MEASURED AND CALCULATED DATA

Measurements were performed with the refrigerant R12 at pressures between 10 and 28 bar, with mass flow rate densities from 400 to 2400 kg/m^{2}s, and with the heat flux densities between 1 and 6 W/cm^2. The experimental results are reported in [25] and [29] in detail. Most of the measurements were performed with uniformly heated tubes, however a part of the investigations was also aimed to see what influence circumferentially non-uniform heating has on the heat transport and whether the opposite cold wall reduces the temperature of the hot side.

Here, only comparisons between measured and calculated data shall be given. Boiler manufacturers are usually not interested in the heat transfer coefficient itself but more in the maximum wall temperature to be expected, because its value limits the design of the steam generating system. Therefore, wall temperatures will be reported here, instead of heat transfer coefficients. The latter comprises also the question of defining the vapour temperature. For practical use and to avoid unnecessary difficulties, it is recommended to choose the saturation temperature of the two-phase mixture - liquid droplets and vapour - in the well-known formulation $\dot{q} = \alpha (T_W - T_{D,s})$.

4.1 Uniformly heated tubes

The post-dryout heat transfer and, by this, the wall temperature of steam generator tubes are influenced by the heat flux and the mass flow rate density of the two-phase mixture, mainly. The tube diameter plays a minor role. Due to the varying density ratio between vapour and liquid the radial transportation of the droplets is depending on the pressure which, therefore, influences the cooling of the wall, too. A much stronger effect has the concentration of droplets in the mixture, however. Low quality, which means a high droplet concentration, guarantees a lower grade of thermodynamic non-equilibrium between the phases than high qualities, because the heat exchanging area between vapour and liquid and the thermal capacity of the liquid phase are larger. Low quality leads to a higher deposition rate of droplets into the superheated boundary layer near the wall. In the following, mainly the influences of the quality $\dot{x}$ and of the heat flux on the wall temperature will be discussed and the agreement between calculation and experiment examined.

The post-dryout heat transfer behaviour is not only quantitatively but also qualitatively different at high and at low qualities. At high qualities with low droplet concentration in the superheated vapour the wall temperature is continuously increasing behind the dryout spot along the flow path of the mixture, as Fig.5 shows. With low qualities, however, the heat transfer downstream of the dryout spot can even improve again, due to strong evaporation of the liquid, as demonstrated in Fig.7. This evaporation enlarges the specific volume of the mixture and, by this, the mean velocity is increased, which again favours the heat transport from the wall. The deposition rate at low qualities - Fig.7 - is much larger than at high qualities (Fig.5).

Both figures (5 and 7) also convey an impression how measured and semi-theoretically predicted wall temperatures agree. The semi-theoretical model, based on the assumption of droplet-splitting and dispersion, fits the experimental data especially in the low quality region, where droplet interaction with the boundary layer near the wall is dominant.

In both figures the temperatures on the inner side of the tube - the wall temperature - is plotted versus the flow path - the length of the tube - with the heat flux as parameter. As discussed in chapter 3, the real vapour temperature $T_{D,tat}$ was measured by a thermocouple, which had a special protection against impinging droplets. The real vapour temperatures, corresponding to the thermo- and fluid dynamic parameters of Fig.5, are presented in Fig.6.

From the measured vapour temperature one can calculate - by a simple energy balance - the real vapour quality, which is also plotted in Fig.6. Comparisions of these experimentally gained data - real vapour temperature and, derived from that, real quality - with predicted ones by the semi-theoretical model give satisfactory agreement.

The tube diameter has a negligible small influence on the heat transfer under post-dryout conditions, as mentioned before and as a comparison of Figs.7 and 8 demonstrates.

The thermo- and fluid dynamic conditions of the experiments reported in these figures were similar, but the tube diameter in Fig.8 is twice of that in Fig.7. The degree of heat-transfer-enhancement downstream of the dryout spot is more and more marked with decreasing steam quality.

To prove that the model presented in chapter 2 is not only valid for the refrigerants but also for water, which is the usual working fluid in power engineering, comparisons were also made with experiments by Kastner [30]. The test section used by Kastner had the same dimensions as that in the R12 - loop presented here. An example of the agreement between correlated and measured data [30] is presented in Fig.9.

Scaling the thermodynamic properties between R 12 and water with the reduced pressure p/p critical, the pressure of 150 bar in the water-data of Fig.9 corresponds to the pressure of 27.9 bar in the R12-data of Fig.7. Comparing these two figures, one realizes with water an even more strongly marked enhancement of the heat transfer downstream of the dryout spot, which can be explained by the larger latent heat of vaporisation in water compared to R12. A more detailed analysis of the semi-empirical correlation with respect to its validity in water is presented in [29].

4.2 Circumferentially non-uniformly heated tubes

In a first evaluation of the data measured in tubes which were only heated from one side, as described in chapter 3, it was assumed that the circumferential heat conduction in the material of the tube wall can be neglected. This means that one half of the tube is uniformly heated and the opposite half would be unheated. Under dryout conditions the temperature of this unheated wall corresponds roughly to the temperature of the superheated vapour. Due to the limited number of heating elements - as described in chapter 3 - the temperature at the inner surface of the heated side of the tube shows a certain waviness, as shown in Fig.10.

In the graphs of the following figures this waviness is smoothed out.

The course of the wall temperatures on the heated and the unheated side of the tube is parallel along the flow path downstream of the dryout spot, as one can see from Fig.11.

This parallelism - but not the temperature difference - is independent from the heat flux. At low vapour qualities the grade of heat transfer enhancement was much smaller with non-uniform heating than it was observed with the uniformly heated tubes before. In this connection, however, it has to be mentioned that the dryout occurred under non-uniform heating conditions much later than under uniform ones at the same thermo- and fluid dynamic conditions. The downstream - shifting of the dryout spot - due to non-uniform heating - can be upto 1 m, as shown in Fig.12.

From this figure we also can see that the wall temperature on the heated side is lower with non-uniform heating than with the uniform one. This is caused by a better cooling, which again is a function of re-distribution of droplets in the vapour.

Finally, an attempt was made to predict the wall temperature under conditions of non-uniform heating with the same semi-empirical model - described in chapter 2 - as for uniform heating.

Fig.13 gives some information about the agreement or disagreement between measured and calculated data. The calculation always over-predicts the wall temperatures of the heated side in this case which, however, easily can be explained by the fact that in these calculations the circumferential heat conduction in the material of the tube wall was not yet taken in to account. Therefore, the over-prediction is larger with high heat fluxes. At low heat fluxes the circumferential heat conduction becomes negligible and then the calculated and measured data agree well. Therefore, the conclusion may be allowed that the semi-empirical model can also be used for non-uniformly heated tubes ; however, it gives conservative data for the wall temperature.

ACKNOWLEDGEMENT

The authors wish to thank the Bundesminister fuer Forschung und Technologie (BMFT) for financially supporting the work reported here.

NOMENCLATURE

- A area
- B Spalding-number
- c_p Specific heat
- D diameter
- h specific enthalpy
- L length
- $\dot{m}$ mass flow rate density
- $\dot{M}$ mass flow rate
- N number of droplets
- $\dot{N}$ flow rate of droplets

Nu Nusselt-number
p pressure
Pr Prandtl-number
q heat flux density
Q heat flux
Re Reynolds-number
r latent heat of evaporation
t time
T temperature
V volume
w velocity
We Weber-number
$\dot{x}$ vapour quality
α heat transfer coefficient
ε local void fraction
η dynamic viscosity
λ heat conductivity
ρ density
σ surface tension

Indices

D," vapour
Do dryout
f film
Fl,' liquid
g gas
G.g equilibrium
ges total
i inner
krit critical
max maximum
rel relative
s saturated
tat real
Tr droplet
W wall

REFERENCES

1. Hewitt, G.F., Hall-Tayler, N.S., Annular two-phase flow, Pergamon Press (1970)

2. Langner, H., Untersuchungen des Entrainment-Verhaltens in stationären und transienten zweiphasigen Ringströmungen, Dissertation, TU-Hannover (1978)

3. Mayinger, F., Strömung und Warmeubergang in Gas-Flüssigkeits Gemischen, Springer-Verlag Wien-New York, (1982)

4. Konkov, A.S., Experimental study of the conditions under which heat exchange deteriorates when a steam-water mixture flows in a heated tube, Teploenergetika 13 (12), 53-57 (1966).

5. Cumo, M., Ferrari,G., and Urbani,C., Prediction of burn-out power with Freon up to the critical pressure, 27th Meeting ATI, Napoli, Sep. (1972)

6. Roko, K, Takitani, K. et al. Dryout characteristics at low mass velocities in a vertical straight tube of a steam generator, Proceedings 6th Int.Heat Trans. Conf. Toronto, (1978).

7. Levitan, L.L., et al., Teploenergetika, (1975) 1.

8. Cumo, M., et al., On the limiting critical quality and the "deposition controlled" burnout, Com.Nat.-Energia Nucleare, RT ING (79)4

9. Bertoletti, S., et al., Heat transfer crisis with steam-water mixtures, Energia Nucleare 12, (1956) 121.

10. Iloeje, O.C., Three step model of dispersed flow heat transfer, General Electric Presentation Post CHF Heat Transfer, Oct. (1975)

11. Miropolski, Z.L., Heat transfer in film boiling of a steam-water mixture in generating tubes, Teploenergetica, No.10, 5 (1963) 49-53.

12. Dougall, and Rohsenow, W.M., Film boiling on the inside of vertical tubes with upward flow of the fluid at low qualities, Techn al Rep.No.9079-26 M.I.T. Dep. of Mech. Engineeri

13. Bishop, A.A., et al., Forced convection heat transfer at high pressure after the critical heat flux, ASME 65-HT-31, (1965)

14. Herkenrath, H., and Mörk-Mörkenstein,P., Wärmeübergang an Wasserbei erzwungener Strömung im Druckbereich von 140-250 bar, EUR-3678d, (1967)

15. Hynek, S.J., Rohsenow, W.M., and Bergles, A.B., Forced convection dispersed flow film boiling, M.I.T. Heat Transfer Lab., Rep.No. DSR 70586-63, (1969)

16. Brevi, R., Cumo, M., et al., Post-dryout heat transfer with steam/water mixtures, Trans.Am.-Nucl.Soc.,12 (1969) 809-811.

17. Lee,D.H., Studies of heat transfer and pressure drop relevant to subcritical once-through evaporators, IAEA-SM-130/56, IAEA Symp.on Progress in Sodium-Cooled Fast reactor Engineering, Monaco (1970)

18. Tong, L , Film boiling heat transfer at low quality subcooled region, Proc.2nd Joint - USAEC-EURATOM Two-Phase Flow Meeting, Germantown, Rep. CONF 640507 (1964) 63.

19. Slaughterbeck, and Mattson, Statistical regression analysis of experimental data for flow film boiling heat transfer, ASME Publ. at the ASME-AIChE Heat Trans. Conf.Atlanta/Ga.,(1973)

20. Laverty,W.F., and Rohsenow, W.M., Film boiling of saturated nitrogen flowing in a vertical tube, ASME 65-WA/HT-26, (1965)

21. Groeneveld, D.C., and Delorme, G.G.J., Prediction of thermal non-equilibrium in the post-dryout regime, Nuclear Engng. and Design 36 (1976) 17-26, North Holland Publ. Company.

22. Plummer, D.N., Post critical heat transfer to flowing liquid in a vertical tube, Ph.D.-Thesis, MIT, (1974)

23. Ganic,E.N., and Rohsenow, W.M., Post critical heat flux heat transfer, Techn.Rep.No.82672-97, Mass.Inst.of Techn.,(1976)

24. Chen, J.C., Sundaram, R.K., and Ozynak, F.T., A phenomenological correlation for post-Chf heat transfer, NUREG 0237,(1977)

25. Schnittger, R., Untersuchungen zum Wärmeübergang bei vertikalen und horizontalen Rohrströmungen im Post-Dryout-Bereich, Dissertation, Univ.Hannover, (1982)

26. Hoffman, T.W., and Ross,L.L., A theoretical investigation of the effect of mass transfer on the heat transfer to an evaporating droplet, Int.J. Heat Mass Trans. 15 (1972) 599 - 617.

27. Ueda, T., Tanaka,H., and Koizumi,Y., Dryout of liquid film in high quality, R-113 upflow in a heated tube, Proc.Int.Heat Transfer Conf.(1978) Toronto/Canada

28. Niihawan,S., Chen, J.C., Sundaram, R.K., and London, E.J.,aMeasurements of vapor superheat in post-critical-heat-flux boiling, J.of Heat Transfer, 102 , (1980) 465-470.

29. Mayinger, F., Schnittger,R., and Scheidt, M.,Untersuchungen zum Wärmeübergang bei vertikaler und horizontaler Rohrstromung, AbschluBbericht BMFT Forschungsvorhaben ET 1412A, (1982)

30. Kastner, W., Köhler, W., Krätzer, W.,, and Hein, D., Wärmeübergang im Post-Dryout-Bereich in vertikalen und horizontalen Rohren bei gleichformiger Beheizung, KWU-AbschluBericht zum-Forderungsvorhaben BMFT ET 1409A (1982)

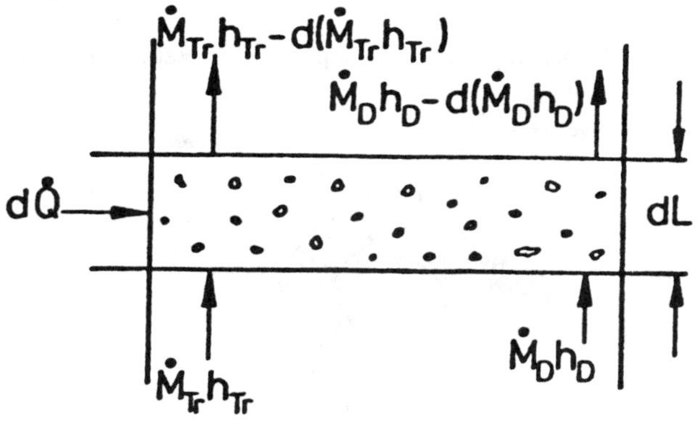

Fig.1 Energy balance of dispersed flow

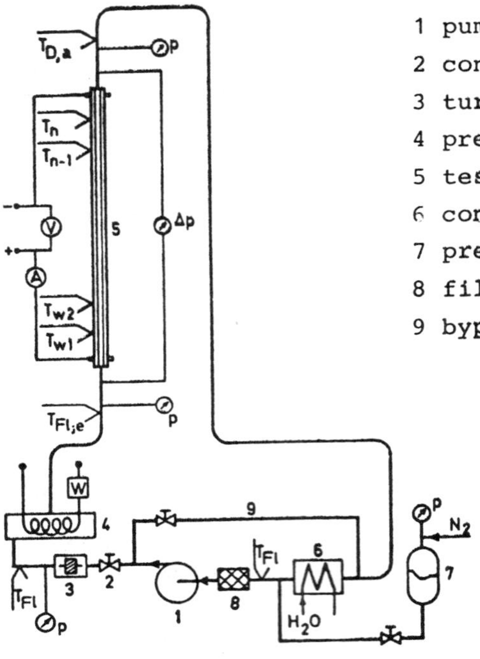

1 pump
2 control-valve
3 turbine flowmeter
4 preheater
5 test section
6 condenser
7 pressurizer
8 filter
9 bypass

Fig.2 Test facility

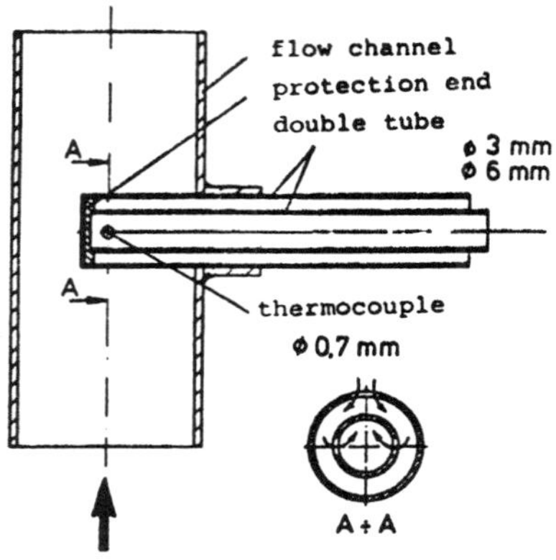

Fig.3 Thermocouple arrangement for superheated vapour in dispered flow

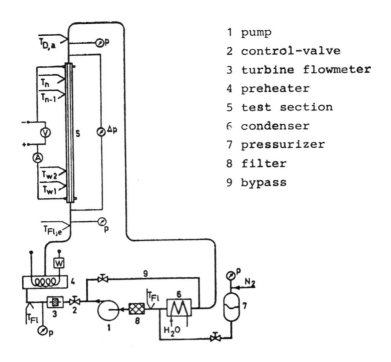

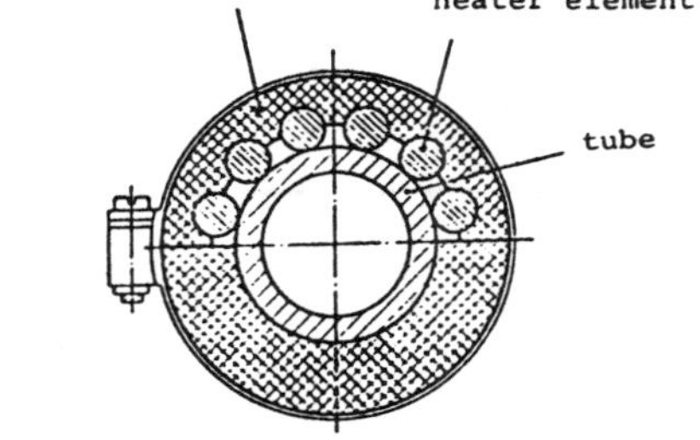

Fig.4 Heating elements on test tube

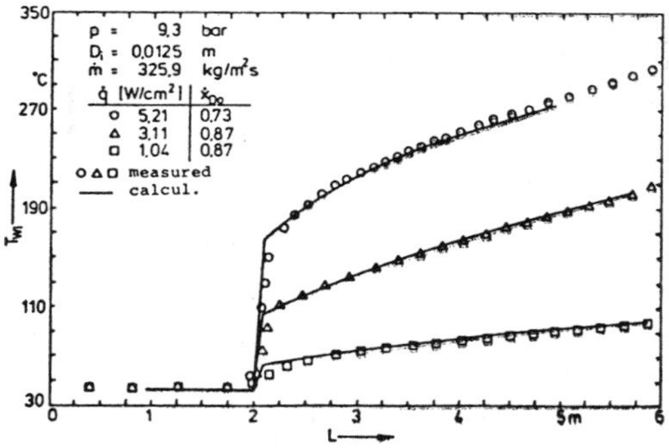

Fig.5 Comparison of measured and calculated wall temperatures with R12; high vapour qualities, no downstream enhancement

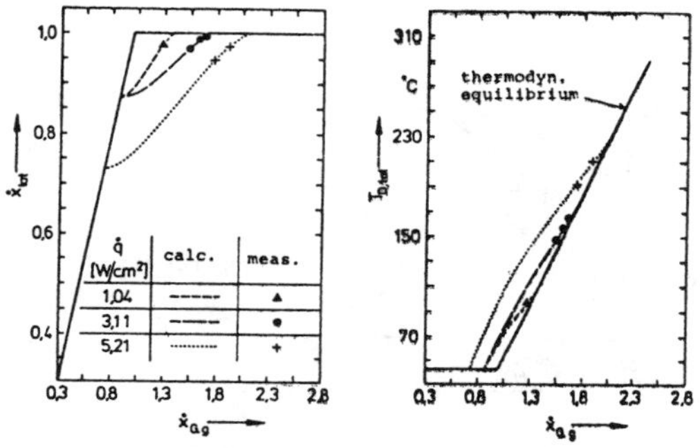

Fig.6 Real Vapour quality and temperatures of the superheated vapour as function of the fictitious equilibrium quality equilibrium quality

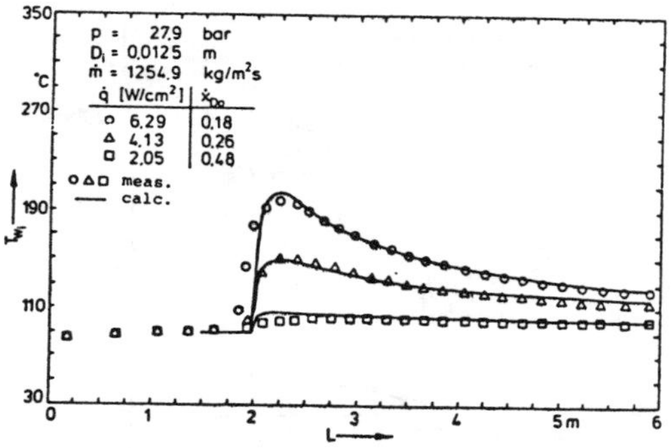

Fig.7 Wall temperatures with R12, low vapour qualities, downstream enhancement

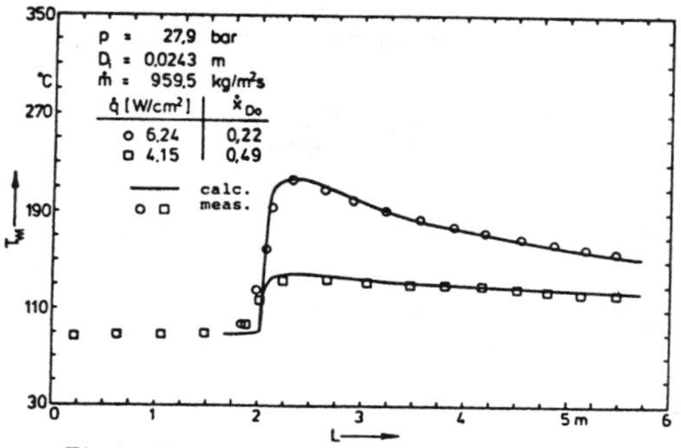

Fig.8 Wall temperatures in large tubes, R12

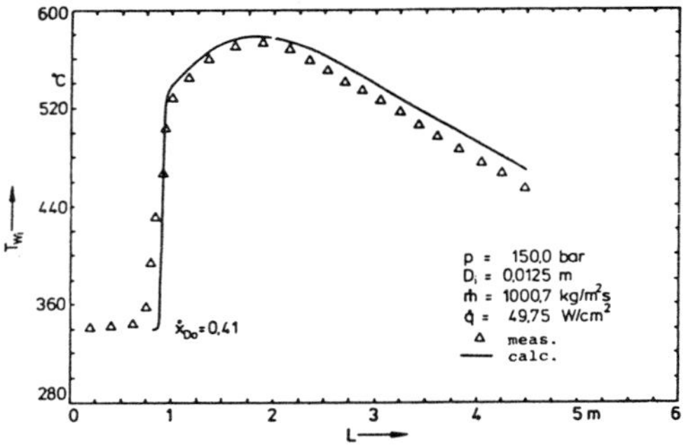

Fig.9 Comparison of measured and calculated wall temperatures; water, low vapour qualities, downstream enhancement

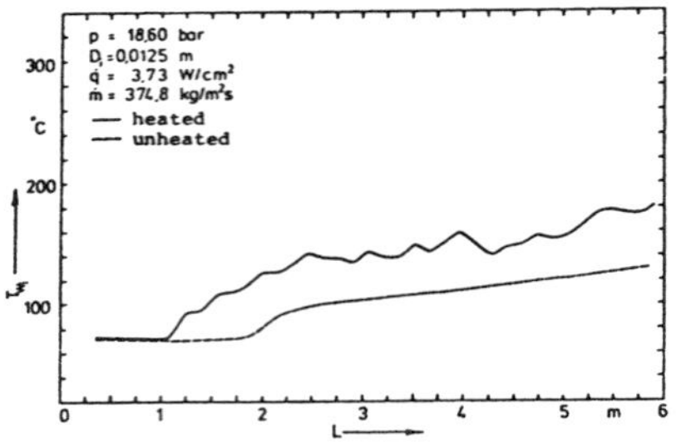

Fig.10 One-sided heated tube, wall temperatures on heated and unheated side

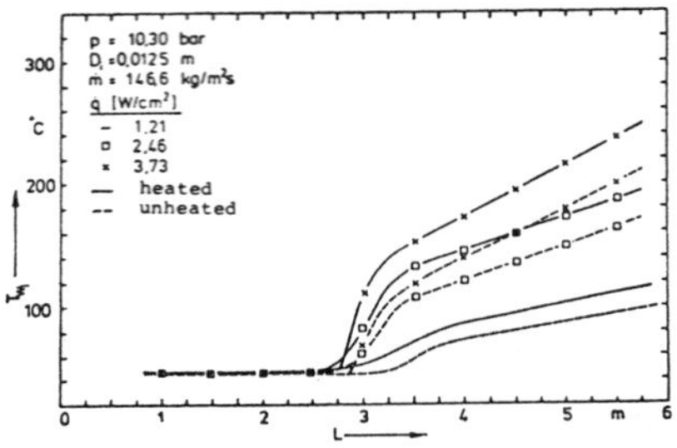

Fig.11 Wall temperatures of one-sided heated tubes, influence of the heat flux

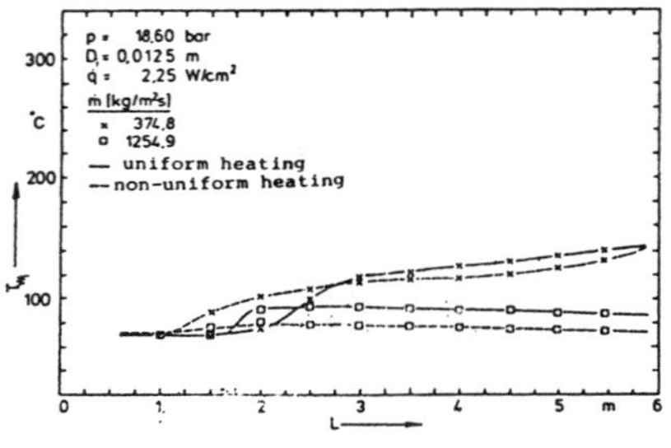

Fig.12 Comparison of wall temperatures with uniform and non-uniform heating

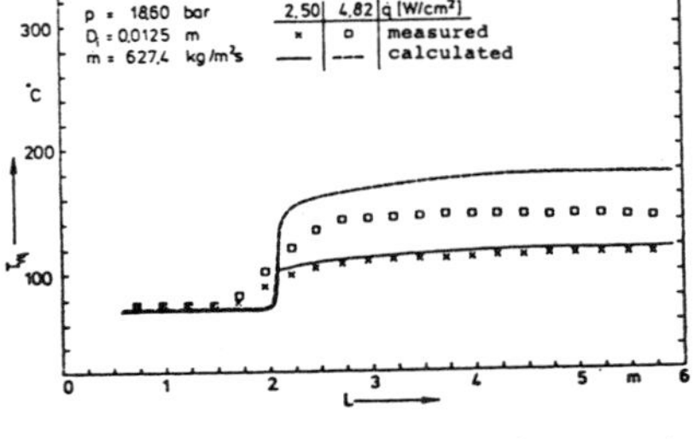

Fig.13 Comparison of calculated and measured wall temperatures for non-uniformly heated tubes (one-sided heating, heated side)

Heat Transfer and Fluid Flow Research Relevant to India's Nuclear Power Program

S. K. MEHTA and V. VENKAT RAJ
Reactor Group
Bhabha Atomic Research Centre
Trombay, Bombay 400 085, India

1. INTRODUCTION

The Indian Nuclear Power Programme envisages three important stages viz., installation of thermal reactors, fast reactors and utilisation of Thorium. By the year 2000 AD, it is proposed to have an installed total capacity of nuclear power of about 10,000 MWe. Starting from the present installed capacity of 1330 MWe, the additional contribution will be mainly made by thermal power reactors of the Pressurized Heavy Water type (PHWR). Apart from the reactors presently under construction about 12 numbers of 235 MWe units are planned to be constructed, which will be based on the standardized design of the reactors at Narora Atomic Power Project (NAPP). In addition, 10 units of 500 MWe capacity each, the design for which is currently under progress, will also be installed. The design, construction and operating agency is the Nuclear Power Board (NPB), while the Bhabha Atomic Research Centre (BARC) is responsible for the research and development work required.

In addition to the programme on thermal power reactors, a thermal research reactor (DHRUVA) of 100 MWth capacity has been designed and constructed by BARC and has been commissioned recently (August 1985). The second phase of the power programme involves the installation of fast reactors. Work in this regard has already been initiated at the Reactor Research Centre (RRC), Kalpakkam. A 15 MWe Fast Breeder Test Reactor (FBTR) has been commissioned recently (October, 1985). Work on the design of a 500 MWe Prototype Fast Breeder Reactor (PFBR) is already in progress. In addition to the nuclear reactors, various associated plants such as Heavy Water Plants, Chemical Process Plants for required raw materials such as Uranium, Zirconium etc, Reprocessing Plants and Waste Management Plants etc. are also required to sustain the nuclear power programme. The design and operation of nuclear reactor systems and associated plants calls for considerable R & D in the fields of heat and mass transfer. Some of the important heat transfer and fluid flow research problems relevant to the Indian nuclear power and research reactors are discussed in this paper.

2. DESCRIPTION OF RELEVANT SYSTEMS

To enable a better appreciation of the R & D problems involved, it may be in order to describe the salient features of the reactor systems and the phenomena involved in these reactors and other associated plants. The details of NAPP type PHWR is given first. The reactor core consists of an array of horizontal pressure tubes (306) housed in a cylindrical vessel called calandria. The pressure tube contains 12 fuel bundles (each of about 495 mm length) which are cooled by pressurised (about 90 bar) heavy water at a mean temperature of 271.2°C (inlet 249°C and outlet 293.4°C). While the fuel bundles used in present day PHWRs have 19 rods arranged in triangular pitch, 22 rod fuel bundles may be used in the 235 MWe units in the future. The Calandria is filled with cool heavy water moderator, at average temperature of 60.5°C, that is insulated from the hot reactor coolant by the calandria tube and the presssure tube with gas space in between.

The Primary Heat Transport (PHT) circuit is arranged in a figure of eight loop with two pumps and two steam generators in each leg of the figure of eight loop. In this circuit (see figure 1) the coolant leaving the reactor core passes through the steam generators and is then pumped back through the reactor and into the other set of steam generators from where it goes to the other set of pumps. The fuel channels in each core pass are individually connected by small diameter feeder pipes (32 mm to 65 mm diameter), to headers at each end. The header is of 400 mm diameter. The reactor is provided with shutdown cooling system for removal of decay heat (generated by fission product decay) under shut down condition and Emergency Core Cooling System (ECCS) to provide cooling under Loss of Coolant Accident (LOCA) conditions. The reactor is housed inside a containment whose purpose is to contain the radioactive fission products in the event of their release from the PHT system. The heat picked up by the primary coolant (heavy water) from the reactor core is transferred to the secondary coolant (light water) in the steam generators (SG). The 500 MWe PHWR being designed is a scaled up version of the 235 MWe unit. As such while there may be some differences, there are many features which will be common to both types of units.

The research reactor DHRUVA (see figure 2), on the other hand is of a completely different design. It has vertical coolant channels with the coolant heavy water flowing from the bottom to top. The fuel is a seven rod cluster, 3050 mm long. The coolant is distributed to the different fuel channels from an inlet plenum. The coolant circuit operates at low temperature (50°C at inlet) and low pressure (about 7 bar at inlet plenum). As against the thermal reactors which use heavy water as coolant, Sodium is used as coolant in fast reactors. In both the 15 MWe FBTR which has been recently commissioned as well as the proposed 500 MWe PFBR which is being designed, the heat removed by the primary sodium from the core is transferred to the secondary sodium in the Intermediate Heat Exchanger. The secondary sodium in turn transfers heat to the water in the Steam Generator. While the primary coolant circuit operates at low pressure, the temperatures are high (about 380°C at core inlet and about 530°C at outlet). The proposed fuel subassemblies consist of 217 slender pins of 6.5 mm diameter.

Apart from the reactor systems described above, other chemical and process plants also form part of the nuclear fuel cycle. The production of fuel, viz. Uranium, and other special raw materials such as Zirconium, involves mining and milling of ores and their subsequent processing to the required materials. The required high recovery of the desired components to the specified levels of high purity is a challenge. Heavy water which forms the coolant and moderator of thermal reactors is to be concentrated from levels of about 150 ppm in light water to purity levels of 99.8%. The spent fuel is reprocessed to recover the fissile materials and separate the highly radioactive fission products. The radioactive waste is further treated for safe disposal. The design of the fuel reprocessing plants and waste treatment plants calls for a knowledge of various heat and mass transfer phenomena. Furthermore as the operation and maintenance of these plants is to be done remotely, detailed studies need to be conducted on prototype equipment to avoid any costly mid-course correction/modification. Apart from the requirements of nuclear fuel cycle, work is also in progress in other areas such as desalination.

The design and safe operation of nuclear reactors and associated systems and plants calls for an understanding of a number of extremely complex heat/mass transfer and fluid flow phenomena. These include: heat transfer and fluid flow in multi-rod bundles involving heat transfer across low conductivity oxide fuels with high temperature gradients, conductance across the gap between the fuel and its clad (this varies with time and operational conditions) and subchannel mixing, thermal hydraulics of horizontal and vertical channels, flow distribution between parallel channels and in plenums, transient behaviour of fuel and heat transfer equipment, blowdown from high pressure systems, condensation induced flow oscillations, rewetting of hot surfaces, heat transfer to liquid metals, natural circulation and thermosyphon cooling phenomena, ore separation and benefication, solvent extraction, calcination and reduction, studies on thermosyphon evaporators, pulsed extraction and molecular sieve columns, exchange processes, vitrification of radioactive wastes, development of profiled tubes, grooved surface tubes for desalination etc. Details of studies in some of these areas are discussed further.

3. HEAT TRANSFER DURING NORMAL OPERATION AND ANTICIPATED OCCURENCES FOR NUCLEAR REACTORS

The design of a nuclear reactor involves interaction between physics and thermal hydraulic parameters. The designer has to cater to various limitations and analysis is carried out to arrive at suitable geometry and appropriate values for the system parameters. The optimisation of major reactor parameters for steady state operation as well as the prediction of reactor behaviour during anticipated transient and accident conditions involves the application of various heat transfer and fluid flow principles. While the analytical methods required for analysis have been developed and improved over the years, there are many areas where analytical methods by themselves are not adequate and recourse is made to experimental investigations.

3.1 Subchannel Analysis of Rod Clusters

One of the important areas of concern in the design is the evaluation of temperature profiles in the fuel bundles used in nuclear reactors. Because of the complicated geometry of the fuel bundles conventional methods of analysis yield results which are at best approximate. For more accurate evaluation of thermal hydraulic parameters in rod cluster fuel elements the subchannel approach has been evolved over the years. The rod cluster cross section is divided into a number of interconnected subchannels (as shown in figure 3 for a 19 rod cluster) for the purpose of analysis. Experimental studies are also required to determine the intersubchannel mixing rates. Much data has been collected on mixing in multi-element bundles and has been analysed to evolve appropriate subchannel mixing correlations. Such correlations are used in subchannel analysis computer codes together with a Dittus-Boelter form of correlation for heat transfer coefficient to evaluate coolant and cladding temperature profiles. For subchannel analysis, a number of computer codes have been developed by various organisations over the years. Codes like COBRA, HECTIC, COBRA-IIIC are available at BARC. These codes have been used on a regular basis for the analysis of rod cluster fuel elements for power reactors are research reactors. Some observations in this regard follow.

In the code HECTIC, the flow distribution between the subchannels is determined on the basis of overall pressure drop across the cluster. The code cannot take into account change in flow area along the flow path. Also diversion cross flow between subchannels due to pressure difference, along the length of the cluster, is not considered. Because of this, difficulties are encountered when this code is used for analysis of rod clusters with spacers at intermediate locations along the length of the cluster. Hence in such cases it is necessary to use a code like COBRA or its later version, COBRA-IIIC. Observations on the use of HECTIC and COBRA for the subchannel analysis of a seven rod cluster fuel element are presented in [1,2]. These studies show that for the geometry of interest the predicted temperatures are only marginally affected by the method of subchannel division. However, the value of mixing parameter has a significant influence.

The 22 element bundles, a cross-section of which is shown in figure 4, have rods of two different sizes. Subchannel analysis, based on assumed values for mixing coefficients, using the code COBRA has been carried out for this fuel; subchannel mixing experiments [3] on such a geometry are being conducted at the Indian Institute of Science (I.I.Sc.), Bangalore for better estimates of subchannel mixing rates. For the proposed 500 MWe unit, it is planned to use 37 rod clusters as fuel elements. Cross-section of a 37 rod cluster is shown in fig. 5. Studies on this design are also in progress. Typical calculated temperature profiles are given in figure 6. Under the auspices of the Engineering Sciences Committee (ESC) of the Board of Research in Nuclear Sciences (BRNS) some research work is being carried out at other Institutes also. Under a project completed at the

Indian Institute of Technology (I.I.T.), Bombay, a generalised computer code [4] has been developed for the numerical prediction of pressure drop and heat transfer charracteristics of fully developed and developing, laminar and turbulent, uniform property single phase flow in rod cluster assemblies consisting of upto 46 rods enclosed in a circular channel. Another project on Combined Free and Forced Convection in Rod-Bundle Assemblies has been completed at I.I.T., Kharagpur [5].

One special feature unique to reactor heat transfer is the presence of spacers in the fuel clusters. The spacers affect flow patterns, improve mixing, and enhance local heat transfer. Downstream of spacers, there is a possibility of recirculation and under such conditions the simple marching technique becomes inadequate. To be able to model such flow situations realistically, elliptic character of the equations has to be retained. Development of a code which can account for such effects and model the flow situations more realistically is of interest.

3.2 Flow Distribution Studies

The various components/equipments/systems of a nuclear reactor involve complex geometries and flow paths. Experimental and analytical studies are required to determine the flow distribution in these. Some illustrative examples follow. The PHWR coolant headers have a number of parallel inflow and outflow connections branching off from them, which results in complex flow patterns in the headers. Scale model studies on a transparent model are planned. The calandria of PHWR is another equipment where complex flow fields are encountered. Studies relevant to NAPP reactor have been carried out at I.I.Sc., Bangalore [6]. Studies pertaining to 500 MWe reactor are to be carried out. Flow distribution studies related to the inlet plenum of DHRUVA which has a number of tubes in cross flow have been carried out on a scale model at BARC. Other problems of interest include flow patterns in the inlet and outlet plenums of the vessel of the fast reactor and the flow distribution in the core.

3.3 BWR Core Thermal Hydraulic Analysis

Core thermal hydraulic analysis forms an essential part of the fuel management service for the Boiling Water Reactors (BWRs) in operation at Tarapur. A distinguishing feature of BWRs is the dependence of core flow distribution on power distribution. Because of the changes in the power distribution from cycle to cycle it is necessary to evaluate the core thermal hydraulic parameters for every reload fuel cycle. A computer code THABNA has been developed for this purpose. The original version of the code developed [7] has been subsequently modified to accommodate 71 channels, i.e. quarter-core of the reactor. The code can be used for calculating the flow distribution, pressure drop, Critical Heat Flux, Minimum Critical Heat Flux Ratio (MCHFR), quality, void fraction etc. in the core of a BWR. Either the total core flow or the core pressure drop can be specified as input for the code. When one of the two is specified as input the other is calculated by the code.

The coolant flow distribution between various fuel assemblies is estimated based on equal pressure drop in all the assemblies. For each channel type, the code carries out a node by node calculation from bottom to top. The following pressure drop components are taken into account in calculating the total pressure drop: friction pressure drop, local pressure drop, acceleration pressure drop and elevation pressure drop. The coolant flow through the various channel types is so adjusted that the core pressure drop or total core flow agrees with the specified input value, within the convergence limits specified. The code has been used to evaluate the various thermal hydraulic parameters for different reload fuel cycles of Tarapur reactors and ensure that specified thermal margin criteria are not violated. Thermal Hydraulic performance of the newly designed 7 x 7 fuel bundles has also been evaluated using this code.

3.4 Thermosyphon/Natural Convection Cooling Phenomena in Reactor Loops

Nuclear reactors continue to generate heat even after they are shut down due to decay of fission products. Unless the heat generated is effectively removed under all possible conditions including abnormal and accidental conditions, fuel overheating will occur leading to undesirable consequences. One such abnormal condition involves the complete loss of power to coolant circulating pumps. Heat removal is effected by the thermosyphon cooling mechanism under such conditions. A phased programme of studies on natural convection in a figure of eight loop with horizontal channels (relevant to PHWR) is being carried out. In the first phase, phenomenological experiments were carried out at atmospheric pressure in a transparent loop [8,9]. The loop consisted of two horizontal channels of annular geometry. The inner stainless steel tube was heated by electric current and outer tube was of glass. The heaters were connected to U-tube coolers through stainless steel headers.

These studies indicate that from a stagnant initial condition, thermosyphon flow initiates in either direction. Below a certain power, flow does not initiate. However, after initiation at a higher power natural convection flow is maintained even if power is reduced to very low values. Noncondensibles hinder natural convection flow by collecting at the top of the inverted U bend in the inner tube of the cooler. When the volume of noncondensibles at the U-bend exceeds a critical value, thermosyphon flow breaks down temporarily and restarts again after reorienting the air bubble (see figure 7). The steady natural convection flow was found to be stable for the entire range of powers tested. Non-dimensional expressions have been developed for flow and heat transfer under natural circulation conditions so as to extend and generalise the validity of experimental data generated. An analytical model has also been developed for the prediction of loop flow rates and temperatures. The buoyancy pressure differential is balanced against the pressure losses. The resulting equation is integrated and further rearranged to obtain the flow rate in terms of non-dimensional parameters (Reynolds number and modified Grashoff number) [8]. Comparison of measured and predicted flow rates gives reasonably good agreement (figure 8). Computer codes for the analysis of natural convection cooling of PHWR and fast reactor cores as well as other closed loop systems have been developed and used to analyse the system behaviour [10-12].

3.5 Studies on Spent Fuel Transfer/Storage and Jacket Piping of Fast Reactors

The spent fuel from the FBTR will be transferred into an Argon filled flask and kept in the storage bay with air cooling. Experimental studies on the decay heat removal from the fuel assembly, under natural and forced cooling conditions, were carried out [13] using an electrically heated full scale mock-up sub-assembly (with 61 pins) simulating the geometry and heat generation rates of the FBTR spent fuel assembly. MgO insulated, stainless steel clad nichrome heater pins were used. The experiments indicated that the temperature excursion caused by loss of forced cooling is not unacceptably high and loss of forced cooling for short periods is not of concern. The problem involves conduction, convection and radiation heat transfer. The calculated temperatures were found to agree well with experimental values when the heater pin emissivity values were taken to be 0.7 to 0.8.

In the 500 MWe PFBR being designed, the spent fuel subassembly (with 217 fuel pins) while being transferred from the internal storage area in reactor vessel to the external storage vessel passes through argon atmosphere. Parametric studies on various designs of transfer pots (used for transferring the fuel) of different shapes have been carried out to evolve a suitable design. A two-dimensional finite element heat conduction code has been used to evaluate the temperatures in the fuel pins during normal transfer operations when the transfer pot will remain in argon atmosphere for about ten minutes as well as under upset conditions when the pot may have to be in argon atmosphere for longer durations. The transfer pot design should be such that under such conditions, the clad temperature at any location should not exceed permissible limits so as to ensure the safety and integrity of the fuel pins. The studies [14] indicated that pots of different shapes and sizes could meet the stipulated temperature limitations.

The primary piping of FBTR (carrying radioactive sodium) is provided with a jacket to contain the active sodium in case of leak. While the outer pipe is insulated, the space between the two is filled with nitrogen gas. A computer code has been developed [15] for calculating the temperature distribution in the double envelope for both concentric and eccentric configurations of the two pipes. It was found that radiation is the predominant mode of heat transfer through the interspace; the eccentricity of the jacket was found to affect the temperature distribution in the inner pipe to a limited extent only. The results of the analysis were found to be in good agreement with the data obtained from experiments on a geometrically similar one-third scale model.

3.6 Transient Analysis of Nuclear Reactors

The transients under operational state are routine and planned. Examples are reactor start up, shutdown, power changes etc. But there are many other transients which are expected to occur once or several times in the reactor life and for which appropriate design provisions are made. Examples are: (i) reactivity changes caused by inadvertent control rod withdrawal, cold water injection from an inactive coolant loop etc.; (ii) failure of equipments such as recirculating pump, feed water pump, turbine trip, loss of power etc. The reactor system should be capable of withstanding such transients. Computer codes have been developed for the analysis of such transients. A discussion follows.

Some of the transients frequently encountered include power transients and flow transients in the primary coolant circuit. A computer code [16,17] has been developed for analysis of such cases given the relevant transients as input. The code can also calculate the flow transient for the case of flow coast down consequent to pump failure. The flow coast down is calculated based on the model given in [18]. In this model, considering the energy balance of the system an expression for transient flow is obtained in terms of steady state energy, rate of energy loss, hydraulic and mechanical losses in pump and motor. The code FLATT considers cylindrical fuel geometry and as such can be used for average flow conditions in rod cluster fuels. Results of an analysis for a 7 rod cluster research reactor fuel for a power transient are presented in figure 9. Computer code COBRA-IIIC can be used for transient subchannel analysis. Computer codes have also been developed for analysis of pump start-up transients as well as transient and stability analysis of heat transfer equipment such as steam generators, heat exchangers etc. [19-21]. Analytical tools for evaluating temperatures in fast reactor coolant channels have also been developed. The applicability of lumped parameter technique vis a vis exact numerical solution, for slow transients, has been examined [22].

While the above mentioned codes model individual equipment/systems, it is necessary to carry out transient analysis of the reactor system as a whole. Computer codes have been developed for this purpose. These include:

i) A Code [23] for the analysis of research reactors like DHRUVA, which models the heat transfer dynamics in the primary, secondary and tertiary coolant circuits involving fuel, pumps, heat exchangers, piping etc.

ii) A Code [24] for the transient analysis of PHWRs. This code has models for fuel pin, pump, steam generator, turbine, etc.

iii) A Code [25] for the dynamic analysis of fast reactors with models for heat transfer in primary system, Intermediate Heat Exchanger, Steam generator, piping etc.

Improvement of the component models built into these codes is an evolving process as also the development of new computer codes.

4. SAFETY STUDIES RELATED TO PHWRs

The Design Basis Accident (DBA) postulated for the Pressurised Heavy Water Reactors (PHWRs) is the Loss of Coolant Accident (LOCA) involving the double ended guillotine rupture of the biggest pipe in the primary heat transport system. Protection against such highly unlikely events as major ruptures of the primary coolant piping system has long been an essential part of the defence-in-depth concept adopted by the nuclear power industry to ensure the safety of nuclear power plants. While in the past attention has been centred mainly on the large break LOCA. the accident at TMI has brought into focus the consequences of a small break LOCA and the need to provide adequate safety features to mitigate the consequences of such an accident. Thus it is necessary to calculate and demonstrate the adequacy of the Engineered Safety Features over a wide range of break sizes.

The computer programs used for analysis are required to evaluate the PHT system transients from the beginning of the blowdown phase to the end of the Emergency Core Cooling (ECC) phase when long term assured cooling is established. This calls for proper modelling of the various phenomena involved viz.

(1) subcooled and saturated blowdown from high pressure systems.
(2) dryout, heat transfer in various two-phase regimes.
(3) emergency coolant injection and distribution and
(4) rewetting of hot horizontal channels.

All these phenomena involve complex two-phase flow and heat transfer and the analytical models need to be validated against experimental data. These data are generated through a 'separate effects' tests programme wherein each of the above mentioned phenomena is investigated individually. The theoretical models so developed for the individual phenomena, have to be integrated into a system code, wherein interactions between the different subsystems and components are accounted for. Finally the predictions of this code have to be validated against Integral System Behaviour experiments which are to be carried out on an appropriately scaled model of the reactor system. The relevant research and development activities involving extensive work in areas such as physical, mathematical and experimental modelling of various transient two-phase flow and heat transfer phenomena, development of appropriate numerical techniques, studies on integral system behaviour etc. have been identified and work is in progress.

4.1 Computer Codes for PHT System Behaviour During LOCA

Two major computer codes, THYNAC and NODE-1, are being developed.

4.1.1. Computer Code THYNAC

The code THYNAC lumps all the 153 core channels connected to one set of reactor headers into one average channel. This channel is divided axially into six segments. The axial variation in heat flux is taken into account while axial conduction is ignored. Variation in thermal properties of UO_2 with temperature are considered. Two phase pressure drop is calculated using Martinelli-Nelson friction multiplier and slip between the two phases is considered. Heat transfer coefficient correlations are based on local coolant conditions. In addition the code takes into account the thermal coupling between PHT system and secondary system in an approximate way. The primary coolant pump behaviour is modelled in a simple way based on single phase flow characteristics. The Zaloudek correlation for subcooled coolant discharge rate and the Moody model for saturated two phase critical flow are used. In its present state of development, the code can perform the thermal hydraulic analysis for a large break LOCA. Additional work to be taken up includes:

(i) Proper modelling of pump under two-phase forward and reverse flow conditions
(ii) Choice of proper heat transfer correlations
(iii) Modelling of clad ballooning and fuel deformation
(iv) Incorporation of a point neutron kinetics subroutine
(v) Modelling of accumulators as proposed for NAPP reactors
(vi) Incorporation of model for metal water reaction

4.1.2 Computer Code NODE-1

The code NODE-1 is essentially a control volume equilibrium code. The system can be divided into any number of specified volumes connected to each other through junctions. Initial conditions such as density, pressure and enthalpy in different volumes and flows in junctions are given as input. The double ended break causing LOCA could occur in any of the junctions. For each volume the conservation equations for mass and energy are solved. The momentum equations include terms for elevation, friction pressure drop and developed pump head. The pump coast-down is calculated allowing the pump to operate in three possible quadrants. The two-phase degradation in pump characteristics is considered. For friction in two-phase flow, Martinelli-Nelson's factor together with mass flux correction is used. The flow through the break is checked and if critical flow is found to occur, Moody's critical flow model is used. The core can be modelled as a number of parallel channels, some of which are lumped together. The heat transfer from the primary to the secondary coolant during the transient is accounted for in an approximate way.

4.1.3 It may be noted that both the codes can handle only the blowdown phase of LOCA for large breaks in PHWRs. The refilling and rewetting phase following emergency coolant injection is not accounted for. Further development work is required before either of these codes can be used for full fledged LOCA analysis for a complete spectrum of break sizes. The codes need to be validated against experimental data. For this purpose separate effects tests as well as integral system behaviour studies have been planned.

In addition to THYNAC and NODE-1, work is also on hand to commission the LOCA analysis code RELAP4-MOD6.

4.2 Separate Effects Tests

4.2.1. Blowdown Studies

4.2.1.1. Analytical Studies

An assessment of the various critical flow models, for both subcooled as well as saturated blowdown is under progress. The subcooled blowdown model is due to Burnell. The saturated blowdown models include the Homogeneous Equilibrium Model and the models due to Fauske, Levy, Moody (with and without friction effects) and Henry-Fauske. A computer code package has been developed for the calculation of critical flow using these models [26]. Presently work is on hand to test the predictions of these models with experimental data available in literature. Subsequently these models will be tested against experimental data to be generated by us simulating PHWR system.

4.2.1.2. Experimental Investigations

Experiments on blowdown from horizontal pipes and pressure tube geometries are planned to be carried out in a phased manner. The experiments planned under the first phase involve generation of data that will enable a better understanding of the phenomenon of blowdown from horizontal pipes of different sizes ranging from 25 mm to 100 mm. These experiments will be followed by experiments with simulated fuel clusters, both without and with heat addition during the transient. Studies are also in progress to develop techniques for measuring transient two-phase flow during blowdown. One technique is to condense the discharge in a pool of water in a vessel and measure its transient weight. Experiments on the use of load cells for measuring the rate of change of weight of a vessel have been carried out [27]. Studies on the use of drag disc in conjunction with gamma-ray densitometer are also in progress. To test its suitability, an experimental facility has been set up. Calibration tests on drag discs are in progress in this set up.

4.2.2. Studies on Emergency Core Cooling System

In PHWRs the emergency coolant is injected into headers, from where it is distributed between the various channels connected in parallel. The relevant problems are discussed further.

4.2.2.1 Steam-Water Interaction in Headers

When the emergency core coolant is injected into headers, the steam-water interaction in the header is likely to cause condensation induced flow and presssure oscillations. Such an oscillatory phenomenon is of great interest because of its impact on ECC delivery to the feeder and in turn to the fuel channels. A phased programme of experiments in this area is in progress. In the first phase, experiments were conducted on a transparent 1/10th scale model of a portion of a typical PHWR header, feeder pipes being simulated by two tubes [28]. The boiler branch was connected to steam supply system and cold water injected through ECC injection line.

The experimental studies conducted at very low steam flow rate (less than 0.3 Kg/min.) gave valuable insight into the phenomena of steam-water interaction. The steam-water interface was found to oscillate as shown in figure 10. The frequency of oscillation varied from about 0.5 Hz to 1.3 Hz depending on steam and water flow rates. Frequency was found to increase with increase in steam flow rate. Flow reversal in the feeder pipes was noted during the upward motion of the interface through the boiler branch. Further studies in this area are in progress.

4.2.2.2. Flow Distribution between Parallel Channels under ECC Conditions

The emergency coolant injected into the coolant header is to be distributed between a large number of parallel channels. In addition to the channel thermal hydraulics, it is likely that the distribution of the injected coolant between the various fuel channels in parallel is affected by the thermal condition of fuel cladding in the various channels. Scale model studies have been planned to investigate the severity of this phenomenon. In this model the hydraulic resistances of the maximum-rated and the minimum-rated channels of the PHWR core are simulated. The remaining channels are lumped into one equivalent channel. The flow distribution between these channels is measured under cold conditions as well as under hot conditions with the surfaces of the different channels maintained at different temperatures before initiating the coolant injection.

4.2.2.3. Studies on Rewetting of Hot Horizontal Channels

By the time emergency coolant is injected into the fuel channels, the cladding may reach very high temperatures due to deterioration in heat transfer. Consequently the emergency coolant will not wet the cladding surface immediately on coming in contact. For effective cooling, rewetting of the fuel clad is essential. The rewetting behaviour is governed by a number of factors such as the coolant flow rate, initial wall temperature, channel orientation etc. A phased programme of analytical and experimental work on the rewetting of hot horizontal channels is carried out [29]. A computer code for conduction controlled rewetting analysis has been developed [30]. Basic studies [31] on the rewetting behaviour of hot horizontal surfaces indicate that

(a) The sputtering temperature is about 200°C at atmospheric pressure.

(b) A transition in the rewetting behaviour occurs when the initial surface temperature is around 350°C (Leidenfrost temperature); above this temperature sustained film boiling occurs ahead of the wet front.

(c) The rewetting velocity increases as the flow rate increases, and decreases as the initial wall temperature is increased.

Experiments on the rewetting of hot horizontal annular channels have been carried out [32]. It is found that stratification effects are significant in the rewetting of hot horizontal channels. At higher flows considerable oscillation of the pressure drop is noted along with corresponding flow oscillations. Visual observations also indicated considerable chugging. Typical rewetting transients are shown in figure 11. The refilling front has been observed to move ahead of the rewetting front. With peaked distribution of the temperature of the inner tube, downstream locations at lower temperature were found to rewet earlier than the upstream locations at higher temperature. At a given axial location the bottom of the inner tube was found to rewet first, the midside next and the top last. The rewetting front is inclined. The rewetting velocity is maximum along the bottom and minimum along the top locations. Heat generation in the tube wall during the rewetting transient causes a reduction in the rewetting velocity.

4.3 Studies on Integral System Effects

The LOCA analysis codes are to be validated against Integral System Behaviour Experiments. These experiments are planned to be carried out in a high pressure, high temperature facility which is presently under advanced stage of construction. It is proposed to incorporate in this facility simulated PHWR fuel channels, with electrically heated rod clusters, for carrying out integral system tests. The rest of the system will be scaled down appropriately.

4.4 Fuel Behaviour under Accident Conditions

During a LOCA, the temperature of the cladding rises at a very fast rate. Due to increase in the temperature, the internal pressure due to fission gases increases; the strength of the clad also comes down. As a result the clad balloons. The ballooning of the clad offers extra resistance to flow and in extreme cases the cladding of adjacent fuel rods may touch each other. The ballooning behaviour of single fuel elements at constant pressure and at constant volume with various rates of increase of temperature is being studied in an out of pile set up. The next phase of studies would include the effect of neighbouring rods and the channel on the ballooning behaviour of fuel rods. The modelling work on this phenomenon is complicated by the fact that as the clad balloons, the clad temperature decreases locally and thus the clad regains part of the lost strength.

Due to the temperature increase, the oxidation rate of the cladding also increases. This reaction being exothermic, in addition to the decay heat, the heat of reaction also heats up the clad. To accurately evaluate the peak cladding temperature, the rate of reaction at various temperatures needs to be studied in detail. The work on the measurement of reaction rate as a function of temperature has already started. During the heat up phase of LOCA, the cladding temperature crosses the alpha to beta transformation temperature for Zircaloy. The oxygen diffusion in various phases and the effect of oxygen on the embrittlement behaviour of cladding needs very detailed study. A computer program to study the diffusion of oxygen in Zircaloy is being developed.

4.5 Containment Behaviour

The reactor containment constitutes the ultimate safety barrier of a nuclear power plant which protects the general public and the environment from unacceptable releases of radioactivity. The design of the PHWR containment has been progressively improved. The present day design uses double containment with a facility to purge the intermediate space. The engineered safety feature to limit the pressure and temperature rise within the containment following a LOCA has also undergone changes. The initial PHWRs had dousing system to condense the vapour in case of an accident. This has been changed to a vapour suppression pool type system to condense the flashing coolant. Considerable effort had to be put in to adopt this system modification for the PHWR. Unlike in LWRs where only steam needs to be condensed following LOCA, in the case of PHWRs, during the initial stages following the accident, a mixture of steam and air flows through the downcomers.

Experiments [33] were conducted to determine condensation of steam and heat transfer from air to suppression pool during its passage through the pool. Condensation of steam was found to be complete. It was observed that the air steam mixture during its passage through different compartments of the containment is far from homogeneous. There is a need for greater understanding of the phenomenon with a view to correctly model it. Experiments to determine the pressure and temperature transients in a scaled containment model are planned to be conducted. This model is 1/1000th of MAPP prototype by volume i.e 1/10 on linear scale. Besides this the following areas have been identified for further investigations:

(i) Experimental studies for evaluation of chugging loads
(ii) Development of analytical model for pool swell and related experiments
(iii) Studies on impact loads during pool swell
(iv) Experimental determination of lateral loads on downcomers due to asymetries in flow during vent clearing.

5. STUDIES ON FAST REACTOR SAFETY

In liquid metal cooled Fast Breeder Reactors (LMFBRs) the core compaction following any accident leading to fuel melting may result in severe reactivity and power excursions and consequent core disruption. The accidents called Hypothetical Core Disruptive Accidents (HCDAs) because of their very low probability are also known as Design Basis Accidents [34]. The unprotected Loss of Flow Accident (LOFA) and the uncontrolled Transient Over Power Accident (TOPA) fall under this category. The HCDA involves (i) the formation of highly pressurised bubble of vaporised fuel and coolant (called core bubble). (ii) Expansion of the core bubble in the liquid sodium. (iii) Propagation in the coolant of pressure waves which induce upward lifting or radial outflow motion of fluid along with the associated loadings on the reactor structures. (iv) Large displacements of the different shell structures leading to eventual plastification and/or buckling of these structures [35].

The modelling of DBAs involves: core neutronics, reactivity feedbacks, thermal hydraulics, sodium boiling, fuel pin mechanics and failure, cladding and fuel slumping and their relocation, fuel-coolant interaction, vaporisation and expansion and post accident heat removal. The various computational models and codes required have been developed. Studies on the importance of nucleate and film boiling mode for heat transfer during Molten Fuel Coolant Interaction (MFCI) [36] indicate that the heat transfer is initially through film boiling. An energetic MFCI is unlikely in LMFBRs since before the conditions for vapour explosion are met, the temperature completely flattens out because of the large thermal conductivity of coolant and also availability of large amount of coolant. Studies on the sensitivity of heat transfer parameters during LOFA [37] indicate that heat transfer coefficients, particularly the gap bond-conductance influences the sodium boiling propagation in LMFBR cores. Larger the bond-conductance, more incoherent is the sodium boiling propagation, whereas fuel slumping propagation is more coherent. Higher the bond conductance lower is the energy release. Work on further improvement and validation of the various models and computer codes is in progress.

6. BASIC STUDIES

Basic studies on a number of topics relevant to nuclear reactor systems have been carried out. Some of these are discussed further.

6.1. Two-Phase Flow and Heat Transfer

Critical Heat Flux (CHF) measurements on tubes have been carried out with water (at about 70 bar pressure). CHF measurements have also been carried out with Freon (at about 10.3 bar pressure) [38], with a view to developing fluid modelling techniques. The scaling factor approach is found to be quite useful in fluid to fluid modelling of CHF in tubular geometries. Mass flux scaling factor F_G found in this investigation i.e. 1.45, is in close agreement with that found by others. Literature review indicates that the scaling factor approach and use of Freon have so far met with only limited success in the context of modelling of CHF in rod clusters. The modelling relations are fairly accurate for simple geometries, but not adequate when applied on global basis to full scale rod bundles. Hence in the design of reactor fuels which are invariably rod bundles, only water data are used. In view of this emphasis needs to be placed on achieving a better understanding of the physical phenomena occurring in the rod cluster.

Using two-phase pressure drop measurements carried out with water in diabatic two-phase flow at about 71 to 75 bar pressure an assessment of the predictive capabilities of various pressure drop correlations (6 in number) was carried out [39]. It was found that the available correlations give reasonably good predictions. The uncertainties can be of the order of 40%, depending on the correlations used. The Baroczy [40] or CNEN [41] correlation for friction pressure drop in combination with Bankoff correlation for void fraction was found to give the best prediction.

Studies on two-dimensional modelling of annular two-phase flow have been carried out [42]. A two-dimensional model for the analysis of annular two-phase flow through tubular geometry based on the solution of continuity, momentum and energy equations has been developed. The numerical method used is an adaptation of Patankar-Spalding method. For the evaluation of Reynolds stresses a mixing length model has been postulated for the case of annular two-phase flow. The model calculates a variety of two-phase annular flow characteristics including drying out for both uniform and non-uniform heat flux distribution, in a single frame work. The model has been validated against available experimental data obtained from air-water, steam-water and Freon-12 systems.

6.2 Flow Visualization

Flow visualization plays an important role in gaining an understanding of the physical processes involved in fluid flow phenomena. It is a useful, sometimes indispensable supplement to theoretical and other experimental studies. A number of techniques suitable for visualising flow patterns in models of nuclear reactor components have been developed [43,44]. Tellurium dye emanating from a tellurium electrode was used to visualize a streak line in natural convection of water as shown in figure 12. Flow pattern covering a larger flow field in natural convection of water was obtained using aluminium particles of size varying from 0.5 micron to 11 micron (Please see figure 13). To study air flow patterns close to a solid surface a tracer paint consisting of titanium dioxide, liquid paraffin and oleic acid was successfully used. A typical flow pattern thus obtained is shown in figure 14. Further work is in progress regarding various methods of flow visualization. Some of the important flow phenomena requiring visualization studies pertain to flow fields in the calandria of a PHWR, vessel of a fast breeder reactor, flow distribution in PHWR header ect. as explained earlier.

6.3 Natural Convection in Annuli

In the design of many nuclear systems one often encounters the problem of restricting natural convection of hot process fluid through a vertical annular passage in order to limit temperatures at locations of temperature sensitive components and equipments. Two possible ways of reducing natural convection through a vertical annulus open to hot fluid at bottom are: (1) reduction in annular gap [45] and (2) introduction of baffles in the annular gap. Studies were carried out to assess the effectiveness of these methods. Heat transfer through the annular gap under consideration involves both conduction as well as natural convection. This combined conduction-cum-convection problem is transformed into a "conduction only" problem by ascribing an effective thermal conductivity, K_{eff}, to the fluid in the annulus. Subsequently, solution is obtained by using available conduction codes like HEATING. In the case of annular gap without baffles, K_{eff} was determined by using available correlations. The temperature distribution calculated using these values of K_{eff} was found to agree quite well with the measured temperature distribution. The studies indicated that the threshold between viscid and inviscid regimes is highly sensitive to the annular gap. Typical flow pattern in a 1 mm annulus is shown in figure 15. For the case of annulus with baffles a correlation for K_{eff} was obtained from experimental data. Temperature distribution predicted using this correlation is compared with measured temperature distribution in figure 16. These studies indicate that both the methods of reduction of natural convection are effective. However, it was observed that for annulus without baffles substantial reduction of convection can be achieved only when the annular gap is made very narrow.

6.4 Pressure Drop in Rod Clusters

Experimental measurement of pressure drop across rod cluster fuel elements has been carried out. Based on studies on longitudinally finned seven-rod cluster bundles [46], the following conclusions have been drawn.

(i) In addition to the P/D ratio, the W/D ratio may also have an important influence on the friction factor for rod clusters and specially so for clusters with fewer rods. (Hence P is the rod to rod distance, D is the rod diameter and W is the wall distance)

(ii) In clusters with finned rods it seems more appropriate to compare rod cluster friction factor with that for smooth tubes on the basis of P/D_R instead of P/D. (Here D_R is the equivalent rod diameter including the effect of fins)

(iii) The length of a spacer also has a significant effect on the pressure drop across it.

To generate data required for safety analysis, experiments have been carried out to measure pressure drop across a PHWR fuel channel containing 19 rod fuel bundles (wire-wrap type) in the Reynolds number range of 70 - 50,000 [47]. The data are plotted in figure 17. It is found that the flow is laminar in character upto Re = 200. Further studies on 22 and 37 element fuel bundles are to be carried out.

6.5 Some Special Techniques, Instrumentation and Equipment

In nuclear reactors and heat transfer research associated with nuclear industry often special measurement techniques and instrumentation are required to be developed. Some examples are given here. In heat transfer experiments nuclear heating is simulated by electrical resistance heating of tubes. Diferent methods of fixing thermocouples to measure temperature of current (A.C) carrying surfaces have been tested to arrive at a suitable technique for accurate measurement [48]. Techniques for recording temperature transients during safety related experiments (rewetting) have been evolved [49]. Response time of special thermocouples to be used for measurement of sodium temperature in FBTR has been determined experimentally [50].

Special equipment for use in high temperature, high pressure heat transfer research facilities have been developed. These include heaters based on direct electrical resistance heating concept and pool boiling type of coolers. The latter offer many advantages when the fluid to be cooled is at a temperature substantially higher than the saturation temperature of water at atmospheric pressure [51]. The fluid to be cooled flows through 'U' tubes immersed in a pool of boiling water. Boiling occurs in the water pool and high heat fluxes are achieved. Cooling water requirements are also small.

7. SOME STUDIES RELATED TO ASSOCIATED PLANTS

As explained earlier various chemical and process plants associated with the production of raw materials such as fuel, cladding, heavy water etc. and reprocessing and waste management plants form part of the nuclear fuel cycle. Some of the heat transfer, mass transfer and fluid flow phenomena associated with these plants are discussed further.

7.1 Heat Transfer Studies Associated with Storage of Radioactive Waste

The radioactive waste produced in nuclear reactors and associated plants is to be treated and stored under monitored conditions. One method of providing monitored storage is to have underground storage vaults. The design and construction of such vaults is a challenging task in view of the special requirements such as arrangements for handling, provision of assured long term cooling, seepage free boundaries etc. The last requirement calls for careful evaluation of temperature gradients in the walls to enable proper structural design. A parametric study of the effect of parameters such as inside air temperature, wall thickness, wall thermal conductivity, heat generation in the wall due to incident radiation, condition of surrounding soil etc. on the wall temperature differentials has been carried out [52]. It is found that the temperature differential across the boundary wall increases with increase in heat generatioan rate, rise in) air temperature, increase in soil thermal conductivity, reduction in thermal conductivity of wall material and increase in wall thickness. Based on the studies, a successful design for monitored storage of radioactive waste has been evolved.

Work related to the dispersion of chemical nuclides in porous soils is being carried out at I.I.T., Bombay under a BRNS sponsored project. The work involves developing a mathematical model of the transportation equation for dispersion of chemical nuclides in porous saturated/unsaturated soils, solving the equation for different boundary conditions using numerical methods and estimating concentration profiles for various distances at different times.

7.2 Studies on Reverse Osmosis (RO) Desalination

R & D work has been carried out on reverse osmosis with a view to develop this technology for commercial scale desalination and other important industrial applications. A compact tubular module based on tube bundle concept (consisting of 14 numbers of tubes) offering 1 m^2 membrane area was designed and fabricated. The module offers 95% salt rejection and 0.700 m^3/m^2 day permeate flux when tested under 40 kg/cm^2 operational pressure for saline water of 5000 ppm TDS. Eight to ten such modules when connected in series provide an unit of 5 m^3/day capacity RO plant. A number of such units make an RO plant of 50 to 100 m^3/day capacity. The operational parameters of such an assembly (i.e applied pressure, feed velocity, percentage recovery etc.) have been optimised. The effect of other process parameters such as feed temperature, pH, turbidity, fouling, scaling potential etc. has been studied for ensuring longer membrane life. The RO systems developed are capable of desalting brackish waters having TDS 10,000 ppm into fresh potable water.

Apart from desalination, the RO plants have been also tested for production of boiler grade water in conjunction with demineralizer and in the treatment of liquid effluents from chemical industry as well as radioactive effluents generated in the nuclear industry.

Work is being carried out at present on increasing the number of tubes and their length to derive a membrane area 4-5 times greater than existing modules. Attempts are also under way to test cheaper material of construction such as hot dip galvanised/electroplated mild steel or FRP as end plates in place of stainless steel. Incorporation of turbulence promoting devices to improve the mass transfer is being studied. These systems are ideal for making small RO units driven at very low feed velocities.

A compact module design based on filter press concept has been recently developed. Overall improvement in economics of RO is noted. The membranes used so far in the RO system are based on cellulosic polymers available locally. Work on development of new polymeric membranes with far superior properties is in progress.

8. CONCLUDING REMARKS

Some of the R & D problems, in the area of thermal hydraulics, relevant to nuclear reactors and associated systems/plants have been discussed in this paper. It may be seen that these problems cover a wide range of topics in single/two phase flow and heat transfer as well as mass transfer. R & D in this area is an evolving and continuous process.

ACKNOWLEDGEMENTS

The authors wish to thank the engineers of Reactor Group, Chemical Engineering Group and Waste Management Division of BARC, Reactor Research Centre and Nuclear Power Board for supplying the required information.

REFERENCES

1. Venkat Raj, V., Grover, R.B., Murthy L.G.K., and Mehta S.K., Thermal Hydraulic Subchannel Analysis of a Seven Rod Cluster Nuclear Fuel Element, Paper No. D-10, Second National Heat and Mass Transfer Conference, Indian Institute of Technology, Kanpur, December (1973).

2. Grover R.B., Koranne S.M., Venkat Raj V., and Mehta S.K., Steady State and Transient Thermal Hydraulic Analysis for the Fuel of a Research Reactor, Part 1 - Steady State Analysis, Paper No.4HMT-82, Fourth National Heat and Mass Transfer Conference, University of Roorkee, November (1977).

3. Das M et. al., Flow Mixing studies on 22 Element Fuel Bundle for Narora Atomic Power Station, HMT-A10-83, Seventh National Heat and Mass Conference, I.I.T., Kharagpur, December (1983).

4. Benodekar R.W., and Date A.W., Prediction of Flow and Heat Transfer in Nuclear Rod-cluster Assemblies, I.I.T., Bombay (1978), (Report on BRNS Project).

5. Das R., and Mohanty A.K., Combined Free and Forced Convection in Rod-Bundle Assemblies, I.I.T., Kharagpur, (1982) (Report on BRNS Project).

6. Sridharan K., Neema L.K., and Rama Prasad, Flow Velocities and Frequencies in the NAPP Calandria, Report No.80 N 9, I.I.Sc., Bangalore, (1980).

7. Venkat Raj V., Anand A.K., and Saha D., THABNA - A Computer Program for the Thermal Hydraulic Analysis of Boiling Nuclear Assemblies, Paper No.G-4, Second National Heat and Mass Transfer Conference, Indian Institute of Technology, Kanpur, December (1973).

8. Vijayan P.K., Venkat Raj V., Mehta S.K., Date A.W., Investigations on Natural Convection in a Figure of Eight Loop, Paper No. B 10, 8th National Heat and Mass Transfer Conference, Visakhapatnam December (1985).

9. Vijayan P.K., Venkat Raj V., Mehta S.K., and Date A.W., Investigations on Natural Convection in a Loop Relevant to a PHWR, ASME Winter Annual Meeting, Florida, Nov.(1985).

10. Ghosh A.K., Bandyopadhyay S.K., Murthy L.G.K. and Ramamoorthy N., Ultimate Heat Sink for PHWRs - Evaluation of Existing Methods and Alternative Proposals, Nuclear Engineering and Design 74 (1982) 145-152.

11. Vaidyanathan G., and Paranjpe S.R., An Analysis of Phenomena Leading to Establishment of Natural Convection in FBTR Paper presented at the Indo-German Workshop on Transient Analysis and Emergency Core Cooling Systems held at BARC from Jan.21 - Feb.1, (1985).

12. Vijayan P.K., Shah D., and Venkat Raj V., Analysis of Thermally Induced Flow and Heat Transfer in a Closed Loop, Paper No.HMT-C3-83, 7th National Heat and Mass Transfer Conference, I.I.T., Kharagpur, (1983).

13. Prahlad B., and Kale R.D., Study of Temperature Evolution During Dry Storage of Spent Fuel Sub-assembly of the Fast Breeder Test Reactor, Reg. J. Energy Heat Mass Transfer No.3, 6, (1984) 229-235.

14. Markandeya S.G., Dutta B.K., and Venkat Raj V., Thermal Design Analysis of Transfer Pot for Handling of Spent Fuel from Fast Reactors Paper No.E 3/5, 8th International Conference on Structural Mechanics in Reactor Technology, August 19-23, Brussels, (1985).

15. Prahlad B., Gunasekaran T.G., and Kale R.D., Study of Temperature Distribution on the Jacket Piping of Fast Breeder Test Reactor, Reactor Research Centre, Kalpakkam.

16. Venkat Raj V., and Koranne S.M., FLATT - A Computer Program for calculating Flow and Temperature Transients in Nuclear Fuels, Paper No. HMT-93-75, Third National Heat and Mass Transfer Conference, Indian Institute of Technology Bombay, December (1975)

17. Koranne S.M., Grover R.B., Venkat Raj V., and Mehta S.K., Steady State and Transient Thermal Hydraulic Analysis for the Fuel of a Research Reactor, Part-II - Transient Analysis, Paper No. 4 HMT-83, Fourth National Heat and Mass Transfer Conference, University of Roorkee, November (1977).

18. Yokomura T., Flow Coast Down in Centrifugal Pump System, Nuclear Engg. Design, 10, (1969), 250-258.

19. Grover R.B. and Koranne S.M., Pump Start-up Transients, Nuclear Engg. Design, 67, (1981), 137-141.

20. Rajakumar A., Development of a Theoretical Model for Stability Analysis of Sodium Heated Once Through Steam Generator, Reactor Research Centre, Kalpakkam.

21. Koranne S.M., Grover R.B., Kulkarni V.P., and Venkat Raj V., Mathematical Simulation of Transients in Shell and Tube Type Heat Exchangers, Paper No. HMT-68-81, 6th National Heat and Mass Transfer Conference, I.I.T., Madras (1981).

22. Singh O.P., Ponpandi S., and Parikh M.V., On the Exact Heat Conduction and Lumped Model of Heat Transfer and the effects of Heat Transfer Parameters on the Temperature Distribution in LMFBR Cooling Channels, REDG, Reactor Research Centre, Kalpakkam

23. Koranne S.M., Grover R.B., and Venkat Raj V., A Computer Code for Thermal Hydraulic Dynamic Analysis of a Nuclear Reactor System, National Systems Conference, I.I.T., Bombay, Dec.(1984)

24. Chamany B.F., Murthy L.G.K., Ray R.N., and Sastry S.R., Digital Simulation of the Dynamics of Rajasthan Atomic Power Station, BARC, Bombay.

25. Vaidyanathan, G., Kothandaraman A.L., Rajakumar A., and Raviprasan G.R., Dynamic Modelling of a LMFBR Plant, Paper No. HMT-64-81, 6th National Heat and Mass Transfer Conference, I.I.T., Madras (1981).

26. Dolas P.K., and Venkat Raj V., Critical Flow Models for Nuclear Reactor Safety Evaluation, Paper No. HMT-C2-83, 7th National Heat and Mass Transfer Conference, Indian Institute of Technology, Kharagpur, (1983).

27. Dolas P.K., Venkat Raj V., Ghosh A.K., Murthy L.G.K., and Muralidhar Rao S., Measurement of Blowdown Flow rates using Load Cells B.A.R.C. 1070, (1980).

28. Venkat Raj V., Saha D., Dolas P.K., Markandeya S.G., Some Separate Effects Tests Pertaining to LOCA, Paper Presented at the Indo-German Workshop on Transient Analysis and Emergency Core Cooling Systems held at BARC, Trombay, from Jan.21 - Feb.1, (1985).

29. Venkat Raj V., Studies on the Rewetting of Hot Horizontal Surfaces, Ph.D., Thesis, I.I.T., Bombay, (1985).

30. Venkat Raj V., and Date A.W., Analysis of Conduction Controlled Rewetting of Hot Surfaces Based on Two Region Model, Paper accepted for presentation at the 8th International Heat Transfer Conference, San Francisco, August, (1986).

31. Venkat Raj V., Sukhatme S.P., and Mehta S.K., Experimental Investigations on the Rewetting of Hot Horizontal Surfaces, Paper No.FB 26, 7th International Heat Transfer Conference, Munich, September (1982).

32. Venkat Raj V., Experimental Investigations on the Rewetting of Hot Horizontal Annular Channels Int. Comm. Heat Mass Transfer, 10, (1983). 299-311.

33. Anil Kakodkar, and Mehra V.K., Vapour Suppression Pool Experiments (Phase-1), BARC Report B.A.R.C./1-171, BARC, (1972)

34. Om Pal Singh, Bhaskar Rao P., Ponpandi S., and Shankar Singh R, Design Basis Accident Analysis Work at RRC, Paper Presented at the Indo-German Workshop on Transient Analysis and Emergency Core Cooling Systems held at BARC, Trombay, Bombay from Jan. 21 - Feb. 1, (1985)

35. Chellapandi P., and Bhaskar Rao P., Consequences of HCDA: Shock-structure Interaction, Paper Presented at the Indo-German Workshop on Transient Analysis and Emergency Core Cooling Systems held at BARC, Trombay, Bombay from Jan. 21 - Feb. 1, (1985).

36. Bhaskar Rao P., Om Pal Singh, and Shankar Singh R., On the Importance of Nucleate and Film Boiling Mode for Heat Transfer in Molten Fuel Coolant Thermal Interaction Studies, Paper No. HMT-F8-83, 7th National Heat and Mass Transfer Conference, I.I.T., Kharagpur, (1983)

37. Om Pal Singh, Bhaskar Rao P., and Shankar Singh R., A Study of the Sensitivity of Heat Transfer Parameters on Sodium Boiling Propogation, Fuel Slumping and Subsequent Effects in LMFBRs Under Unprotected Loss of Flow Accidents, Paper No. HMT-F20-83, 7th National Heat and Mass Transfer Conference, I.I.T., Kharagpur, (1983).

38. Koranne S.M., Grover R.B., and Venkat Raj V., Experimental Investigations on Modelling of Critical Heat Flux in Water Using Freon-12, Report B.A.R.C./1-707 (1982).

39. Vijayan P.K., Prabhakar B., and Venkat Raj V., Experimental Measurement of Pressure Drop in Diabatic Two-phase Flow and Comparison of predictions of Existing Correlations with the Experimental Data, Paper No. HMT-71-81, 6th National Heat and Mass Conference, I.I.T., Madras (1981).

40. Baroczy C.J., Chem. Engg. Progr. Symp. Series, No.64, 62 (1964) 232-249.

41. Marinelli V., and Pastori L., AMLETO - A Pressure Drop Computer Code for LWR Fuel Bundles, RT/ING(73)11, CNEN, August (1973).

42. Grover, R.B., Two-Dimensional modelling of Annular Two-Phase Flow, Ph.D. thesis, I.I.Sc., Bangalore (1981).

43. Vijayan P.K., Saha D., and Venkat Raj V., Some Techniques for Flow Visualisation In Air and Water, Paper No. HMT-70-81, 6th National Heat and Mass Transfer Conference, I.I.T., Madras, (1981)

44. Saha D., Vijayan P.K., and Venkat Raj V., Visualisation of Natural Convection Flow Using Aluminium Particles Measurement Techniques in Heat and Mass Transfer, Proceedings of the International Centre for Heat and Mass Transfer; 18, (1985) 91-102.

45. Vijayan P.K., Saha D., and Venkat Raj V., Investigations on Natural Convection Heat Transfer in Narrow Vertical Annuli, Paper No. HMT - 69-81, 6th National Heat and Mass Transfer Conference, I.I.T., Madras, (1981).

46. Grover R.B., and Venkat Raj V., Pressure Drop Along Longitudinally-Finned Seven-Rod Cluster Nuclear Fuel Elements, Nuclear Engineering and Design, 58, 79-83.

47. Vijayan P.K., Saha D., and Venkat Raj V., Measurement of Pressure Drop in PHWR Fuel Channel with 19 Rod Fuel Bundles (Wire-Wrap Type) at Low Reynolds Numbers, Report No. B.A.R.C./1-811, (1984)

48. Saha D., Grover R.B., and Venkat Raj V., Development of a High Rating Uniform Flux Heater and Temperature Measurement Techniques for some Out-of Pile Experiments Simulating Nuclear Heat Transfer, Paper No.HMT-72-75, 3rd National Heat and Mass Transfer Conference, I.I.T., Bombay, Dec.11-13, (1975)

49. Venkat Raj V., and Kalibhat M.B., Studies on PHWR Safety - Development of Instrumentation for Experiments on Rewetting of Hot Surfaces, Symposium on Process Instrumentation and Control for Power Plants and Nuclear Facilities, BARC, Bombay, April (1983).

50. Prahlad B., Swaminathan K., and Kale R.D., Determination of Response Time of Core Thermocouples of the Fast Breeder Test Reactor, Paper No.HMT-38-31, 6th National Heat and Mass Transfer Conference, I.I.T., Madras (1981).

51. Venkat Raj V., Anand A.K., and Ramamurthy H. Experimental Studies on Heat Transfer from Vertical 'U' Tubes Immersed in a Pool of Boiling Water, Paper No.HMT-51-75, 3rd National Heat and Mass Transfer Conference, I.I.T., Bombay, Dec. (1975)

52. Markandeya S.G., and Venkat Raj V., A Parametric Study of the Temperature Differential Across the Wall of an underground Storage Vault for Monitored Storage of Radioactive Wastes, Paper No.4HMT-84, 4th National Heat and Mass Transfer Conference, University of Roorkee, November (1977).

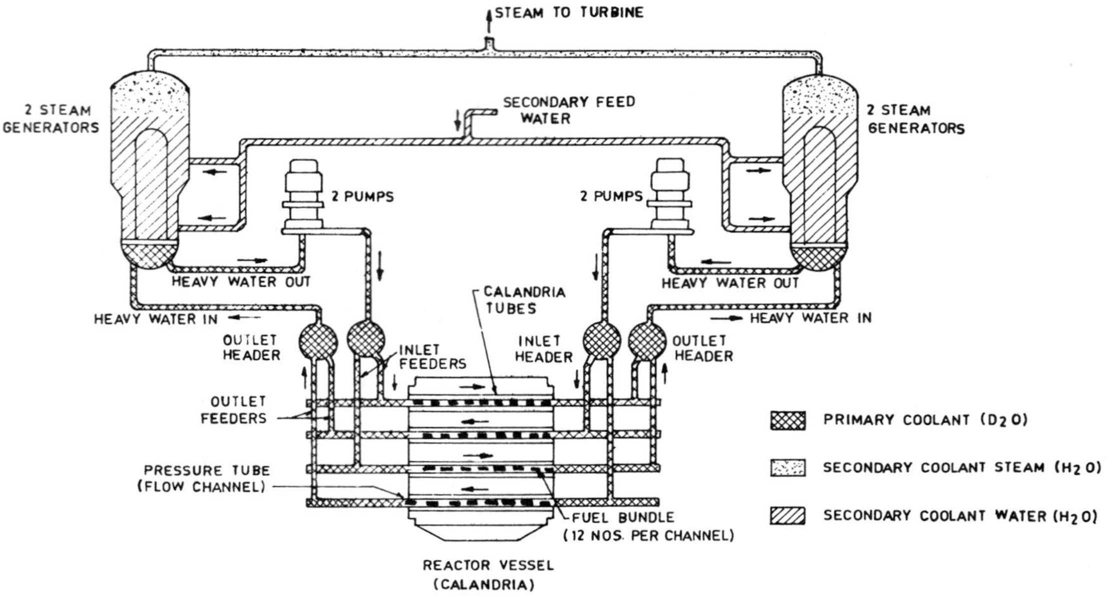

Fig.1 Napp Reactor Primary Heat Transport System

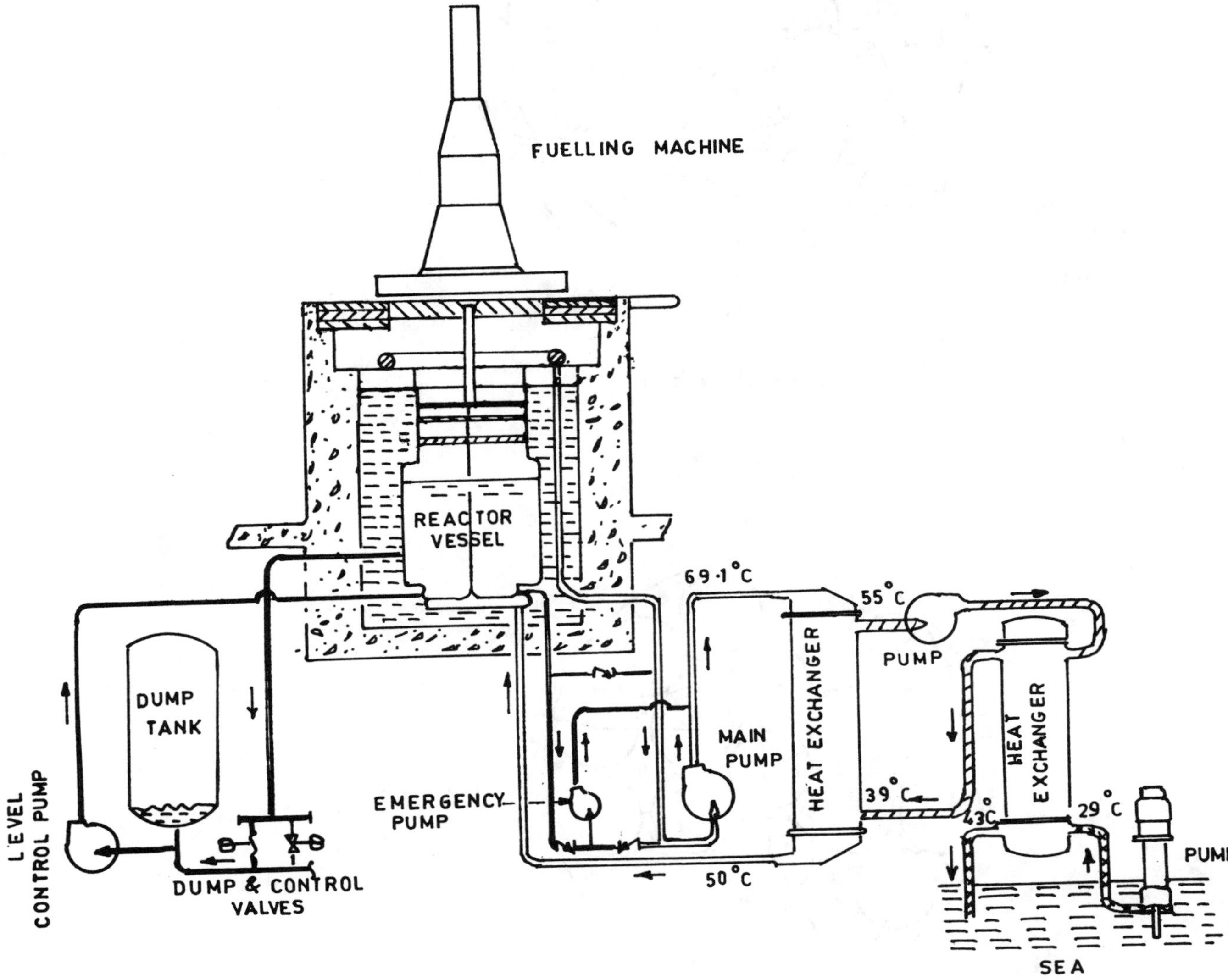

Fig.2 Simplified Flow Diagram of Dhruva Reactor

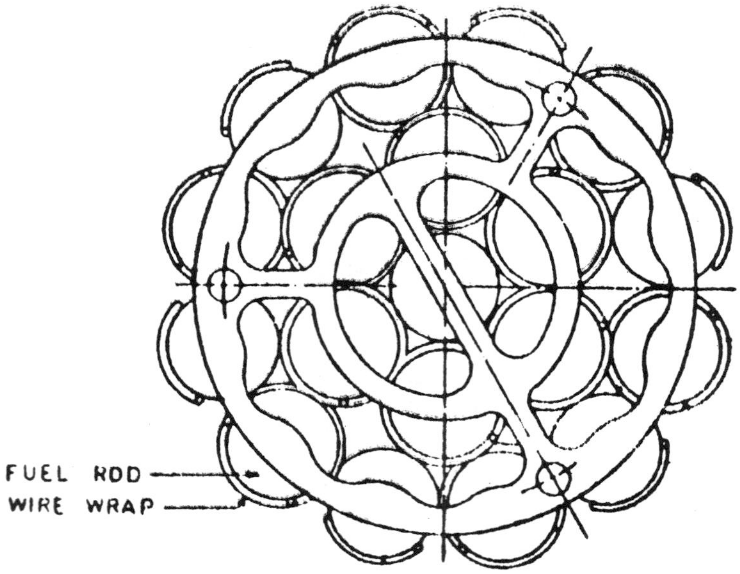

a) END VIEW

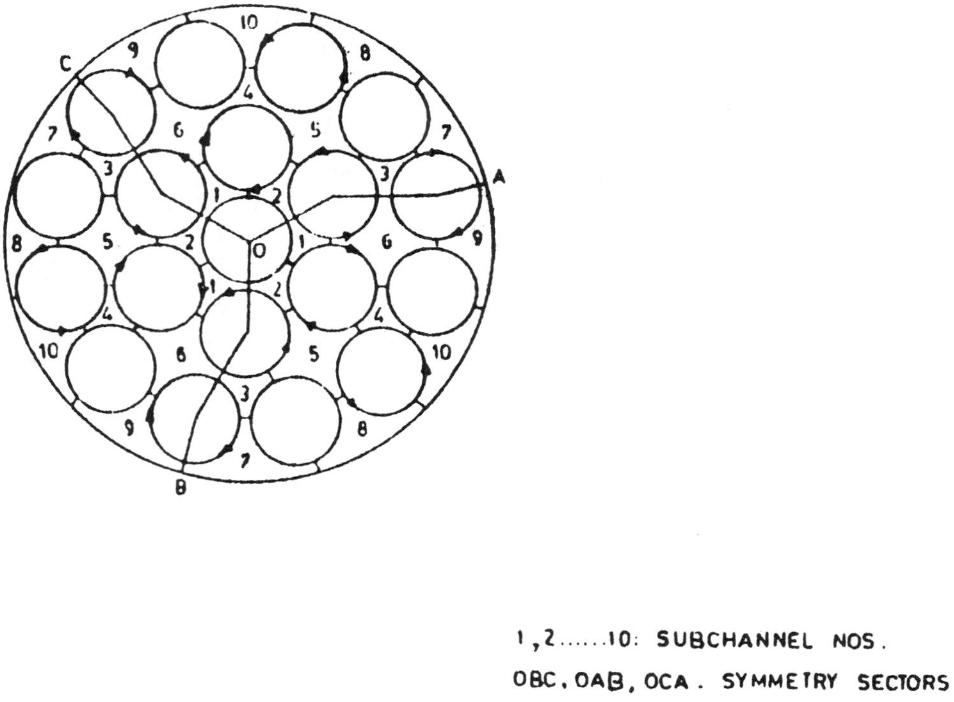

b) SUBCHANNELS

Fig.3 19 Rod Cluster

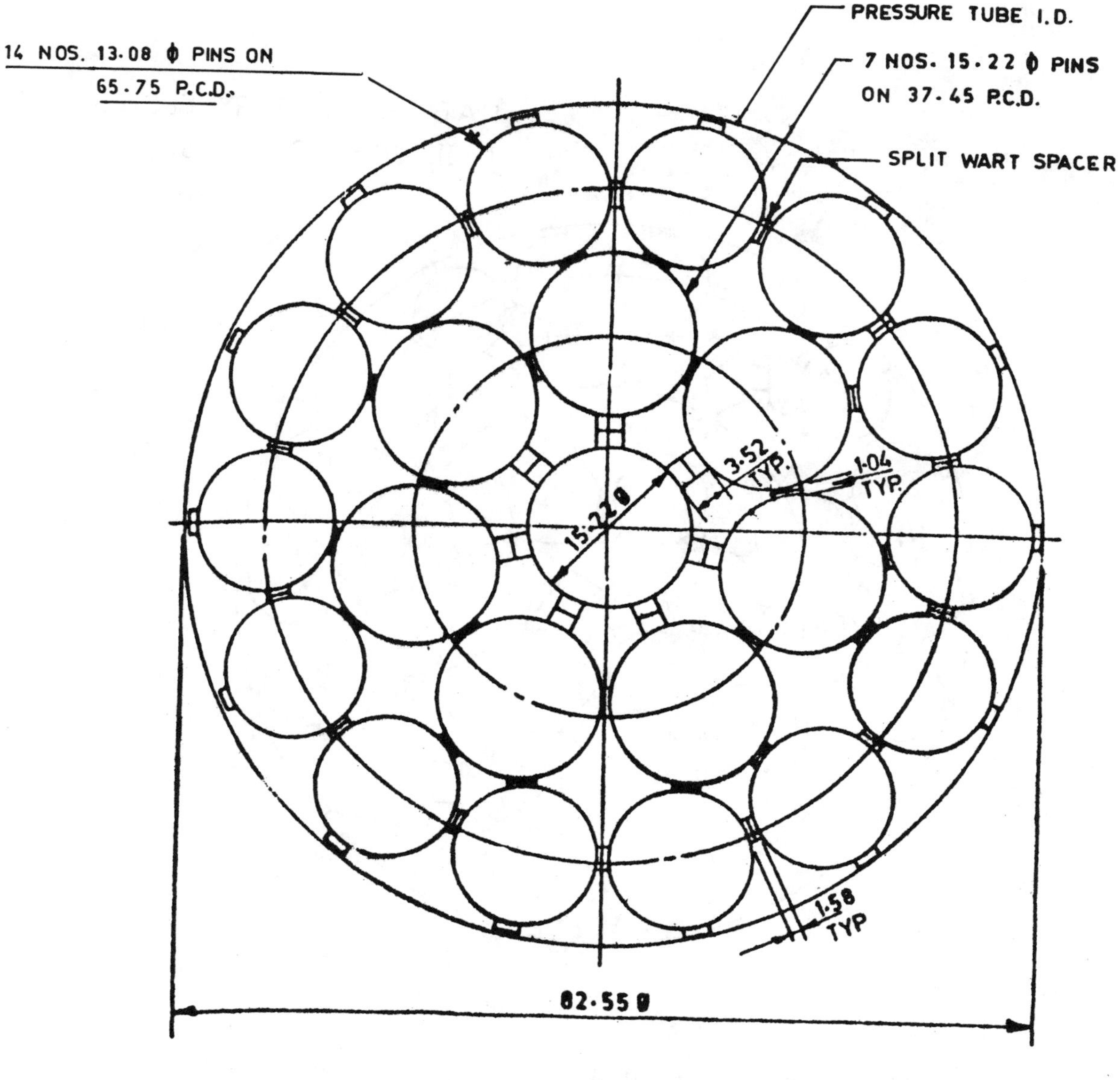

Fig.4 Cross Section of a 22 Element cluster

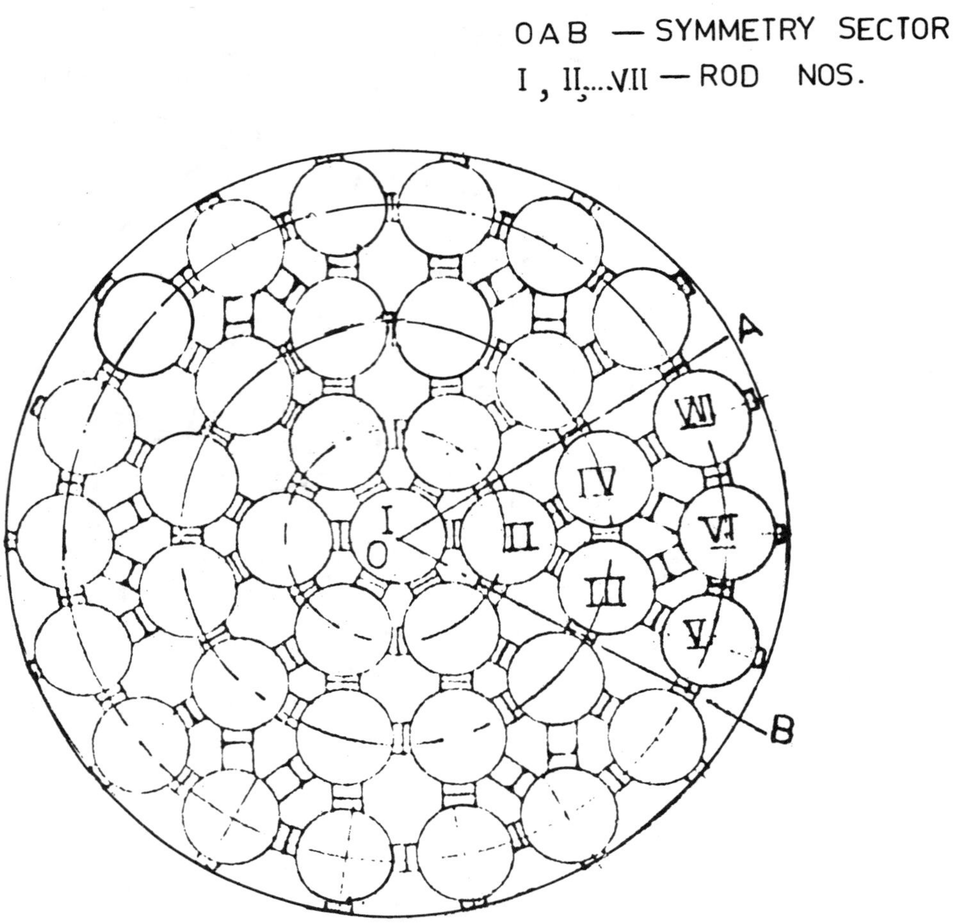

Fig.5 Cross Section of A 37 Element Fuel Bundle

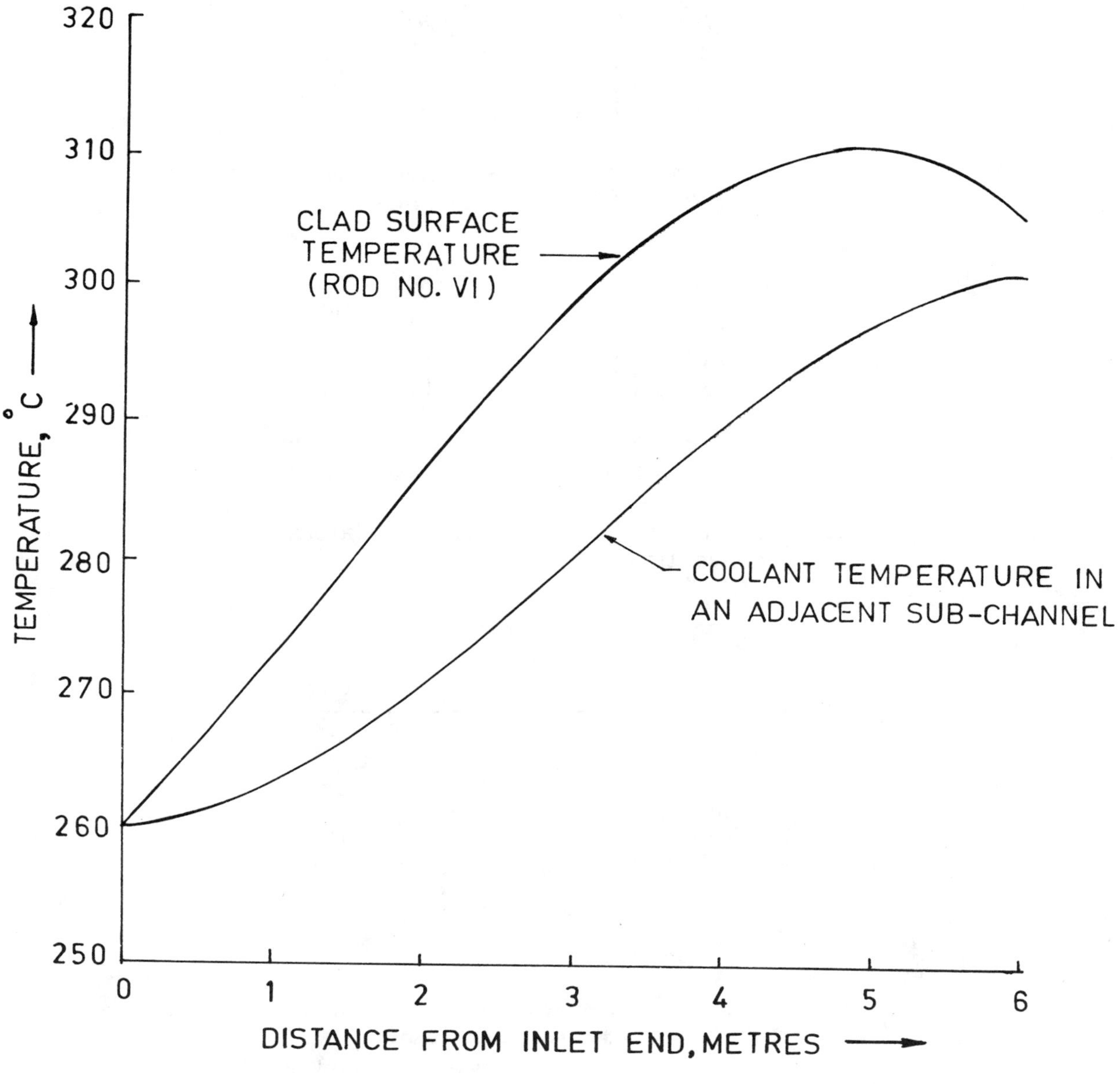

Fig.6 Axial variation of clad surface and coolant temperatures (37 Element Cluster)

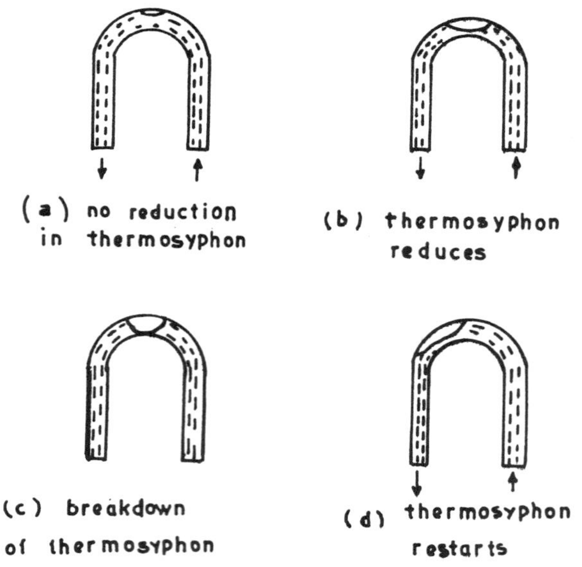

Fig.7 Effect of Noncondensibles on Natural convection Flow

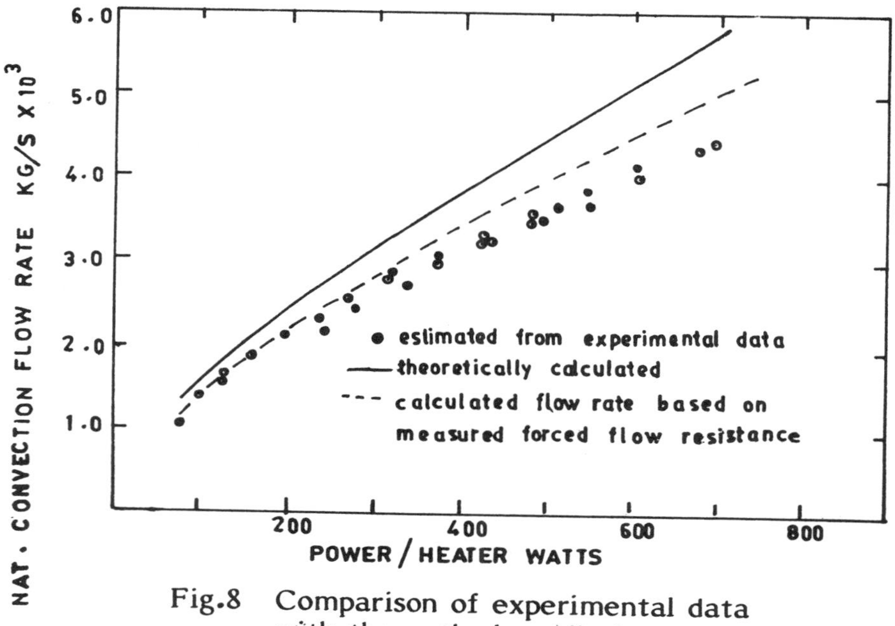

Fig.8 Comparison of experimental data with theoretical predictions

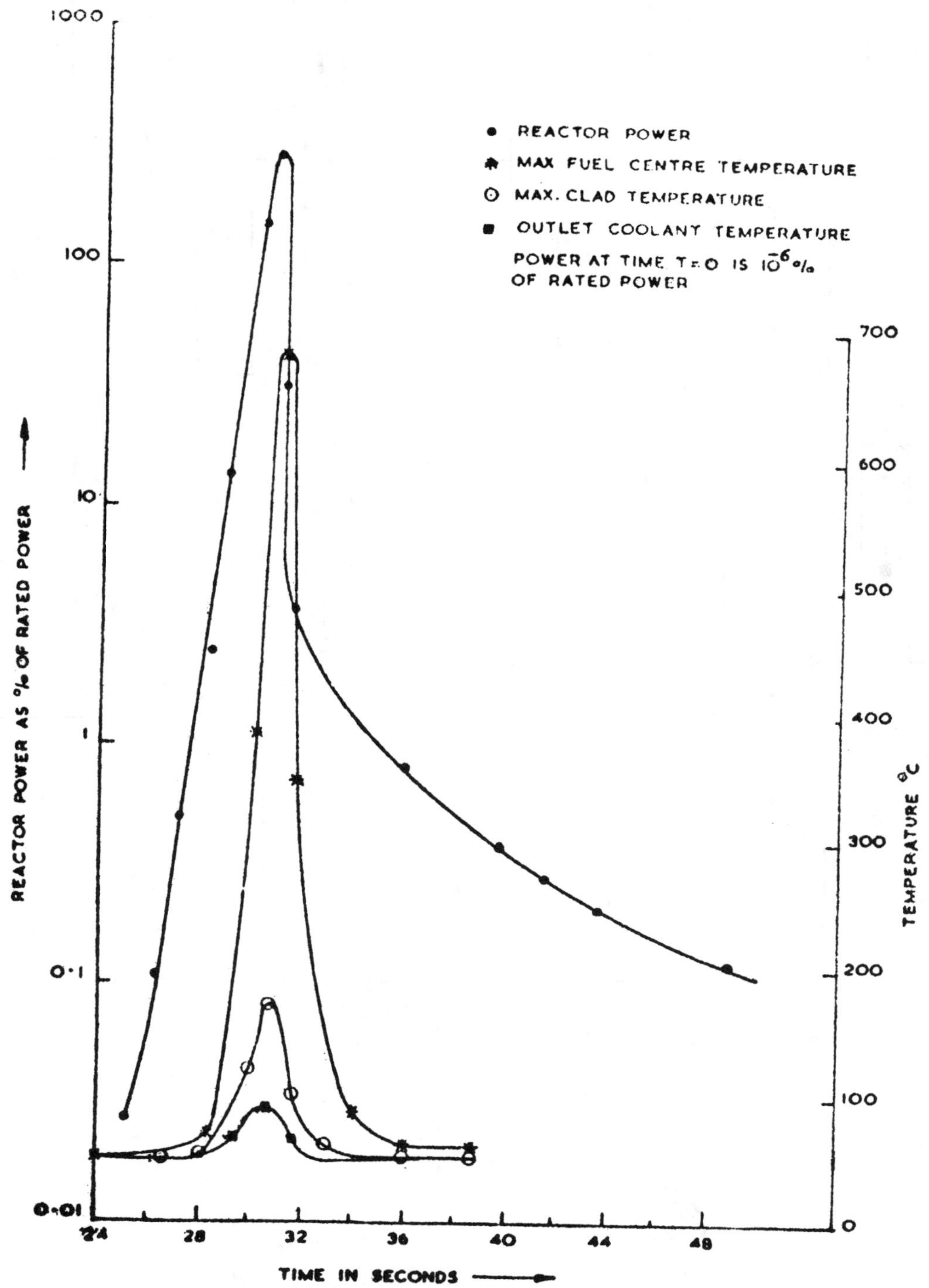

Fig.9 Temperature Transients During A Moderator Pump-Up Transient

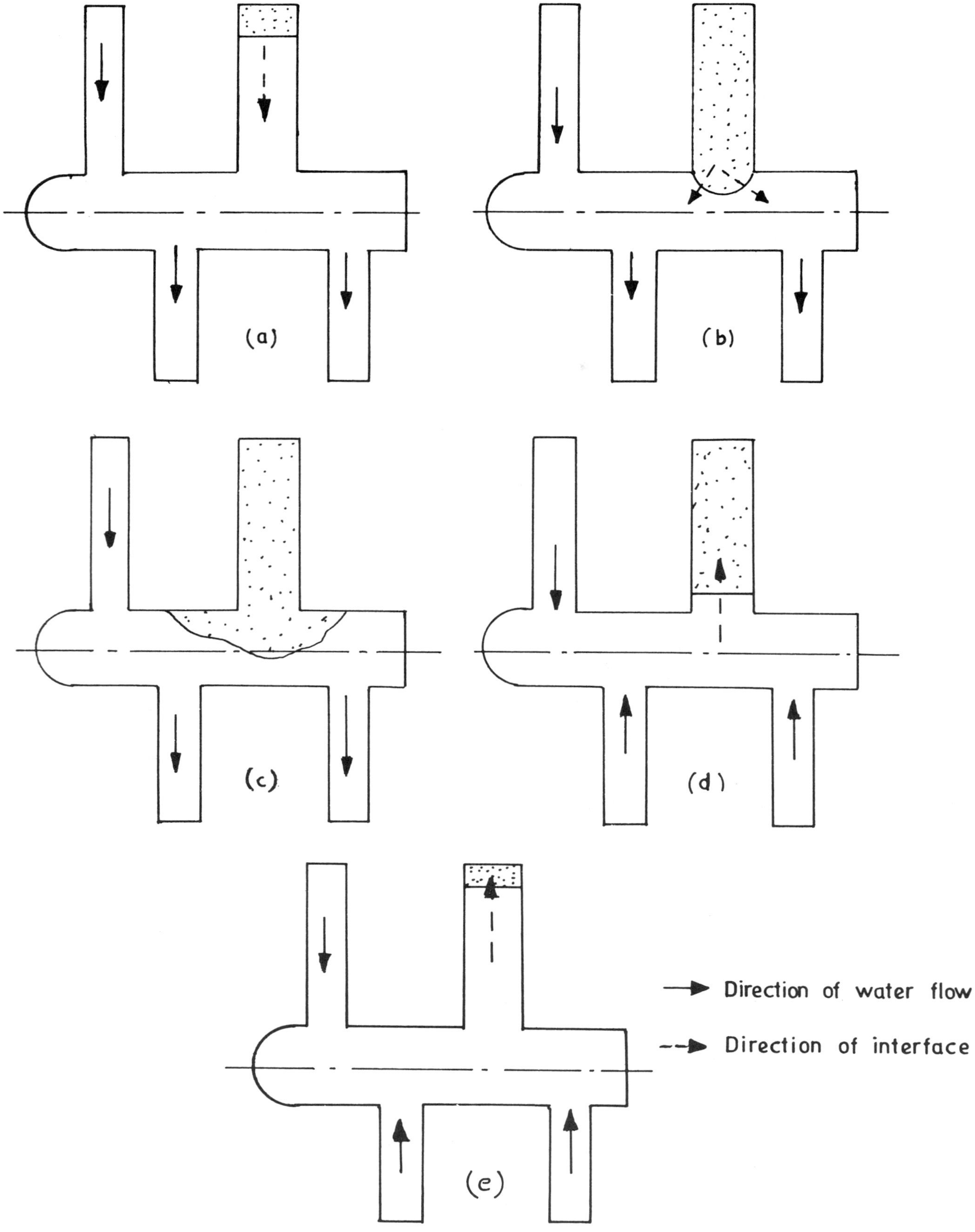

Fig.10 Sequential position of steam-water interface

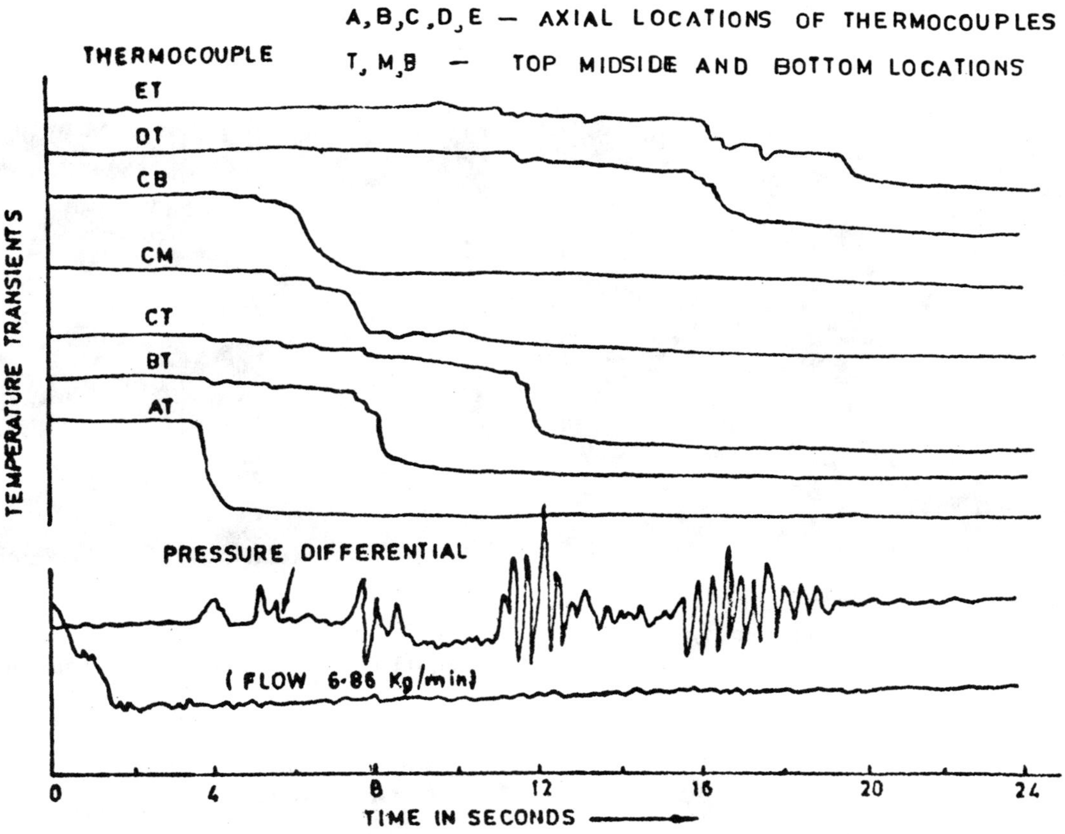

Rewetting Transients (Initial Temperature - 500°C)

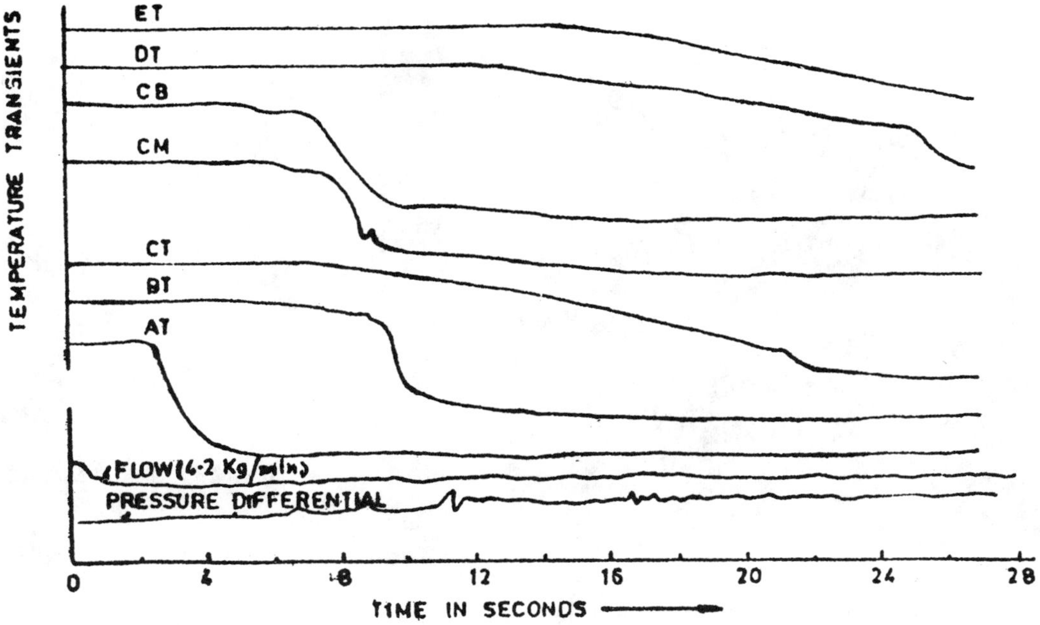

Rewetting Transients (Initial Temperature 400°C)

Fig.11

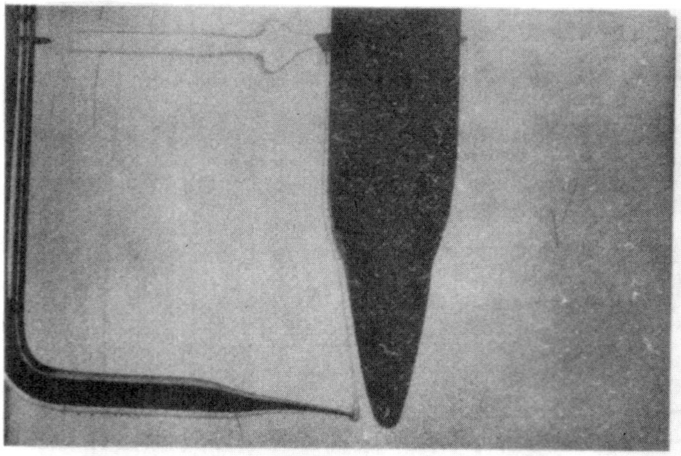

Fig.12 Flow visualisation using tellurium dye

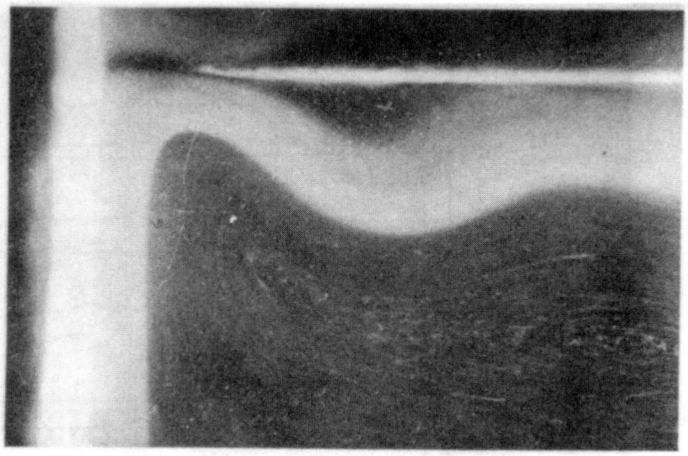

Fig.13 Flow visualisation using aluminium particles

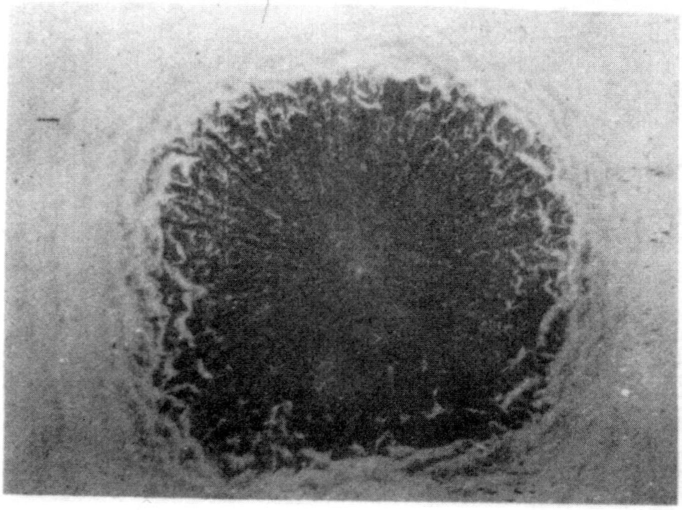

Fig.14 Flow visualisation using tracer paint

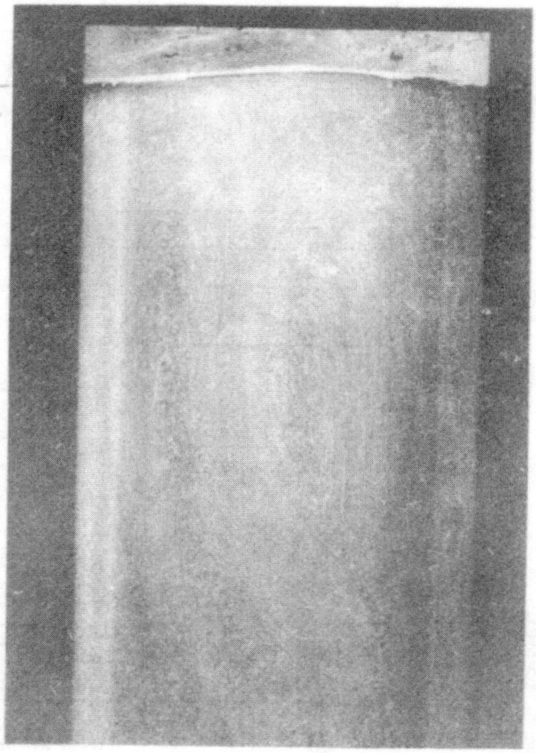

Fig.15 Flow pattern in A 1 MM. Annular Gap

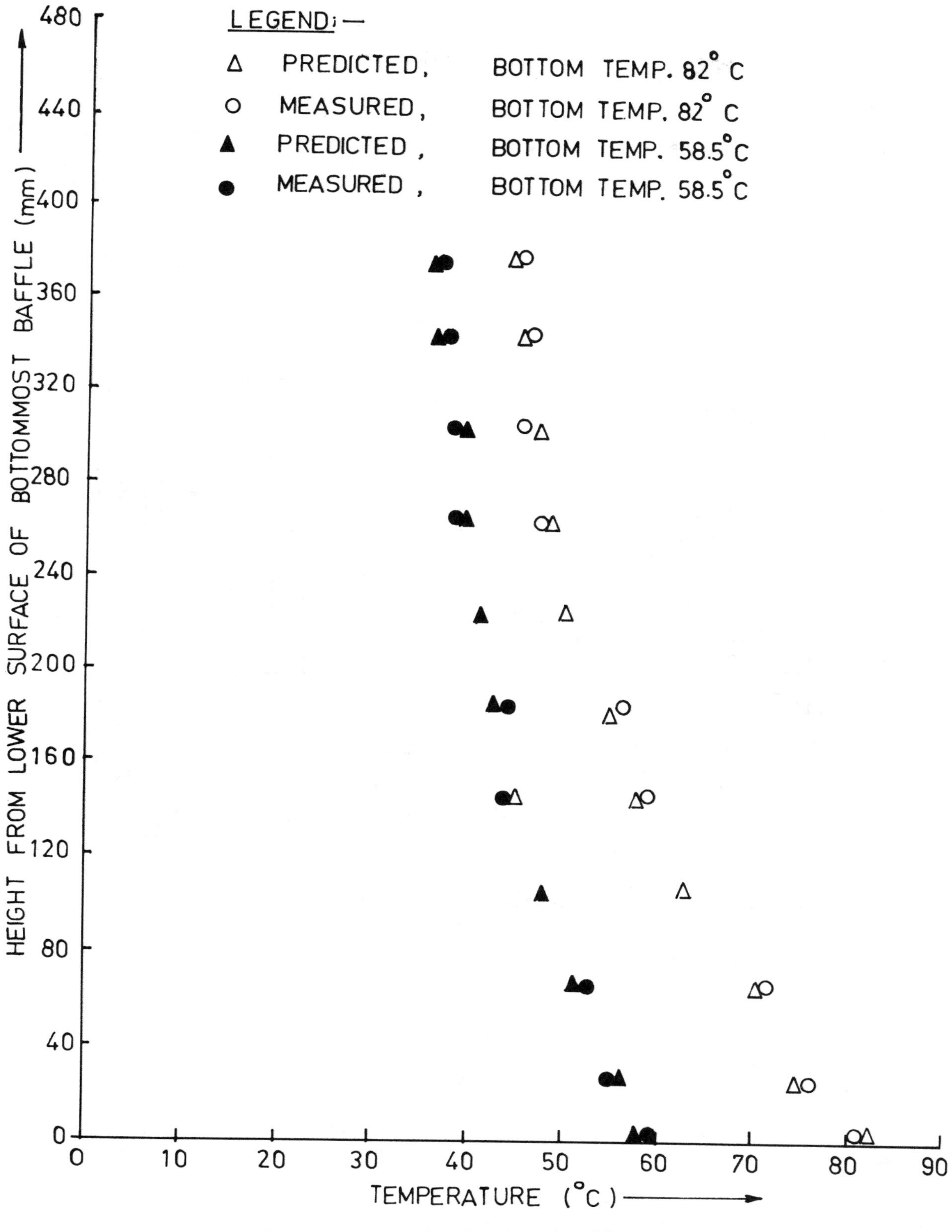

Fig.16. Temperature Distribution for 30 mm Baffle spacing

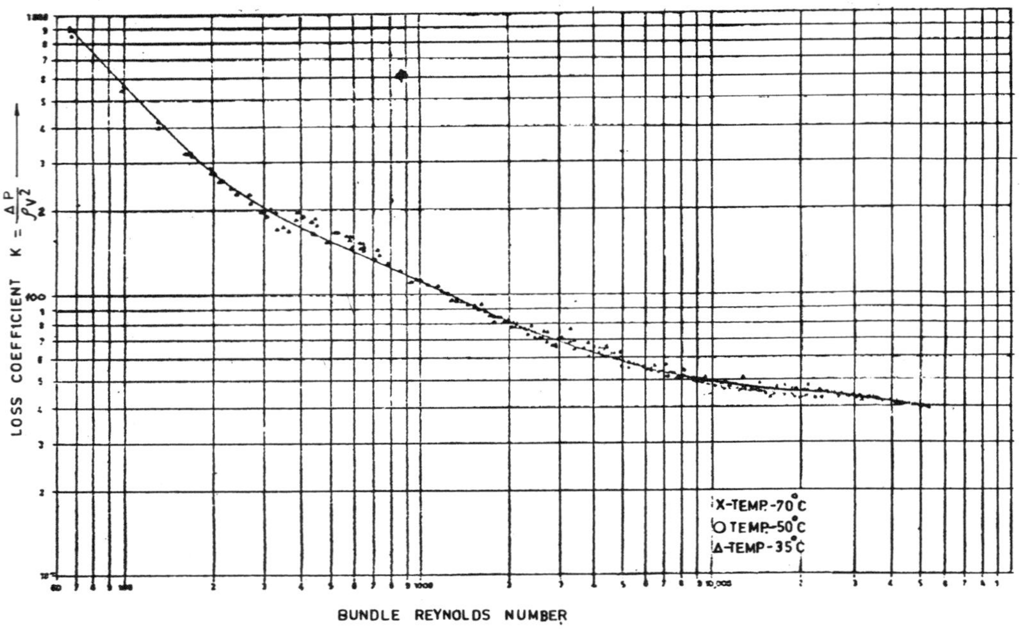

Fig.17 Fuel Channel Pressure Drop Data

Heat Transfer Experiments for an under Platform Ventilation System

A. K. MOHANTY
Department of Mechanical Engineering
I. I. T., Kharagpur, India

An under platform ventilation system is adopted to remove significant amount of traction heat from the coaches of an underground railway system. Heat transfer experiments conducted on a 1:10 scale model, typical for the Calcutta Tube Railway, and the estimated values of the efficiency of the UPV system are presented.

1. INTRODUCTION

An underground railway system is now considered to be one of the most viable means of inter and intra-city communications. Such a system can handle a large passenger traffic, does not occupy the on-surface free space, does not use fossil fuel nor adds to the pollution level, being electricity-driven. Advances in magnetic levitation are expected to reduce the tractive power considerably.

A major technological problem in an underground railway system, however, relates to the removal of heat that is generated due to the operation of traction equipment, passenger and auxiliary services. The heat load is typically estimated at 6 million kilojoules/km route/hour for the Calcutta tube railway.

A part of the generated heat leaks to the sub-soil through three thermal resistances in series offered by: the air-film on the tunnel wall, the tunnel wall material and the depth of thermal penetration in the soil. The heat flow direction can be alternated from or to the sub-soil, depending on whether the tunnel air temperature is above or below the sub-soil temperature, by matching the schedule of operation of the ventilation plants with the passenger load, hourly temperature variation of the ambient and seasonal effects. The thermal design aims at, through such alternations, attaining a zero net heat deposit to the sub-soil over a time period of one year. The concept applied specifically to the Calcutta sub-soil and tube railway may be noted in references 1 and 2.

2. VENTILATION LAYOUT

A typical arrangement of mechanical ventilation for an underground system consists of : delivery of a quantity of conditioned air to the station, washed air at the inlet to the tunnel and exhaust of the hot air at the mid-section of a tunnel (MTE). Additionally an under platform ventilation system (UPVS) is adopted to remove a fraction of the traction, wheel and braking heat load at the wheel level, thereby reducing thermal discomfort to the passengers and personnel in a station. The typical arrangement for a representative station of the Calcutta tube railway is given in figure 1.

A knowledge of the quantity of heat removed below the platform level through the UPVS is essential for the design of the ventilation system, its ratings and operating schedule. This paper is devoted to an estimation of the effectiveness of a typical under platform ventilation system.

3. MODEL STUDY

The traction motor, wheels and the braking rheostats of an underground railway coach are the sources of heat below the platform level. These under-carriage equipment are primarily cooled by convection due to train motion as well as by the air flow caused by the mid-tunnel exhaust fan in train running condition. When the train stops at a station, the UPVS superimposes an induced draft parallel to the wheel axis. At stationary conditions the hot bodies attain temperature of the order of 300°C because of which radiative loss to the surrounding surfaces also becomes significant.

It is easy to visualize that geometry and flow phenomena are too complex to allow a purely theoretical estimation of the heat loss from the under-carriage components.

A 1 : 10 scale model study was therefore undertaken.

The arrangement for the model study of heat transfer during the residence of a train in a station is shown in figure 2. The station air supply was achieved by means of a forced flow centrifugal fan delivering through overhead grills. The UPVS was simulated by connecting an induced draft fan to a header that was communicating with the under platform hoods. For the purpose of experimentation, only one axial flow fan was used to simulate the station conditions. A circular connecting duct was taken from the far end of the station to join with the fan-end, figure 2. Throttle adjustments were made for attaining equal flow rates in both the main and connecting ducts.

The normal running section was a twin-box tunnel with intermediate pillars for separating the up and down tracks. Representative dimensions may be noted from figure 3.

The model coaches were each 2 metre long with 40 mm coupling distance between the coaches. The station model was 13.5 metres long suitable for a 6 coach train.

For a representative coach, there are eight wheels, four motors and two rheostats, half number of each being on the UPVS side and the other half on the wall side. Electric resistance heaters made up of nichrome wire elements mounted on mica sheets and sandwiched between mild steel plates to fit on to the wheel, the motor or the rheostat were made and clipped on to the respective elements. Each component was provided with iron-constantan thermocouples for temperature measurement.

A set of similar heaters for a coach were connected electrically in series. For example, the heaters for the four wheels on the UPVS side were in series with one supply connection, whereas the other four on the wall side were connected to a separate supply. Similar were the connections for the motor and rheostat heaters. By such arrangement it was possible to estimate the cooling rates separately on the UPVS and the wall sides. Heat input to similar elements on both the UPVS and wall sides were adjusted for attaining equal temperature.

In order to account for differential cooling of the components due to varying distances from UPVS openings, readings were taken with the train moved over a distance of 90 cm in 15 cm intervals.

The heated coach assembly was kept on rails with an asbestos gasket sheet to avoid heat flow from the wheels to the rails during the long time required to carry out the experiments. In practice, a train halts at a station for about 1 minute and heat conduction to the rails is negligible through the small contact area of the wheels.

Measurement of the station air temperature at three heights equidistant between the platform and the mezanine was carried out by three thermocouple probes held in position by a wooden support, and the readings averaged for heat transfer calculations.

Air flow through the station inlet, UPVS and tunnel exhaust systems were measured by pitot traverse. Temperature and power input measurements were taken under thermally steady state conditions.

The UPVS fan outlet area was 410 x 300 mm, and the tunnel crossectional area 0.3094 m^2 and hydraulic diameter 505 mm. The surface area and hydraulic diameter of the model hot bodies were as below.

Wheel	: 0.009747 m^2 ;	91 mm
Motor	: 0.02644 m^2 ;	17.8 mm
Rheostat	: 0.038 m^2 ;	21.6 mm

4. HEAT TRANSFER RESULTS

Experimental results indicated that the cross flow created by the UPVS improved the near side heat transfer from the wheels marginally over that due to the pure tunnel flow. The wall side wheels were unaffected by the UPV system. The extrapolated values of Nusselt number for wheel heat transfer at Re_t 2.5 x 10^5 are of the order of 91 and 98 for the wall and the UPVS sides. The corresponding values for rheostat are 13 and 12, and for the motors Nu = 11.5.

All Reynolds numbers are defined on the basis of tunnel hydraulic diameter, and the Nusselt numbers on the basis of the component hydraulic diameter.

The motors being aligned parallel to the tunnel flow, heat transfer from these components were practically unaffected by the under platform cross flow.

Sample experimental results for heat transfer from the wheels are presented in figure 4. Heat transfer from the rheostat on the platform side was also improved marginally by the UPVS. These transport rate values are primarily convective with marginal radiation component since the maximum temperature of the hot surfaces was limited to 100°C. In practice, however, the motors and the rheostats shall be at temperatures of the order of 180 and 300°C respectively, and the wheels may remain at about 100°C. For the design calculation, therefore, additional heat transport by radiation exchange were calculated for rheostats and motors assuming a surface emissivity of 0.6 appropriate for cast iron construction.

Calculations using the above values indicated that the traction heat removable by the ventilation system at surface temperatures of 100, 180 and 300°C for the respective components with station air temperature at 31°C was 42.11 kW per coach. This value works out to 19.56 kJ for a 8 coach train during 58 seconds residence in a station.

5. EFFICIENCY OF THE UPV SYSTEM

The foregoing discussions indicate that the contribution of the UPVS in enhancing the rate of heat transfer from a traction component is insignificant compared to when the mid tunnel exhaust works alone. This is to be expected since the UPVS being an induced flow arrangement has a limited 'catchment' area to improve heat transfer from components locally, especially when the under-carriage flow is highly recirculating and turbulent.

The true advantage of the UPVS, however, occurs from the fact that it exhausts a bulk of the hot under-carriage air, thereby reducing thermal infiltration to the platform level.

The efficiency of the UPVS in removing a portion of the traction heat is estimated from its capacity to remove heat in excess to what could have been, had the temperature above and below the platform were the same. This was estimated by the following set of experiments.

In order to create a high temperature in the under-platform air, the traction heaters in all the six coaches were heated and the total power input was estimated as the traction heat. The temperature at the platform level and at the entry to the UPVS duct were measured. The heat removed by the UPVS was estimated by additionally measuring the mass flow rate through this system.

A sample calculation for a set of experiments is given below for ease of visualization.

Total traction heat input = 4.45 kW

Mass flow rate of air through the UPVS,

$$\dot{m} = 0.671 \text{ kg/s}$$

Rise in temperature of the UPVS inlet air over the station air, $\Delta T = 3.74°C$

Heat carried away by the UPVS air

$$Q = \dot{m}\, Cp\ \Delta T = 0.671 \times 1.0 \times 3.74 = 2.518 \text{ KW}$$

Hence the percentage of traction heat removed by the UPV system is

$$X_q = \frac{2.518}{4.45} = 56.6\ \%$$

The balance heat would mix with the station air and finally be exhausted by the mid tunnel fan.

The above value of X_q resulted when the flow rate through the UPVS and the MTE corresponded to:

$$Re_{t_{UPVS}} = 6.26 \times 10^4 \text{ and } Re_{t_{MTE}} = 8.71 \times 10^4$$

based on a common hydraulic diameter of the tunnel.

In otherwords, the percentage mass flow of air through the UPVS was

$$X_m = \frac{6.26}{6.26 + 8.71} = 41.8\ \%$$

Had the UPVS been of zero efficiency, it would have removed heat equal to the mass ratio. We therefore define the efficiency of the under platform ventilation system as : $X_q = X_m\,[1 + \eta_{UPVS}]$ or

$$\eta_{UPVS} = \frac{X_q}{X_m} - 1 = 34\ \%$$

The efficiency of the UPVS was estimated experimentally for several mass flow ratios, X_m, and traction heat input. The results are presented in figure 5. The measurement accuracy is estimated at ± 10%.

6. CONCLUSIONS

We have outlined the ventilation system adopted typically for the Calcutta underground railway. Aerodynamic and heat transfer tests on a 1:10 scale model were conducted at IIT Kharagpur to establish the value of different parameters, which were then used for design calculations leading to plant ratings and operating schedules. In this paper we have discussed the experiments pertinent to the under platform ventilation system. For the prototype mass flow ratio of $X_m = 0.456$, the UPVS efficiency is estimated at 0.625. In otherwords 62.5 % of the traction heat is removed by the UPVS with corresponding reduction in the platform heat load.

ACKNOWLEDGEMENT

This work was part of a study on a consultancy project assigned by the Calcutta Metro Railways. Professors G.B.Misra and M.S.Sastry were also involved in the study.

NOMENCLATURE

c_p	specific heat of air at constant pressure	kJ/kg K
D_h	hydraulic diameter of the model tunnel	m
d	hydraulic diameter of the heated bodies	m
h	convective heat transfer coefficient	W/m^2 K
k	thermal conductivity of air	W/m K
MTE	mid-tunnel exhaust	
$\dot{m}$	mass flow rate of air through the UPVS	kg/s
Nu	Nusselt number, Nu = hd/k	
Re_t	Reynolds number based on tunnel hydraulic diameter	
U_{av}	average velocity	m/s
UPVS	under platform ventilation system	
X_m	ratio of mass flow	
X	ratio of heat removal	
η	efficiency	
ν	kinematic viscosity	m^2/s

REFERENCES

1. Misra, G.B., Mohanty, A.K., and Sastry, M.S., "Report on the Ventilation System Design for the Calcutta Tube Railway", IIT Kharagpur, February, (1978).

2. Mohanty, A.K., and Misra, G.B., "Temperature and Thermal Conductivity of Calcutta Sub-Soil". Journal of the Institution of Engineers (I), September, 62 (1981) 81-84.

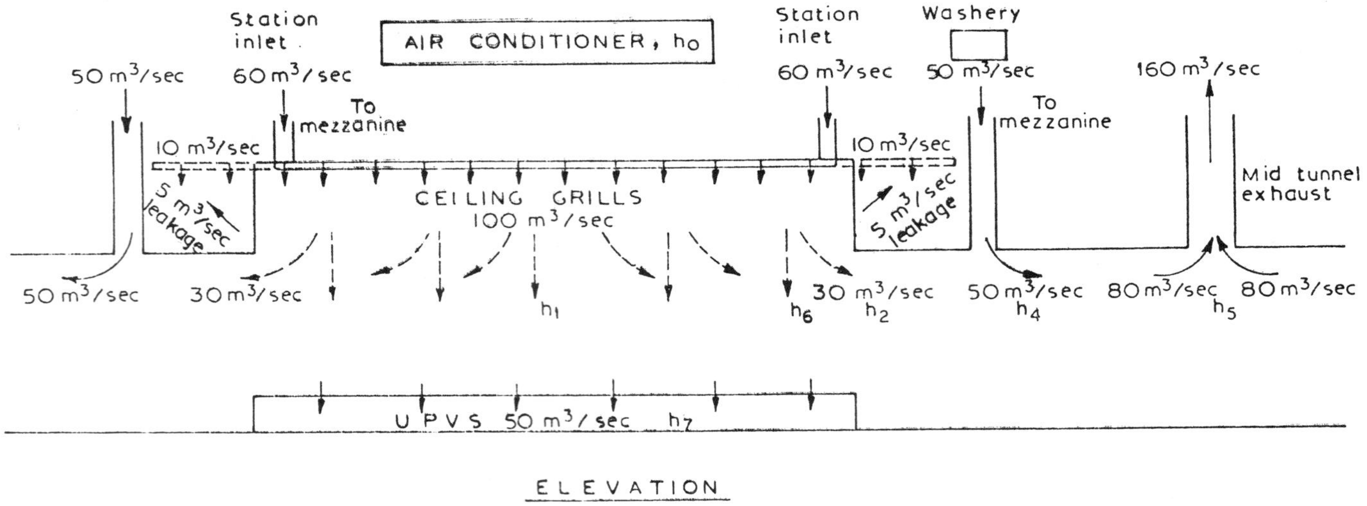

Fig.1 Outline of the Ventilation System

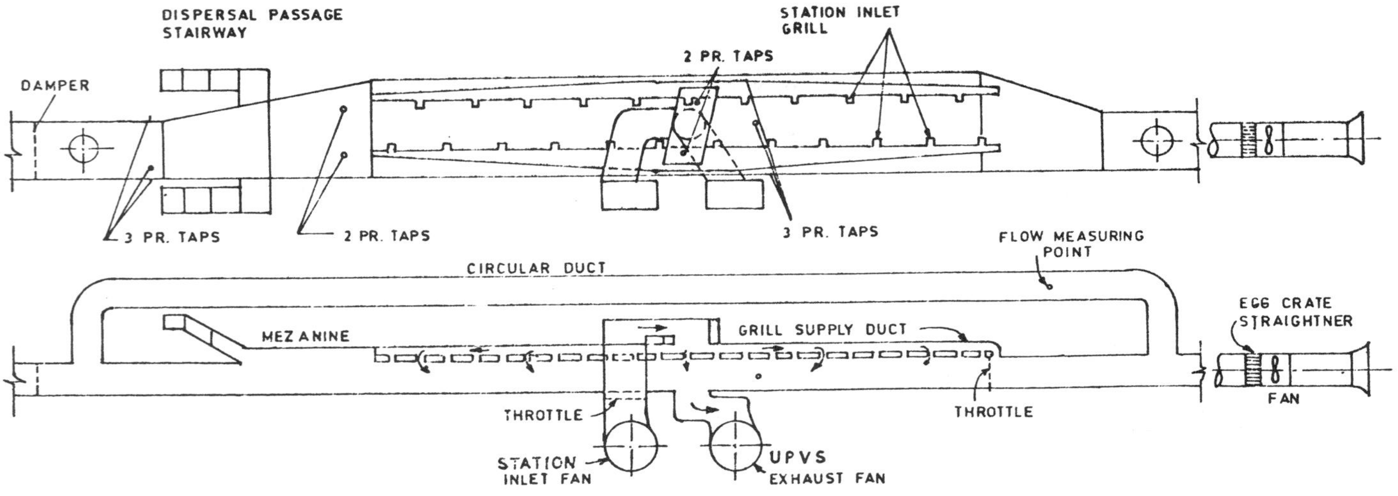

Fig.2 Schematic Diagram of a Station

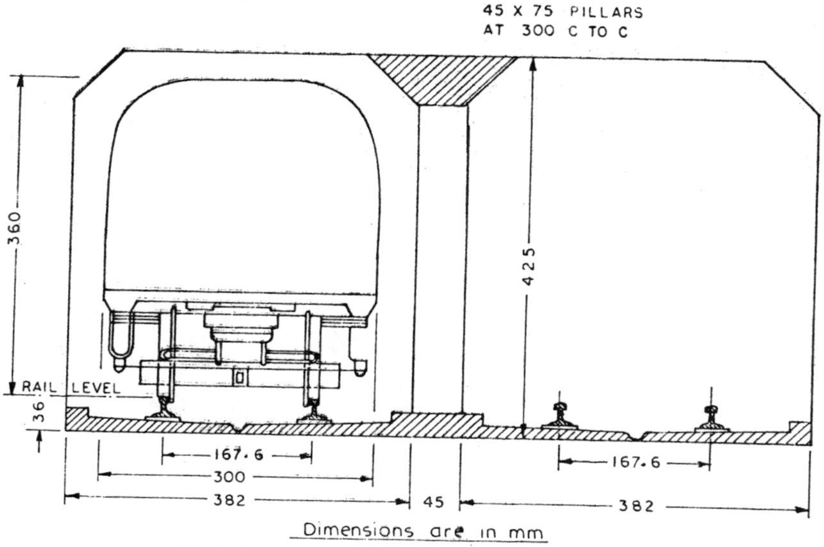

Fig.3 Scale Model of a Twin-Box Tunnel and Train

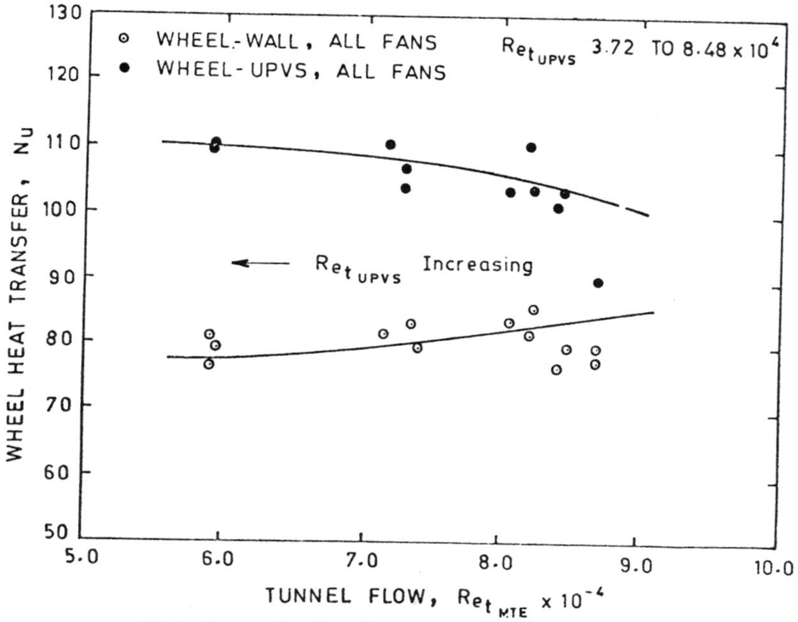

Fig.4 Convective Heat Transfer from Wheels,All Fans Running

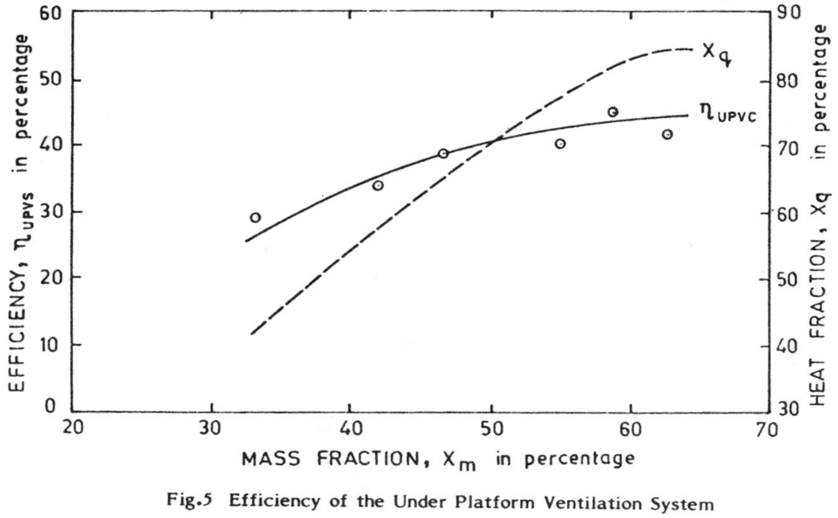

Fig.5 Efficiency of the Under Platform Ventilation System

Binary Diffusion and Heat Transfer in Laminar Free Convection Boundary Layers around Spheroids

A. S. MUJUMDAR and M. HASAN
Department of Chemical Engineering
McGill University
Montreal, PQ, Canada H3A 2A7

1. INTRODUCTION

Free convection in a fluid often arises because of the density changes within the fluid. For a single component or constant composition fluid, density varies inversely with temperature. In a system of varying composition, however, density is a function of both composition as well as temperature. A variation in composition within the fluid then will either enhance or retard free convection. Since the variation of density in fluids due to thermal and concentration gradients is very common in natural as well as technological processes there is a considerable body of research related to the analysis, measurement, prediction and understanding of external and internal heat and mass transfer by free convection.

Most of the studies on external natural convection have been devoted to plate, cylinders and spheres. Very few studies exist on free convective transfer from spheroids. All such studies, theoretical or experimental, related to spheroids were concerned with either heat or mass transfer in free convection. A thorough search of heat and mass transfer literature has failed to unearth any prior work on coupled heat and mass transfer in free convection around spheroids. In the case of single mode transfer, mass transfer at low mass transfer rates is analogous to the corresponding heat transfer phenomenon. Thus, the Sherwood number results can be interpreted as Nusselt number results. This is not true in the case of coupled mode of transfer since transfer of one affects the other.

The knowledge of combined heat and mass transfer is important for a correct understanding of the transfer phenomena. For example, it has been shown (1) that the presence of natural convection on droplet combustion can severely distort the combustion process from spherical symmetry and therefore significantly modify the combustion behaviour. Understanding of the free convective heat and mass transfer processes is also important for a proper design of some problem in chemical reactors, atomizers, combustion chambers, in prediction of the vaporization and evaporation of mists and fog in meteorology, etc. Although heat and mass transfer in most of the above practical processes operate under the combined mode of free and forced convection, free convective transfer is one of the limiting situations and therefore merits study in its own right. Moreover, in most of the practical circumstances perfect spherical bubbles, drops or particles are rarely encountered. Both theoretical and photographic evidence indicates that drops of a dispersed phase often assume non spherical shapes (2). The latter justifies the practical importance of the present study on spheroidal bodies.

Due to their practical and engineering importance, effects of mass diffusion on natural thermal convection flow have been widely investigated for flat plates which are vertical (Gill et al., 1965; Gebhart and Pera, 1971), horizontal (Pera and Gebhart, 1972; Chen and Strobel,1980), or inclined (Chen and Yuh, 1979; Strobel and Chen, 1980), and cones (Hasan and Mujumdar,1984) among other simple geometries. In contrast to the plates with different orientations, the analysis of natural convection around spheroids have been confined to flow induced solely by temperature (Saville and Churchill, 1967; Lin and Chao, 1974; Raithby et al., 1976) or solely by concentration variations (Weber et al., 1983). Due to the lack of closely relevant work, literature which is of peripheral relevance to coupled free convective heat and mass transfer around spheroids, is discussed below.

Saville and Churchill (1967) appear to be the first to analise laminar free convection boundary layer flow along two-dimensional planar and axisymmetric isothermal bodies of fairly arbitrary contour. They used Görtler-type series and showed that for cases of horizontal cylinders and spheres, the series converge much faster than the corresponding Blasius type series used earlier by Chiang and Kaye (1963) who studied the same problems. Later, Saville and Churchill (1970) analyzed simultaneous mass and thermal transport in natural convection caused by buoyancy forces for a very low concentration level of the diffusing species. Attention was restricted to asymptotic processes in terms of Prandtl (Pr) and Schmidt (Sc) numbers. The parametric dependence of the transport processes was reported for $Pr \to 0$ and $Sc \to 0$, for $Pr = Sc \to \infty$, for $Pr \to 0$ and $Sc \to \infty$ and Pr and $Sc \to \infty$ with $Sc >> Pr$. The asymptotic nature of the results shed interesting light for $Pr = Sc$ but are of only peripheral interest for the ranges of Sc and Pr under which most of the practical processes operate. Moreover, the above authors did not provide any results on coupled heat and mass transfer for any axisymmetric bodies.

Watson and Poots (1972) studied the three-dimensional free convective boundary-layer over a heated ellipsoid of revolution. Full numerical method of solution was not used because of the excessive computing time and storage required. An expansion scheme was adopted instead. The results were reported in graphical form and were described as preliminary. Lin and Chao (1974, 1978) also studied laminar free convection heat transfer around two-dimensional and axisymmetric bodies but unlike Saville and Churchill (1967) they adopted a series expansion method to solve their nonsimilar boundary-layer equations. The adopted expansion method was of the form originally proposed by Merk (1958) and which was later corrected by Chao and Fagbenle (1973). The analysis was carried out employing a suitable coordinate transformation. The expanded form of the governing momentum and energy equations contained sequence of universal functions which were dependent on Prandtl number and the configuration variable.

Raithby et al. (1976) measured free convective heat transfer in air for a wide range of Rayleigh numbers from isothermal prolate and oblate spheroids. The experimental conditions encompassed both laminar and turbulent regimes. Based on their experimental findings, they proposed correlations from which average free convective heat transfer rates can be calculated for a wide range of eccentricity of the spheroids. These authors also provided an approximate method of calculating average free convective heat transfer from spheroids of moderate eccentricity. Later, Raithby and Hollands (1976) reported an approximate but modified correlation for prolate spheroids with very small ratio of minor to major axis based on Langmuir-type correction.

Very recently Weber et al.(1983) measured the rates of natural convective mass transfer to a variety of non-spherical objects in a number of orientations. By using the limiting current density technique (depositing copper from aqueous solution of copper sulphate containing sulphuric acid as a supporting electrode) they obtained free convective mass transfer rates for high Schmidt numbers. They provided a useful correlation for calculating the Sherwood number from Schmidt and Rayleigh numbers for objects of any shape with sphericities larger than 0.6. Later they extended theoretically the validity of their correlation for a wide range of Schmidt numbers through Churchill and Churchill (1975) approach.

The object of the present work is to describe the most fundamental features of the free convection laminar boundary-layer flow driven by temperature and mass fraction variation around prolate spheroids. A novel aspect of this research is the focus on the mutual effects of the two buoyancy forces that drive the flow, namely, the density difference caused by temperature variations and the density variations caused by species mass fraction variations. The present analysis is therefore a useful extension of the work of Lin and Chao (1974). In this work consideration is given to the situation in which the surface of the spheroid is maintained at a uniform temperature and a uniform concentration. Following previous investigators (4-9) the mass fraction, C, of the diffusing species in the binary mixture is assumed to be very small in comparison to the other inert chemical species present. As a result of this assumption, the secondary effects, namely (often referred to as thermodynamic coupling effects) diffusion-thermo (i.e., Dufour) and thermo-diffusion (i.e.,Soret) effects can be neglected. The governing laminar binary boundary layer equations which do not admit similarity solutions are solved by the local nonsimilarity method. Numerical solutions of the governing transformed boundary-layer equations are carried out for convection of water vapor and naphthalene about prolate spheroids in air.

2. ANALYSIS

Consider a spheroid whose meridian plane is an ellipse with a vertical major axis of length 2a and a minor-axis of length 2b as in Fig.1. The spheroid is situated in an otherwise quiescent fluid at temperature T_∞, and concentration, C_∞. The surface of the spheroid is kept at uniform temperature, T_w, and uniform mass fraction, C_w. Let the local orthogonal coordinates be chosen such that, x measures the distance from the lower stagnation point along the surface of the spheroid and y measures the distance normal to the surface into the fluid. The steady, laminar boundary-layer equations, neglecting curvature effects for the spheroid can be written as:

$$\partial (ru)/ \partial x + \partial (rv)/ \partial y = 0 \quad (1)$$

$$u(\partial u/\partial x) + v(\partial u/\partial y) = \nu(\partial^2 u/\partial y^2) + g\,\beta(T - T_\infty) \sin(\alpha^*) + g\beta^*(C - C_\infty) \sin(\alpha^*) \quad (2)$$

$$u(\partial T/\partial y) + v(\partial T/\partial y) = \alpha(\partial^2 T/\partial y^2) \quad (3)$$

$$u(\partial C/\partial y) + v(\partial C/\partial y) = D\,(\partial^2 C/\partial y^2) \quad (4)$$

The approporiate boundary conditions are

$$u = 0,\ V = V_w,\ T = T_w,\ C = C_w \text{ at } y = 0$$
$$u \to 0,\ T \to T_\infty,\ C \to C_\infty \text{ as } y \to \infty \quad (5)$$
$$u = 0,\ T = T_\infty,\ C = C_\infty \text{ at } x = 0 \text{ and } y > 0$$

The above equations are dependent upon the assumptions that the only body force operating is that of gravity and that the temperature and mass fraction variations within the fluid are not large, so that Boussinesq's approximation can be applied thus enabling the density to be treated as a constant in all terms of the transport equations except the buoyancy term. Other fluid and transport properties, i.e. viscosity, specific heat, thermal conductivity and binary mass diffusivity, are taken as constant. Also viscous dissipation as well as volumetric heat generation effects are neglected.

The conservation equations can be recast into their dimensionless form by introducing the length of the major-axis, 2a, as the reference length, the appropriate stretched coordinates and the nominalized velocity, temperature and concentration are as defined below:

$$q = (T - T_\infty) /(T_w - T_\infty), \qquad \omega = (C - C_\infty) / (C_w - C_\infty)$$

$$X = x/L,\ Y = Gr_L^{1/4}\, Y/L, \qquad U = u/ (\nu/L)\, Gr_L^{1/2}, \quad V = v/(\nu/L)\, Gr_L^{1/4} \quad (6)$$

Where L = 2a and Grashof number, $Gr_L = g\beta(T_w-T_\infty)L^3/\nu^2$

To facilitate the analysis the following geometric relationships are derived from Fig.1.

$$X = x/2a = \frac{1}{2} \int_0^{\theta} (1 - e^2 \cos^2 \theta)^{1/2} d\theta$$

$$R = r/2a = \frac{1}{2} \frac{b}{a} \sin\theta \qquad \Phi = \frac{\sin\theta}{\sqrt{(1 - e^2 \cos^2 \theta)}} \qquad (7)$$

Now introduce the stream function Ψ (X,Y) such that

$$U = \frac{1}{R} \frac{\partial \Psi}{\partial Y} \quad \text{and} \quad V = -\frac{1}{R} \frac{\partial \Psi}{\partial X} \qquad (8)$$

In Eqn. (7), θ is the eccentric angle and e is the eccentricity of the meridian ellipse of the spheroid and is given by

$$e^2 = 1 - (b/a)^2$$

In forced flows the streamwise pressure gradient in the momentum boundary layer equation assumes the role of buoyancy in Eqn. (2). This observation led Lin and Chao (1974) to define a 'hypothetical' or equivalent outer stream function such that

$$u^* = U/(\nu Gr_L^{1/2} / L) = (2 \int_0^X \Phi dX)^{1/2} \qquad (9)$$

Thus, following Lin and Chao (1974), it can be shown that for the spheroid

$$u^* = (1 - \cos\theta)^{1/2} \qquad (10)$$

The (X,Y) coordinate system can be transformed into the Goertler-Meksyn system of stretched coordinate ξ and psuedo-similarity variable η using the following transformation

$$\xi = \int_0^X R^2 u^* dX = \frac{1}{8} \frac{b^2}{a^2} \int_0^{\theta} \sin^2 \theta (1 - \cos\theta)^2 \cdot (1 - e^2 \cos^2 \theta)^{1/2} d\theta \qquad (11)$$

$$\eta = \frac{Ru^* Y}{(2\xi)^{1/2}} = \frac{1}{2} \frac{b}{a} \frac{(1 - \cos\theta)^{1/2} Y}{(2\xi)^{1/2}} \sin\theta \qquad (12)$$

Eqs. (1) - (4), become

$$f''' + ff'' + \Lambda(\xi)[-(f')^2 + q + N\omega] = 2\xi \frac{\partial(f',f)}{\partial(\xi,\eta)} \qquad (13)$$

$$q'' + fq'Pr = 2\xi \frac{\partial(q,f)}{\partial(\xi,\eta)} Pr \qquad (14)$$

$$\omega'' + f\omega' Sc = 2\xi \frac{\partial(\omega, f)}{\partial(\xi,\eta)} Sc \qquad (15)$$

and the boundary conditions become

$$f'(\xi,0) = q(\xi,0) - 1 = \omega(\xi,0) - 1 = 0 \qquad (16)$$

$$f(\xi,0) + 2\xi \frac{\partial f}{\partial \xi} = 0 \qquad (17)$$

$$f'(\xi,\infty) = q(\xi,\infty) = \omega(\xi,\infty) = 0 \qquad (18)$$

where the function $\Lambda(\xi)$ is defined by

$$\Lambda = 2\Phi\xi/R^2 (u^*)^3 \qquad (19)$$

$$\text{and } N = Gr_{L,c}/Gr_L = \beta^*(C_w - C_\infty)/\beta(T_w - T_\infty) \qquad (20)$$

At the stagnation point:

$$X = \frac{x}{2a} = \frac{1}{2} \int_0^{\theta} (1 - e^2 \cos^2 \theta)^{1/2} d\theta, \quad R = X$$

$\theta \to 0$ (as $\theta \to 0$)

$$= \frac{1}{2}\theta\left(\frac{b}{a}\right) \text{ or } \theta = \left(\frac{a}{b}\right) 2X$$

$$\Phi = \frac{\sin\theta}{\sqrt{1 - e^2 \cos^2 \theta}} = \left(\frac{a}{b}\right)\theta = \left(\frac{a}{b}\right)^2 2X \qquad (21)$$

$\theta \to 0$ $\theta \to 0$

$$u^* = \{2\int_0^X \left(\frac{a}{b}\right)^2 2X\, dX\}^{1/2} = 2^{1/2}\left(\frac{a}{b}\right)X, \quad \xi = \int_0^X R^2 u^* dX = \frac{X^4}{2^{3/2}}\left(\frac{a}{b}\right)$$

$$\Lambda = \frac{2\xi\Phi}{R^2 (u^*)^3} = 0.5 \text{ (for all ratios of a/b)}$$

In writing the boundary condition, Eq.(17), it is assumed that the species mass fraction is low and hence the normal component of interfacial velocity is neglected. The condition for the neglect of the component of the interfacial velocity can be shown to be

$$V (2\xi)^{1/2} / [(\upsilon Gr^{1/4} / L) Ru^*] << 1 \qquad (22)$$

Using the condition of no mass transfer of the inert component at the interface ana applying Fick's law, it can be shown that the transverse velocity at the wall is given by

$$v = v_w = -D(\partial C / \partial y)_w / (1 - C_w) \qquad (23)$$

Using Eqn. (22) and Eqn. (23) the following alternative condition can be derived for the neglect of the contribution of the transverse velocity at the surface to the interface mass transfer

$$(C_w - C_\infty)[-\omega'(\xi,0)] / [Sc(1 - C_w)] << 1 \qquad (24)$$

Often in natural convection processes with low mass fraction the condition shown in equation (24) is fully satisfied and the neglect of the interfacial velocity effect is justified [6].

The problem defined by Eqs. (13) - (18) could be solved bv a finite difference scheme, for example, by a Keller Box type scheme [8]. In the present analysis the above equations have been solved by local nonsimilarity approaches. The latter approaches have been sucessfully used by various investigators for solving nonsimilar boundary layer equations and are also well documented in the literature [23 - 26]. The local nonsimilarity method of solution possesses, in addition to providing reasonably accurate results, the two attractive features, namely, the governing equations can be treated as ordinary differential equations and their solution at a certain value of the streamwise coordinate ξ can be obtained independently of the solutions from other ξ values.

The governing equations for the local nonsimilarity models will now be briefly derived. To begin with, the following three new functions are defined

$$G(\xi,\eta) = \partial f/\partial\xi, \quad H(\xi,\eta) = \partial q/\partial\xi, \quad M(\xi,\eta) = \partial\omega/\partial\xi \qquad (25)$$

After substituting these new functions into Eqs.(13)-(18), as set of ordinary differential equations are obtained. However, the introduction of the three new dependent variables in the problem require three additional equations with appropriate boundary conditions. These equations were obtained by differentiating Eqs. (13) - (18), with respect to ξ. Using the foregoing results, the governing equations and the boundary conditions for the 2-eqation local nonsimilarity were obtained in the following form:

$$f''' + ff'' + \Lambda(\xi)[-(f')^2 + q + N\omega] = 2\xi\,(f'G' - f''G) \qquad (26)$$

$$q'' + fq'\,Pr = 2\xi\,(f'H - q'G)\,Pr \qquad (27)$$

$$\omega'' + f\omega' Sc = 2\xi\,(f'M - \omega'G)Sc \qquad (28)$$

$$G''' + fG'' - 2f'G' + 3f''G + (d\Lambda/d\xi)[-(f')^2 + q + N\omega] + \Lambda(\xi)[-2f'G' + H + NM] - \underline{2\xi\,(G'G' - G''G)} = 2\xi\,[f'(\partial G'/\partial\xi) - f''(\partial G/\partial\xi)] \qquad (29)$$

$$H''/Pr - 2f'H + 3Gq' + fH' - \underline{2\xi\,(G'H - H'G)} = 2\xi\,[f'(\partial H/\partial\xi) - q'(\partial G/\partial\xi)] \qquad (30)$$

$$\omega''/Sc - 2f'M + 3G\omega' + fM' - \underline{2\xi\,(G'M - M'G)} = 2\xi\,[f'(\partial M/\partial\xi) - \omega'(\partial G/\partial\xi)] \qquad (31)$$

It is to be noted that the terms in the RHS as well as the underlined terms of Eqs. (29) - (31) have been neglected in parlance with the 2-equation model local nonsimilarity approach as they are assumed to be very small (23 - 27). This assumption is also valid for the present problem since the 3-equation model (not shown) of the local nonsimilarity approach applied to a specific case of the problem only marginally improved the results over those from the 2-equation model. Hence this approach was discarded in favour of the latter due to the excessive computing cost required to solve the 3-equation set. The boundary conditions become:

$$f(\xi,0) = f'(\xi,0) = G(\xi,0) = G'(\xi,0) = H'(\xi,0) = M(\xi,0) = 0$$

$$q(\xi,0) = \omega(\xi,0) = 1 \qquad (32)$$

$$f'(\xi,\infty) = G'(\xi,\infty) = q(\xi,\infty) = H(\xi,\infty) = \omega(\xi,\infty) = M(\xi,\infty) = 0$$

For the pure thermal convection, the two-equation local nonsimilarity model simplifies Eqs.(26) - (31) to

$$f''' + ff'' + \Lambda[-(f')^2 + q] = 2\xi\,(f'G' - f''G') \qquad (33)$$

$$q'' + fq'Pr = 2\xi(f'H - q'G)Pr \qquad (34)$$

$$G''' + fG'' - 2f'G' + 3f''G + (d\Lambda/d\xi\,[-(f')^2 + q] + \Lambda(\xi)[-2f'G' + H] - 2\xi\,(G'G' - G''G) = 2\xi\,[f'(\partial G'/\partial\xi) - f''(\partial G/\partial\xi)] \qquad (35)$$

$$H''/Pr - 2f'H + 3Gq' + fH' - 2\xi(G'H - H'G) = 2\xi[f'(\partial H/\partial\xi) - q'(\partial G/\partial\xi)] \qquad (36)$$

and the boundary conditions in Eq.(32) to

$$f(\xi,0) = f'(\xi,0) = G(\xi,0) = G'(\xi,0) = H(\xi,0) = 0$$

$$q(\xi,0) = 1, \quad f'(\xi,\infty) = G'(\xi,\infty) = q(\xi,\infty) = H(\xi,\infty) = 0 \qquad (37)$$

The coupled highly nonlinear 14th order set of ordinary differential equations obtained from the 2-equation local nonsimilarity approach for the combined heat and mass transfer cases and the 10th order set for the pure heat transfer cases constitute two-point boundary value problem. These equations were solved by the Runge-Kutta-Gill integration scheme along with the modified multivariable Newton-Raphson iteration scheme. Initial values of $f''(\xi,0)$, $q'(\xi,0)$, $\omega'(\xi,0)$, $G''(\xi,0)$, $H'(\xi,0)$ and $M'(\xi,0)$ for the Eqn.(26)-(37) and $f''(\xi,0)$, $q'(\xi,0)$, $G''(\xi,0)$ and $H'(\xi,0)$ for the Eqs. (33)-(37) were unknown. A forward integration procedure requires that all of the boundary conditions be specified at the starting of integration. Therefore initial values of the above qualities were guessed along with the step size and a suitable upper bound of integration for the independent variable η. Subsequent initial values were obtained by modified Newton-Raphson method. Initial values were refined until the conditions at the edge of the boundary-layer (i.e., at $\eta = \eta_\infty$) $|f'(\xi\eta)|$, $|f''(\xi\eta)|$, $|q(\xi\eta)|$, $|q'(\xi,\eta)|$, $|\omega(\xi,\eta)|$, $|\omega'(\xi,\eta)|$, $|G'(\omega,\eta)|$, $|G''(\xi,\eta)|$, $|H(\xi,\eta)|$, $|H'(\xi,\eta)|$, $|M(\xi,\eta)|$ and $|M'(\xi,\eta)|$ for the Eqs. (26) - (32) and $|f'(\xi,\eta)|$, $|f''(\xi,\eta)|$, $|q(\xi,\eta)|$, $|q'(\xi,\eta)|$, $|G'(\xi,\eta)|$, $|G''(\xi,\eta)|$, $|H(\xi,\eta)|$ and $|H'(\xi,\eta)|$ for the Eqs.(33)-(37) were each of the order of $10^{-5} \sim 10^{-6}$. Because of the strongly nonlinear nature of the above equations, some difficulty in converging the values at $\eta = \eta_\infty$ was experienced if initial guesses of the boundary values at the surface were far from the solution values. This difficulty was overcome by starting the computations with a relatively small values of $\eta = \eta_\infty$ then its value was successively increased until preassigned tolerances on the values of the variables at η_∞ were satisfied. For all the cases initial η_∞ were selected as 2.0 and were progressively increased with a step size of 1.0 or 2.0. A uniform integration step size of 0.04 was found to provide sufficiently accurate numerical results. The final upper limit of integration of $\eta_\infty=25$ satisfied the error condition fully for all cases studied. It should be pointed out that, the solutions of the velocity, thermal and concentration fields provide information for f, q, ω, G, H and M functions for the combined heat and mass transfer cases and f, q, G and H in the case of pure thermal convection. Only f, q, and ω and their derivatives for the former case and f and q along with their derivatives for the latter case respectively, as will be seen latter are physically relevant.

The parameters of the problem are Prandtl number, Pr, Schmidt number, Sc, and the ratio of the species to the thermal buoyancy force, N. Computations were carried out for the diffusion of water vapour and napthalene into air. Both of these diffusing species have practical as well as technological applications. Calculations were carried out both when the thermal buoyancy force was opposing and aiding the concentration buoyancy force. The case N = 0, corresponds to the pure thermal convection; when $N < 0$, the thermal and concentration buoyancy forces oppose each other, while for $N > 0$ they assist each other in driving the flow. The values of N may assume any positive or negative value since the quantities, β, β^*, $(T_w - T_\infty)$ and $(C_w - C_\infty)$ may be positive or negative.

The physical quantities of interest include, the local Sherwood number, Sh_L, the local Nusselt number, Nu_L, and the local wall shear stress, τ_w, defined, respectively, by

$$Sh_L = \overset{o}{m}_w L / [\rho D^* (C_w - C_\infty)],\ Nu_L = q_w L/[k(T_w - T_\infty)]$$

$$\text{and } \tau_w = \mu(\partial u / \partial y)_{y=0} \tag{38}$$

From Fick's law, $\overset{o}{m}_w = \rho D^* (\partial C/\partial y)_{y=0}$ and from Fourier's law, $q = -k(\partial T/\partial y)_{y=0}$ along with Eqns.(6)-(12) can be shown that

$$Sh_L (Gr_{L,c})^{-1/4} = Ru^* [-\omega'(\xi, 0)] / [(2\omega)^{1/2} N^{1/4}] \tag{39}$$

$$Nu_L (Gr_L)^{-1/4} = Ru^* [-q'(\xi,0)] / (2\xi)^{1/2} \tag{40}$$

$$\tau_w L^2 / [\rho \nu^2 (Gr_L)^{3/4}] = \Phi^{1/3} \xi^{1/4} f''(\xi, 0) / R^{1/3} \tag{41}$$

The dimensionless tangential velocity distributions is given by,

$$uL/\{\nu[1 - \cos\theta) Gr_L^{1/4}\} = f'(\xi, \eta) \tag{42}$$

The general nondimensional form of boundary-layer thickness is defined by

$$\delta X = y_x Gr_L^{1/4} / L = \eta_x [2\xi / (1 - \cos\theta)]^{1/2} / R \tag{43}$$

where x corresponds to c, t or v and η_x's corresponds to the distances inside the boundary-layers such that

$$\omega(\xi, \eta_c) = q(\xi, \eta_t) = f'(\xi, \eta_v) = 0.01 \tag{44}$$

3. RESULTS AND DISCUSSION

The parameters which govern the transformed non-dimensional boundary-layer equations are : Prandtl number (Pr), Schmidt number (Sc), ratio of mass fraction buoyancy force to thermal buoyancy force (N) and the geometric parameter which defines the shape of the spheriods (aspect ratio = a/b). For convenience the calculations have been restricted to Pr=0.72 (air) for Sc = 0.63 (water vapor) and Sc = 2.57 (napthalene). R was varied from -0.5 to 2.0 and two aspect ratios (A.R), 4.0 and 2.0 were considered. Although solutions have been reported over limited ranges of parameters the equations and the general trends hold for other values of the parameters as well.

In order to test the accuracy of the present solution scheme the pure free convective heat transfer solutions around spheroids were calculated and compared with the solutions given in [11]. The present solution agreed with those obtained in [11] within 1.0% . For space limitations the comparison is not reported here; rather the present pure free convective heat transfer results are plotted in Figs.2 and 4 for the sake of comparison and interpretation of the coupled heat and mass transfer solutions.

An interesting feature observed during the course of numerical solutions was that the local similarity scheme (obtained by deleting the right hand sides Jacobian terms from Eqs. (13) - (15)) provided results which were within 1.0% of the local nonsimilarity scheme. To conserve space those results are also not reported here. This aspect is somewhat unexpected. Usually, local nonsimilarity solution [23-27] provides more accurate results than the local similarity solution. This unexpected result by the lower order solution scheme can be credited to the fact that with the present form of selection of the transformed variables the rate of change of the streamwise derivative of the configuration function $\Lambda(\xi)$ is very small and the streamwise coordinate ξ itself also have very low values, thereby contributing almost nothing towards the local similarity results when solved with local nonsimilarity scheme.

The local Nusselt number distribution at the spheroidal surfaces is shown in Fig.2. From the figure it is seen that the maximum value of the local Nusselt number is obtained at the frontal stagnation point which increases with the increase of the A.R. Similar results were also observed in forced convection around prolate spheroids [22]. From Fig.2 it is also seen that for prolate spheriods Nu_L falls very rapidly from the high value at the frontal stagnation point as the flatter part of the spheriod is approached and the rate of fall increases with the increase of A.R. Figure 2 further reveals that with the meridional distance from the stagnation point there is a changeover in predicting the heat transfer results with the change of A.R.; lower A.R spheroids predicting constantly higher values in comparison with higher

Table 1. Results for the local Nusselt number, the local Sherwood number, the local wall shear stress and the cumulative tangential mass flow rate (R = 2, Pr = 0.72).

a/b	Sc	x/2a	$Nu_L/Gr_L^{1/4}$	$Sh_L/Gr_{L,c}^{1/4}$	$\tau_w L^2/\rho\nu^2 Gr_L^{3/4}$	$\dot{m} / 2L\mu\pi Gr_L^{1/4}$
2	0.63	0	1.03644	0.81466	0	0
		0.1463	0.90249	0.70938	0.91922	0.01493
		0.3525	0.73175	0.57518	1.44071	0.06798
		0.6055	0.60522	0.47570	1.78742	0.14527
		0.8586	0.48725	0.38292	0.04444	0.21818
		1.0032	0.38626	0.30349	2.14314	0.25891
		1.0647	0.31959	0.25104	2.12359	0.27934
	2.57	0	0.91585	1.40433	0	0
		0.1463	0.79724	1.22197	0.78862	0.01180
		0.3525	0.64642	0.99080	1.23604	0.05370
		0.6055	0.53496	0.82063	1.53719	0.11471
		0.8586	0.44723	0.66344	1.77061	0.17217
		1.0032	0.34303	0.52979	1.87760	0.20407
		1.0647	0.24249	0.44149	1.88000	0.21992
4	0.63	0	1.46574	1.15210	0	0
		0.0962	1.01760	0.79989	0.99802	0.03367
		0.2853	0.75327	0.59211	1.44927	0.13995
		0.5361	0.60959	0.47914	1.77975	. 2 9 2
		0.7869	0.49256	0.38710	2.07931	0.44117
		0.9230	0.39666	0.31164	2.28900	0.53202
		1.0179	0.33251	0.26114	2.38204	0.58172
	2.57	0	1.29520	1.98602	0	0
		0.0962	0.89861	1.37670	0.85499	0.02660
		0.2853	0.66528	1.0.943	1.24228	0.11055
		0.5361	0.53876	0.82631	1.52984	0.23108
		0.7869	0.43619	0.67080	1.80138	0.34814
		0.9230	0.35250	0.54499	2.01059	0.41917
		1.0179	0.29651	0.46117	2.12044	0.45771

A.R spheroids although the spread is not very significant. The above results can be attributed to curvature effects. A spheroid with A.R. = 4.0 is more slender than one with A.R = 2.0 ; the former having higher curvature at the stagnation point than the latter. Since in the presently deployed local orthogonal coordinate system the buoyancy force is a variable quantity as it is dependent on the local streamwise coordinate, x and the surface curvature requires that the boundary-layers must continually adjust to the changing strength of the surface parallel component of the buoyancy force therefore higher curvature favors more energy transfer. With the increase of the distance from the stagnation point spheroids with A.R of 4 become more flatter with respect to the spheroids with lower A.R and hence predicts lower heat transfer results.

Heat transfer results displayed in the form of $Nu_L/(Gr_L)^{1/4}$ in Fig.2 and Table 1 portrays that, $Nu_L/(Gr_L)^{1/4}$ increases for aiding buoyancy forces i.e for $N > 0$ and decreases for opposing buoyancy forces i.e., for $N < 0$. Observe that $Nu_L/(Gr_L)^{1/4}$ increases and decreases beyond the values for N = 0 when the buoyancy force from species diffusion assists and opposes, respectively, the thermal buoyancy force. Figure 2 along with the results in Table 1 also indicates that larger departures of $Nu_L/(Gr_L)^{1/4}$ from N = 0 are associated with smaller values of Sc's for both $N > 0$ and $N < 0$. This trend can be interpreted physically in terms of the fact that diffusing species with smaller Sc possess a larger binary diffusion coefficient which exerts a greater effect on the flow and hence also thermal fields.

Figure 2 and Table I display the fact that for a fixed A.R and constant Pr, Nu_L decreases with increasing Sc for aiding flows i.e. for $N > 0$, while Nu_L increases with increasing Sc, for counter-buoyant flows i.e $N < 0$. The reason for this reversed effect of Sc on the thermal diffusion mechanism can be explained from the relative changes of the boundary layer thicknesses observed during numerical computations. During computations it was observed that for $N > 0$ with the increase of Sc, δ_c decreases due to the lower diffusion coefficient associated with the higher Sc. Since in a coupled convective flow concentration and thermal fields are intimately related with the flow field, the adverse effect of one tends to be balanced by the other two. This reduction in δ_c causes the momentum boundary-layer thickness δ_v to adjust itself by expanding its depth. Since in this problem the thermal field is intimately coupled with the flow field, δ_v also forces δ_t to follow its expansion. This results in a greater δ_t thereby reducing the thermal gradients at the wall and hence also the Nu_L.

The results for the local Sherwood number in the form $Sh_L/(Gr_{L,c})^{1/4}$ are shown partly in graphical form in Fig.3 and partly in Table 1. The results show that for a given buoyancy similar to the heat transfer results maximum mass transfer occurs at the lower stagnation point and that the rate decreases progressively with angular distance from the lower stagnation point. The attributes of the mass transfer results with respect to the A.R are similar to that of heat transfer solutions and they can be explained similarly.

Figure 3 and Table 1 further reveals that large $Sh_L/(Gr_{L,c})^{1/4}$ is associated with large Sc. A larger Sc corresponds to a smaller binary diffusion coefficient in a given binary mixture and hence a thinner δ_c relative to the momentum boundary-layer thickness, δ_v. This results in a larger concentration gradient at the wall which in turn enhances the mass transfer rate.

For a fixed A.R dependence of $Sh_L/(Gr_{L,c})^{1/4}$ on N is quite interesting. From Fig.2 it is seen that in the case of diffusion of napthalene (Sc = 2.57) into air with N = -0.5 or 1.0, local mass transfer rates over the entire sphere are higher than the corresponding mass transfer rates for N = 2.0. From the calculations it was seen that the δ_c were thinner for N = 2.0 than those for N = -0.5 or 1.0. Thus it is expected that the concentration gradient will be higher at the surface for the former case than for the latter cases. This is indeed true. From the numerical calculations it is seen that for a spheroid with A.R of 4.0 and at a meridional distance x/2a = 0.787 in the case of diffusion of naphthalene into air the mass fraction gradients at the surface of the spheroid with N = 2, 1.0 and -0.5 are -0.76318, -0.70599 and -0.57387, respectively. However, $Sh_L/(Gr_{L,c})^{1/4}$ apart from the concentration gradient at the wall, $-\omega(\xi,0)$ also depends on $|N|^{1/4}$ (note that the absolute value of $|Gr_{L,c}/N|$ is needed to ensure that Sh_L is positive since $Gr_{L,c}$ and N can assume positive as well as negative values). For a small value of N, either positive or negative, the $|1/N|^{1/4}$ takes on a very high value and hence results in higher $Sh_L/(Gr_{L,c})^{1/4}$ For a fixed spheroid depending on Sc, $Sh_L/(Gr_{L,c})^{1/4}$ For a fixed spheroid depending on Sc, $Sh_L/(Gr_{L,c})^{1/4}$ at small N may be higher than that at high values of N. This may appear somewhat surprising but physical interpretation can elucidate the computed results. Physically, for a low value of N the flow condition is induced almost entirely by the thermal buoyancy force and the species diffusion mechanism becomes very effective at very low concentration levels. For a fixed $Gr_{L,c}$ a small value of N implies a large value of Gr_L. Thus a large $Sh_L/(Gr_{L,c})^{1/4}$ results when Gr_L is large compared to $Gr_{L,c}$

The results of the local wall shear stress reported in the form $\tau_w L^2[\rho\nu^2(Gr_L)^{3/4}]$ are partly portrayed in Fig.4 and partly tabulated in Table 1. The limited results shown reveal that with the increase of A.R local surface shear stress around the spheroids increases the difference in results with the variations of slenderness of A.R is quite significant at the frontal stagnation region and near the rear end. In the middle section the change of wall shear stress is not appreciable with the change of A.R. The results of wall shear stress also show trends similar to those for the heat transfer results, i.e., the wall shear stress increases beyond the values for N = 0 when the buoyancy force from mass diffusion acts in the same direction as the thermal buoyancy force to assist the flow and decreases beyond N = 0 when the two buoyancy forces are counter-buoyant. The results of Fig.4 and Table 1 also depict that smaller Sc (associated with larger binary diffusion coefficient) exerts a greater influence on the flow field ; as a result a larger departure of wall shear stress from N = 0 occurs for both N > 0 and N < 0.

For the purpose of illustration representative dimensionless tangential relative velocity, temperature and mass fraction for the spheriod with A.R of 4.0 and at two meridional locations of x/2a = 0.287 and 0.787 are shown through Figs.5 - 10 for various combinations of parameters.

From Figs.5 and 6 it is seen that the velocity gradient at the wall shows an increase with increasing N but with decreasing values of Sc. This trend agrees with the wall shear stress results shown in Fig.4. Indeed, if the combined buoyancy force effect is greater than the effect due to pure thermal buoyancy viz., if $q + \omega N > q$, the velocity gradient at the wall increases. Conversely, the velocity gradient at wall decreases when the combined effect is less than the effect due to pure thermal buoyancy force, viz., for $q + N\omega < q$. The velocity profiles show similar patterns to those of single buoyancy flow viz., the velocity for no-slip condition at the surface starts from zero at the surface and increases to a maximum and then gradually decreases to zero at an infinite distance from the surface of the spheroids. Depending on the values of the parameters, in the case of simultaneous transfer, this maximum is seen to shift towards the surface of the spheroids or away from the surface. For positive N the shift is towards the surface while for negative N it moves away from the surface.

The dimensionless scaled temperature profiles in Figs. 6 and 7 show that the surface temperature gradient increases with N and decreases with Sc. This behaviour is consistent with Nu_L results shown in Fig. 2. As in the case of the velocity gradient the temperature gradient at the wall increases above that for N = 0 when $q + N\omega > q$ and decreases below it when $q + \omega N < q$.

The scaled mass fraction profiles in Figs.9 and 10 exhibit trends that are somewhat different from those of the velocity and temperature profiles. While the mass fraction gradient at the wall increases as N increases, a larger increase is associated with a larger Sc. It should be noted that a higher mass fraction gradient at the surface does not necessarily mean higher Sh_L since, as explained earlier, apart from the mass fraction gradient Sh_L also depends on the multiplicative factor $|1/N|^{1/4}$.

4. CONCLUSIONS

From the present formulation of the problem of coupled free convective mass and heat transfer around spheroids, it is observed that the two-equation local nonsimilarity approach provides reasonable results for the major portion of the spheroids. During numerical computations it was observed that very little improvement in results was obtained by using local nonsimilarity approach over that of local similarity solution. Therefore, for the present set of transformed boundary-layer equations simple local similarity solution can provide reasonably accurate solution compared to the computationally many times more demanding solution scheme of local nonsimilarity.

Analyses testify that as in pure thermal and pure species transfer in free convection, the maximum local mass and heat transfer rates in the case of simultaneous transfer around spheroids occur at the frontal stagnation point. With the increase of the aspect ratio both heat and mass transfer rates increase at the stagnation point but sharply decrease some distance from the neighborhood of the stagnation point. With the progress from the stagnation point along the meridian plane the heat and mass transfer rates decrease and their respective difference with respect to the aspect ratio diminishes.

Both Nu_L and the T_w increase and decrease from their respective pure thermal free convection values as the buoyancy forces from species diffusion assist and oppose, respectively, the thermal buoyancy force. The departures of these two quantities from the pure thermal convection results are more pronounced at low Schmidt numbers. The local heat transfer rate decreases with increasing Schmidt number for aiding flows while it increases with Schmidt numbers for counterbuoyant flows. For a fixed $Gr_{L,c}$ the mass transfer rate is found to increases as the thermal buoyancy force increases. While the local surface heat transfer and local wall shear stress are enhanced for diffusion of lower molecular weight gases and vapors, the surface mass transfer rate increases. for high molecular weight gases and vapors. The results obtained from the present analysis have demonstrated and quantified the effect of heat and mass transfer on each other in free convective boundary-layer transfer around spheroids.

NOMENCLATURE

a = semi-major axis, m

b semi-minor axis, m

C mass fraction, dimensionless

D binary diffusion coefficient, m^2/s

f reduced stream function, dimensionless

g = accelaration due to gravity, m/s^2

Gr_L = local thermal Grashof number ($= \beta(T_w - T_\infty) L^3/\nu^2$), dimensionless

$Gr_{L,c}$ = local Grashof number due to mass diffusion ($= \beta^* (C_w - C_\infty) L^3 / \nu^2$)

k = thermal conductivity of fluid, W/mK

$\dot{m}$ = mass flow rate, kgm/m^2s

$\dot{m}_w$ = **mass flux of diffusing species, kgm/m^2s**

N = ratio of concentration to thermal Grashof nubers ($Gr_{L,c}/Gr_L$), dimensionless

q = dimensionless temperature ($= (T - T_\infty) / T_w - T_\infty$

q_w = local wall heat flux, W/m^2

R = dimensionless radial distance from axis of symmetry to a surface element (= r/2a), dimensionless

r = radial distance from symmetrical axis, to surface, m

Sc = Schmidt number ($=\nu/D$), dimensionless

Sh = local Sherwood number ($=m_w L/\rho D (C_w - C_\infty)$), dimensionless

T = Temperature of the fluid, K

u,v = velocity components in x and y directions, m/s

u,V = dimensionless velocity components

x, y = local streamwise and normal coordinates, m

X,Y = dimensionless streamwise and normal coordinates

Greek Letters

α = thermal diffisivity, $m^2 s$

β = volumetric coefficient of thermal expansion, $(-(\partial\rho/\partial T)_{p,c} / \rho)$, K^{-1}

β^* = volumetric coefficient of expansion with mass $(-(\partial\rho/\partial C)_{p,T}/\rho)$

θ = angular distance from the stagnation point, rad

δc = dimensionless concentration boundary-layer

δ = dimensionless thermal boundary-layer

δ = dimensionless velocity boundary-layer

η = pseudo similarity variable, dimensionless

η_∞ = final limit of integration, dimensionless

μ = dynamic viscosity, Ns/m^2

ν = kinematic viscosity of fluid, m^2/s

ω = dimensionless concentration ($=(C - C_\infty)/(C_w - C_\infty)$)

ρ = density of fluid, kg/m^3

ξ = transformed streamwise coordinate, dimensionlesss

Λ = configuration function defined in equation (13), dimensionless

τ = shear stress, N/m^2

ψ = stream function, dimensionless

Super Scripts

= denotes differentiation with respect to η

Subscripts

w = condition at the wall

∞ = condition at the edge of the thickest boundary layers among concentration, thermal and velocity

REFERENCES

1. Fernandez-Pello, A.C., and Law, C.K., A Theory for the Free Convective Burning of a Condensed Fuel Particle, Combustion and Flame, 44 (1982) 97-112.

2. Beg, S.A., Forced Convective Mass Transfer Studies from Spheroids, Warme-und-Stoffubertragung, 8 (1975) 127-135.

3. Gill, W.N., Del Casal, E., and Zeh, D.W., Binary Diffusion and Heat Transfer in Laminar Free Convection Boundary Layers on a Vertical Plate, 8 (1965) 1131-1151.

4. Gebhart, B., and Pera, L., The Nature of Vertical Natural Convection Flows Resulting from the Combined Buoyancy Effects of Thermal and Mass Diffusion, Int. J. Heat Mass Transfer, 15 (1971) 2025-2050.

5. Pera, L., and Gebhart, B., Natural Convection Flows adjacent to Horizontal surfaces resulting from Combined Buoyancy effects of Thermal and Mass Diffusion, Int. J. Heat Mass Transfer, 15 (1972) 269-278.

6. Chen, T.S., and Strobel, F.A., Combined Heat and Mass Transfer in Mixed Convection over a Horizontal Flat plate, Trans. ASME, J. Heat Transfer 102 (1980) 538-543.

7. Chen, T.S., and Yuh, C.F., Combined Heat and Mass Transfer in Natural Convection on Inclined Surfaces, Numer. Heat Transfer, 2 (1979) 233-250.

8. Strobel, F.A., and Chen, T.S., Buoyancy Effects on Heat and Mass Transfer in Boundary Layers Adjacent to Inclined, Continuous, Moving Plate, Num. Heat Transfer, 3 (1980) 461-481.

9. Hasan, M., and Mujumbdar, A.S., coupled Heat and Mass Transfer in Natural convection under Flux Condition Along a Vertical cone, Int. Comm. Heat Transfer, No.2, 2 (1984) 157-172.

10. Saville, D.A., and Churchill, S.W., Laminar Free Convection in Boundary Layers Near Horizontal Cylinders and Vertical Axisymmetric Bodies, J. Fluid Mech., 29 (1967) 391-399.

11. Lin, F.N., and Chao, B.T., Laminar Free Convection Over Two-Dimensional and Axisymmetric Bodies of Arbitrary Contour, Trans. ASME, J. Heat Transfer, 94 (1974) 435-442.

12. Raithby, G.D., Pollard, A., Hollands, K.G.T., and Yavanovich, M.M., Free Convection Heat Transfer from Spheroids, ASME, J. Heat Transfer, 98, (1967) 452.

13. Weber, M.E., Astrauskas, P., and Petsalis, S., Natural Convection Mass Transfer at High Rayleigh Number, The Cand. J. Chem./Eng., 62 (1983) 68-72.

14. Chiang, T., and Kaye, J., On Laminar Free Convection from a Horizontal Cylinder, Proceedings of the Fourth National Congress of Applied Mechanics, (1962) 1213-1219.

15. Saville, D.A., and Churchill, S.W., Simultaneous Heat and Mass Transfer in Free Convection Boundary Layers, AICh E Jl (1970) 268-273.

16. Watson, A., and Poots, G., On steady Laminar Free Convection due to a Heated Elliposid of Revolution, Int. J. Heat Mass Transfer, (1972) 1467-1475.

17. Lin, F.N., and Chao B.T., Predictive Capabilities of Series Solutions for Laminar Convection Boundary Layer Heat transfer, Trans.ASME, J. Heat Transfer, 100 (1978) 160-163.

18. Merk, H.J., Rapid Calculations for Boundary-Layer Transfer Using Wedge Solutions and Asymptotic expansions, J. Fluid Mech., 5 (1959) 460-480

19. Chao, B.T., and Fagbenle, R.O., On Merk's Method of Calculating Boundary-Layer Transfer, Int. J. Heat Mass Transfer, 17 (1974) 223-240.

20. Raithby, G.D. and Hollands, K.G.T., Free Convection Heat Transfer from Vertical Neddles, J. Heat Transfer, 98 (1976) 552-523.

21. Churchill, S.W., and Churchill, R.U., Comprehensive correlating Equation for Heat and Component Transfer by Free Convection, AIChE J., (1975) 604-606.

22. Masliyah, J.H., and Epstein, N., Numerical Solution of Heat and Mass Transfer from Spheroids in Steady Axisymmetric Flow, Progress in Heat and Mass Transfer, 6 (1972) 613-632.

23. Minkowycz, W.J., and Sparrow, E.M., Local Nonsimilar Solutions for Natural Convection on a Vertical Cylinder, Trans. ASME, J. Heat Transfer, 96 (1974) 178-183.

24. Minkowcyz, W.J., and Sparrow, E.M., Interaction Between surface mass Transfer and Transverse Curvature in Natural Convection Boundary-layers, Int. J. Heat Mass Transfer, 22 (1979) 1445-1454.

25. Chen, T.S., and Lohman M.E., Axial Heat Conduction Effects in Forced Convection Along a Cylinder, Trans. ASME, J. Heat Transfer, 97, (1975) 185-190.

26. Hasan, M., and Mujumdar, A.S., Simultaneous Heat and Mass Transfer in Mixed Convection Along a Vertical Cone Under Uniform Heat and Uniform Mass Fluxes, Chem. Engg. Commn., (in press).

27. Hasan, M., and Mujumdar, A.S., Coupled Heat and Mass Transfer in Free Convection Along a Rotating Plate under Uniform Heat and Uniform Mass Fluxes, Chem. Engg. Commns. (in press).

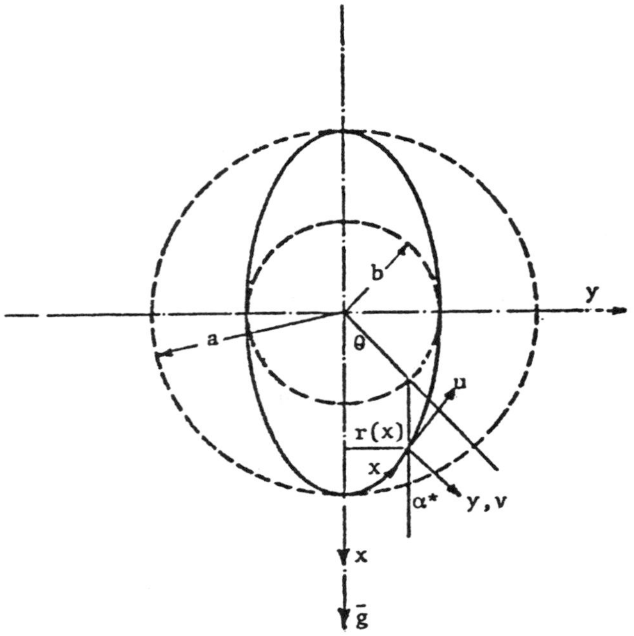

Fig.1 Vertical cross-section of the physical model and coordinate system

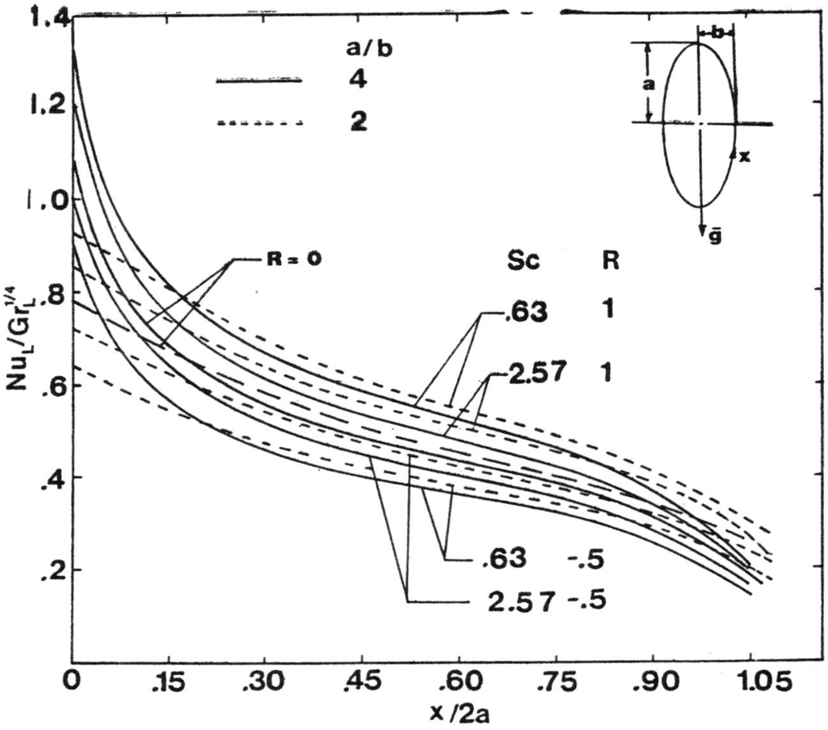

Fig.2 Angular distributions of local Sherwood number, Pr=0.72

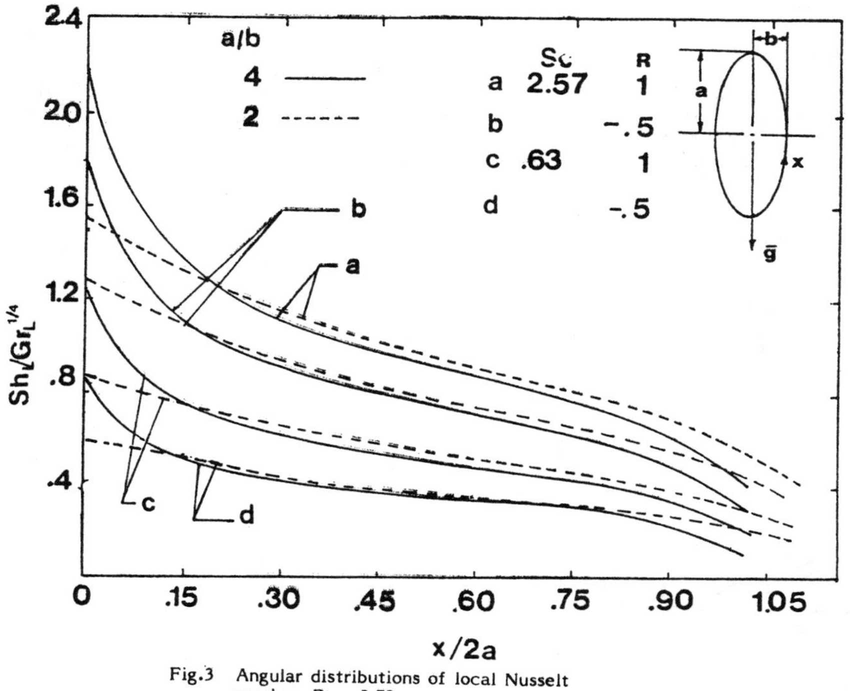

Fig.3 Angular distributions of local Nusselt number, Pr = 0.72

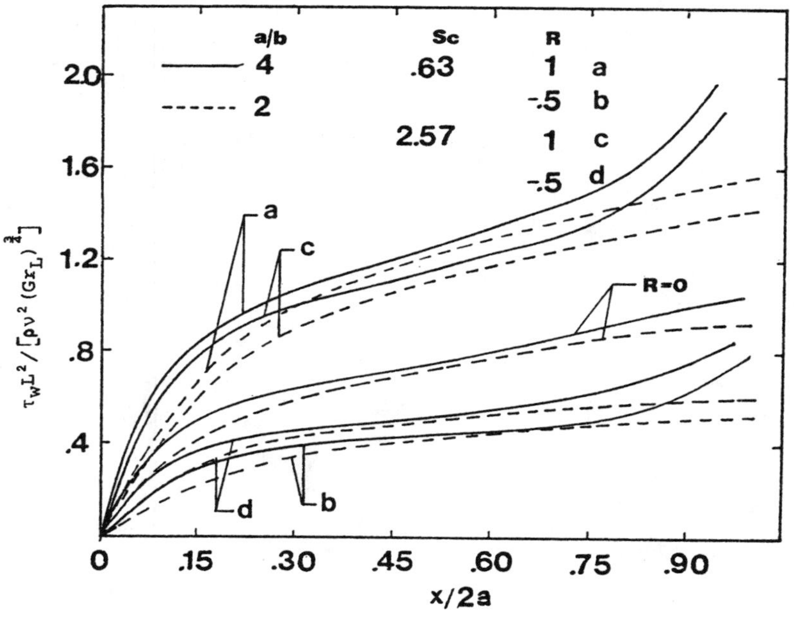

Fig.4 Angular distributions of local wall shear stress, Pr = 0.72

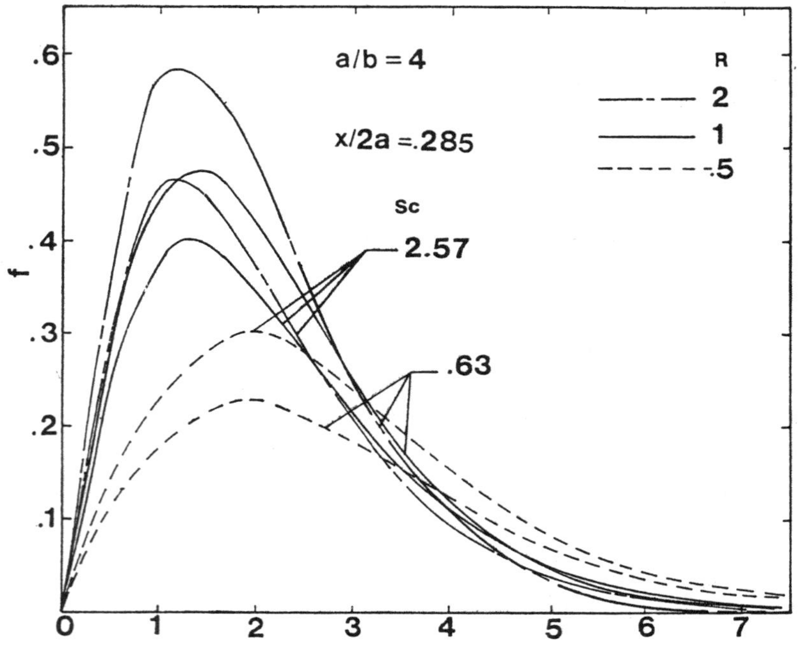

Fig.5 Variations of the dimensionless streamwise velocity for various conditions of buoyancy force effects, x/2a = 0.285

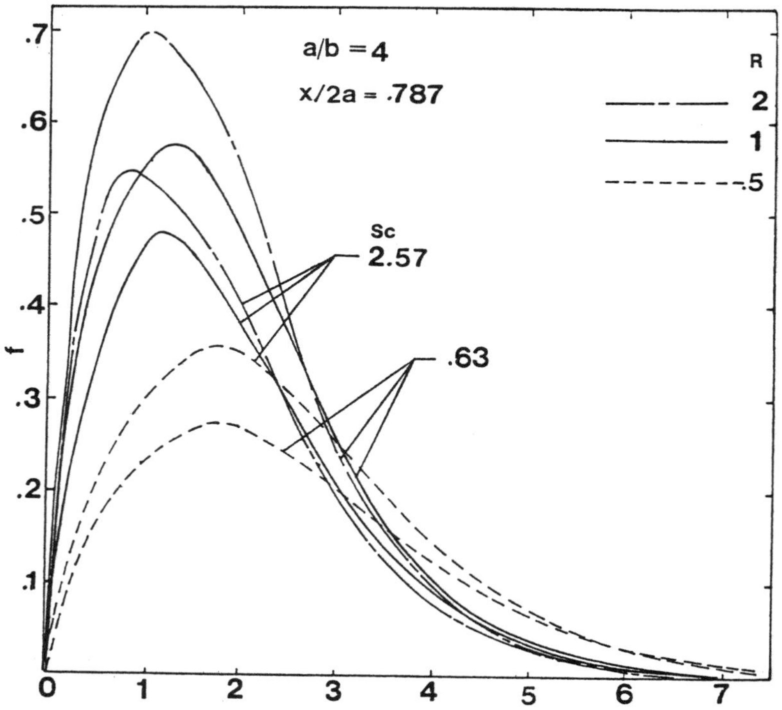

Fig.6 Variations of the dimensionless streamwise velocity for various conditions of buoyancy force effects, x/2a = 0.787

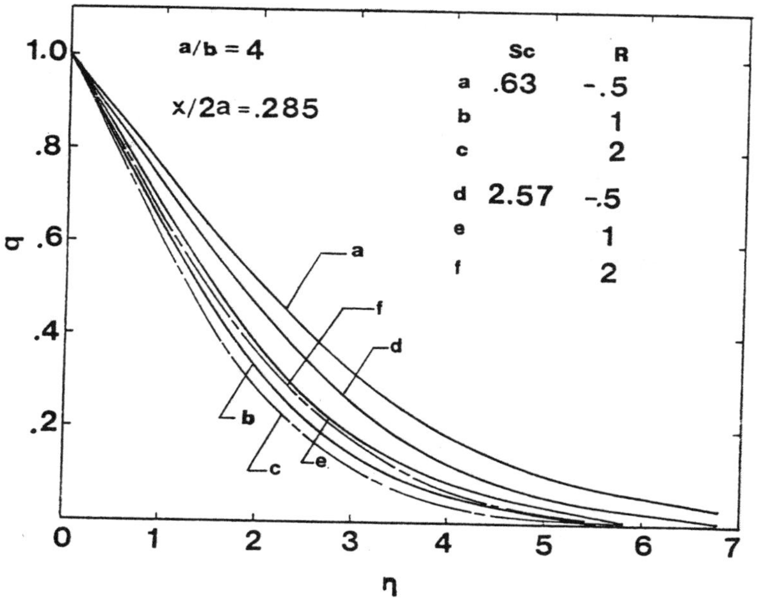

Fig.7 Dimensionless temperature profiles, x/2a = 0.285, a/b=4.0

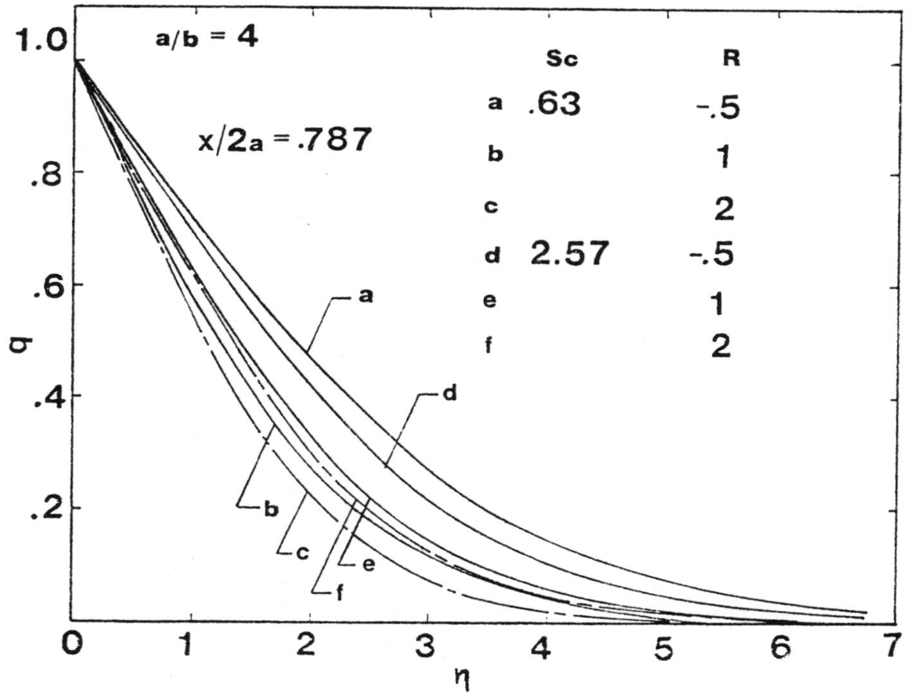

Fig.8 Dimensionless temperature profiles, x/2a=0.787, a/b=4.0

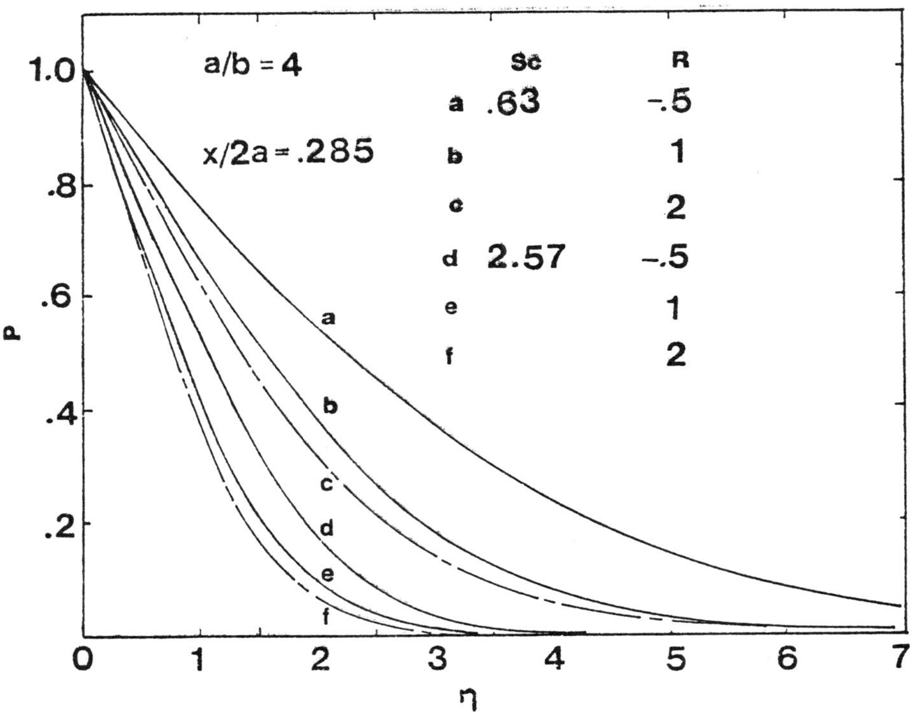

Fig.9 Dimensionless mass fraction profiles, x/2a = 0.285, a/b = 4.0

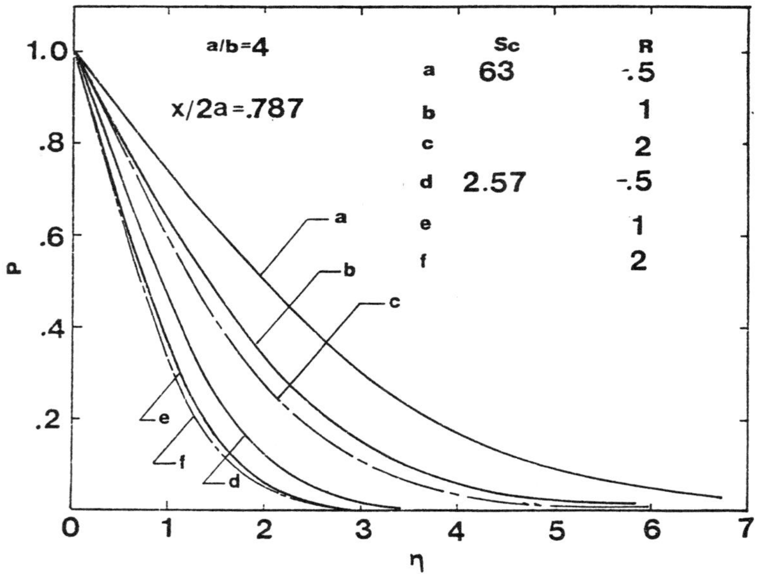

Fig.10 Dimensionless mass fraction profiles, x/2a = 0.787, a/b = 4.0

The Methods of Solution of the Nonlinear Finite Element Modelled Transient Heat Transfer Problems

BHIMAVARAPU S. REDDY and ANAND M. SHARAN
Faculty of Engineering
Memorial University of Newfoundland
St. John's, Newfoundland, A1B 3X5, Canada

In this work, the transient temperature distribution within a solid subjected to nonlinear boundary conditions has been obtained using the nonlinear finite element analysis. The nonlinear system of algebraic equations are solved using three techniques.

1. INTRODUCTION

There are a large number of industrial processes where one has to solve the transient nonlinear finite element modelled heat transfer equations [1]. The nonlinearity in these problems arises due to two reasons; the first reason is the variation of the material properties with the temperature and the second one is due to the radiative heat flux where higher powers of temperatures are involved. The first type of nonlinearity can be taken care of by evaluating the elemental matrices at each time-step thus these matrices are updated at each time-step. The second type of the nonlinearity can be analysed by formulating the problem using the variational principles which is the subject of discussion in this paper.

The present work involved the formulation based on the nonlinear finite element analysis of the transient heat transfer problems and solution of these problems using several techniques.

2. THE MATHEMATICAL FORMULATION

The differential equation of the heat conduction process is solids can be written as

$$K_x \frac{\partial^2 T}{\partial x^2} + K_y \frac{\partial^2 T}{\partial y^2} + K_z \frac{\partial^2 T}{\partial z^2} + Q = \rho c \frac{\partial T}{\partial t} \qquad (1)$$

with the boundary conditions

$$T = T_B \text{ on } S_1, \text{ and} \qquad (2)$$

$$K_x \frac{\partial T}{\partial x} \ell_x + K_y \frac{\partial T}{\partial y} \ell_y + K_z \frac{\partial T}{\partial z} \ell_z + q$$

$$+ h (T - T_\infty) + \sigma \varepsilon (T^4 - T_\infty^4) = 0 \text{ on } S_2. \qquad (3)$$

The union of both the surfaces S_1 and S_2 forms the complete boundary of the solid having a volume V. The functional formulation which is equivalent to Eqns. (1) and (2) can be written as [1]

$$X = \int_V \frac{1}{2} [k_x(\frac{\partial T}{\partial x})^2 + k_y (\frac{\partial T}{\partial y})^2$$

$$+ k_z(\frac{\partial T}{\partial z})^2 + 2 \rho c T \frac{\partial T}{\partial t}] dV$$

$$\int_S [q t + \frac{h}{2} (T - T_\infty)^2 + \sigma \varepsilon \frac{T^5}{5} - T_\infty^4 T)] d S \qquad (4)$$

The contribution of each of the elments can be added up and the resulting global matrices can be written as [1]

$$[C^G] \frac{\partial \{T^G\}}{\partial t} + [K^G] \{T^G\}$$

$$= \{F_Q^G\} - \{F_q^G\} - \{F_c^G\} + \{F_r^G\} \qquad (5)$$

where the expressions for the matrices $[C^e]$, $[K^e]$ and the vectors $\{F_Q^e\}$, $\{F_q^e\}$, $\{F_c^e\}$, $\{F_r^e\}$ are given in the Appendix A.

The vector $\{F_r^G\}$ is the global force vector due to the radiative heat transfer. This vector contains the terms of higher powers of the nodal temperatures. Since Eqn. (5) represents a system of nonlinear first order differential equations, it must be solved by the nonlinear methods. Some of the methods are:

(1) The iteration method.
(2) The Newton-Raphson method, and
(3) The nonlinear optimization method.

3. THE METHODS OF SOLUTION

3.1 The Transformation of the Simultaneous Differential Equations

There are two commonly known methods for solving the nonlinear set of transient temperature equations which is Eqn. (5). It is a system of first order nonlinear differential equation. The first method of solving these equations is by using the finite element method defined in the time domain [2]. The second method [3] for solving these equations is by approximating the time derivative using a finite difference scheme. However, the number of computations involved in the first method are very large. On the other hand, the Crank-Nicholson central finite difference method [3,4] is unconditionally stable and can be easily used in the present investigation for obtaining the temperature distribution. Using this method, the first derivative of the nodal temperatures between two points in the time domain can be approximated as

$$\frac{d \{T^G\}_t}{dt} = \frac{\{T^G\}_{t+\frac{\Delta t}{2}} - \{T^G\}_{t-\frac{\Delta t}{2}}}{\Delta t} \qquad (6)$$

where Δt is the time step.

Similarly, we can also express $\{T^G\}_t$ and $\{F^G\}_t$ as

$$\{T^G\}_t = \frac{\{T^G\}_{t+\frac{\Delta t}{2}} + \{T^G\}_{t-\frac{\Delta t}{2}}}{2} \tag{7}$$

and

$$\{F^G\}_t = \frac{\{F^G\}_{t+\frac{\Delta t}{2}} + \{F^G\}_{t-\frac{\Delta t}{2}}}{2} \tag{8}$$

Substitution of Eqns. (6), (7), (8) into Eqn. (5) gives

$$\begin{aligned}
&\frac{1}{\Delta t}[C^G]\{T^G\}_{t+\frac{\Delta t}{2}} - \frac{1}{\Delta t}[C^G]\{T^G\}_{t-\frac{\Delta t}{2}} \\
&+ \frac{1}{2}[K^G]\{T^G\}_{t+\frac{\Delta t}{2}} \\
&+ \frac{1}{2}[K^G]\{T^G\}_{t-\frac{\Delta t}{2}} = \frac{1}{2}\{F_Q^G\}_{t+\frac{\Delta t}{2}} \\
&\quad + \frac{1}{2}\{F_Q^G\}_{t-\frac{\Delta t}{2}} \\
&- \frac{1}{2}\{F_q^G\}_{t+\frac{\Delta t}{2}} - \frac{1}{2}\{F_q^G\}_{t-\frac{\Delta t}{2}} \\
&+ \frac{1}{2}\{F_c^G\}_{t+\frac{\Delta t}{2}} \\
&+ \frac{1}{2}\{F_c^G\}_{t-\frac{\Delta t}{2}} + \frac{1}{2}\{F_r^G\}_{t+\frac{\Delta t}{2}} \\
&+ \frac{1}{2}\{F_r^G\}_{t-\frac{\Delta t}{2}}
\end{aligned} \tag{9}$$

The above equation can be rearranged as

$$\begin{aligned}
&([K^G] + \frac{2}{\Delta T}[C^G])\{T^G\}_{t+\frac{\Delta t}{2}} \\
&= (\frac{2}{\Delta t}[C^G] - [K^G])\{T^G\}_{t-\frac{\Delta t}{2}} \\
&+ \{F_Q^G\}_{t+\frac{\Delta t}{2}} + \{F_Q^G\}_{t-\frac{\Delta t}{2}} \\
&- \{F_q^G\}_{t+\frac{\Delta t}{2}} - \{F_q^G\}_{t-\frac{\Delta t}{2}} \\
&+ \{F_c^G\}_{t+\frac{\Delta t}{2}} + \{F_c^G\}_{t-\frac{\Delta t}{2}} \\
&+ \{F_r^G\}_{t+\frac{\Delta t}{2}} + \{F_r^G\}_{t-\frac{\Delta t}{2}}
\end{aligned} \tag{10}$$

In the above equation, the force vectors due to radiation heat transfer, $\{F_r^G\}_{t-\frac{\Delta t}{2}}$ and $\{F_r^G\}_{t+\frac{\Delta t}{2}}$ are nonliner. They contain the higher powers of some of the nodal temperatures of the elements along the surface of the ingot. Here, the nodal temperature vector $\{T^G\}$ at time $t+\frac{\Delta t}{2}$ is unknown whereas the other quantities (vectors and matrices) are known. The substitution of $\{T^G\}_{t-\frac{\Delta t}{2}}$ on the right side of Eqn. (10) reduces to

$$\begin{aligned}
([K^G] + \frac{2}{\Delta t}[C^G])\{T^G\}_{t+\frac{\Delta t}{2}} &= \{A_1\} + \{F_Q^G\}_{t+\frac{\Delta t}{2}} \\
&- \{F_q^G\}_{t+\frac{\Delta t}{2}} + \{F_c^G\}_{t+\frac{\Delta t}{2}} + \{F_r^G\}_{t+\frac{\Delta t}{2}}
\end{aligned} \tag{11}$$

where $\{A_1\}$ is known.

The Eqn. (11) has vectors $\{F_r^G\}_{t+\frac{\Delta t}{2}}$ and $\{F_c^G\}_{t+\frac{\Delta t}{Z}}$ on the right hand side and these vectors contain the unknown nodal temperatures at the time $t+\frac{\Delta t}{2}$. Therefore, this equation can only be solved by iteration techniques.

3.2 The Iteration Method

In this method, [1] the nodal temperatures at time $t+\frac{\Delta t}{2}$ are initially assumed to be the same as the temperatures at time $t-\frac{\Delta t}{2}$, and used on the right hand side of Eqn. (11). Then the force vectors, $\{F_Q^G\}$, $\{F_q^G\}$, $\{F_c^G\}$, **and** $\{F_r^G\}$ in Eqn. (11) are evaluated with the assumed nodal temperatures. At this stage Eqn. (11) reduces to

$$([K^G] + \frac{2}{\Delta t}[C^G])\{T^G\}_{t+\frac{\Delta t}{2}} = \{A_2\} \tag{12}$$

where $\{A_2\}$ is known.

The linear equation, Eqn. (12) can be easily solved for $\{T^G\}_{t+\frac{\Delta t}{2}}$. The calculated nodal temperatures are then compared with those assumed ones. If they do not meet the convergence criteria then, the average of the assumed and calculated temperatures are used in the next iteration.

3.3 The Newton-Raphson Method

It is a commonly known method for solving the

nonlinear equations [3,5]. In this method the unknown temperature vector $\{T^G\}$ at the ith stage of iteration is substituted in Eqn. (11) and rearranging this equation one can write

$$\{f\}^i = ([K^G] + \frac{2}{\Delta t}[C^G])\{T^G\}^i_{t+\frac{\Delta t}{2}} - \{A_1\} - \{F_Q^G\}^i_{t+\frac{\Delta t}{2}} + \{F_q^G\}^i_{t+\frac{\Delta t}{2}} - \{F_c^G\}^i_{t+\frac{\Delta t}{2}} - \{F_r^G\}^i_{t+\frac{\Delta t}{2}} \tag{13a}$$

In this equation $\{f\}^i$ is the residual vector whose all the elements are zero only when the assumed vector $\{T^G\}^i$ converges. Clearly, $\{f\}^i$ is a function of all the nodal temperatures, therefore, one can evaluate the matrix $[A]^i$, the Jacobian matrix whose general term A_{jk}^i can be defined as

$$A_{jk}^i = \frac{\partial f_j^i}{\partial T_k} \tag{13b}$$

The correction vector $\{\Delta T^G\}^i$ can be obtained using

$$[A]^i \{\Delta T^G\}^i = -\{f\}^i$$

Then, the improved temperature vector can be known using the relation

$$\{T^G\}^{i+1}_{t+\frac{\Delta t}{2}} = \{T^G\}^i_{t+\frac{\Delta t}{2}} + \{\Delta T^G\}^i_{t+\frac{\Delta t}{2}} \tag{13c}$$

The iteration can be terminated if

$$|f_j| < f_{tol} \quad (j = 1,2, \dots n)$$

Alternately, the iteration can also be terminated if

$$|\Delta T_j^G| < \Delta T^G_{tol} \quad (j = 1, 2, \dots, n)$$

3.4 The Nonlinear Optimization Method

The nonlinear optimization methods can also be used for solving Eqn. (11) for the unknown nodal temperature vector $\{T^G\}_{t+\frac{\Delta t}{2}}$ [6]. As an example, the solution of Eqn. (11) can be obtained by the Davidon-Fletcher-Powell method [7,8] which is a very stable method and converges quadratically.

If we denote the global temperature vector at the ith iteration stage of the optimization as $\{To^i\}_{t+\frac{\Delta t}{2}}$ and substitute on the right hand side of Eqn. (11) then we can compute the force vectors $\{F_c^G\}_{t+\frac{\Delta t}{2}}$ and $\{F_r^G\}_{t+\frac{\Delta t}{2}}$. Then, this equation reduces to system of linear equations which can be solved for the unknown temperature vector at time $t + \frac{\Delta t}{2}$. This calculated vector can be denoted as $\{Tc^i\}_{t+\frac{\Delta t}{2}}$. So we can define the objective function Θ for the minimization as

$$\Theta = \sum_{i=1}^{n} (Tc^i - To^i)^2_{t+\frac{\Delta t}{2}} \tag{14}$$

where n denotes the total number of nodes.

Therefore, the global temperature vector $\{T^G\}_{t+\frac{\Delta t}{2}}$ is obtained by minimizing the objective function Θ. The iterative procedure of this method can be stated as follows [7]:

(1) Start with an initial assumed vector $\{To^i\}$ and a positive definite symmetric matrix $[H^1]$. Usually we consider $[H^1]$ as the identity matrix. Set iteration number i to 1.

(2) Compute the gradient of the objective function $\nabla\Theta^i$, at $\{To^i\}$, and evalute $\{G^i\}$ such that

$$\{G^i\} = -[H^i]\{\nabla\Theta^i\} \tag{15a}$$

(3) Obtain the optimal step length λ_i^* in the direction $\{G^i\}$ and set

$$\{To^{i+1}\}_{t+\frac{\Delta t}{2}} = \{To^i\}_{t+\frac{\Delta t}{2}} + \lambda_i^* \{G^i\} \tag{15b}$$

Substitute $\{To^{i+1}\}_{t+\frac{\Delta t}{2}}$ in $\{F_c^G\}_{t+\frac{\Delta t}{2}}$ and $\{F_r^G\}_{t+\frac{\Delta t}{2}}$ Solve for $\{Tc^{i+1}\}_{t+\frac{\Delta t}{2}}$. Compute the value of objective function $\Theta^{i+1}_{t+\frac{\Delta t}{2}}$ and test for optimality. If the objective function attains an optimum value, then terminate the iterative procedure. The temperature vector corresponding to optimum objective function represents the global temperature vector at time $t + \frac{\Delta t}{2}$ Otherwise go to the next step.

(5) Update the matrix [H] using the relation

$$[H^{i+1}] = [H^i] + [M^i] + [N^i] \tag{15c}$$

where

$$[M^i] = \frac{\lambda_i^* \{G^i\}\{G^i\}^T}{\{G^i\}^T \{p^i\}} \tag{15d}$$

$$[N^i] = \frac{-[[H^i]\{p^i\}][H^i]\{p^i\}]^T}{\{p^i\}^T [H^i]\{p^i\}} \tag{15e}$$

and

$$\{p^i\} = \nabla\Theta(\{To^{i+1}\}) - \nabla\Theta(\{To^i\}) = \nabla\Theta^{i+1} - \nabla\Theta^i \tag{15f}$$

(6) Set the new iteration number i=i+1 and go to Step 2.

4. NUMERICAL EXAMPLE

The transient heat transfer process of the body shown in Fig.1 was studied using the finite element method [1]. It is a ceramic body whose sides were maintained at 300°C and the bottom surface was insulated. The top surface exchange heat with the surrounding fluid which was at 50°C. This heat transfer process was by convection and radiation mechanisms. Initially the whole body was at 300°C. This problem was analyzed as a two-dimensional problem. Because of the symmetry about the vertical axis, the heat transfer analysis was carried out for the right half of the body only. The linear finite triangular elements were used in this study. The finite element discritization of the system is shown in Fig.2. The unsteady heat transfer process of this body was represented by Eqn. (11). The Eqn. (11) was solved for $\{T^G\}_{t+\frac{\Delta t}{2}}$ by the three methods mentioned earlier i.e. the iteration method, the Newton-Raphson method and the optimization method.

Fig. 3 shows the temperature distribution of the solid using the iteration technique. This figure shows the continuous decline of temperature for all the nodes with time. The temperatures of various nodes oscillate in the beginning, and then die out after some time. At a given time, the oscillation amplitudes are higher at the top surface nodes compared to the nodes at the insulated boundary. The rate of decrease of the temperature at the top surface is very high, whereas the temperatures at the nodes which are on insulated boundary decrease very slowly. At any given time, node 21 is at the lowest temperature. This is because it lies on the symmetrical axis losing heat to the surrounding fluid by convection as well as radiation. Moreover, it is the farthest point from insulated and isothermal surfaces. Figs. 4 - 6 show the time-temperature plot of the nodes 1,13,23 respectively. It is clear from these figures that the temperature distributions obtained by different numerical methods are quite close. As expected, the finite element results show oscillations in the initial time period, whereas finite difference results do not exhibit such a behaviour. However, one has to take care of the stability criteria in the finite difference method. The finite element equation (11) is unconditionally stable.

In order to solve the heat transfer equation (11) by Newton-Raphson method, one has to compute the Jacobian matrix, $[A]^i$. The computation of Jacobian matrix requires more computer memory storage and involves more number of computations as compared to the iteration method. Similarly, the solution of Eqn. (11) by the nonlinear optimization method required more computation time [6]. The iteration method was found to be the most efficient among the three methods for solving nonlinear heat transfer equations. Using this method the solution was obtained in the least CPU time compared to the Newton-Raphson method or the nonlinear optimization method.

5. CONCLUSIONS

In this paper, the equations for the transient nonlinear temperature distribution within a solid due to the convective and radiative heat flux from the surroundings were obtained using a combination of finite element and finite difference methods. The nonlinearities in this analysis were due to two reasons; the first reason was due to the variation of the material properties with the temperature and the second one was the radiative heat flux. The resulting nonlinear heat transfer equations were solved by (a) the iteration method, (b) the Newton-Raphson method, and (c) the nonlinear optimization method. Based on this study it can be concluded that the iteration method is the most efficient method among the three methods used for solving transient heat transfer problem.

ACKNOWLEDGEMENT

The authors would like to thank the Natural Sciences and Engineering Research Council of Canada for their grant A5549.

NOMENCLATURE

[A]	Jacobian matrix
[B]	thermal gradient matrix
c	specific heat
[C]	capacitance matrix
[D]	thermal material property matrix
$\{f^i\}$	residual vector
$\{F_c\}$	force vector due to convection
$\{F_q\}$	force vector due to the heat flux
$\{F_Q\}$	force vector due to heat generation within the body
$\{F_r\}$	force vector due to radiation
h	convection heat transfer coefficient
k_x, k_y, k_z	thermal conductivities in the x, y and z directions respectively
[K]	thermal conduction matrix
ℓ_x, ℓ_y, ℓ_z	direction cosines normal to the surface
n	total number of nodes
[N]	shape function matrix
q	heat flux
Q	heat generated within the body
S_1	surface experiencing heat flux
S_2	surface experiencing convection and radiation heat transfer
t	time
Δt	time increment
{T}	nodal temperature vector
T_∞	fluid temperature
{To}	optimization temperature vector
{Tc}	surface nodal temperatures
ε	emissivity of the body
Θ	objective function to be minimized for the solution of nonlinear heat transfer equations
ρ	density of the material
σ	stefen-Boltzmann Constant

REFERENCES

1. Reddy, B.S., and Sharan, A.M., "The Transient Heat Transfer Analysis of Solids with Radiative Boundary Condition Using Finite Element Analysis" Int. Comm. Heat Mass Transfer, 12 (1985), 169-178.

2. Segerlind, L.J., "Applied Finite Element Analysis", John Wiley and Sons, Inc., New York, (1976), 212-222.
3. Huebner, K., and Thronton, E.A., "The Finite Element Method for Engineers", John Wiley and Sons, Inc., New York, (1982) 427-428; 431-435.
4. Rao, S.S., "The Finite Element Method in Engineering", Pregamon Press, Oxford, (1982) 493-494.
5. Gerald, C.F., "Applied Numerical Analysis",. Addison-Wesley Publishing Company, Inc., Reading, (1978) 15-26.
6. Sharan, A.M., and Reddy, B.S., "Nonsteady Temperature Distribution in Solids Using the Optimization Principles", AIAA-85-1015, Williamsburg, Virginia.
7. Rao, S.S., "Optimization theory and applications", Wiley Eastern Limited, New Delhi, (1978) 315-327.
8. Siddall, J.N., "Optimal Engineering Design", Mercel Dekker, Inc., New York, (1982) 161-167.

APPENDIX A

The following are the expressions for the elemental matrices:

$$[C^e] = \int_{v^e} \rho c \, [N^e]^T \, [N^e] \, dV$$

$$[K^e] = \int_{v^e} [B^e]^T \, [D^e] \, [B^e] \, dV + \int_{S_2^e} h \, [N^e]^T \, [N^e] \, dS$$

$$\{F_Q^e\} = \int_{v^e} Q \, [N^e]^T \, dV$$

$$\{F_q^e\} = \int_{S_1^e} q \, [N^e]^T \, dS$$

$$\{F_c^e\} = \int_{S_2^e} h \, T_\infty \, [N^e]^T \, dS$$

$$\{F_r^e\} = \int_{S_2^e} \sigma \varepsilon \, T_\infty^4 \, [N^e]^T \, dS$$

$$\int_{S_2^e} \sigma \varepsilon \, [N^e]^T ([N^e]\{T^e\})^4 \, dS$$

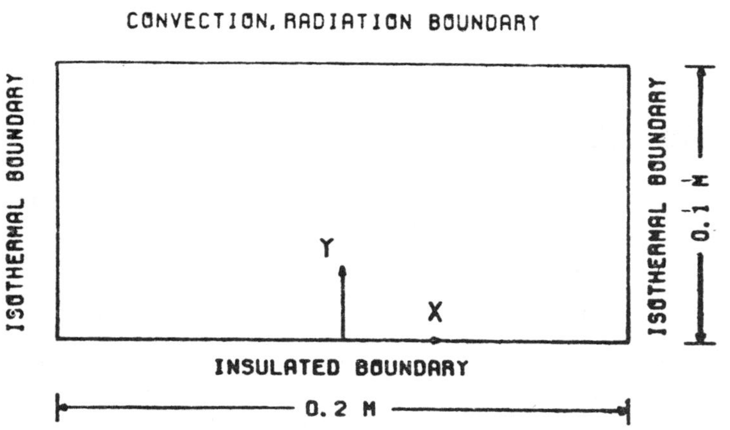

Fig. 1 The system configuration with various boundary conditions

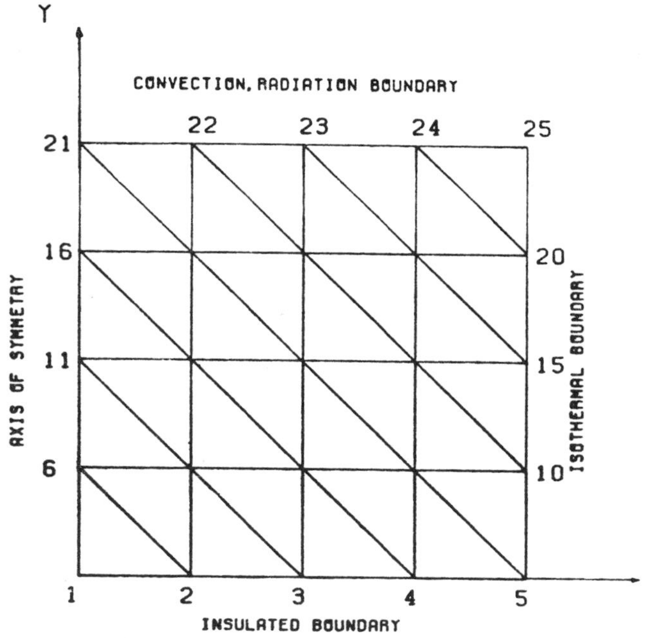

Fig.2 The descretization of the system into finite elements.

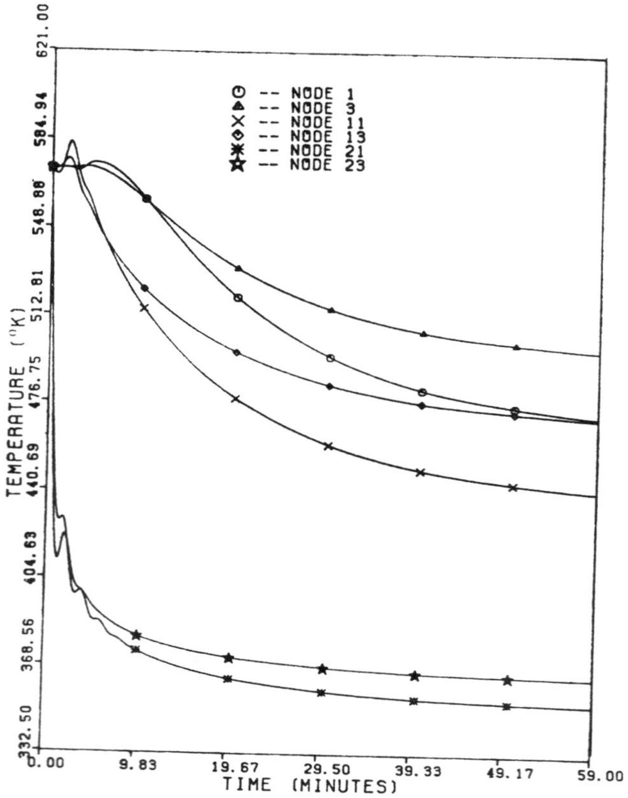

Fig.3 The time-temperature plot of the ceramic body

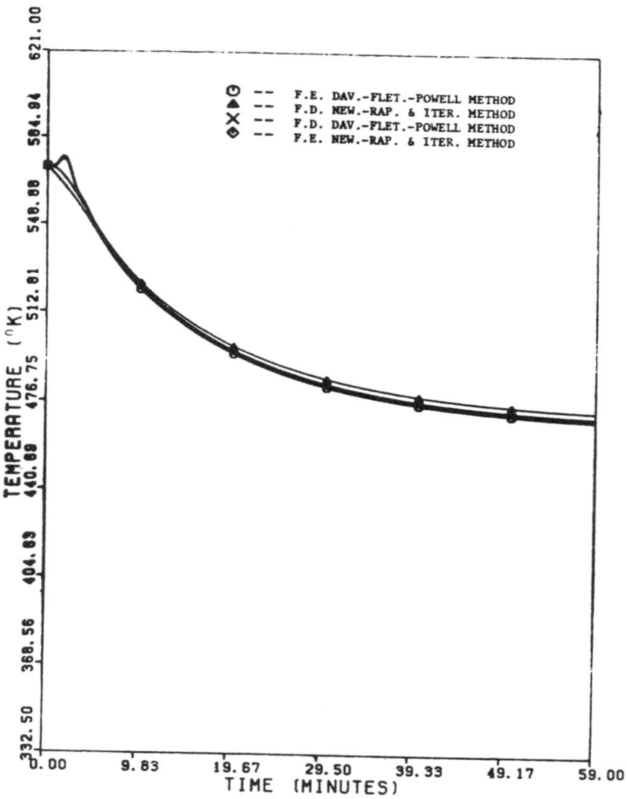

Fig.4 The temperature-time plot at the node-1.

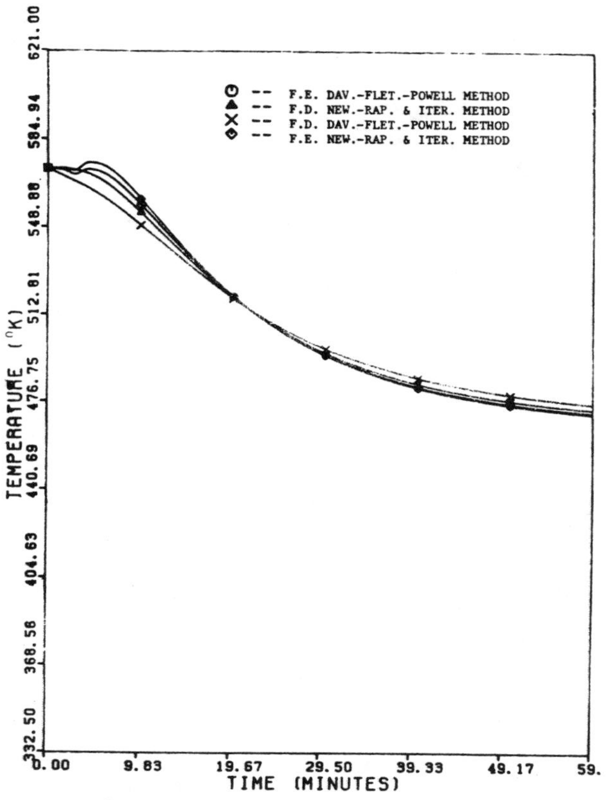

Fig.5 The temperature-time plot at the node 13.

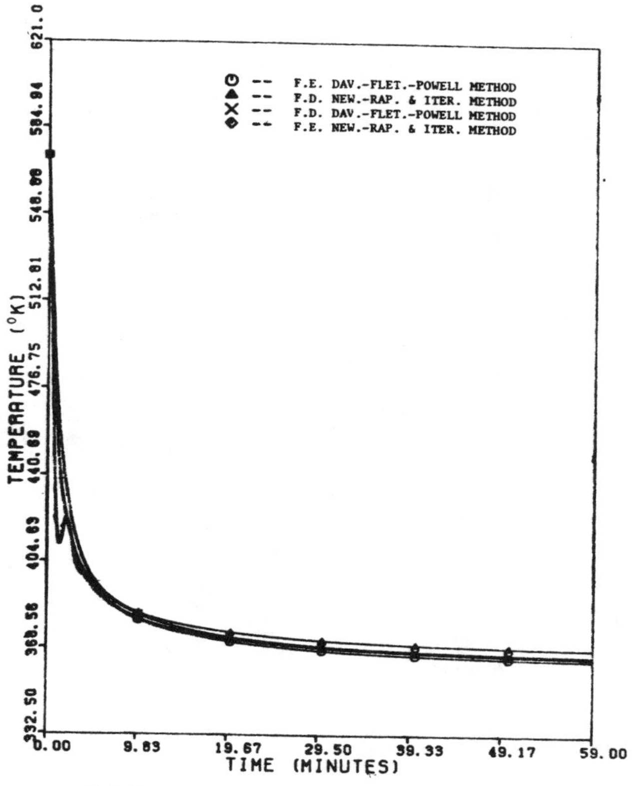

Fig.6 The temperature-time plot at the node 23

Studies on Heat Transfer in Fluidized Beds

V. M. K. SASTRI and N. S. RAO
Indian Institute of Technology
Madras, India

Heat transfer data have been obtained for different tube bundle configuration and particle sizes in a fluidized bed. Results are compared with those in literature. The modified Vreedenberg's correlation overpredicts the present data. For lesser compact tube bundles, however, the values of the heat transfer coefficient agree closely with the correlation suggesting the limitations of the original Vreedenberg's correlation, which was proposed for single tube being extended to tube bundles.

1. INTRODUCTION

Heat transfer to immersed tubes in fluidized beds has been a subject of intensive research in recent years. The nearly isothermal nature of the bed and the high heat transfer rates are the main advantages of fluidized bed heat transfer. The mechanism of heat transfer between the bed and an immersed surface is, however, very complicated because of the large number of variables involved. Such variables include particle size, shape and size distribution, particle and gas thermal properties, bed geometry, type of gas distributor etc. in addition to the size, shape, spacing, pitch and material of the heat transfer tubes.

Numerous investigations have been made with laboratory scale beds on tube-to-bed heat transfer. Correlations were empirically developed on the basis of the observed data with limited attention to the mechanisms of the heat transfer process. Therefore, the scope of the correlations is restricted to the data for the specific test conditions. This implies the need for further tests for the existing correlations and investigation of the effects of some of the least investigated parameters such as particle size distribution in a mixed particle bed, before the existing correlations can be applied with greater confidence.

The performance of a fluidized bed combustor is a function of several design and operating conditions such as pressure, temperature, velocity, particle size and tube geometry. It is essential, therefore, to study the effect of these variables on the combustor performance and the interrelation between these effects. It should also be understood that an individual parameter can influence the various bed phenomena in different ways, thus demanding an optimum value to be determined, perhaps as a compromise.

The effects of two of the more important parameters viz., particle size and tube geometry, on heat transfer have been studied and discussed in the following sections.

2. PREVIOUS WORK

2.1 Effect of particle size on bed dynamic and heat transfer

Particle size is the most influential parameter in the overall combustor performance. In view of the fact that the bed material in a fluidized combustor is characterised by a fairly wide size distribution, which must be taken into account for meaningful interpretation of data, most of the research so far has relied on the use of some statistical mean diameter for design and performance study of a combustor.

The bed particle size influences every single bed phenomenon, directly or indirectly, e.g., the sulphur adsorption process occuring in outer layers of the sorbent particle irrespective of the size, or the fluid mechanics in the bed, where small particles are generally in the fast bubble regime where the gas flow through the bubble is faster than that through emulsion, or the heat transfer to immersed heat exchangers which occurs in four modes viz., unsteady conduction through the emulsion, steady conduction across the gas layer between emulsion and surface, gas convection through the emulsion and bubbles, and radiation.

The total heat transfer coefficient, which is the sum of the individual components including the bubble fraction effect, is found to decrease with increasing particle size as evidenced in the coal fired tests by Zakkay [1], Cherrington [2] experimentally found that the heat transfer coefficient to an immersed tube varied as $d_p^{0.36}$. However, Golan et al [3] in their experiments with limestone of different size distribution concluded that d_p is important only for a bed of narrow size distribution. In a mixture of wide blend, smaller particles control the heat transfer and the d_{pw} does not significantly affect the heat transfer. While Mclaren and Williams [4], and Andeen and Glicksman [5] indicated a considerable effect on h_w, Rice and Coates have reported that the presence of 7% oversize had no effect on the value of h_w. Pitts et al [6] in their recent study found that their heat transfer data was underpredicted up to 25% by three of the existing correlations.

Geldart [7] has indicated that the principal effect of adding fines to a bed of powder is reduction of the mean particle size. At equal values of $U-U_{mf}$ this results in increased bed expansion and solid circulation rates but no decrease in mean bubble size.

2.2 Effect of Tube Geometry on Bed Dynamics and Heat Transfer

Newby and Keairns [8] have concluded that a) Improved bubble distribution, and reduced bed slugging and elutriation is characterisitic of fluidization with compact tube-bundles, b) The totally fluidized bed pressure drop and bed expansion are not significantly influenced by the presence of compact tube bundles and c) h_w for a single, central subject-tube locations considered, changes very little for the 1.59 cm diameter tube, but caused a significant increase in the coefficient for the 3.18 cm diameter heaters (10 to 20%).

Mclaren and Williams [4] measured the total heat transfer coefficient between a water cooled copper tube bundle and a fluidized bed of boiler ash. The value of the total heat transfer coefficient averaged 465 W/m^2k, as compared to 530 W/m^2k for a straight, stainless steel tube under the same operating conditions. Experiments were performed to study the effects of tube diameter and orientation, air fluidization velocity, bed height, position of the tube bundle in the bed and particle size distribution.

Little apparent variation in the heat transfer coefficient around the circumference was found, possibly because of the high thermal conductivity of the tube material. The value of the total heat transfer coefficient increased with the increase in fluidization velocity. Raising the tube bundle with a relative horizontal pitch of 4 from the bottom of the bed to the upper half of the bed resulted in about 20% increase in the value of h_w. Experimental data showed no change in the heat transfer coefficients for the tube bundle near the bottom of the bed, when the bed height was varied. It was also found that the heat transfer coefficient decreased with decrease in relative horizontal pitch. The heat transfer coefficient was reduced by about 25% as the gap between the tube surface was reduced from 282 mm to 15 mm with sudden decrease occuring indicating a marked decrease in the solid-particle mobility.

Gelperin and Einstein [9] reported experimental investigations conducted by Korotyanskaya [10] and Gelperin et al [11] on two horizontal rows of tubes arranged, one above the other with various angles between the centre-line of the rows. It was found that the heat transfer coefficient increased 3-5% when the very closely arranged tube rows were changed gradually from an in-line to a crossed arrangement.

Gelperin et al [12,13] studied heat transfer from bundles of horizontal tubes in in-line and staggered arrays. Bed particle size, fluidizing velocity, and relative horizontal and vertical pitches were varied to study their effects on the heat transfer coefficient.

It was observed that for the in-line arrangement of the tube bundle, the total heat transfer coefficient decreased with decrease in horizontal pitch. However, the effect of decrease of vertical pitch was noticeable only when the tubes were nearly touching each other. For the staggered tube bundles, both horizontal and vertical pitches influenced the value of h_w. The total heat transfer coefficient was reduced with a decrease in horizontal pitch. This was more pronounced at small values of vertical pitches. Further, it was found that the value of h_w for S_v/D_t > 2 was independent of the vertical pitch (S_v) even when S_h/D_t = 2-3. This decrease in h_w with decreases in tube spacing was attributed to hindrance of particle movement.

Bartel and Genetti [14,15] investigated the rate of heat transfer from a bundle of electrically heated horizontal tubes in a staggered array in an air fluidized bed. It was found that total heat transfer coefficient for the bundle increased when the tube spacing was increased from 11 mm to 38 mm. However, with further increases in tube surface spacing to 119 mm there was no additional increase in the total heat transfer coefficient. The authors report that the above result was true for all the particle sizes.

From these various experimental results, it may be concluded that:

1. With increase in D_t the h_w diminishes, when D_t > 10 mm the dependence of h_w on D_t disappears. Two main causes for the increase of h_w with increase in D_t [16 - 18] are:

(i) The regularities of particle packet heating at cylinder and plane surfaces are different if the thickness of the solid particle layer that is heated during the time of contact with the surface is commensurate with the radius of curvature of the latter.

(ii) Packets of solid particles are thrown aside from a heat transfer surface of small dimension D_t by the action of gas bubbles, and total replacement of the packets causes a significant increase in h_w. If the heat transfer surface is sufficiently large, some of the gas bubbles (particularly those that do not flow directly over the surface) only move the packet along the surface so that the increase in h_w is less appreciable.

2. According to [19-21], h_w to vertical tubes is usually somewhat (5 to 15%) higher than to horizontal ones, because the latter have inferior conditions of contact with the fluidized bed.

3. For vertical tubes inside bed, gas distribution within the bed becomes more uniform with increase in the number of tubes in the bundle, and values of h_w for different tubes in the bundle, and values of h_w for different tubes in the bundle tend to the same value. For more closely packed bundles h_w can reduce by 35 to 50%.

4. In most cases reported [12,13] there is a small (5 to 7%) drop in h_w in the direction from the outside to the centre of the bundle. For this reason increase in the number of tubes rows in the bundle above 2 to 4 usually causes some reduction (2 to 4%) in the heat transfer rate for the bundle as a whole.

5. It was found for in-line tube bundles [22], S_v has a significant influence on h_w only at small values of S_v, with diminution in S_h, h_w diminishes appreciably.

6. In staggered bundles the influence of S_v on h_w is most marked at relatively small values of S_h and then only at small values of S_v. When S_v/D_t > 2 to 3, h_w is independent of S_v even for S_h/D_t = 2 to 3.

3. EXPERIMENTAL SET-UP

The general arrangement of the test facility is shown in figure 1. The various components are: 1) the main test section, 2) air supply system, 3) tube bundle and 4) instrumentation.

3.1 Test Section

The 0.2 m x 0.2 m test section is constructed of 15 mm thick, clear perspex to allow visual observation of the bed when fluidized. It is square in cross section and about 1 m high. A square cross section was chosen because it is easier to fabricate the tube assembly and to position it in the bed. Also, all the tubes would be in the same relative position to the walls. The test section is positioned over a conical diffuser, about 0.5 m high, in order to allow for sufficient smoothing of the air flow before entering the particle bed.

The air distributor, located 100 mm above the base of the column, is made of a stainless steel plate, 2.5 mm thick. 686 holes of 3 mm diameter are drilled in 8 mm triangular pitch, which yielded a free flow area of 11.8%. The top of the distributor is covered with 200 mesh brass cloth in order to prevent the solid particles leakage.

Pressure taps are located 50 mm below the distributor and at heights of 5 mm, 105 mm, 205 mm, 305 mm & 405 mm respectively above the distributor. The gas exit port, covered with fine wiremesh to prevent particle carry over, is removable to position the tube bundle in the bed and also to pour in the solid particles. A particle drain is provided just above the air distributor. The column is supported by a steel structure fabricated with one inch diameter mild steel tubes and slotted angles. The entire assembly is then bolted to the floor.

3.2 Air Supply System

Air is supplied to the bed from a compressor (LG bros.; displacement volume = 5 m^3/mm; pressure range = 0 to 15 atm.). An orifice plate (D-D/2 tapping) is installed in the feeding line. The air flow rate is controlled by means of a needle valve and constant pressure is maintained by a pressure regulator. The pressure drop across the orifice plate is measured with an inclined (30° to the horizontal) water manometer. Earlier, the calibration of the air supply system was performed with the help of a vane type gas flow meter (SCHIEBERLUNG MIT SPEZIALOL, Range: 2 m^3/hr to 300 m^3/hr). The mass flow can be calculated and plotted against the pressure drop.

3.3 Tube Assembly

Two perspex plates, held apart by means of 4 tie-rods, are used to position the tubes. For a given pitch, in one plate, holes are drilled completely through and fitted with check nuts, through which the heaters extended. The other plate has U-clamps in corresponding positions to hold the other ends of the tubes. This arrangement held the tubes firmly in place. The lead wires are taken out of the assembly through the top of the test section. The staggered arrangement of the tubes is shown in figure 2.

3.4 Tube Heater

Two heaters, made of copper are placed symmetrically, 230 mm above the air distributor. Three grooves, 2 mm wide and 2.5 mm deep, were cut 90° apart on the copper tube surface. The thermocouples are fixed with copper cement and the heater surface is polished with fine emery. Ceramic rod of 9 mm dia. with nichrome wire wound on the grooves, could be inserted into the copper tube. Thin mica sheets are used to prevent electrical contact between the heater and the copper-tube. The lead wires are taken through an asbestos plug, screwed at one end of the tube. The other end is also sealed with an asbestos plug, using high-temperature araldite.

3.5 Calibration and Placement of Thermocouples

The thermocouples used in the present study are 24 WG, copper-constantan. Thermocouples are calibrated in a constant temperature waterbath. The e.m.f. values of each thermocouple are plotted against a standard thermocouple and the agreement between them was within $\pm$ 5%.

Apart from the six thermocouples placed on the heater surface, another six are placed in the bed at different locations. One each for measuring inlet and exit gas temperatures are located at 5 mm below the air distributor and 805 mm above it respectively. The remaining four are located at 4 mm, 100 mm, 190 mm and 295 mm above the air distributor respectively and are used to measure bed temperature.

3.6 Power Supply System

Each heater is given power supply by means of an adjustable autotransformer (GLOLITE Electricals; model - VT/8 1/p - 240 V; 50; o/p 0.270 V) an ammeter (GK, range 0 - 5 V/10 V) and a voltmeter (Automatic Electricals class 0.5, range 0-65 V/125 V/250 V).

3.7 Digital Voltmeter and Thermal Printer

The thermocouples are connected to the digital voltmeter and thermal printer through a selector switch. The digital voltmeter is Hewlett Packard; model 3455A, and has a resolution of 1μV in the 0-100 m V range. The e.m.f. values developed are displayed and a thermal printer attached (Hewlett Packard; model 5150A) is used to record the displayed values. The thermal printer is capable of printing up to 20 characters per line at a maximum rate of 3 lines per second. The data print interval could be selected by means of a clock.

The interval used in the present work is one line per second and a maximum of six values are recorded for a given reading.

4. RESULTS AND DISCUSSION

4.1 Bed Temperature

The temperatures were measured using thermocouples shielded by 200 mesh brass wire cloth in order to prevent direct particle contact. Thus it was ensured that the temperature measured was the bulk bed temperature. The bed temperature, measured at 4 different heights are shown plotted in figure 3 for the tube bundles, 1.5 x 1.5 and 4 x 4 at air velocities 0.073, 0.099, and 0.15 m/s. The temperature variations is found to be less than 0.5°C over the velocity and particle size range covered.

It was also observed in the present study that after a few hours of operation with the thermocouple inside the bed, very fine dust accumulated inside the shielded bulb. Care must be taken while using shielded probes as the dust accumulated might effect the response time of the thermocouple. In the present work, the average of the four temperature readings was used for further analysis.

4.2 Effect of heat flux

It was anticipated that, while conducting the experiments fluctuations would occur in heat input to the heaters. As a preliminary study, the effect of heat flux variation on h_w for two tube bundles, 1.5 x 1.5 and 4 x 4, were conducted with sand of 0.141 mm. The heat flux was varied using an autotransformer. The minimum heat flux used was limited by the temperature difference, ΔT, realised between heater surface and bed. As the ΔT, gets too small, say less than 10°C, errors in h_w can become large. The results of this study are plotted h_w versus Q in figure 4. The third parameter in the plots is the air velocity, U, which is varied from 0.087 m/s to 0.146 m/s.

From the above figure, it can be seen that there is negligible dependence of h_w on the heat flux, Q. However, it was observed that h_w tends to drop at the lowest value of Q = 120 W. The effect is relatively small, being less than 4% over the entire range of heat fluxes and air velocities covered. The variation in operating heat input of the present study was less than 4% and this could have resulted in less than 3% variation in h_w. Since this is rather small compared to the possible experimental errors, it has been neglected for the remainder of the study.

4.3 Effect of Pitch

The effect of tube pitch on h_w is shown in figures 5 and 6. As P increases the h_w also increased. For a given value of P, the coefficient increases with increase in U and, decreases with increase in d_p. The lower value of h_w for compact tube bundles may be due to restricted particle movement.

Data for four tube bundles, 1.5 x 1.5, 2 x 2, 3 x 3, and 4 x 4 and for three sample, d_p = 0.112, 0.141 and 0.165 mm are shown compared in figure 7 with some of the heat transfer correlations reported in literature. The present data is shown plotted in their coordinates along with the modified Vreedenberg's correlation, proposed by Andeen and Glicksman. They reported that this correlation predicted their experimental data well for two rows of tubes arranged in a staggered pattern with an electrically heated aluminium tube and the modified correlation also predicted original Vreedenberg's data better than his own correlation.

From the above figure it is seen that the correlation over predicts the present data for more compact tube bundles. As the pitch is increased, it is expected that h_w values tend to single tube h_w values and hence expected to be predicted well by the modified Vreedenberg's correlation. For example, the data for d_p = 0.165 mm is predicted well, whereas the data for lower sizes suggest that the d_p effect is not properly accounted for by the correlations.

5. CONCLUSIONS

1. The bed temperature remains essentially constant and the variation is less than 0.5°C for constant air flow rate for all tube bundles.

2. Changing surface flux of the heater tube does not alter h_w significantly (less than 4%)

3. h_w increase with increase in tube pitch, thus suggesting particle movement is restricted in case of more compact tube bundles.

The modified Vreedenberg's correlation overpredicted the present data for lesser compact bundles. However, h_w, for 4 x 4 tube bundle closely agreed with the correlation.

NOMENCLATURE

d_p	diameter of particles
D	diameter of orifice plate
D_t	tube diameter
h_w	heat transfer coefficient
p	tube pitch
Q	heat flux
S_h	horizontal pitch
S_v	vertical pitch
ΔT	temperature difference, $T_w - T_b$
T_b	bed temperature
T_w	immersed tube wall temperature
U	superficial air velocity
U_{mf}	U-valve at incipient fluidization

REFERENCES

1. Zakkay, V et al, Interim Report, DOE/MC/14322 - 168, U.S. Dept. of Energy, Morgantown, (1981).

2. Cherrington, D.C., Golan, L.P., and Hammitt, F.G., "Industrial Applications of FBC - Single Tube Heat Transfer Studies", Proc. 5th Int. Conf. Fluidized Bed Combusion, Dec. 3 (1977)

184-209.

3. Golan, L.P., Diener, R., Scarborough, C.E. and Weiner S.C., Chem. Eng. Progr., 75 (1979) 63.

4. McLaren, J., and Williams, D.F., J.Inst.Fuel, 42 (1969) 303.

5. Andeen, B.R., and Glicksman, L.R., ASME - AIChE Heat Transfer Conf., 1976 Paper 76 - HT - 67 (1976).

6. Pitts D.R., Figliola, R.S., and Hamlyn, K.M., ASME Jl of Heat Transfer, 104 (1982) 563.

7. Geldart, D., Powder Technology, 6 (1972) 201.

8. Newby, R.A., and Keairns, D.L., in "Fluidization", Eds. Davidson J.F., and Keairns, D.L., 320.20), Cambridge Univ.Press, Cambridge, (1978)

9. Gelperin, N.I. and Einstein, V.G., in "Fluidization", Eds. Davidson J.F., and Harrison D.. Academic Press, New York, (1971) 517 - 540.

10. Korotyanskaya, L.A., Diss., Inst.Fine Chemical Technology, Moscow, (1968)

11. Gelperin, N.I., Einstein, V.G., and Korotyanskaya, L.A., Khim. Prom. 6 (1968) 427.

12. Gelperin, N.I., Einstein, V.G., and Zaikouski, A.V., Khim. Mashinostr (Moscow), 3 (1968) 17.

13. Gelperin, N.I., Einstein, V.G., and Korotyanskaya L.A., Int. Chem. Eng., 9 (1969) 137.

14. Bartel, W.J. and Genetti, W.E., Chem. Eng. Prog., Symp. Ser., 69 (1973) 85.

15. Bartel, W.J., Ph.D. Thesis, Montana State University, Bozeman, (1971)

16. Baskakov, A.P., InSh. - Fiz. Zh., Akad. Nauk, Belorussk SSR 6, 11 (1963) 20.

17. Baskakov, A.P., "High Speed non - oxidative heating and heat treatment in a fluidized bed", IZd. "Mettalurgia", Moscow, (1968)

18. Baskakov, A.P. and Berg, B.V. Insh - Fiz. Zh., Akad. Nauk Belorussk. SSR, 10 (1966) 739.

19. Gelperin, N.I., Einstein, V.G. and Romanova, N.A., Khim. Prom. 11 (1963) 823.

20. Gelperin, N.I., Einstein, V.G., and Kwasha, U.B., "Fluidization Technique Fundamentals" IZd. "Khimia", Moscow, (1967)

21. Kofman, P.L., Gelperin, N.I., and Einstein, V.G., in "Processes and Equipment for Chemical Technology", Inst. Fine Chem. Technol., Moscow, (1967) 166-174.

22. Malukovich, S.A., Diss., Inst. Heat - Mass Transfer, Akad, Nauk Belorussk, SSR, Minsk, (1968)

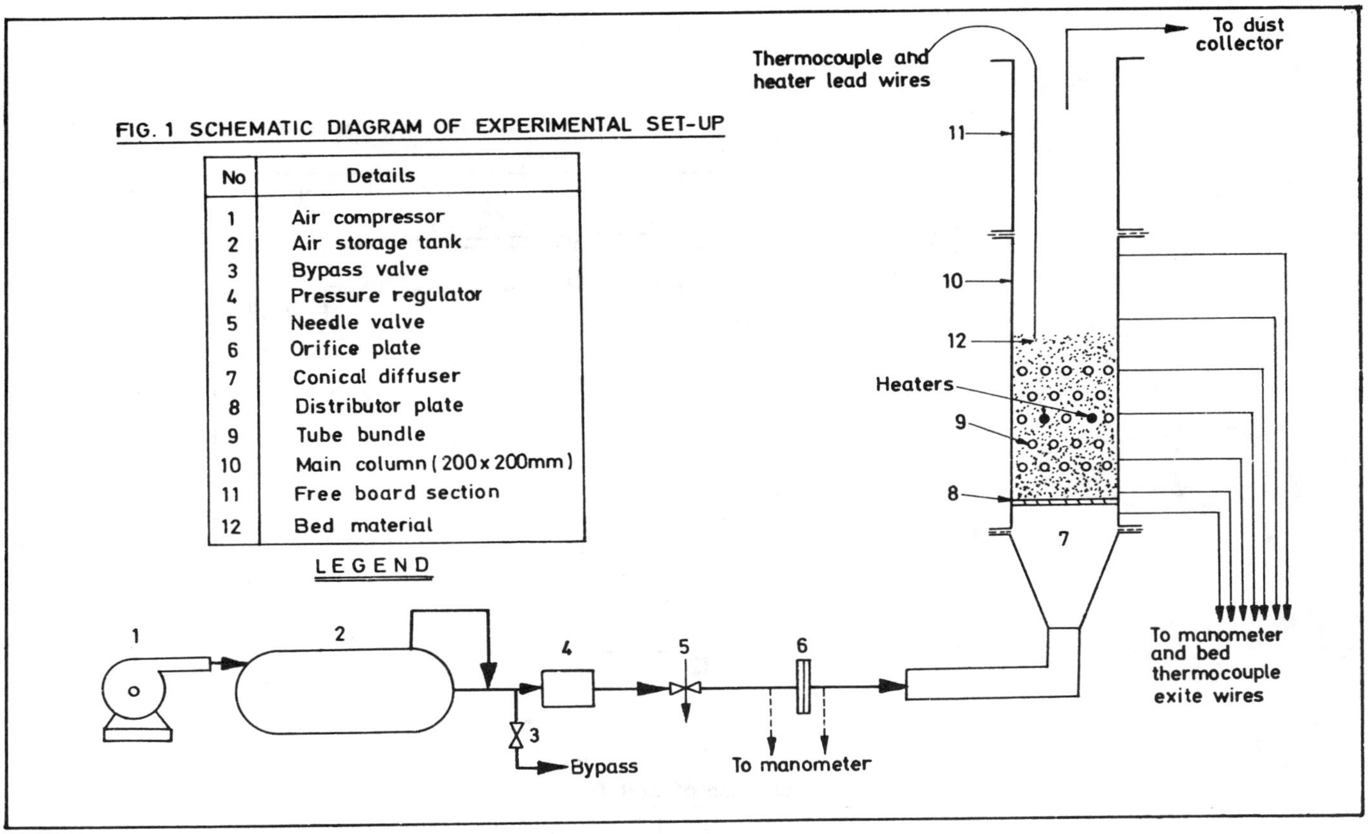

Fig.1 Schematic Diagram of Experimental Set-up

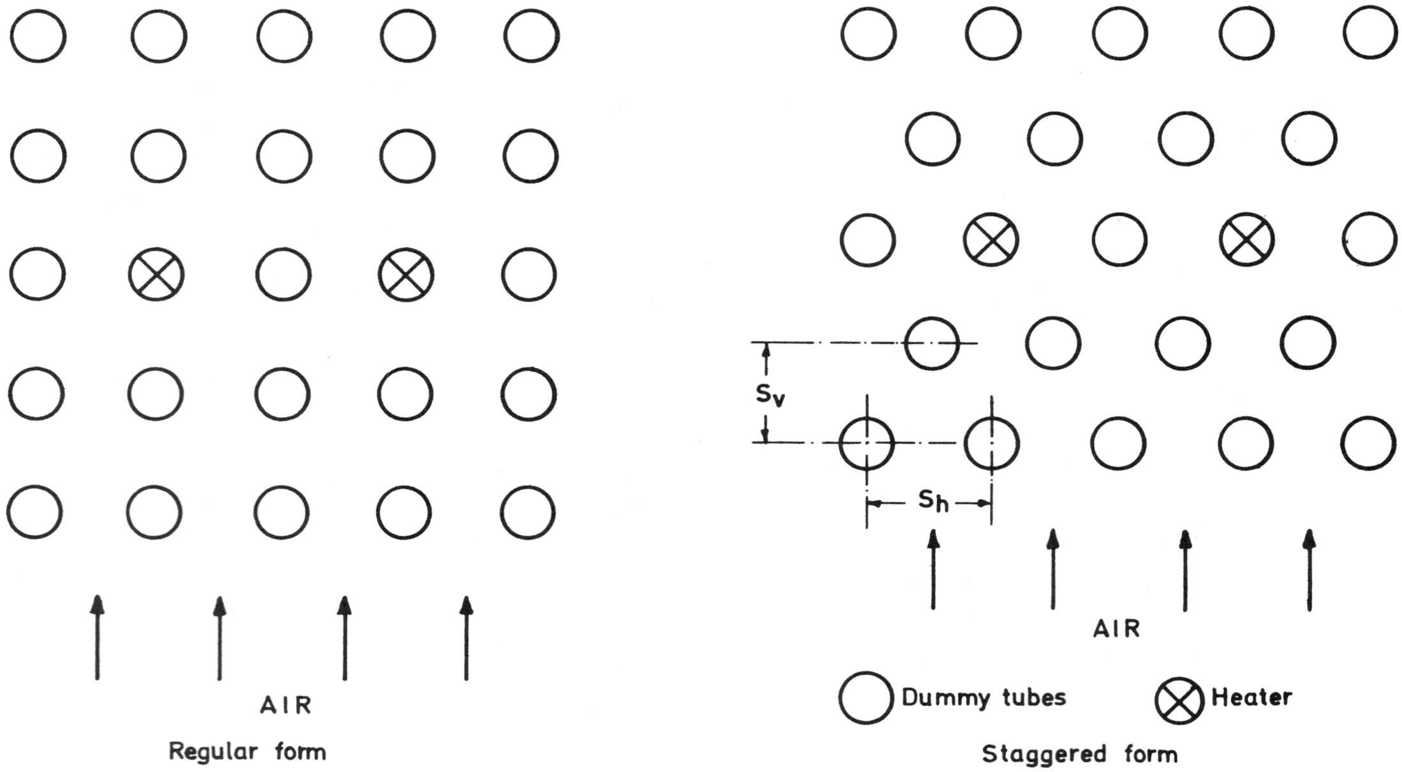

Fig.2 Arrangement of Heaters in Tube Bundle

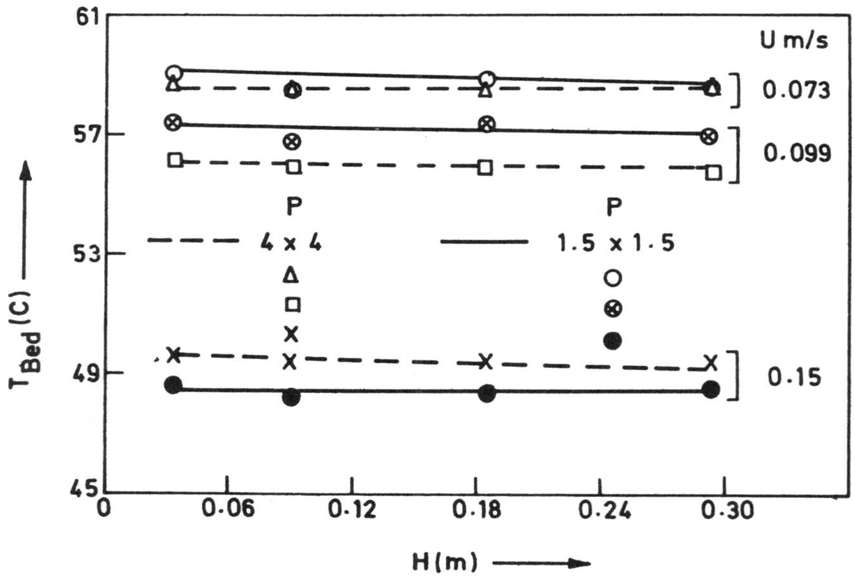

Fig.3 Variation of Bed Temperature with Height

d_p = 0.141, 1.5 x 1.5 & 4 x 4

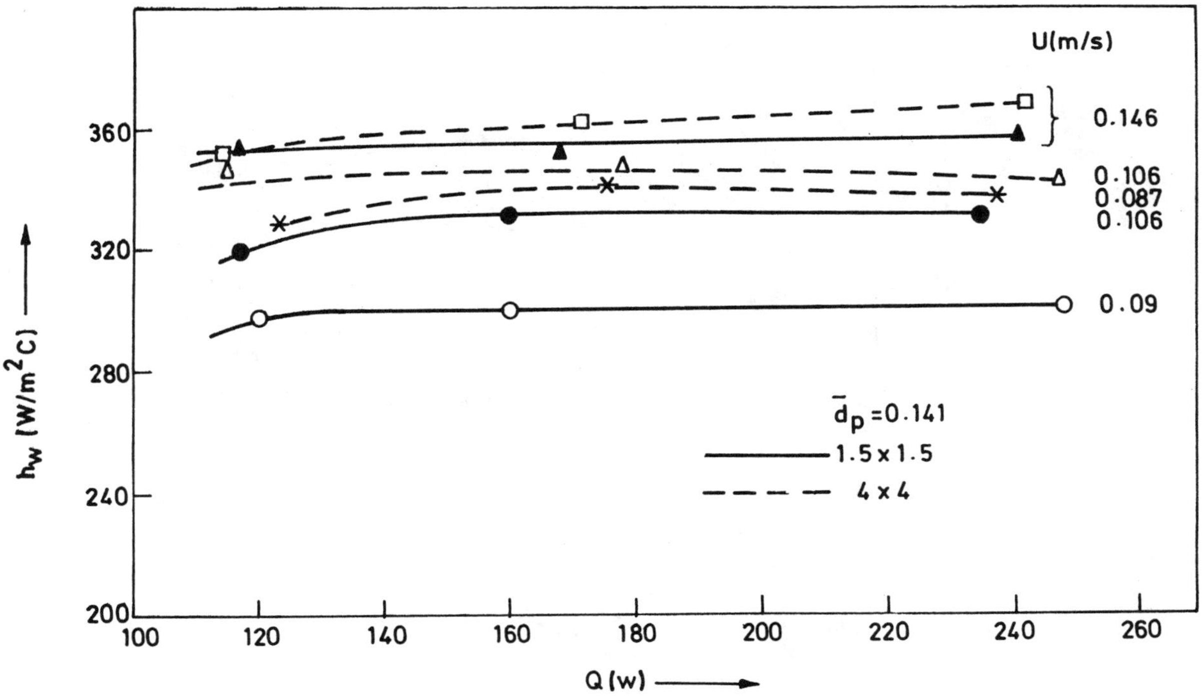

Fig.4 Variation of h_w with Heat Flux for Different Pitches (1.5 x 4) Parameter U.

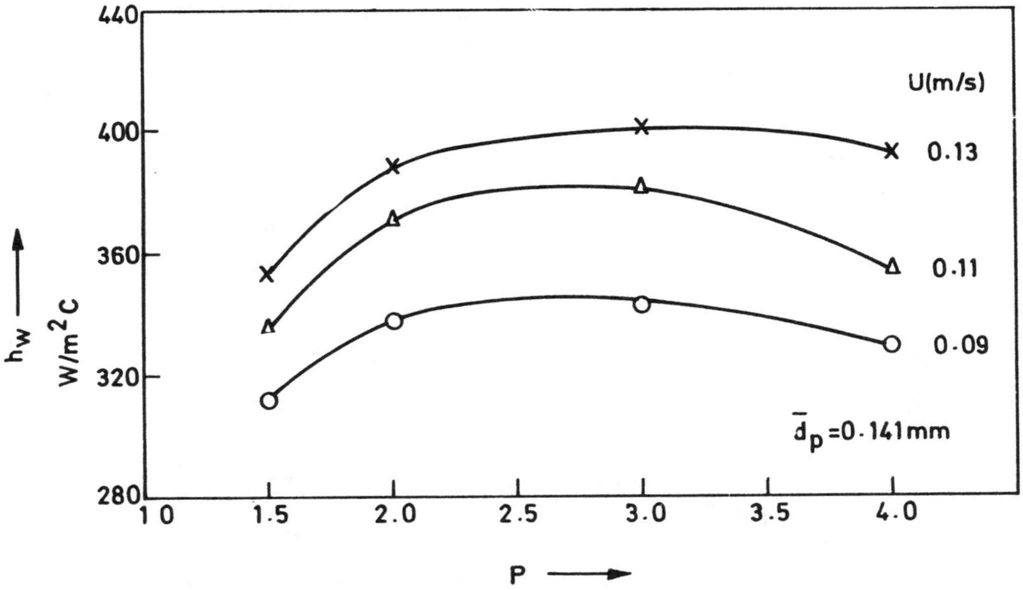

Fig. 5 Variation of h_w with P, d_p = 0.141 mm, Q = 80W, Parameter U.

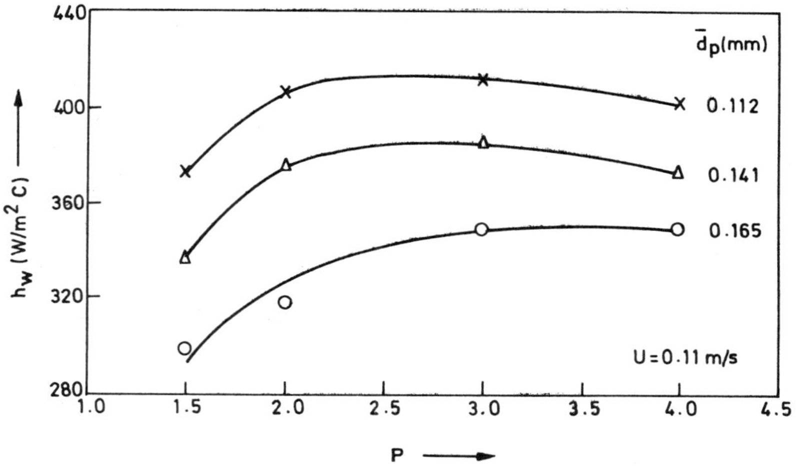

Fig.6 Variation of h_w with P, U = 0.11 m/s, Q = 80 w, Parameter d_p

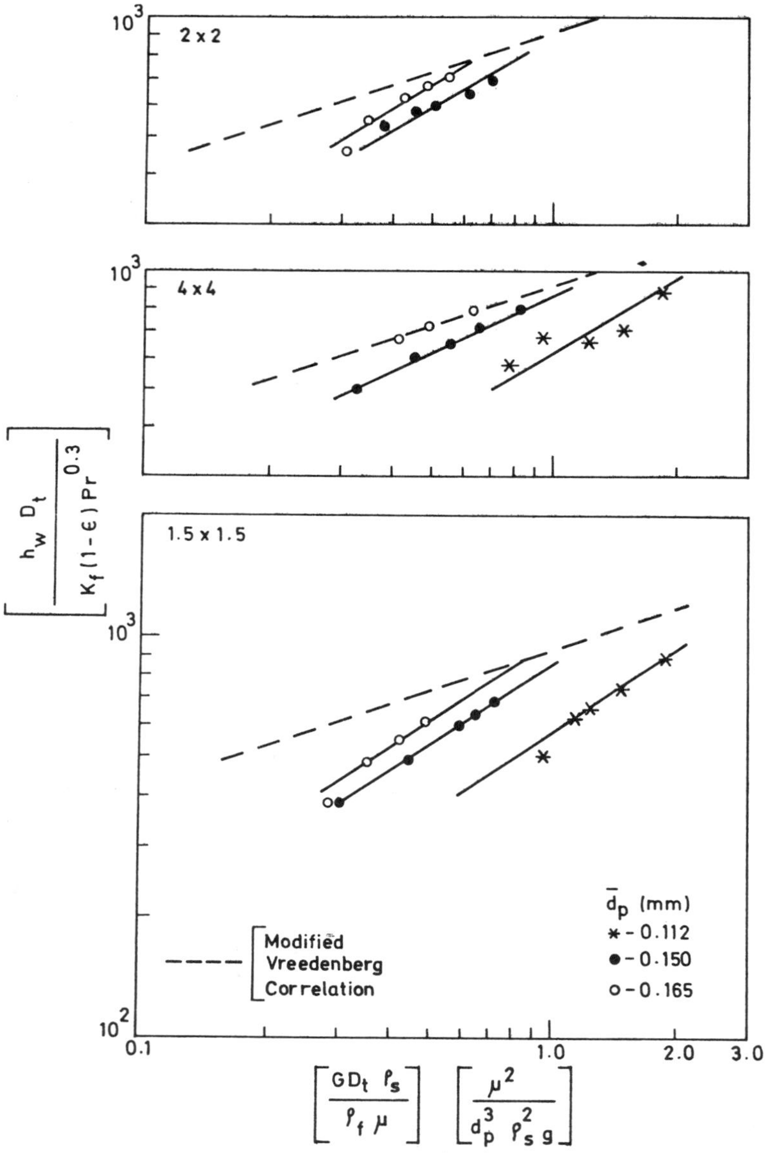

Fig.7 Comparison of Experimental Data with the Modified Vreedenberg Correlation.

The Influence of the Number of Thermal Plates on Plate Heat Exchanger Performance

RAMESH K. SHAH* and SATISH G. KANDLIKAR†

*Harrison Radiator Division, General Motors Corporation, Lockport, N.Y. 14094, USA

†Mechanical Engineering Department, Rochester Institute of Technology, Rochester, N.Y. 14623, USA

The performance of plate heat exchangers has been analyzed numerically by a finite difference method for 1 pass-1 pass, 2 pass-1 pass, and 3 pass-1 pass flow arrangements. Extensive tabular results for the exchanger temperature effectiveness and the log-mean temperature difference correction factor are presented for a wide range of the number of transfer units, heat capacity rate ratio and the number of thermal plates. The "end effect" due to a small number of thermal plates is important only for N < 40. In most plate exchangers the number of thermal plates is greater than 40. For this case, the influence of the number of thermal plates is found to be negligible on the performance, i.e., the performance approaches that of an exchanger with an infinite number of thermal plates. Based on the results, design guidelines are also presented for the selection of a 1-1, 2-1 or 3-1 pass exchanger when the heat capacity rate ratio is about two and higher.

1. INTRODUCTION

The plate and frame or gasketed plate heat exchanger (PHE) consists basically of a number of rectangular thin metal plates held together in a frame as shown in Fig. 1. A large number of flow arrangements of the working fluids are possible in plate exchangers. Some of them for a two-fluid exchanger are shown in Fig. 2. In a plate exchanger, the two outer plates serve as end plates and ideally do not transfer heat, while the remaining plates, known as thermal plates, transfer heat.

Plate heat exchangers are commonly used in hygienic applications as well as in chemical processing and other industrial applications. Extensive information and literature on the PHE has been summarized by Focke [1], Cooper and Usher [2], and Raju and Bansal [3,4] among many others.

The overall design methodolgy for PHE's would include thermal design, mechanical design, and manufacturing considerations for a specific application. This methodology will be somewhat similar to that for a compact heat exchanger [5], but the details vary.

One of the key requirements in the design of a heat exchanger is to perform the heat transfer and pressure drop analyses. Several nondimensional methods exist for the heat transfer analysis that relate to the basic dimensional variables associated with the problem [6]. The two most commonly used methods for plate heat exchanger design are the ε-NTU (or P-NTU) and LMTD (long-mean temperature difference) methods. The results are available for most basic flow arrangements [7-12] and the literature is reviewed in the next section.

When a number of thermal plates is greater than one, the heat is transferred from one fluid to the other fluid from both sides of an interior channel while the heat is transferred from only one side of an end channel. Thus the heat transfer rate from the end channel is different from an interior channel of the same fluid. This "end effect" reduces the exchanger effectiveness (performance) over that having an infinite number of thermal plates. As a result, the influence of the number of thermal plates has been investigated from the first publications on the subject. However, the results available in the open literature are incomplete, and the objective of this study is to present a complete set of results, for 1 pass-1 pass, 2 pass-1 pass and 3 pass-1 pass flow arrangements from a designer's point of view.

2. LITERATURE REVIEW

The ε-NTU results or the LMTD correction factors F have been obtained for various flow arrangements [7-12] for a two-fluid single-phase plate exchanger.

Buonopane et al. [7] experimentally determined the F factors for 1 pass-1 pass flow arrangement with up to 17 thermal plates and multipass series flow arrangements with up to 11 thermal plates. No attempt was made to present the F factors as a function of P (or NTU) and R.

Jackson and Troupe [8] investigated the 1 pass-1 pass and 2 pass-2 pass flow arrangements for overall counterflow and parallel flow, and 4 pass-4 pass arrangement with overall counterflow. They analyzed the problem using the Runge-kutta method treating the problem as an initial value problem. They presented ε-NTU results for $0.3 \le \text{NTU} \le 12$, and $C_{min}/C_{max} = 0.25, 0.5, 0.75$ and 1.0, while the number of thermal plates was considered as 1, 2, 3, 4 and 5 for 1 pass-1 pass arrangements, and only one thermal plate was considered for each pass of multipass arrangements. An analog method was used for higher number of thermal plates. The ε-NTU results for 1 and 2 thermal plates of 1 pass-1 pass arrangements are identical to those for an infinite number of thermal plates; those results for one thermal plate in each pass of an M pass- M Pass (M > 1) arrangement also correspond to those for an M pass-M pass exchanger with an infinite number of thermal plates. For 1 pass-1 pass counterflow (and parallel flow) arrangements, Jackson and Troupe considered a maximum of 5 thermal plates, while the number of thermal plates may be at least 20-50 in a real plate exchanger. As will be shown, the exchanger effectiveness with 40 or less number of thermal plates may not approach that for the infinite number of thermal plates at high NTU values. Hence, more detailed results than those by Jackson and Troupe are required by an engineer. Also because of the "end effects", the ε-NTU results are not symmetric with respect to the heat capacity rate ratio C_1/C_2. Hence, the results should be presented for

C_1/C_2 varying from 0 to ∞ rather than 0 to 1. This unsymmetric case is similar to the crossflow exchanger with one fluid mixed [13]. Since Jackson and Troupe [8] present the results for C_1/C_2 varying from 0 to 1, we have extended their results for the values of C_1/C_2 up to 3.

Foote [9] investigated 2 pass-1 pass, 3 pass-1 pass and 4 pass-1 pass plate exchangers with $2 \leq R \leq 10$ and an infinite number of thermal plates. Also for 2 pass-1 pass flow arrangement, N = 3, 7 and 11 were considered, and for a specific 3 pass-1 passs system (parallel flow at inlet), N = 5 was investigated. The results were presented in terms of the LMTD correction factor F as a function of $P_1R_1/(1-P_1R_1)$ with R_1 as a parameter. The results were obtained by a numerical method that was not described in the report. Again the results presented by Foote are restricted to only a small number of thermal plates.

Usher [10] presented the F factor as a function of the number of thermal plates (N up to 10) for a single-pass counter-flow exchanger (one curve only); however, no mention is made of the value of R or NTU.

Marriott [11] presented empirical F factors for 1-1, 2-1, 2-2, 3-1, 3-3, 4-1 and 4-4 pass flow arrangements as a function of NTU (≤ 11) which he mentioned should be valid for $0.7 \leq R \leq 1.4$. However, no details are presented for the basis of his empirical curves.

Kandlikar [12] numerically analyzed 1-1, 2-1, 2-2 and 3-1 pass counterflow arrangements and 1-1 and 2-2 pass parallel flow arrangements. Most of the results are presented for 3 or 5 thermal plates. As a result, an attempt is made in this paper to provide comprehensive results for 1-1, 2-1 and 3-1 pass overall counterflow arrangements covering the practical range of the thermal plates N and heat capacity rate ratio R.

3. ANALYSIS

In this section, the details are provided for numerical analysis and how the temperature effectiveness P_1 and the LMTD correction factor F are computed for various flow arrangements.

The plate exchanger is analyzed under the following

- The heat exchanger operates under steady-state conditions.
- Heat loss to the surroundings is negligible. This idealization is reasonable since there are dead air spaces between the cover and end plates, a situation which closely resembles an adiabatic wall. Heat losses to ambient are usually less than 1 percent.
- The overall heat transfer coefficient is constant throughout the exchanger. This usually implies that fluid properties are independent of temperature and that the temperature and velocity fields are fully developed.
- The specific heat of each fluid is constant throughout the exchanger so that the heat capacity rate C on each side is treated as constant.
- Temperature and fluid flow are uniform across the channel width and at the inlet of the channel for each fluid stream.
- The fluid flow rate is uniformly distributed through the exchanger on each fluid side in each pass, and the fluid is perfectly mixed at any section inside each flow channel.
- Heat is not conducted in the direction of fluid flow either by the fluid iteslf or by the channel walls.
- Either there is no phase change (condensation or boiling) or the phase change occurs under one of the following conditions: (i) the temperature is constant such as for a single component fluid at constant pressure; the effective specific heat for the phase changing fluid is infinity in this case, and hence $C_{max} \rightarrow \infty$; (ii) the temperature varies linearly with heat transfer during consensation or boiling ideally occuring uniformly over the entire surface. In this case, the effective specific heat is constant and finite.

Each channel is divided into n number of steps of equal area, from the inlet to the outlet end of the channel. The step i = 1 corressponds to the inlet section of all channels. This means i = 1 is at the lower end for Fluid 1 in Fig. 3 and i = 1 is at the upper end for Fluid 2. Referring to the plate arrangement shown in Fig. 3, for any section i in channel j, the energy equation may be written as

$$C_j dT_{i,j} = U\, dA \left[(T_{i,j-1} - T_{i,j}) + (T_{i,j+1} - T_{i,j}) \right] \tag{1}$$

For the end channels, heat is transferred from only one adjacent channel, and Eq.(1) is modified accordingly as follows. For the left end channel,

$$C_j\, dT_{i,j} = U\, dA\, (T_{i,j+1} - T_{i,j}) \tag{2}$$

For the right end channel,

$$C_j\, dT_{i,j} = U\, dA\, (T_{i,j-1} - T_{i,j}) \tag{3}$$

All temperatures are nondimensionalized as follows

$$\theta = \frac{T - T_{c,\,in}}{T_{h,\,in} - T_{c,\,in}} \tag{4}$$

This will result in the initial conditions $\theta_{h,\,in} = 1$ and $\theta_{c,\,in} = 0$ for hot and cold fluid inlet temperatures respectively.

Dividing Eq. (1) by nondimensional $dy = \delta = 1/(n-1)$ = step size and rearranging,

$$\left.\frac{d\theta}{dy}\right|_{i,j} = NTU_j\, (\theta_{i,j+1} + \theta_{i,j-1} - 2\theta_{i,j}) \tag{5}$$

$$\text{where } NTU_j = \frac{UA_j}{C_j} = \frac{UA/N}{C/m} = NTU \frac{m}{N} \tag{6}$$

where NTU = number of transfer units on the particular side, i.e. $NTU_1 = UA/C_1$ and $NTU_2 = UA/C_2$; m is the number of flow channels in one pass on the particular side (1 or 2) of concern, and N is the number of thermal plates. Note that A_j in the definition of NTU_j corresponds to the heat transfer surface area on one side of a thermal plate.

Equations (2) and (3) are nondimensionalized in the same way as follows.

$$\left.\frac{d\theta}{dy}\right|_{i,j} = NTU_j\,(\theta_{i,j+1} - \theta_{i,j}) \qquad (7)$$

$$\left.\frac{d\theta}{dy}\right|_{i,j} = NTU_j\,(\theta_{i,j-1} - \theta_{i,j}) \qquad (8)$$

The slope $d\theta/dy$ in Eqs. (5), (7) and (8) is expressed in terms of backward difference formulas. For the first step at the inlet sections, i.e. $i = 2$, a first order backward difference formula is employed.

$$\left.\frac{d\bar{\theta}}{d\,y}\right|_{i,j} = \frac{\theta_{i,j} - \theta_{i-1,j}}{\delta} \qquad (9)$$

For the second subsequent steps, a second order backward difference formula is used.

$$\left.\frac{d\theta}{dy}\right|_{i,j} = \frac{3\,\theta_{i,j} - 4\,\theta_{i-1,j} + \theta_{i-2,j}}{2\,\delta} \qquad (10)$$

The solution is started first with the input/initialization. For specified values of NTU_1 and $R_1 = C_1/C_2$, first $NTU_2 = NTU_1 R_1$ is calculated and individual channel NTU_j's are computed. A step size δ is chosen, and an initial temperature distribution throughout the exchanger is determined as follows. The heat exchanger effectiveness ε on the C_{min} side is guessed as 0.80 for counterflow and multipass flow arrangements. The temperature effectiveness P_1 or P_2 (whichever is appropriate) on the C_{max} side is then equal to $\varepsilon\, C_{min}/C_{max}$. If the exchanger is multipass on a particular fluid side, the temperature distribution is assumed to be linear in the initial guess to determine the second (and third) pass inlet temperature. For example, with the initial guess of $\varepsilon = 0.8$ for the 2 pass-1 pass arrangement with the two pass side being cold and also the C_{min} side, the dimensionless θ of the two pass side at the inlet is zero, at the inlet to the second pass 0.4, and at the outlet is 0.8. With this initial guess of ε, the outlet temperatures for both fluids are computed. The temperature distribution from the inlet to the outlet of a fluid in a given pass is then computed linearly for each nodal point.

The foregoing finite difference equations are then solved by the Gauss-Seidel method by a marching process from $i = 2$ to n. First, the temperature distribution at $i=2$ step are obtained for all channels regardless of the exchanger being single-pass or multipass. In this case, a first order backward difference formula and initial guessed temperature distribution is used to compute $\theta_{2,j}$'s. In the next sweep, all the temperatures $\theta_{3,j}$'s are computed using the second order backward difference formula, with all θ_j's from the previous iteration. The calculations are similarly continued until $i = n$. During the calculations for each node, the difference between the newly computed temperature and the temperature from the previous iteration is recorded. If the maximum difference between these temperatures at any node in the exchanger is greater than 0.00001, the iterations are restarted with $i = 2$ and continued until $i = n$, again the maximum difference between two consecutive temperatures of each node are compared, and the iterations are stopped when this difference at all nodes is found to be less than 0.00001.

For 1 pass-1 pass arrangement, the mean outlet temperature on Fluid 1 (or Fluid 2) is the algebraic sum of the outlet temperatures of Fluid 1 (or Fluid 2) divided by the number of respective flow channels. For multipass arrangements, the same approach is used for the mean outlet temperature from each pass; the mean outlet temperature from the previous pass is then used as the inlet temperature for each channel in the next pass. The mean outlet temperatures from the exchanger are then related to the temperature effectiveness of each fluid side as

$$P_1 = \begin{cases} \theta_{1,\,out} & \text{for Fluid 1 cold} \\ 1 - \theta_{1,\,out} & \text{for Fluid 1 hot} \end{cases} \qquad (11)$$

$$P_2 = \begin{cases} 1 - \theta_{2,\,out} & \text{for Fluid 2 hot} \\ \theta_{2,out} & \text{for Fluid 2 cold} \end{cases} \qquad (12)$$

The temperature effectivenesses P_1 and P_2 are defined in Eq. (13).

The number of steps tried was 10, 100 and 1000. For the last case, double precision was employed. It was observed that 100 steps were sufficient, as going to 1000 steps did not change the results up to fourth decimal place in the nondimensional temperature values. The order of magnitude of error in second order equations is δ^2, i.e. $(1/100)^2$ or 0.00001 with 100 steps. This is well within the desired accuracy limit of 0.0001.

Also several initial guesses of ε from 0.3 to 0.8 were taken, and the final solution was found totally independent of the initial guessed value of ε.

In the computation of the temperature at any node, the surrounding temperatures from the previous itertion (rather than the latest values) were used. Even then the convergence was very rapid. The number of iterations varied from 3 to 12 for 1 pass-1 pass counterflow for convergence depending upon the value of NTU_1 and R_1. The number of iterations for convergence were 15 to 24 for 2 pass-1 pass exchangers, and 20 to 44 for 3 pass-1 pass exchangers.

Some further details on the computer program may be found from [12].

4. RESULTS

Extensive checks were made on the computer program before generating the results. The ε-NTU results obtained from the computer program for 1 pass-1 pass counterflow with 1 and 2 thermal plates were found to be identical to those from the closed-form formula for the single-pass counterflow exchanger [13]. The ε-NTU results for 3, 4 and 5 thermal plates were found to be in excellent agreement with the graphical results by Jackson and Troupe [8] for $C_{min}/C_{max} \le 1$. The F factors for 2 pass-1 pass for 3,

7, and 11 thermal plates were found to be in excellent agreement with the graphical results of Foote [9] for $R_1 = 4$. The present F factors for the 3 pass-1 pass exchanger with parallel flow at the inlet for N = 5 and R = 3 were found to be in fair agreement with the graphical results of Foote. However, it is believed that the present results are more accurate due to the known accuracy of the modified Gauss-Seidel method. Thus the current computer program has been thoroughly checked out for accuracy. Subsequently, the detailed results have been obtained for several flow arrangements as summarized next.

In all of the results presented in this paper, Side 1 is considered as the 1 pass side, and Side 2 is considered as the multipass (2 pass or 3 pass) side. In the case of the 1 pass-1 pass counterflow exchanger with an even number of thermal plates, the results are different depending upon which fluid is in the two outermost channels. Hence, for the 1 pass-1 pass counterflow exchanger with an even number of thermal plates, Side 1 is considered to be the fluid in the two outermost channels. Explicitly P_1, NTU_1, R_1 and F are defined as

$$P_1 = \frac{T_{1,in} - T_{1,out}}{T_{1,in} - T_{2,in}} \qquad P_2 = \frac{T_{2,out} - T_{2,in}}{T_{1,in} - T_{2,in}} \tag{13}*$$

$$NTU_1 = \frac{UA}{C_1} \qquad R_1 = \frac{C_1}{C_2} \tag{14}$$

$$F = \frac{1}{NTU_1 (1-R_1)} \ln\left[\frac{1 - P_1 R_1}{1 - P_1}\right] = \frac{1}{NTU_2 (1-R_2)} \ln\left[\frac{1 - P_2 R_2}{1 - P_2}\right] \tag{15}$$

Equation (15) is valid for $R_1, R_2 \neq 1$. For $R_1 = R_2 = 1$,

$$F = \frac{P_1}{NTU_1 (1-P_1)} = \frac{P_2}{NTU_2 (1-P_2)} \tag{16}$$

Note that P_2, NTU_2 and R_2 are then given by

$$P_2 = P_1 R_1 \qquad NTU_2 = NTU_1 R_1 \qquad R_2 = 1/R_1 \tag{17}$$

For 1 pass-1 pass counterflow exchanger, the temperature effectiveness P_1 and the LMTD correction factor F have been calculated for even and odd number of thermal plates for extensive ranges of NTU_1 and R_1. All 1 pass-1 pass exchanger results are presented in Tables 1-7. The influence of even and odd number of thermal plates is shown in Fig.4.

For 2 pass-1 pass exchanger, there are four possible flow arrangements as shown in Fig.5. The flow arrangement of Fig. 5a has the highest thermal effectiveness for given values NTU_1 and R_1. Hence, this arrangement is analyzed. The computer results are presented in Table 8 and the partial results are presented in Fig. 6. The 1 pass-1 pass and 2 pass-1 pass arrangements are compared in Fig. 7 for two values of R_1.

For 3 pass-1 pass exchanger, two possible flow arrangements are shown in Fig. 8. The arrangement of Fig. 8b was analyzed by Foote [9], while the arrangement of Fig. 8a is also analyzed here for the first time. The detailed results are presented in Tables 9 and 10, and the partial results are compared in Figs. 9 and 10.

5. DISCUSSION

Some useful design guidelines and interesting observations can be made from the tabular and graphical results presented here.

For 1 pass-1 pass counterflow exchanger, if the "end effect" is negligible, the F factors should be unity for all values of NTU_1 and R_1. In this case, the temperature effectiveness P_1 is related to NTU_1 and R_1 as

$$P_1 = \frac{1 - \exp[-NTU_1 (1-R_1)]}{1 - R_1 \exp[-NTU_1 (1-R_1)]} \tag{18}$$

In many plate exchangers, the number of thermal plates is usually greater than 40. Figure 4 indicates that the "end effect" is negligible (F close to unity) for N > 40 and $NTU_1 \leq 1$. However, the end effect becomes important with decreasing number of thermal plates and/or increasing values of NTU_1, particularly for $NTU_1 > 3$. For a fewer number of thermal plates, the strong end effect is dependent upon whether the number of thermal plates is odd or even. A significant reduction in F or P_1 may result for N < 10, particularly for odd number of thermal plates.

If the flow arrangement is symmetric about R = 1, then the exchanger effectiveness ε is a unique function of $NTU = UA/C_{min}$ and $C^* = C_{min}/C_{max}$. In that case, ε, NTU and C^* are calculated as follows.

For $R_1 < 1$,

$$\varepsilon = P_1 \qquad NTU = NTU_1 \qquad C^* = R_1 \tag{19}$$

For $R_1 > 1$,

$$\varepsilon = P_1 R_1 \qquad NTU = NTU_1 R_1 \qquad C^* = 1/R_1 \tag{20}$$

Note that in both cases F remains the same for a given NTU and C^*.

For a counterflow exchanger with an odd number of thermal plates, the number of flow channels are identical for Fluids 1 and 2, and the flow arrangement is symmetric with respect to R = 1. This has also been confirmed with numerical results for $R_1 > 1$ and an odd number of thermal plates. Hence to conserve the

* All temperatures in this equation are bulk mean temperatures.

space, the tabular results for R_1 = 1.25, 1.5, 2 and 3 for an odd number of thermal plates are not included.

However, for a counterflow exchanger with an even number of thermal plates, the number of flow channels for one fluid is one greater than that for the other fluid, and hence the exchanger effectiveness is dependent upon whether the heat capacity rate C of the fluid in the end channels is lower or higher than that for the fluid with one less flow channel. This "end effect" is now calculated for two specific cases. From Table 4, $\varepsilon = P_1 = 0.8426$ for $NTU = NTU_1 = 4$, $N = 4$ and $C^* = R_1 = 0.8$. From Table 6, $P_1 = 0.6779$ at $NTU_1 = 3.2$, $N = 4$ and $R_1 = 1.25$. Then using Eq.(17), $\varepsilon = P_1 R_1 = 0.8474$, $NTU = NTU_1 R_1 = 3.2 \times 1.25 = 4$, and $C^* = 1/R_1 = 0.80$. Thus at NTU = 4 and $C^* = 0.8$, $\varepsilon = 0.8426$ if the fluid in the end channels has the lower heat capacity rate, and $\varepsilon = 0.8434$ if the fluid in the end channels has the higher heat capcity rate. Thus this end effect or asymmetrical condition results in $\Delta\varepsilon/\varepsilon = 0.57\%$. Using Tables 2 and 7, a similar case for NTU = 4 and $C^* = 0.5$ will yield $\Delta\varepsilon/\varepsilon = 1.3\%$ at N = 4. The corresponding $\Delta\varepsilon/\varepsilon$ for N = 40 are 0.39% and 0.76% respectively. These typical points clearly indicate the end effect that results in asymmetry may be important for higher vlaues of NTU's and intermediate values of C^* (=0.5). A slightly higher ε would result if the fluid in the outermost channel has a heat capacity rate higher than that for the other fluid for 1 pass-1 pass counterflow exchanger with an even number of thermal plates.

For the 2 pass-1 pass exchanger (Fig. 5a), the F factors, are shown in Fig. 6 for NTU_1 = 1,2 and 5 for R_1 = 2, 3 and 4. The F factors are found to be considerably lower than unity due to the flow in one pass being parallel flow, and they are not as sensitive to N for low values of N as for the 1 pass-1 pass counterflow exchanger.

When there is a significant imbalance in the flow rate of the two fluids, a multipass arrangement may be used. As an example, if the flow rate on Side 1 is twice the flow rate on Side 2 with both fluids having approximately the same specific heats, a design question arises as to whether to use a 1 pass-1 pass or 2 pass-1 pass arrangement with two passes on Side 2. In a 2 pass-1 pass arrangement, the fluid on Side 2 is counterflow in one pass and in parallel flow in other pass with respect to the fluid on Side 1. So half of the exchanger is in parallel flow which reduces the exchanger overall effectiveness and heat transfer capability compared to the 1 pass-1 pass exchanger at the same N, R_1 and NTU_1. However, the heat transfer coefficient h on Side 2 in two pass may be considerably higher than that on Side 1 due to the double fluid velocity. This will result in a higher UA and NTU compared to the 1 pass-1 pass exchanger at the same N and R_1. This will increase ε and compensate for the reduction in ε in one pass due to parallel flow.

Thus when there is an imbalance in the flow, the optimum design may be 1 pass-1 pass or 2 pass-1 pass depending upon the values of R_1 and NTU_1 (i.e.,the type of plates and their nondimensional heat transfer coefficient versus Reynolds number characteristics). In Fig. 7, the temperature effectiveness P_1 is compared as a function of NTU_1 for 1 pass-1 pass and 2 pass-1 pass exchangers for R_1 = 2 and 3. As mentioned above, for the same values of N and R_1, the 2 pass-1 pass exchanger will have higher NTU_1 than that for the 1 pass-1 pass exchanger. From the review of the results of Tables 7, 8, and Fig. 7, a preliminary observation may be made that for low values of NTU_1, 2 pass-1 pass arrangement will have effectively higher P_1 due to higher NTU_1 compared to the 1 pass-1 pass conterflow with the same number of thermal plates at the same R_1. However, at high values of NTU_1, the 2 pass-1 pass arrangement will have effectively lower P_1 compared to the 1 pass-1 pass counterflow at the same N and R_1; this is because the increase in NTU_1 (due to high h) may not compensate for the reduction in ε due to parallel flow in one pass. The following procedure is recommended to find an optimum 1 pass-1 pass versus 2 pass-1 pass arrangement. (with no considerations on the pressure drops) when there is a significant imbalance in the flow rates of two fluids. First determine the heat transfer coefficients on both fluid sides considering 1 pass-1 pass and 2 pass-1 pass arrangement, and accordingly determine the NTU_1's for both arrangements. Go to the appropriate P_1-NTU_1 tables and find the appropriate temperature effectiveness P_1 for the specified R_1 and N. The arrangement which yields a higher value of P_1 is then desired from the heat transfer point of view. This process of determining the better flow arrangement (1-1 vs. 2-1 passes) from the heat transfer point of view can be computerized if the dimensionless heat transfer coefficient (Colburn factor j or Nusselt number Nu) versus Reynolds number characteristics or similar design data are available for various plate configurations under consideration and a computer program similar to the one used for generating the results of this paper is available.

Since the 2 pass-1 pass system is used when there is a flow imbalance, the results are presented only for R_1 = 2, 3 and 4. For $R_1 < 2$ most probably the 1 pass-1 pass arrangement may be most desirable from the heat transfer point of view and for $R_1 > 4$, 3 pass-1 pass arrangement may be desirable as discussed later.

For the 3 pass-1 pass case, two possible flow arrangements shown in Fig. 8 are investigated. P_1 of Fig. 8a arrangement (counterflow at inlet is always higher than P_1 of Fig. 8b arrangement (parallel flow at inlet) except for low values of NTU_1 for N = 5 and and R_1 = 3 (see Tables 9 and 10)*. The difference in P_1 increases with increasing values of NTU_1, decreasing

* N=5 represents the minimum possible number of thermal plates in a 3 pass-1 pass arrangement. The "end effect" is the strongest for this case.

values of N, and R_1 decreasing from 6 to 3. A typical comparison of the results for N = 41 is shown in Fig. 10 While P_1 of Fig 8a arrangement always increases monotonically with increasing values of NTU_1, P_1 of Fig. 8b has a severe temperature cross and P_1 first increases and then decreases with increasing values of NTU_1 for a low number of thermal plates.

An important observation may be made for the choice of 2 pass-1 pass versus 3 pass-1 pass arrangements. A review of Tables 8 and 9 indicate that P_1's are almost identical for these two flow arrangements for $R_1 = 3$ and $N = \infty$. This means for exchangers with the number of thermal plates greater than 40, the exchanger effectiveness will be approximately the same for 2-1 and 3-1 pass exchangers for the same R_1 and NTU_1. If the pressure drop is tolerable, a selection of 3 pass over 2 pass for a given fluid side will result in the higher fluid velocity, higher h and subsequently higher NTU_1 and P_1.

6. SUMMARY AND CONCLUSIONS

Extensive numerical results are presented for 1 pass-1 pass, 2 pass-1 pass, and 3 pass-1 pass plate exchangers. These results are presented in terms of the exchanger temperature effectiveness and the log-mean temperature difference correction factor as functions of the number of transfer units, heat capacity rate ratio, and the number of thermal plates. The following conclusions are based on the results presented.

For a 1 pass-1 pass exchanger, even versus odd number of thermal plates have a strong influence on P_1 or F for N < 10, and a negligible influence for $N \gtrsim 40$.

For a 1 pass-1 pass exchanger with an even number of thermal plates, the fluid in the two outermost channels is the same. The exchanger effectiveness ε is slightly higher if the outer fluid has the higher heat capacity rate compared to that for the other fluid having one less flow channel. However, this increase in ε is negligible for engineering purposes except possibly for the high NTU case.

The performance (P_1 or F) of a 2 pass-1 pass exchanger may be higher, equal or lower than that for a 1 pass-1 pass exchanger for the same N and R_1. It depends upon the heat transfer characteristics of the plates and the number of thermal plates. A procedure is outlined to decide whether 1 pass-1 pass or 2 pass-1 pass arrangement will be better.

For the 3 pass-1 pass exchanger, the flow arrangement with counterflow at the inlet yields a slightly higher effectiveness, particularly at high NTU and $R_1 = 3$, compared to the flow arrangement with parallel flow at the inlet.

NOMENCLATURE

A	heat transfer surface area on one fluid side of an exchanger, m^2.
C	Flow stream heat capacity rate, Wc_p, W/°C
C^*	the ratio of minimum to maximum heat capacity rates, C_{min}/C_{max}, dimensionless.
c_p	specific heat of fluid at constant pressure, J/kg °C
F	log-mean temperature difference correction factor, dimensionless
h	heat transfer coefficient, W/m^2 °C
m	number of flow channels on one fluid side in one pass
N	number of thermal plates
n	total number of finite difference nodes along the plate length (flow length)
NTU	number of heat transfer units for the whole exchanger, $NTU = UA/C_{min}$, $NTU_1 = UA/C_1$, $NTU_2 = UA/C_2$, dimensionless
P	temperature effectiveness of a given fluid stream see Eq.(13) for the definitions of P_1 and P_2, dimensionless
R	heat capacity rate ratio, $R_1 = C_1/C_2$, $R_2 = C_2/C_1$, dimensionless
U	overall heat transfer coefficient, W/m^2°C
W	fluid mass flow rate, kg/s
y	distance along the plate length normalized to the plate length, dimensionless
δ	step size along the plate length, $\delta = 1/(n-1)$ dimensionless
ε	heat exchanger effectiveness, larger of P_1 and P_2, dimensionless
θ	temperature defined by Eq. (4), dimensionless

Subscripts

i	node number along the flow length, see Fig. 3
in	inlet
min	minimum
max	maximum
out	outlet
1	single-pass side
2	multipass side

REFERENCES

1. Focke, W.W., Plate heat exchangers: review of transport phenomena and design procedures, CSIR Report CENG 445, CSIR, Pretoria, South Africa (1983).
2. Cooper,A., and Usher,J.D., Plate heat exchangers, in Heat Exchanger Design Handbook, E.U. Schlunder, Editor-in-Chief, Vol.3, Section 3.7, Hemisphere Publishing Corp., Washington, D.C. (1983).
3. Raju,K.S.N., and Bansal J.C., Plate heat exchangers and their performance, in Low Reynolds Number Flow Heat Exchangers, S. Kakac, R.K. Shah, and A.E. Bergles, Editors, pp. 899-912, Hemisphere Publishing Corp., Washington, D.C. (1983).

4. Raju, K.S.N., and Bensal, J.C., Design of plate heat exchangers, in Low Reynolds Number Flow Heat Exchangers, S. Kakac, R.K. Shah, and A.E. Bergles, Editors, pp. 913-932, Hemisphere Publishing Co., Washington, D.C. (1983).

5. Shah R. K., Heat exchanger design methodology - an overview, in Low Reynolds Number Flow Heat exchangers, Kakac, S., Shah, R.K., and Bergles, A.E., Editors, PP. 15-19, Hemisphere Publishing Corp., Washington, D.C. (1983).

6. Shah R.K., Heat exchanger basic design methods, in Low Reynolds Number Flow Heat Exchangers, S. Kakac, R.K. Shah, and A.E. Bergles, Editors, Hemisphere Publishing Corp., Washington, D.C. (1983) 27-72.

7. Buonopane, R.A., Troupe, R.A., and Morgan, J.C., Heat transfer design method tor plate neat exchangers, Chem. Engg. Prog., No.7, 59 (1961) 57-61.

8. Jackson, B.W., and Troupe,R.A., Plate exchanger design by ε-NTU method, Chem. Engg. Prog. Symp. Series No.64, 62 (1966) 185-190.

9. Foote,M.R., Effective mean temperature differences in multipass heat exchangers, NEL Report 303, National Engineering Lab, Glasgow, U.K. (1967).

10. Usher,J.D., The plate heat exchanger, in Compact Heat Exchangers, NEL Report 482, National Engineering Lab, Glasgow, U.K.(1969).

11. Marriott,J., Where and how to use plate heat exchangers, Chem. Engg. No.8, 78 (1971) 127-133

12. Kandlikar, S., Performance curves for different plate heat exchanger configurations, ASME Paper No.84-HT-26 (1984).

13. Kays,W.M., and London,A.L., Compact Heat Exchangers, Third Edition, McGraw-Hill, New York (1984).

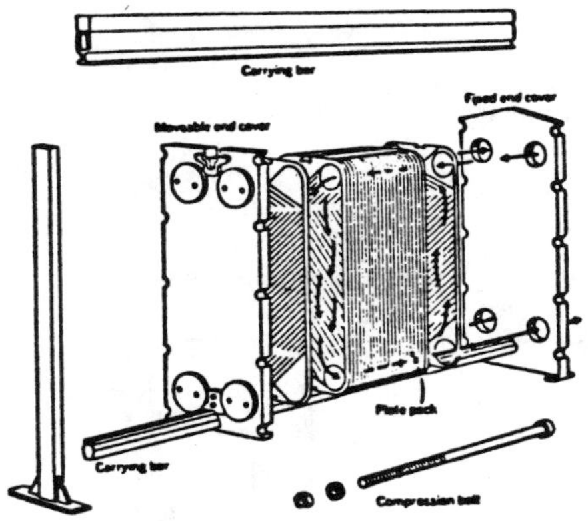

Fig.1 A Plate Heat Exchanger Assembly

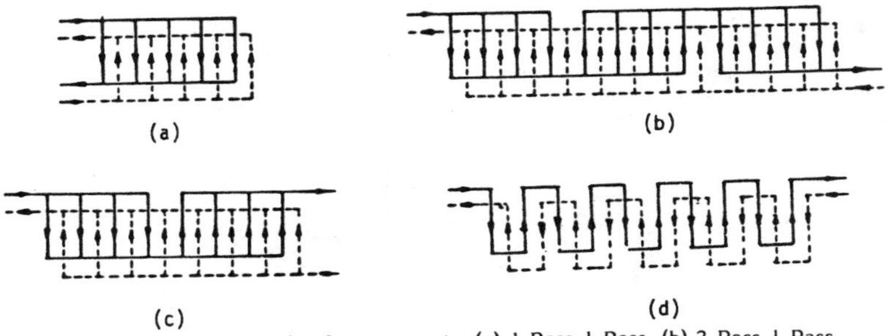

Fig.2 Some Plate Exchanger Arrangements. (a) 1 Pass-1 Pass, (b) 2 Pass-1 Pass, (c) 3 Pass-1 Pass and (d) Series Arrangement, M Pass-M Pass with one Channel Per Pass. Only the Fluirl Streams are Shown, No Thermal Plates are Shown.

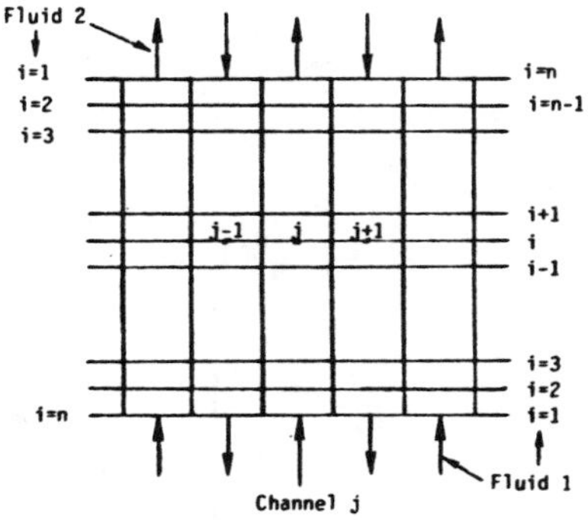

Fig.3 The Grid Structure for the Finite Difference Analysis.

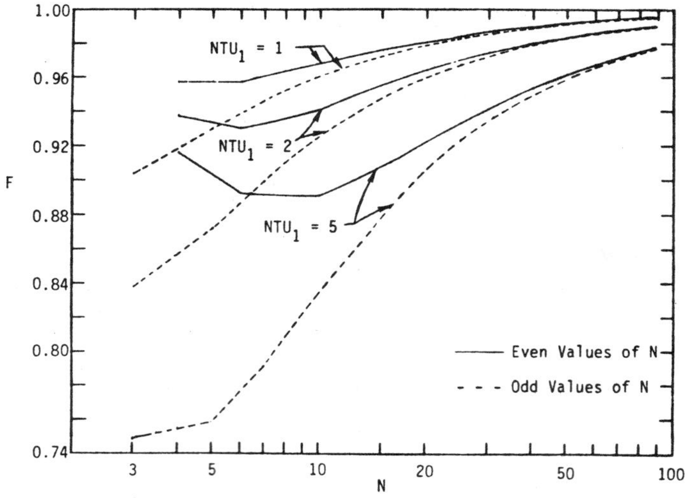

Fig.4 The Influence of Odd and Even Number of Thermal Plates for 1 Pass-1 Pass Counterflow Exchanger Performance for $R_1^I = 1$. All Curves should Approach to Unity as $N \to \infty$.

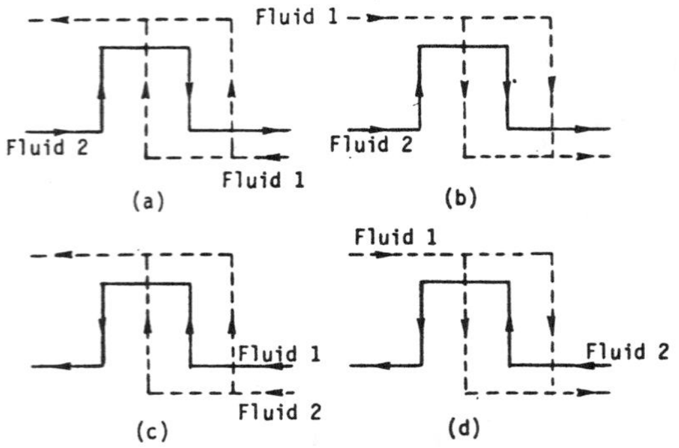

Fig.5 Four possible flow arrangements for the 2 pass-1 pass exchanger.

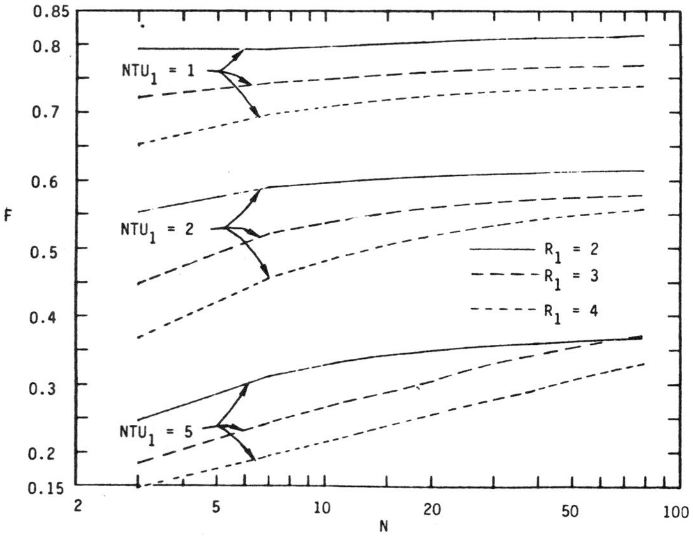

Fig.6 The influence of the number of thermal plates on the 2 pass-1 pass exchanger F factors.

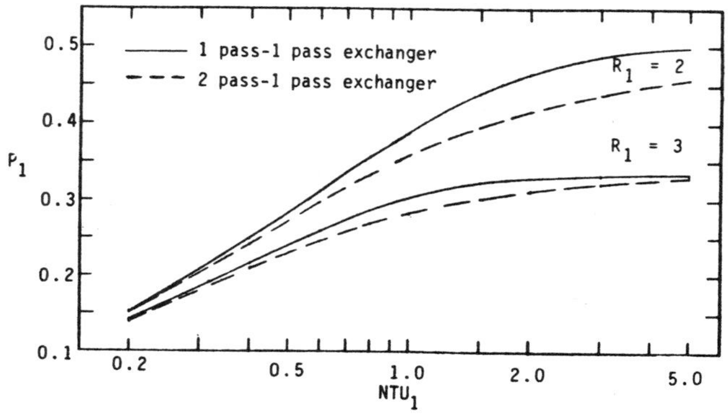

Fig.7 A comparison of the temperature effectivenesses of 1 pass-1 pass and 2 pass-1 pass exchangers with $N \to \infty$.

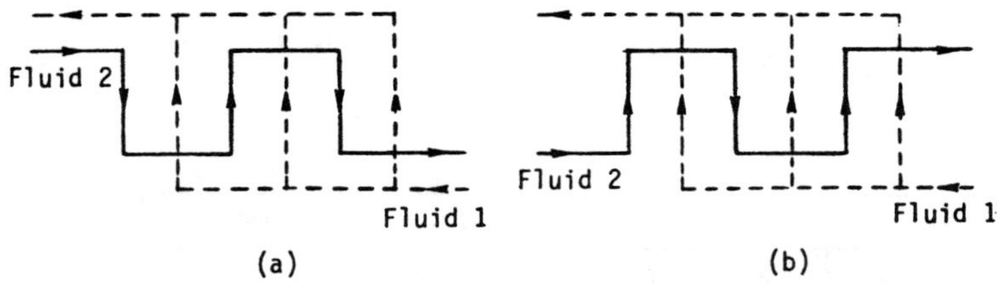

Fig.8 Two 3 pass-1 flow arrangements: (a) Counterflow at inlet, (b) Parallel flow at inlet.

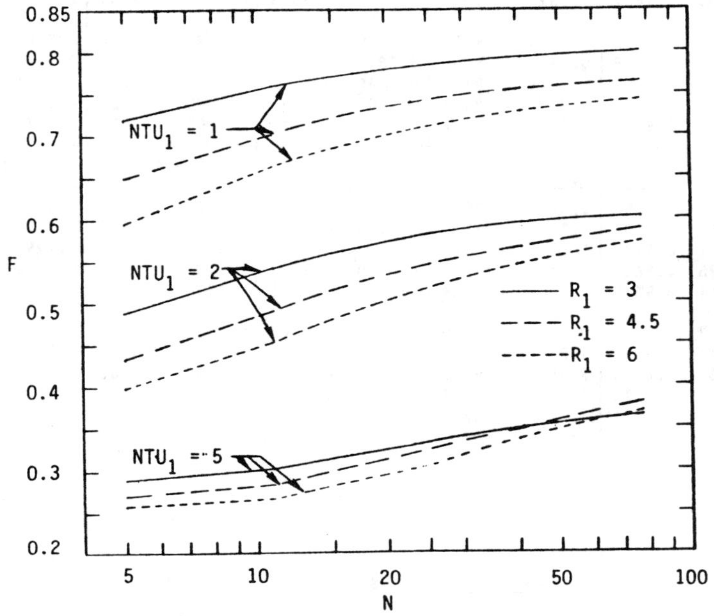

Fig.9 The influence of the number of thermal plates on the 3 pass-1 pass exchanger (counterflow at inlet) F factors.

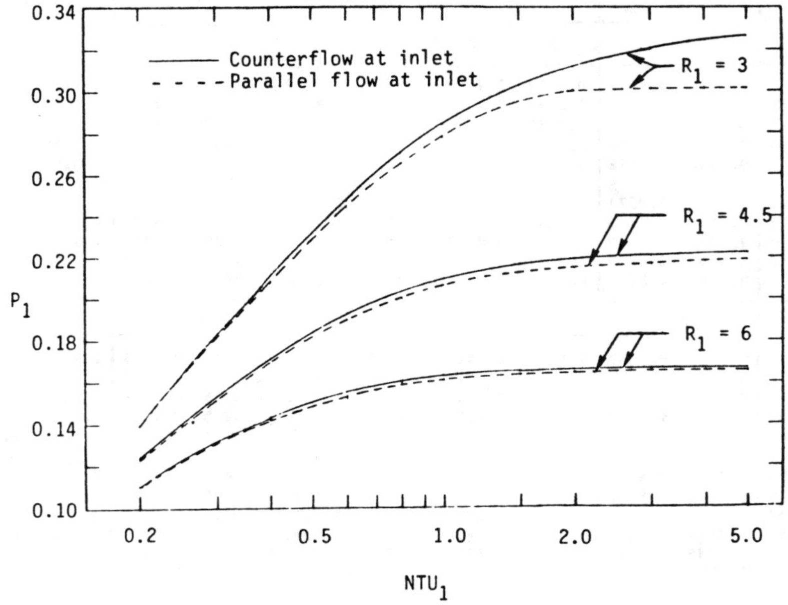

Fig.10 P versus NTU_1 for the two 3 pass-1 pass arrangements of Fig.8 at N = 41.

Table 1. 1 pass - 1 pass counterflow exchanger: P_1 and F as functions of NTU_1 and N for R_1 = 0.25

NTU_1	N=4		N=6		N=8		N=10		N=20		N=40		N=80		N=∞
	P_1	F	P_1	F	P_1	F	P_1	F	P_1	F	P_1	F	P_1	F	P_1
0.2	0.1756	0.9880	0.1758	0.9890	0.1760	0.9906	0.1762	0.9920	0.1768	0.9954	0.1771	0.9975	0.1773	0.9986	0.1775
0.4	0.3122	0.9766	0.3127	0.9785	0.3134	0.9815	0.3141	0.9841	0.3158	0.9908	0.3168	0.9950	0.3175	0.9975	0.3181
0.6	0.4207	0.9661	0.4213	0.9682	0.4226	0.9724	0.4238	0.9762	0.4268	0.9861	0.4288	0.9926	0.4299	0.9961	0.4311
0.8	0.5082	0.9564	0.5089	0.9583	0.5107	0.9635	0.5122	0.9682	0.5167	0.9813	0.5196	0.9899	0.5212	0.9947	0.5229
1.0	0.5798	0.9474	0.5803	0.9487	0.5825	0.9547	0.5845	0.9603	0.5901	0.9764	0.5939	0.9872	0.5960	0.9933	0.5983
1.5	0.7104	0.9276	0.7098	0.9262	0.7124	0.9333	0.7150	0.9406	0.7229	0.9636	0.7284	0.9799	0.7315	0.9894	0.7350
2.0	0.7959	0.9115	0.7941	0.9059	0.7964	0.9130	0.7991	0.9214	0.8081	0.9501	0.8147	0.9719	0.8185	0.9850	0.8228
3.0	0.8946	0.8873	0.8906	0.8716	0.8918	0.8761	0.8940	0.8849	0.9026	0.9216	0.9096	0.9538	0.9139	0.9748	0.9188
4.0	0.9440	0.8708	0.9394	0.8451	0.9394	0.8451	0.9407	0.8522	0.9475	0.8921	0.9536	0.9330	0.9576	0.9622	0.9622
5.0	0.9698	0.8593	0.9656	0.8248	0.9649	0.8197	0.9655	0.8241	0.9702	0.8629	0.9751	0.9100	0.9783	0.9472	0.9823
NTU_1	N=3		N=5		N=7		N=11		N=19		N=39		N=79		N=∞
0.2	0.1753	0.9862	0.1759	0.9901	0.1763	0.9924	0.1766	0.9947	0.1770	0.9968	0.1772	0.9983	0.1774	0.9992	0.1775
0.4	0.3112	0.9728	0.3131	0.9803	0.3143	0.9849	0.3155	0.9897	0.3165	0.9937	0.3173	0.9968	0.3177	0.9984	0.3181
0.6	0.4187	0.9597	0.4221	0.9707	0.4241	0.9773	0.4263	0.9845	0.4282	0.9906	0.4296	0.9952	0.4304	0.9976	0.4311
0.8	0.5050	0.9470	0.5098	0.9611	0.5128	0.9699	0.5160	0.9793	0.5187	0.9873	0.5208	0.9936	0.5219	0.9968	0.5229
1.0	0.5753	0.9348	0.5814	0.9517	0.5852	0.9624	0.5893	0.9741	0.5928	0.9841	0.5955	0.9919	0.5969	0.9960	0.5983
1.5	0.7026	0.9063	0.7107	0.9285	0.7160	0.9436	0.7219	0.9607	0.7270	0.9756	0.7309	0.9875	0.7329	0.9937	0.7350
2.0	0.7856	0.8809	0.7941	0.9060	0.8002	0.9247	0.8071	0.9467	0.8131	0.9665	0.8178	0.9827	0.8203	0.9912	0.8228
3.0	0.8820	0.8392	0.8887	0.8643	0.8946	0.8876	0.9017	0.9175	0.9081	0.9465	0.9133	0.9716	0.9160	0.9854	0.9188
4.0	0.9320	0.8079	0.9361	0.8278	0.9407	0.8524	0.9467	0.8873	0.9524	0.9242	0.9571	0.9584	0.9596	0.9782	0.9622
5.0	0.9599	0.7845	0.9618	0.7971	0.9650	0.8205	0.9696	0.8574	0.9741	0.9000	0.9780	0.9427	0.9801	0.9692	0.9823

Table 2. 1 pass - 1 pass counterflow exchanger: P_1 and F as functions of NTU_1 and N for R_1 = 0.50

NTU_1	N=4		N=6		N=8		N=10		N=20		N=40		N=80		N=∞
	P_1	F	P_1	F	P_1	F	P_1	F	P_1	F	P_1	F	P_1	F	P_1
0.2	0.1720	0.9881	0.1722	0.9892	0.1724	0.9907	0.1726	0.9922	0.1731	0.9955	0.1734	0.9975	0.1736	0.9986	0.1738
0.4	0.3015	0.9773	0.3020	0.9791	0.3027	0.9820	0.3033	0.9845	0.3048	0.9911	0.3058	0.9952	0.3063	0.9976	0.3069
0.6	0.4025	0.9676	0.4030	0.9695	0.4042	0.9735	0.4052	0.9771	0.4079	0.9866	0.4097	0.9929	0.4106	0.9963	0.4117
0.8	0.4832	0.9588	0.4837	0.9604	0.4852	0.9653	0.4866	0.9698	0.4905	0.9823	0.4930	0.9904	0.4944	0.9950	0.4959
1.0	0.5491	0.9509	0.5493	0.9518	0.5512	0.9574	0.5529	0.9626	0.5577	0.9778	0.5609	0.9880	0.5628	0.9937	0.5647
1.5	0.6700	0.9343	0.6693	0.9320	0.6714	0.9384	0.6736	0.9452	0.6804	0.9666	0.6851	0.9815	0.6878	0.9903	0.6908
2.0	0.7516	0.9213	0.7495	0.9148	0.7514	0.9209	0.7538	0.9287	0.7618	0.9551	0.7675	0.9749	0.7709	0.9867	0.7746
3.0	0.8518	0.9030	0.8477	0.8870	0.8485	0.8902	0.8506	0.8980	0.8589	0.9316	0.8657	0.9603	0.8698	0.9786	0.8744
4.0	0.9082	0.8912	0.9030	0.8662	0.9027	0.8650	0.9040	0.8711	0.9114	0.9078	0.9182	0.9443	0.9225	0.9692	0.9274
5.0	0.9419	0.8833	0.9366	0.8507	0.9356	0.8447	0.9362	0.8482	0.9420	0.8844	0.9482	0.9271	0.9523	0.9586	0.9572
NTU_1	N=3		N=5		N=7		N=11		N=19		N=39		N=79		N=∞
0.2	0.1713	0.9836	0.1720	0.9883	0.1724	0.9910	0.1728	0.9937	0.1732	0.9963	0.1735	0.9980	0.1736	0.9991	0.1738
0.4	0.2993	0.9678	0.3015	0.9770	0.3028	0.9824	0.3041	0.9880	0.3052	0.9927	0.3060	0.9963	0.3065	0.9982	0.3069
0.6	0.3982	0.9526	0.4020	0.9660	0.4043	0.9739	0.4067	0.9822	0.4086	0.9892	0.4101	0.9945	0.4109	0.9973	0.4117
0.8	0.4766	0.9381	0.4821	0.9553	0.4853	0.9656	0.4887	0.9766	0.4915	0.9857	0.4937	0.9928	0.4948	0.9964	0.4959
1.0	0.5402	0.9243	0.5471	0.9449	0.5512	0.9575	0.5555	0.9710	0.5591	0.9823	0.5619	0.9910	0.5633	0.9955	0.5647
1.5	0.6560	0.8928	0.6652	0.9199	0.6711	0.9376	0.6774	0.9571	0.6826	0.9736	0.6867	0.9866	0.6887	0.9932	0.6908
2.0	0.7336	0.8659	0.7436	0.8963	0.7506	0.9184	0.7582	0.9432	0.7646	0.9648	0.7695	0.9819	0.7721	0.9909	0.7746
3.0	0.8299	0.8234	0.8386	0.8535	0.8462	0.8815	0.8550	0.9155	0.8625	0.9465	0.8684	0.9721	0.8714	0.9858	0.8744
4.0	0.8860	0.7931	0.8919	0.8173	0.8989	0.8475	0.9075	0.8880	0.9151	0.9273	0.9211	0.9613	0.9243	0.9801	0.9274
5.0	0.9216	0.7712	0.9249	0.7874	0.9307	0.8170	0.9384	0.8613	0.9454	0.9074	0.9511	0.9494	0.9542	0.9736	0.9572

Table 3. 1 pass - 1 pass counterflow exchanger: P_1 and F as functions of NTU_1 and N for R_1 = 0.75.

NTU_1	N=4		N=6		N=8		N=10		N=20		N=40		N=80		N=∞
	P_1	F	P_1	F	P_1	F	P_1	F_1	P_1	F	P_1	F	P_1	F	P_1
0.2	0.1685	0.9883	0.1686	0.9894	0.1689	0.9909	0.1691	0.9923	0.1695	0.9955	0.1698	0.9976	0.1700	0.9986	0.1702
0.4	0.2913	0.9780	0.2916	0.9797	0.2923	0.9825	0.2928	0.9850	0.2942	0.9913	0.2951	0.9953	0.2956	0.9976	0.2961
0.6	0.3849	0.9690	0.3854	0.9707	0.3864	0.9746	0.3873	0.9780	0.3897	0.9872	0.3912	0.9932	0.3920	0.9964	0.3930
0.8	0.4588	0.9611	0.4592	0.9624	0.4605	0.9670	0.4617	0.9713	0.4650	0.9832	0.4672	0.9909	0.4684	0.9953	0.4697
1.0	0.5187	0.9542	0.5188	0.9546	0.5203	0.9599	0.5218	0.9648	0.5259	0.9792	0.5287	0.9887	0.5302	0.9942	0.5319
1.5	0.6283	0.9401	0.6275	0.9374	0.6293	0.9432	0.6311	0.9495	0.6368	0.9692	0.6407	0.9831	0.6429	0.9912	0.6454
2.0	0.7030	0.9297	0.7011	0.9229	0.7026	0.9283	0.7046	0.9353	0.7112	0.9596	0.7160	0.9775	0.7187	0.9880	0.7218
3.0	0.7979	0.9157	0.7942	0.9006	0.7949	0.9031	0.7966	0.9101	0.8038	0.9404	0.8097	0.9658	0.8132	0.9816	0.8171
4.0	0.8553	0.9072	0.8505	0.8848	0.8502	0.8834	0.8514	0.8889	0.8583	0.9218	0.8645	0.9538	0.8685	0.9748	0.8730
5.0	0.8930	0.9017	0.8879	0.8736	0.8869	0.8680	0.8875	0.8713	0.8934	0.9040	0.8998	0.9415	0.9039	0.9675	0.9088

NTU_1	N=3		N=5		N=7		N=11		N=19		N=39		N=79		N=∞
0.2	0.1674	0.9809	0.1682	0.9865	0.1687	0.9896	0.1691	0.9928	0.1696	0.9957	0.1698	0.9977	0.1700	0.9989	0.1702
0.4	0.2879	0.9627	0.2903	0.9735	0.2917	0.9798	0.2931	0.9863	0.2943	0.9916	0.2952	0.9958	0.2957	0.9979	0.2961
0.6	0.3787	0.9455	0.3829	0.9612	0.3852	0.9702	0.3877	0.9798	0.3898	0.9877	0.3914	0.9938	0.3922	0.9969	0.3930
0.8	0.4496	0.9292	0.4554	0.9492	0.4588	0.9611	0.4623	0.9736	0.4652	0.9839	0.4674	0.9919	0.4686	0.9960	0.4697
1.0	0.5066	0.9139	0.5138	0.9377	0.5181	0.9521	0.5225	0.9674	0.5262	0.9801	0.5290	0.9899	0.5304	0.9950	0.5319
1.5	0.6099	0.8798	0.6194	0.9102	0.6255	0.9305	0.6320	0.9525	0.6372	0.9708	0.6413	0.9852	0.6433	0.9925	0.6454
2.0	0.6798	0.8515	0.6900	0.8848	0.6974	0.9099	0.7053	0.9379	0.7118	0.9617	0.7168	0.9805	0.7193	0.9902	0.7218
3.0	0.7694	0.8087	0.7783	0.8399	0.7868	0.8714	0.7965	0.9097	0.8046	0.9437	0.8108	0.9710	0.8140	0.9854	0.8171
4.0	0.8252	0.7795	0.8314	0.8031	0.8396	0.8368	0.8501	0.8827	0.8591	0.9258	0.8659	0.9612	0.8695	0.9803	0.8730
5.0	0.8636	0.7591	0.8670	0.7737	0.8744	0.8064	0.8848	0.8571	0.8941	0.9080	0.9013	0.9512	0.9051	0.9751	0.9088

Table 4. 1 pass - 1 pass counterflow exchanger: P_1 and F as functions of NTU_1 and N for R_1 = 0.80.

NTU_1	N=4		N=6		N=8		N=10		N=20		N=40		N=80		N=∞
	P_1	F	P_1	F	P_1	F	P_1	F	P_1	F	P_1	F	P_1	F	P_1
0.2	0.1678	0.9884	0.1679	0.9894	0.1682	0.9909	0.1684	0.9923	0.1688	0.9955	0.1691	0.9975	0.1693	0.9986	0.1695
0.4	0.2892	0.9781	0.2896	0.9798	0.2902	0.9826	0.2908	0.9850	0.2921	0.9914	0.2930	0.9953	0.2935	0.9977	0.2940
0.6	0.3815	0.9693	0.3819	0.9709	0.3829	0.9747	0.3838	0.9782	0.3861	0.9873	0.3876	0.9932	0.3884	0.9965	0.3893
0.8	0.4540	0.9615	0.4544	0.9628	0.4556	0.9674	0.4568	0.9715	0.4600	0.9834	0.4621	0.9910	0.4633	0.9953	0.4645
1.0	0.5127	0.9548	0.5128	0.9552	0.5143	0.9603	0.5157	0.9652	0.5197	0.9794	0.5223	0.9889	0.5238	0.9942	0.5254
1.5	0.6199	0.9411	0.6191	0.9384	0.6208	0.9441	0.6226	0.9503	0.6280	0.9698	0.6318	0.9835	0.6339	0.9914	0.6363
2.0	0.6929	0.9312	0.6911	0.9245	0.6925	0.9297	0.6944	0.9366	0.7008	0.9605	0.7054	0.9781	0.7080	0.9885	0.7109
3.0	0.7860	0.9179	0.7824	0.9032	0.7830	0.9056	0.7847	0.9123	0.7917	0.9422	0.7973	0.9671	0.8006	0.9825	0.8043
4.0	0.8426	0.9100	0.8381	0.8884	0.8378	0.8870	0.8389	0.8922	0.8456	0.9247	0.8517	0.9559	0.8554	0.9761	0.8597
5.0	0.8804	0.9049	0.8755	0.8780	0.8745	0.8727	0.8750	0.8756	0.8809	0.9080	0.8871	0.9444	0.8911	0.9695	0.8957

NTU_1	N=3		N=5		N=7		N=11		N=19		N=39		N=79		N=∞
0.2	0.1666	0.9804	0.1675	0.9861	0.1679	0.9893	0.1684	0.9926	0.1688	0.9956	0.1691	0.9976	0.1693	0.9989	0.1695
0.4	0.2857	0.9617	0.2881	0.9728	0.2895	0.9793	0.2909	0.9859	0.2921	0.9914	0.2931	0.9957	0.2935	0.9979	0.2940
0.6	0.3749	0.9441	0.3791	0.9602	0.3815	0.9695	0.3840	0.9793	0.3861	0.9874	0.3877	0.9936	0.3885	0.9969	0.3893
0.8	0.4444	0.9274	0.4502	0.9480	0.4536	0.9601	0.4572	0.9729	0.4601	0.9835	0.4623	0.9917	0.4634	0.9959	0.4645
1.0	0.5001	0.9118	0.5073	0.9362	0.5116	0.9510	0.5161	0.9666	0.5197	0.9797	0.5225	0.9897	0.5240	0.9949	0.5254
1.5	0.6008	0.8773	0.6103	0.9082	0.6164	0.9290	0.6229	0.9515	0.6282	0.9703	0.6322	0.9849	0.6342	0.9924	0.6363
2.0	0.6690	0.8487	0.6791	0.8823	0.6865	0.9081	0.6944	0.9367	0.7009	0.9611	0.7059	0.9802	0.7084	0.9900	0.7109
3.0	0.7566	0.8059	0.7654	0.8371	0.7739	0.8691	0.7837	0.9083	0.7919	0.9430	0.7980	0.9707	0.8012	0.9851	0.8043
4.0	0.8116	0.7769	0.8177	0.8003	0.8260	0.8343	0.8366	0.8812	0.8457	0.9251	0.8527	0.9611	0.8562	0.9801	0.8597
5.0	0.8498	0.7569	0.8531	0.7710	0.8606	0.8040	0.8712	0.8556	0.8808	0.9075	0.8882	0.9514	0.8920	0.9751	0.8957

Table 5. 1 pass - 1 pass counterflow exchanger: P_1 and F as functions of NTU_1 and N for R_1 = 1.

NTU_1	N=4		N=6		N=8		N=10		N=20		N=40		N=80		N=∞
	P_1	F	P_1	F	P_1	F	P_1	F	P_1	F	P_1	F	P_1	F	P_1
0.2	0.1651	0.9885	0.1652	0.9895	0.1654	0.9910	0.1656	0.9924	0.1661	0.9956	0.1663	0.9976	0.1665	0.9986	0.1667
0.4	0.2813	0.9786	0.2817	0.9802	0.2822	0.9830	0.2827	0.9854	0.2840	0.9916	0.2848	0.9954	0.2853	0.9977	0.2857
0.6	0.3680	0.9703	0.3683	0.9719	0.3692	0.9755	0.3700	0.9789	0.3721	0.9877	0.3735	0.9935	0.3742	0.9966	0.3750
0.8	0.4352	0.9632	0.4355	0.9642	0.4366	0.9686	0.4376	0.9727	0.4405	0.9840	0.4423	0.9914	0.4433	0.9955	0.4444
1.0	0.4890	0.9571	0.4891	0.9573	0.4904	0.9622	0.4916	0.9668	0.4951	0.9804	0.4973	0.9894	0.4986	0.9945	0.5000
1.5	0.5864	0.9451	0.5856	0.9422	0.5870	0.9475	0.5885	0.9534	0.5931	0.9718	0.5963	0.9846	0.5981	0.9920	0.6000
2.0	0.6520	0.9367	0.6504	0.9302	0.6516	0.9350	0.6531	0.9413	0.6584	0.9637	0.6621	0.9799	0.6643	0.9895	0.6667
3.0	0.7353	0.9260	0.7325	0.9126	0.7329	0.9147	0.7342	0.9209	0.7399	0.9483	0.7444	0.9709	0.7471	0.9846	0.7500
4.0	0.7863	0.9198	0.7828	0.9008	0.7826	0.8997	0.7835	0.9046	0.7889	0.9343	0.7937	0.9621	0.7967	0.9798	0.8000
5.0	0.8208	0.9159	0.8170	0.8928	0.8163	0.8887	0.8168	0.8917	0.8217	0.9215	0.8266	0.9536	0.8298	0.9749	0.8333

NTU_1	N=3		N=5		N=7		N=11		N=19		N=39		N=79		N=∞
0.2	0.1636	0.9783	0.1645	0.9846	0.1650	0.9882	0.1655	0.9918	0.1660	0.9951	0.1663	0.9974	0.1665	0.9987	0.1667
0.4	0.2770	0.9577	0.2795	0.9700	0.2810	0.9771	0.2825	0.9845	0.2838	0.9906	0.2847	0.9952	0.2852	0.9976	0.2857
0.6	0.3602	0.9384	0.3646	0.9562	0.3670	0.9664	0.3696	0.9772	0.3717	0.9862	0.3733	0.9930	0.3742	0.9965	0.3750
0.8	0.4241	0.9204	0.4300	0.9428	0.4334	0.9562	0.4370	0.9703	0.4399	0.9819	0.4422	0.9909	0.4433	0.9955	0.4444
1.0	0.4747	0.9037	0.4819	0.9300	0.4862	0.9462	0.4907	0.9634	0.4944	0.9777	0.4972	0.9887	0.4986	0.9944	0.5000
1.5	0.5654	0.8672	0.5744	0.8997	0.5805	0.9224	0.5869	0.9470	0.5920	0.9675	0.5960	0.9835	0.5980	0.9916	0.6000
2.0	0.6263	0.8379	0.6356	0.8721	0.6428	0.8998	0.6506	0.9310	0.6570	0.9575	0.6618	0.9784	0.6642	0.9891	0.6667
3.0	0.7046	0.7952	0.7122	0.8248	0.7203	0.8582	0.7298	0.9005	0.7379	0.9382	0.7439	0.9683	0.7470	0.9839	0.7500
4.0	0.7543	0.7673	0.7590	0.7874	0.7668	0.8219	0.7772	0.8719	0.7862	0.9194	0.7931	0.9583	0.7965	0.9787	0.8000
5.0	0.7892	0.7486	0.7913	0.7584	0.7982	0.7910	0.8086	0.8451	0.8184	0.9012	0.8258	0.9484	0.8296	0.9736	0.8333

Table 6. 1 pass - 1 pass counterflow exchanger: P_1 and F as functions of NTU_1 and even N's for R_1 = 1.25 and 1.50.

R_1	NTU_1	N=4		N=6		N=8		N=10		N=20		N=40		N=80		N=∞
		P_1	F	P_1	F	P_1	F	P_1	F	P_1	F	P_1	F	P_1	F	P_1
1.25	0.16	0.1345	0.9908	0.1346	0.9917	0.1348	0.9929	0.1349	0.9940	0.1352	0.9966	0.1354	0.9982	0.1355	0.9991	0.1356
	0.20	0.1617	0.9887	0.1619	0.9897	0.1621	0.9913	0.1622	0.9925	0.1627	0.9958	0.1629	0.9978	0.1631	0.9989	0.1632
	0.32	0.2322	0.9828	0.2325	0.9842	0.2329	0.9865	0.2332	0.9884	0.2341	0.9934	0.2346	0.9965	0.2349	0.9982	0.2352
	0.40	0.2718	0.9793	0.2720	0.9808	0.2726	0.9835	0.2730	0.9858	0.2742	0.9918	0.2749	0.9957	0.2753	0.9978	0.2757
	0.48	0.3066	0.9761	0.3069	0.9776	0.3075	0.9807	0.3081	0.9833	0.3095	0.9904	0.3104	0.9949	0.3109	0.9974	0.3114
	0.60	0.3517	0.9716	0.3520	0.9729	0.3528	0.9765	0.3534	0.9797	0.3553	0.9882	0.3565	0.9937	0.3571	0.9968	0.3578
	0.64	0.3651	0.9702	0.3654	0.9715	0.3662	0.9752	0.3670	0.9785	0.3689	0.9875	0.3702	0.9934	0.3709	0.9966	0.3716
	0.80	0.4125	0.9651	0.4127	0.9660	0.4137	0.9702	0.4145	0.9740	0.4170	0.9849	0.4185	0.9919	0.4194	0.9958	0.4203
	1.00	0.4604	0.9597	0.4604	0.9597	0.4615	0.9643	0.4625	0.9687	0.4654	0.9816	0.4673	0.9901	0.4683	0.9948	0.4694
	1.20	0.4992	0.9552	0.4989	0.9541	0.5000	0.9589	0.5011	0.9638	0.5044	0.9785	0.5065	0.9884	0.5077	0.9940	0.5090
	1.50	0.5451	0.9496	0.5445	0.9467	0.5455	0.9515	0.5467	0.9570	0.5504	0.9741	0.5528	0.9860	0.5542	0.9927	0.5557
	1.60	0.5580	0.9480	0.5572	0.9445	0.5583	0.9493	0.5594	0.9549	0.5632	0.9727	0.5657	0.9851	0.5672	0.9923	0.5687
	2.00	0.6005	0.9427	0.5993	0.9366	0.6002	0.9409	0.6014	0.9468	0.6053	0.9672	0.6082	0.9821	0.6098	0.9906	0.6115
	2.40	0.6327	0.9387	0.6311	0.9302	0.6318	0.9337	0.6329	0.9396	0.6369	0.9621	0.6399	0.9790	0.6416	0.9890	0.6435
	3.00	0.6683	0.9344	0.6664	0.9226	0.6667	0.9246	0.6677	0.9303	0.6717	0.9550	0.6748	0.9749	0.6766	0.9868	0.6785
	3.20	0.6779	0.9333	0.6758	0.9206	0.6761	0.9221	0.6769	0.9275	0.6809	0.9528	0.6840	0.9736	0.6858	0.9861	0.6878
	4.00	0.7077	0.9298	0.7055	0.9139	0.7055	0.9134	0.7061	0.9179	0.7097	0.9445	0.7128	0.9686	0.7146	0.9833	0.7166
	5.00	0.7329	0.9268	0.7308	0.9079	0.7305	0.9053	0.7308	0.9086	0.7339	0.9354	0.7368	0.9628	0.7386	0.9801	0.7405
1.50	0.20	0.1585	0.9888	0.1586	0.9899	0.1588	0.9914	0.1589	0.9926	0.1594	0.9958	0.1596	0.9978	0.1597	0.9989	0.1599
	0.40	0.2625	0.9800	0.2628	0.9813	0.2632	0.9840	0.2636	0.9862	0.2647	0.9921	0.2653	0.9958	0.2657	0.9978	0.2661
	0.60	0.3361	0.9728	0.3363	0.9740	0.3370	0.9774	0.3376	0.9805	0.3392	0.9887	0.3402	0.9940	0.3408	0.9969	0.3414
	0.80	0.3908	0.9669	0.3909	0.9675	0.3917	0.9715	0.3925	0.9752	0.3945	0.9855	0.3958	0.9922	0.3966	0.9960	0.3974
	1.00	0.4330	0.9621	0.4330	0.9619	0.4338	0.9662	0.4347	0.9705	0.4370	0.9826	0.4386	0.9906	0.4395	0.9951	0.4404
	1.50	0.5054	0.9534	0.5049	0.9506	0.5057	0.9550	0.5066	0.9602	0.5094	0.9761	0.5112	0.9869	0.5123	0.9931	0.5134
	2.00	0.5505	0.9478	0.5496	0.9421	0.5502	0.9460	0.5511	0.9516	0.5540	0.9701	0.5560	0.9836	0.5571	0.9914	0.5584
	3.00	0.6020	0.9413	0.6008	0.9309	0.6010	0.9330	0.6017	0.9384	0.6042	0.9604	0.6061	0.9779	0.6072	0.9882	0.6084
	4.00	0.6287	0.9378	0.6275	0.9240	0.6276	0.9242	0.6280	0.9290	0.6299	0.9523	0.6316	0.9732	0.6325	0.9854	0.6336
	5.00	0.6438	0.9356	0.6427	0.9193	0.6427	0.9178	0.6429	0.9222	0.6444	0.9457	0.6457	0.9690	0.6465	0.9831	0.6474

Table 7. 1 pass - 1 pass counterflow exchanger: P_1 and F as functions of NTU_1 and even N's for R_1 = 2 and 3.

R_1	NTU_1	N=4 P_1	N=4 F	N=6 P_1	N=6 F	N=8 P_1	N=8 F	N=10 P_1	N=10 F	N=20 P_1	N=20 F	N=40 P_1	N=40 F	N=80 P_1	N=80 F	N=∞ P_1
2.0	0.2	0.1522	0.9892	0.1523	0.9902	0.1525	0.9916	0.1526	0.9928	0.1530	0.9959	0.1532	0.9978	0.1533	0.9989	0.1535
	0.4	0.2450	0.9810	0.2452	0.9823	0.2456	0.9848	0.2459	0.9869	0.2468	0.9925	0.2473	0.9960	0.2476	0.9979	0.2479
	0.6	0.3069	0.9748	0.3071	0.9758	0.3076	0.9790	0.3080	0.9818	0.3093	0.9895	0.3100	0.9944	0.3105	0.9971	0.3109
	0.8	0.3505	0.9700	0.3506	0.9704	0.3512	0.9741	0.3517	0.9775	0.3531	0.9869	0.3541	0.9929	0.3546	0.9963	0.3551
	1.0	0.3825	0.9663	0.3825	0.9659	0.3830	0.9698	0.3836	0.9736	0.3851	0.9845	0.3861	0.9917	0.3867	0.9957	0.3873
	1.5	0.4328	0.9598	0.4325	0.9573	0.4329	0.9611	0.4335	0.9657	0.4350	0.9794	0.4360	0.9888	0.4366	0.9942	0.4372
	2.0	0.4601	0.9560	0.4597	0.9513	0.4600	0.9548	0.4604	0.9595	0.4617	0.9752	0.4626	0.9865	0.4632	0.9929	0.4637
	3.0	0.4852	0.9518	0.4848	0.9439	0.4849	0.9460	0.4851	0.9509	0.4860	0.9689	0.4865	0.9829	0.4869	0.9908	0.4872
	4.0	0.4943	0.9495	0.4941	0.9393	0.4941	0.9404	0.4942	0.9450	0.4947	0.9639	0.4950	0.9798	0.4952	0.9889	0.4954
	5.0	0.4978	0.9482	0.4977	0.9363	0.4977	0.9364	0.4977	0.9415	0.4979	0.9604	0.4981	0.9779	0.4982	0.9879	0.4983
3.0	0.2	0.1405	0.9898	0.1406	0.9907	0.1407	0.9921	0.1408	0.9932	0.1411	0.9962	0.1413	0.9980	0.1414	0.9990	0.1415
	0.4	0.2140	0.9830	0.2141	0.9841	0.2144	0.9863	0.2146	0.9882	0.2151	0.9933	0.2155	0.9964	0.2157	0.9981	0.2159
	0.6	0.2567	0.9783	0.2568	0.9791	0.2571	0.9819	0.2574	0.9843	0.2580	0.9910	0.2585	0.9952	0.2587	0.9975	0.2589
	0.8	0.2831	0.9750	0.2831	0.9753	0.2834	0.9784	0.2837	0.9812	0.2843	0.9891	0.2847	0.9942	0.2850	0.9971	0.2582
	1.0	0.3000	0.9726	0.3000	0.9723	0.3002	0.9754	0.3004	0.9785	0.3010	0.9875	0.3014	0.9933	0.3016	0.9965	0.3018
	1.5	0.3210	0.9690	0.3209	0.9673	0.3210	0.9703	0.3211	0.9738	0.3215	0.9847	0.3218	0.9918	0.3219	0.9958	0.3221
	2.0	0.3287	0.9669	0.3286	0.9640	0.3287	0.9669	0.3287	0.9706	0.3289	0.9825	0.3219	0.9907	0.3292	0.9953	0.3292
	3.0	0.3327	0.9649	0.3326	0.9604	0.3326	0.9624	0.3327	0.9662	0.3327	0.9794	0.3327	0.9893	0.3328	0.9943	0.3328
	4.0	0.3332	0.9624	0.3332	0.9590	0.3332	0.9595	0.3332	0.9622	0.3332	0.9758	0.3333	0.9883	0.3333	0.9916	0.3333
	5.0	0.3333	0.9683	0.3333	0.9661	0.3333	0.9603	0.3333	0.9708	0.3333	0.9802	0.3333	0.0034	0.3333	1.0084	0.3333

Table 8. 2 pass - 1 pass exchanger: P_1 and F as functions of NTU_1 and N for R_1 = 2, 3 and 4.

R_1	NTU_1	N=3 P_1	N=3 F	N=7 P_1	N=7 F	N=11 P_1	N=11 F	N=19 P_1	N=19 F	N=39 P_1	N=39 F	N=79 P_1	N=79 F	N=∞ P_1	N=∞ F
2	0.2	0.1511	0.9801	0.1510	0.9789	0.1512	0.9812	0.1515	0.9835	0.1518	0.9856	0.1519	0.9866	0.1520	0.9878
	0.4	0.2394	0.9452	0.2388	0.9411	0.2394	0.9446	0.2400	0.9486	0.2406	0.9522	0.2409	0.9542	0.2412	0.9563
	0.6	0.2944	0.8997	0.2934	0.8941	0.2941	0.8983	0.2950	0.9032	0.2958	0.9079	0.2962	0.9104	0.2967	0.9132
	0.8	0.3300	0.8481	0.3292	0.8434	0.3300	0.8479	0.3310	0.8535	0.3320	0.8588	0.3325	0.8618	0.3331	0.8650
	1.0	0.3540	0.7938	0.3538	0.7930	0.3547	0.7978	0.3558	0.8038	0.3569	0.8095	0.3574	0.8126	0.3581	0.8161
	1.5	0.3866	0.6631	0.3899	0.6796	0.3911	0.6853	0.3924	0.6921	0.3936	0.6983	0.3943	0.7018	0.3950	0.7055
	2.0	0.4009	0.5531	0.4090	0.5890	0.4106	0.5965	0.4122	0.6041	0.4136	0.6109	0.4144	0.6147	0.4151	0.6186
	3.0	0.4112	0.3995	0.4281	0.4602	0.4311	0.4725	0.4334	0.4827	0.4353	0.4913	0.4363	0.4958	0.4373	0.5006
	4.0	0.4139	0.3062	0.4368	0.3737	0.4415	0.3908	0.4448	0.4040	0.4474	0.4146	0.4487	0.4203	0.4500	0.4261
	5.0	0.4147	0.2465	0.4412	0.3118	0.4476	0.3324	0.4519	0.3481	0.4551	0.3607	0.4567	0.3674	0.4583	0.3744
3	0.2	0.1391	0.9757	0.1387	0.9721	0.1390	0.9744	0.1392	0.9772	0.1395	0.9797	0.1396	0.9810	0.1398	0.9824
	0.4	0.2077	0.9288	0.2069	0.9223	0.2074	0.9260	0.2079	0.9305	0.2084	0.9348	0.2087	0.9371	0.2090	0.9397
	0.6	0.2443	0.8664	0.2438	0.8625	0.2444	0.8673	0.2451	0.8732	0.2457	0.8787	0.2461	0.8818	0.2465	0.8852
	0.8	0.2647	0.7958	0.2654	0.8010	0.2661	0.8070	0.2669	0.8141	0.2677	0.8206	0.2681	0.8243	0.2685	0.8282
	1.0	0.2766	0.7236	0.2789	0.7422	0.2798	0.7499	0.2807	0.7583	0.2816	0.7658	0.2820	0.7699	0.2825	0.7743
	1.5	0.2896	0.5631	0.2964	0.6165	0.2980	0.6300	0.2993	0.6424	0.3004	0.6529	0.3010	0.6585	0.3016	0.6644
	2.0	0.2937	0.4453	0.3043	0.5197	0.3067	0.5398	0.3085	0.5575	0.3100	0.5719	0.3107	0.5796	0.3115	0.5876
	3.0	0.2956	0.3047	0.3106	0.3853	0.3143	0.4145	0.3173	0.4421	0.3195	0.4663	0.3206	0.4797	0.3217	0.4941
	4.0	0.2959	0.2293	0.3125	0.2995	0.3172	0.3309	0.3210	0.3639	0.3239	0.3967	0.3253	0.4168	0.3267	0.4406
	5.0	0.2959	0.1836	0.3131	0.2425	0.3184	0.2720	0.3227	0.3060	0.3261	0.3441	0.3278	0.3704	0.3295	0.4059
4	0.2	0.1283	0.9712	0.1278	0.9655	0.1281	0.9681	0.1283	0.9712	0.1286	0.9742	0.1287	0.9757	0.1288	0.9774
	0.4	0.1815	0.9117	0.1809	0.9049	0.1812	0.9091	0.1817	0.9144	0.1822	0.9194	0.1824	0.9222	0.1827	0.9253
	0.6	0.2055	0.8310	0.2058	0.8343	0.2063	0.8405	0.2069	0.8477	0.2074	0.8544	0.2077	0.8581	0.2081	0.8623
	0.8	0.2169	0.7412	0.2187	0.7636	0.2194	0.7725	0.2202	0.7820	0.2208	0.7904	0.2211	0.7951	0.2215	0.8001
	1.0	0.2226	0.6534	0.2261	0.6973	0.2270	0.7100	0.2279	0.7222	0.2286	0.7326	0.2290	0.7383	0.2293	0.7444
	1.5	0.2276	0.4786	0.2345	0.5588	0.2361	0.5820	0.2374	0.6029	0.2383	0.6204	0.2388	0.6297	0.2393	0.6395
	2.0	0.2287	0.3670	0.2376	0.4551	0.2397	0.4862	0.2414	0.5159	0.2427	0.5420	0.2433	0.5569	0.2439	0.5729
	3.0	0.2290	0.2464	0.2393	0.3202	0.2422	0.3542	0.2445	0.3923	0.2462	0.4334	0.2470	0.4610	0.2479	0.4979
	4.0	0.2290	0.1849	0.2397	0.2427	0.2428	0.2721	0.2453	0.3082	0.2473	0.3533	0.2483	0.3903	0.2492	0.4580
	5.0	0.2290	0.1479	0.2397	0.1946	0.2429	0.2190	0.2456	0.2503	0.2477	0.2923	0.2487	0.3309	0.2497	0.4334

Table 9. 3 pass - 1 pass exchanger with counterflow at inlet: P_1 and F as functions of NTU_1 and N for R_1 = 3, 4.5 and 6.

R_1	NTU_1	N=5		N=11		N=23		N=41		N=77		N=∞	
		P_1	F	P_1	F	P_1	F	P_1	F	P_1	F	P_1	F
3	0.2	0.1389	0.9741	0.1393	0.9774	0.1396	0.9807	0.1398	0.9824	0.1399	0.9835	0.1400	0.9849
	0.4	0.2068	0.9216	0.2081	0.9322	0.2090	0.9397	0.2095	0.9435	0.2097	0.9460	0.2101	0.9489
	0.6	0.2429	0.8552	0.2454	0.8761	0.2469	0.8887	0.2476	0.8947	0.2481	0.8987	0.2486	0.9034
	0.8	0.2634	0.7852	0.2672	0.8167	0.2693	0.8347	0.2702	0.8431	0.2708	0.8485	0.2715	0.8550
	1.0	0.2760	0.7183	0.2808	0.7592	0.2833	0.7818	0.2845	0.7927	0.2852	0.7995	0.2860	0.8078
	1.5	0.2922	0.5820	0.2985	0.6348	0.3018	0.6668	0.3033	0.6819	0.3042	0.6919	0.3053	0.7043
	2.0	0.3004	0.4892	0.3069	0.5419	0.3106	0.5782	0.3122	0.5962	0.3133	0.6084	0.3145	0.6237
	3.0	0.3100	0.3817	0.3151	0.4217	0.3188	0.4581	0.3205	0.4785	0.3216	0.4930	0.3229	0.5122
	4.0	0.3163	0.3239	0.3195	0.3499	0.3227	0.3819	0.3243	0.4023	0.3245	0.4179	0.3267	0.4395
	5.0	0.3206	0.2882	0.3224	0.3028	0.3250	0.3295	0.3265	0.3489	0.3275	0.3642	0.3287	0.3878
4.5	0.2	0.1231	0.9661	0.1234	0.9695	0.1237	0.9736	0.1239	0.9757	0.1240	0.9771	0.1241	0.9789
	0.4	0.1696	0.8958	0.1707	0.9107	0.1715	0.9207	0.1719	0.9257	0.1721	0.9291	0.1724	0.9331
	0.6	0.1893	0.8100	0.1915	0.8412	0.1927	0.8588	0.1932	0.8673	0.1936	0.8728	0.1940	0.8796
	0.8	0.1988	0.7248	0.2018	0.7713	0.2032	0.7970	0.2039	0.8091	0.2043	0.8172	0.2048	0.8269
	1.0	0.2040	0.6495	0.2074	0.7069	0.2090	0.7396	0.2097	0.7553	0.2102	0.7656	0.2107	0.7784
	1.5	0.2104	0.5144	0.2139	0.5790	0.2156	0.6233	0.2164	0.6459	0.2168	0.6613	0.2174	0.6814
	2.0	0.2138	0.4334	0.2166	0.4916	0.2183	0.5404	0.2189	0.5672	0.2194	0.5867	0.2199	0.6132
	3.0	0.2176	0.3454	0.2192	0.3853	0.2204	0.4331	0.2209	0.4640	0.2212	0.4887	0.2215	0.5276
	4.0	0.2196	0.2986	0.2204	0.3236	0.2212	0.3661	0.2216	0.3976	0.2218	0.4254	0.2220	0.4767
	5.0	0.2207	0.2695	0.2210	0.2835	0.2216	0.3198	0.2219	0.3512	0.2220	0.3797	0.2221	0.4428
6	0.2	0.1098	0.9583	0.1100	0.9623	0.1103	0.9671	0.1104	0.9697	0.1106	0.9715	0.1107	0.9737
	0.4	0.1416	0.8710	0.1426	0.8916	0.1433	0.9047	0.1436	0.9112	0.1438	0.9155	0.1440	0.9207
	0.6	0.1526	0.7677	0.1543	0.8109	0.1551	0.8345	0.1555	0.8458	0.1558	0.8535	0.1561	0.8627
	0.8	0.1571	0.6723	0.1592	0.7329	0.1602	0.7669	0.160[illegible]	0.7836	0.1608	0.7947	0.1612	0.8084
	1.0	0.1595	0.5947	0.1616	0.6641	0.1626	0.7066	0.1630	0.7279	0.1633	0.7425	0.1636	0.7610
	1.5	0.1625	0.4687	0.1642	0.5362	0.1650	0.5902	0.1653	0.6201	0.1655	0.6420	0.1658	0.6727
	2.0	0.1641	0.3979	0.1652	0.4540	0.1658	0.5104	0.1661	0.5447	0.1662	0.5730	0.1664	0.6158
	3.0	0.1656	0.3220	0.1660	0.3573	0.1664	0.4086	0.1665	0.4458	0.1666	0.4795	0.1666	0.5501
	4.0	0.1662	0.2817	0.1663	0.3024	0.1665	0.3455	0.1666	0.3820	0.1666	0.4180	0.1667	0.5665
	5.0	0.1664	0.2568	0.1665	0.2672	0.1666	0.3024	0.1666	0.3369	0.1667	0.3688	0.1667	0.4532

Table 10. 3 pass - 1 pass exchanger with parallel flow at inlet: P_1 and F as functions of NTU_1 and N for R_1 = 3,4.5 and 6.

R_1	NTU_1	N= 5		N= 11		N= 23		N= 41		N= 77		N= ∞	
		P_1	F	P_1	F	P_1	F	P_1	F	P_1	F	P_1	F
3.00	0.20	0.1389	0.9738	0.1390	0.9746	0.1392	0.9771	0.1394	0.9785	0.1395	0.9794	0.1396	0.9806
	0.40	0.2066	0.9195	0.2069	0.9223	0.2075	0.9271	0.2078	0.9297	0.2080	0.9315	0.2083	0.9337
	0.60	0.2421	0.8484	0.2431	0.8567	0.2440	0.8641	0.2445	0.8680	0.2448	0.8705	0.2452	0.8737
	0.80	0.2616	0.7704	0.2637	0.7873	0.2650	0.7976	0.2656	0.8028	0.2660	0.8061	0.2665	0.8103
	1.00	0.2727	0.6926	0.2761	0.7197	0.2778	0.7332	0.2785	0.7396	0.2790	0.7437	0.2796	0.7488
	1.50	0.2837	0.5236	0.2911	0.5740	0.2939	0.5952	0.2950	0.6046	0.2957	0.6106	0.2966	0.6179
	2.00	0.2855	0.4013	0.2969	0.4654	0.3008	0.4925	0.3024	0.5045	0.3034	0.5121	0.3045	0.5214
	3.00	0.2823	0.2574	0.3005	0.3267	0.3068	0.3608	0.3092	0.3762	0.3107	0.3863	0.3124	0.3988
	4.00	0.2776	0.1829	0.3007	0.2458	0.3093	0.2826	0.3125	0.2999	0.3144	0.3116	0.3167	0.3268
	5.00	0.2733	0.1396	0.2998	0.1941	0.3106	0.2312	0.3145	0.2493	0.3168	0.2621	0.3194	0.2793
4.50	0.20	0.1231	0.9656	0.1231	0.9655	0.1233	0.9683	0.1234	0.9700	0.1235	0.9711	0.1236	0.9726
	0.40	0.1693	0.8924	0.1696	0.8967	0.1701	0.9028	0.1704	0.9061	0.1706	0.9084	0.1708	0.9113
	0.60	0.1885	0.7983	0.1897	0.8147	0.1904	0.8251	0.1908	0.8304	0.1910	0.8340	0.1913	0.8383
	0.80	0.1970	0.6987	0.1993	0.7316	0.2003	0.7475	0.2008	0.7550	0.2011	0.7599	0.2014	0.7659
	1.00	0.2009	0.6050	0.2043	0.6544	0.2056	0.6758	0.2062	0.6857	0.2066	0.6920	0.2070	0.6997
	1.50	0.2031	0.4239	0.2094	0.4992	0.2115	0.5325	0.2124	0.5478	0.2129	0.5578	0.2135	0.5702
	2.00	0.2021	0.3108	0.2110	0.3926	0.2138	0.4333	0.2149	0.4529	0.2156	0.4662	0.2163	0.4837
	3.00	0.1986	0.1924	0.2114	0.2652	0.2156	0.3113	0.2171	0.3359	0.2180	0.3547	0.2191	0.3830
	4.00	0.1954	0.1356	0.2107	0.1947	0.2162	0.2400	0.2180	0.2661	0.2191	0.2880	0.2205	0.3277
	5.00	0.1929	0.1034	0.2098	0.1511	0.2163	0.1935	0.2185	0.2191	0.2197	0.2420	0.2212	0.2929
6.00	0.20	0.1097	0.9578	0.1097	0.9570	0.1099	0.9603	0.1100	0.9623	0.1101	0.9636	0.1102	0.9655
	0.40	0.1413	0.8659	0.1418	0.8740	0.1422	0.8818	0.1424	0.8860	0.1425	0.8889	0.1427	0.8926
	0.60	0.1517	0.7491	0.1530	0.7776	0.1536	0.7924	0.1539	0.7996	0.1540	0.8044	0.1543	0.8103
	0.80	0.1554	0.6315	0.1575	0.6834	0.1583	0.7062	0.1587	0.7169	0.1589	0.7238	0.1592	0.7323
	1.00	0.1567	0.5287	0.1597	0.5988	0.1606	0.6293	0.1611	0.6434	0.1613	0.6527	0.1616	0.6642
	1.50	0.1566	0.3515	0.1615	0.4394	0.1629	0.4827	0.1635	0.5043	0.1638	0.5195	0.1642	0.5398
	2.00	0.1554	0.2522	0.1619	0.3371	0.1637	0.3854	0.1644	0.4117	0.1648	0.4321	0.1653	0.4630
	3.00	0.1527	0.1544	0.1616	0.2216	0.1642	0.2693	0.1651	0.2975	0.1656	0.3234	0.1662	0.3789
	4.00	0.1506	0.1089	0.1610	0.1605	0.1643	0.2038	0.1653	0.2296	0.1658	0.2551	0.1665	0.3350
	5.00	0.1490	0.0834	0.1603	0.1236	0.1643	0.1626	0.1653	0.1856	0.1659	0.2082	0.1666	0.3082

Effect of Association on Diffusion of Diethylamine in Toluene at Low Solute Concentrations

S. SRINIVASAN and G. S. LADDHA
A. C. College of Technology, Madras, India

Predicted diffusivities of diethylamine in toluene are in good agreement with experimental data at low solute concentrations taking into account the preferential formation of pentamer complexes of the solute molecules due to association.

1. INTRODUCTION

The semi-empirical correlations available in literature for the prediction of diffusivity often leads to large errors for systems involving association due to hydrogen bonding. Kuppuswamy and Laddha [1] have reported that diffusivity data may be predicted for simple solute-solvent complexes. Kumanan et al [2] have observed that diffusivity of acetic acid in inert solvents may be predicted by taking into account the association of acetic acid as dimer. A study of diffusion of inert and hydrogen bonding solutes in aliphatic alcohols has been reported by Lusis and Ratcliff [3] indicating that solute-solvent association due to hydrogen bonding influences diffusivity values. The association of alcohols in nonpolar solvents has been postulated with the formation of linear and cyclic n-mers due to hydrogen bonding. The experimental data on the diffusivity of ethanol in carbon tetrachloride indicates the preferential formation pentamer complexes of the solute molecules due to association [4]. Amines as solutes have a tendency to associate through N-H bonding. Liddel and Wuff [5] have reported IR spectra data for various aliphatic and aromatic amines for the presence of N-H groups but there is no agreed physical picture for the association of amines. This investigation reports the effect of association on the diffusivity of diethyl amine in toluene at low solute concentrations.

2. LIQUID DIFFUSIVITY MEASUREMENT

Diaphragm cell technique as reported by Laddha et al [6-8] was used to measure diffusivities experimentally by the following relationship

$$D_{AB} = \frac{1}{\beta t} \ln \Delta C^{\circ}_{A} \; / \; \Delta C^{F}_{A} \qquad (1)$$

where ΔC°_{A} and ΔC^{F}_{A} are the initial concentration and final differences of the solute across the diaphragm in the two compartments of the diaphragm cell, β is the cell constant and t the time of diffusion.

The diffusion coefficient defined by Eqn.1 represents the time averaged integral diffusion coefficient. Gordon [9] has shown that if the original concentration is small, the integral diffusion coefficient calculated by using a diaphragm cell is approximately equal to the true or differential diffusion coefficient if bulk transport due to volume changes or mixing can be neglected. The use of Eqn.1 is subject to the following conditions: (i) quasi-steady state in the diaphragm (ii) cell constant β remaining constant (iii) solution in the two compartments being homogeneous and (iv) mechanism of transport is by diffusion and not by surface transport. These requirements were assumed to be met by the following considerations. The use of a undirectional form of diffusion equation has been shown by Toor [10] to be rigorously valid. The assumption of quasi or pseudo-steady state condition in the pores - i.e., a linear concentration gradient was proved to be valid by Barnes [11] as long as the following two conditions are met: (i) the total pore volume is less than 1% of the total cell volume and (ii) the preliminary diffusion period has lasted long enough for the effect of the original concentration in the diaphragm to disappear which takes about 4 to 5 hours with a D value of 2×10^{-5} sq.cm. per second. The first condition is usually met by using a diaphragm of normal dimensions and the second condition by proper experimentation. The constancy of the cell factor, β, was checked by a number of experiments with 0.1N HCl diffusing into pure water at a temperature of 30°C. The solutions in the two compartments were assumed to be homogeneous, since with the denser solution in the top chamber, the solutions in the two compartments tended to be density -stirred. It was, however, found necessary to have the diaphragm placed perfectly horizontal. The assumption of diffusion controlled transport will be invalidated by a defective diaphragm having a relatively enormous area of the pores, which may permit streaming, the possibility of such transport can be neglected if a proper diaphragm is used.

3. EXPERIMENTAL

The cells consisted of two equal compartments separated by a No.4 sintered glass diaphragm (porosity 5 to 10 microns) about 30 mm in diameter and 2 to 3 mm. thick. Each compartment was provided with a filling and draining capillary tube with stopcocks. In order to measure the diffusion coefficient, the upper chamber of the vertical cell and the diaphragm were filled with the denser solution and the lower chamber with the less dense solution. The whole cell was provided with an outside jacket with a narrow annulus for flow of water at constant temperature (at 30°C from a thermostat) during the experimental runs. The initial concentration of the solute in the chamber from which diffusion occured was C^{0}_{1A}. The initial concentration of the solute in the other compartment was taken to be zero. Diffusion was then allowed to occur for a preliminary period of 4 to 5 hours to establish the concentration gradient across the diaphragm.

Then the cells were emptied and refilled with fresh solutions to begin the experiment. It was assumed that the solutions in the compartments were density-stirred, so that the concentration was uniform throughout each compartment. After a known time the solutions from each compartment were drained and analysed.

The cell constant, β, was found by calibrating the cells with 0.1N HCl diffusing into pure water at a temperature of 30°C, using the diffusivity value of 3.078×10^{-5} cm^2/sec. from Stokes data [12] Amine in solution was estimated by titration with standard HCl solution using Alizarin red as indicator. The Alizarin red indicator was prepared by dissolving the dye alizarin red in methanol ad 1% solution.

4. EFFECT OF ASSOCIATION OF THE SOLUTE

Consider the diffusion of an associating solute A into an inert Solvent B. The solute in pure state consists of monomers A and polymer complexes $A_2, A_3 \ldots\ldots A_n$ (n indicating the number of monomer units in complex A_n). The binary mixture will, therefore, consist of solute monomers, solute polymers and the inert solvent. The molar flux of the solute is related to the experimentally measured diffusion coefficient D_{AB} by the following Fick's relationship

$$J_A = -D_{AB} \quad \Delta C_A \tag{2}$$

Since the binary mixture consists of solute monomers and various polymers, J_A in Eqn.2 may be expressed as

$$J_A = J_{A_1} + J_{A_2} + J_{A_3} + \ldots\ldots\ldots + J_{A_n} \tag{3}$$

The above relationship gives the following expression for the mutual diffusion coefficient D_{AB}

$$D_{AB} = D_{A_1} \frac{dX_{A_1}}{dX_A} + D_{A_2} \frac{dX_{A_2}}{dX_A} + \ldots D_{An} \frac{dX_{A_n}}{dX_A} \tag{4}$$

where $X_{A_1}, X_{A_2} \ldots\ldots X_{A_n}$ are the concentrations and $D_{A_1}, D_{A_2} \ldots D_{A_n}$ are the diffusivities of the monomer, dimer and n-mer respectively. A mass balance on the solute gives

$$X_A = X_{A_1} + X_{A_2} + \ldots\ldots\ldots X_{A_n} \tag{5}$$

We may further assume that n-mer is formed when a monomer comes into contact with (n-1) th order of the polymer as indicated below

$$A_1 + A_{n-1} \rightleftarrows A_n \tag{6}$$

We may assume that all degrees of association are possible and the association constant k for the dimer and n-mer are related as follows

$$X_{A_2} = k_2 (X_{A_1})^2 \text{ and } X_{An} = k_n (X_{A_1}) (X_{A_n} - 1) \tag{7}$$

In such a case it can be seen that

$$X_{An} = (k_2 k_3 \ldots\ldots k_n) (X_{A_1})n \tag{8}$$

$$dX_{An} = n(k_2 k_3 \ldots\ldots k_n) (X_{A_1})^{n-1} dX_{A_1} \tag{9}$$

$$dX_A = dX_{A_1} [1 + 2 k_2 X_{A_1} + \ldots + nk_2 k_3 \ldots\ldots k_n x^{n-1}_{A_1}] \tag{10}$$

In the light of Eqns. (8), (9) and (10) Eqn. (4) gives the following relationship for D_{AB}

$$D_{AB} = \frac{D_{A_1} + 2 k_2 X_{A_1} D_{A_2} + \ldots\ldots\ldots + nk_2 k_3 \ldots k_n X^{n-1}_{A_1} D_{A_n}}{1 + 2 k_2 X_{A_1} + \ldots\ldots + nk_2 k_3 \ldots k_n X^{n-1}_{A_1}} \tag{11}$$

where X_{A_1} is related to the original concentration X_A by

$$X_A = X_{A_1} [1 + k_2 X_{A_1} + k_2 k_3 X^2_{A_1} + \ldots\ldots + k_2 k_3 \ldots k_n X^{n-1}_{A_1}] \tag{12}$$

If we now postulate that one of the polymers, n-mer is formed preferentially, then the concentrations of dimer, trimer etc., will be negligible as compared to the concentration X_{An}. In such a case Eqns. (11) and (12) may be simplified as indicated below

$$D_{AB} = \frac{D_{A_1} + n k_n X^{n-1}_{A_1} D_{A_n}}{1 + n k_n X^{n-1}_{A_1}} \tag{13}$$

$$\text{where } k_n = (k_2 k_3 \ldots\ldots\ldots k_n) \tag{14}$$

$$\text{and} \quad X_A = X_{A_1} [1 + K_n X^{n-1}_{A_1}] \tag{15}$$

5. ANALYSIS OF THE EXPERIMENTAL DATA

The experimental data as obtained in this investigation on the diffusivity of diethyl amine in toluene were analysed according to the relationships 13 and 15 to examine the possibility of preferential formation of pentamer, hexamer and heptamer polymers with diethyl amine as the solute. Diffusivity D_A of diethyl amine solute monomer was estimated by relationship given by Kuppuswamy and Laddha[1].

$$D_{A_1} = 2.0 \times 10^{-7} \frac{T}{\eta_B} (1/VA)^{1/3} (VB/VA)^{0.16} \tag{16}$$

D_{A_n}, the diffusivity of the solute polymer was obtained by Eqn. (16) by replacing V_A by nV_A and is therefore related to D_{A_1} as indicated below

$$D_{A_n} = D_{A_1} \left(\frac{1}{n}\right)^{0.493} \tag{17}$$

where n is the order of the solute polymer formed. The values of the association constant K_n were calculated alongwith monomer concentration X_{A_1} at different solute concentrations in the range of 0.010 to 0.10 mole fraction. The results are summarized in Table 1. It may be observed that the calculated values of the association constant K_5 are approximately constant over the solute concentration range of 0.01 to 0.10 mole fraction as compared to the values of K_6 and K_7 assuming hexamer and heptamer complexes of the solute diethyl amine. This indicates the preferential formation of pentamer complexes due to association of diethyl amine molecules while diffusing into the inert solvent toluene. Taking the average value of K_5 as 2.0 x 10^{-5} at which the predicted values of diffusivities show the least average deviation δ_{av}, the calculated values of diffusivity D_{AB} using Eqns. 13 and 15 are compared with the experimental data at various solute concentrations in Table.2

The average deviation δ_{av} of the experimental data from the predicted values was calculated as follows:

$$\delta_{av} = \sqrt{\frac{\Sigma \delta^2}{N}}$$

where $$\delta = \frac{D_{exp} - D_{cal}}{D_{cal}} \times 100$$

and N = No of data points.

The value of δ_{av} thus estimated was found to b $\pm$ 3.02% indicating that the predicted values ar in good agreement with experimental data.

Table 1. Estimated values of Association constants at 30°C

[At 30°C

$D_{A_1} = 2.4350 \times 10^{-5}$; $D_{A_5} = 1.1013 \times 10^{-5}$

$D_{A_6} = 1.006 \times 10^{-5}$; $D_{A_7} = 0.9330 \times 10^{-5}$

$\eta_B = 0.521$ cp. $V_A = 159.9$ $V_B = 118.2$]

Solute Concentration X_A mole fraction	Assumed Pentamer formation		Assumed hexamer formation		Assumed heptamer formation	
	Association Constant K_5	Calculated X_{A_1}	Association Constant K_6	Calculated X_{A_1}	Association Constant K_7	Calculated X_{A_1}
0.0102	17.8×10^5	0.0101	1.34×10^8	0.0101	1.06×10^{10}	0.0101
0.0205	2.23×10^5	0.0198	8.4×10^6	0.0200	3.29×10^8	0.0200
0.0307	2.75×10^5	0.0269	6.49×10^6	0.0278	1.63×10^8	0.0283
0.0410	2.67×10^5	0.0320	4.27×10^6	0.0342	7.63×10^7	0.0355
0.0512	1.47×10^5	0.0386	1.83×10^6	0.0417	2.56×10^7	0.0435
0.0614	1.21×10^5	0.0432	1.17×10^6	0.0477	1.63×10^7	0.0495
0.0819	1.17×10^5	0.0490	0.69×10^6	0.0573	5.19×10^6	0.0625
0.1023	3.11×10^5	0.0450	0.88×10^6	0.0602	4.19×10^6	0.0695

Table 2: Comparison of Experimental values with calculated values assuming Pentamer Association complexes at 30°C with $K_5 = 2.0 \times 10^5$

mole fraction diethyl amine	D_{AB} Experimented value x 10^5	D_{AB} calculated value x 10^5	error % $\frac{D_{exp} - D_{cal}}{D_{cal}} \times 100$
0.0102	2.322	2.420	-4.03
0.0205	2.238	2.255	-0.75
0.0307	1.876	1.947	-3.65
0.0410	1.670	1.708	-3.00
0.0512	1.608	1.561	-2.96
0.0614	1.530	1:468	4.16
0.0819	1.407	1.363	3.25
0.1023	1.282	1.305	-1.77

NOMENCLATURE

C = molal concentration mol/cm^3
D = diffusion coefficient
J = total molal flux mol cm^{-2} s^{-1}
k = association constant mol $fraction^{-1}$
K_n = association constant as defined by Eq.14
X = concentration of solute, mol fraction
t = time of diffusion, s
V = Le Bas molar volume at normal boiling point, cm^3/mol
η = viscosity, cp.
β = cell constant

Subscripts

A,B = solute and solvent species respectively
AB = of solute in solvent
A_1 = solute monomer
A_n = solute n-mer

Superscripts

O,F = initial and final respectively

REFERENCES:

1. Kuppuswamy, R, and Laddha G.S., Indian J Technol 17 (1979) 35.
2. Kumanan, V.N., Karthiyayini, R., and Laddha G.S., Chem. Engg. Communications 19 (1983) 7.
3. Lusis, M.A., and Ratcliff, G.A., AIChE.J. 17 (1971) 1492.
4. Karthiyayini, R., Raghunathan, V.C., and Laddha G.S., Indian J. Technol 22 (1984) 121.
5. Urner Liddel and Oliver R.Wuff., J.Am.Chem. Soc. 55 (1933) 3574.
6. Krishnan, K.S., and Laddha, G.S., Indian Chem. Engs. 7 (1965) 83.
7. Amourdam, M.J. and Laddha, G.S., J.Chem.Eng. Data 12 (1967) 389.
8. Laddha, G.S., current Science No.16, 51 (1982) 763.
9. Gordon, A.R., J.Chem Phys.5 (1937) 522.
10. Toor, H.L., J.Phys. Chem.64 (1960) 1580.
11. Barnes, C., Physics 5 (1934) 5.
12. Stokes, R.H., Dunlop, P.J., and Hall, J.R., Trans Faraday Soc. 49 (1953) 886.

Heat Transfer from Rectangular Vertical Fin Arrays

S. P. SUKHATME
Department of Mechanical Engineering
Indian Institute of Technology, Bombay, India

1. INTRODUCTION

Rectangular fin arrays are used in a variety of applications to dissipate heat to the surroundings. Studies on the heat transfer and fluid flow patterns associated with such arrays are therefore of considerable engineering significance. In a previous article, [1] attention was primarily focussed on the natural convection heat transfer from rectangular fin arrays on a horizontal base. In this paper, we will be concerned with the problem of analyzing the laminar natural convection heat transfer and the radiative heat transfer from rectangular fin arrays fixed on a vertical face (Fig.1). We will first consider the effects separately (Sections 2 and 3) and then simultaneously (Section 4).

2. NATURAL CONVECTION HEAT TRANSFER FROM RECTANGULAR VERTICAL FIN ARRAYS

2.1 Experimental Investigations

The first investigation on natural convection was due to Starner and McManus [2] who showed that heated arrays gave rise to a single chimney pattern in each channel, inflow occuring from the bottom and from the side, and outflow occurring through the top. Welling and Woolridge [3] presented results for Nusselt numbers based on the hydraulic radius of the fin arrays. Chaddock [4] made a comparison of the work done by earlier investigators and his own experimentally determined Nusselt numbers for fin arrays of different geometries. He reported data for modified Rayleigh numbers ranging from 1 to 10^4 and proposed a correlation between $\overline{Nu}_S$ and Ra_S^*. Aihara [5-7] also carried out extensive work on vertical fin arrays. The data and correlations of the above mentioned investigators are shown in Fig.2. It will be noted that there is a fairly wide variation in the results of the various investigators.

2.2 Analysis

The first simplified numerical solution of the problem was due to Saikhedkar [8,9]. A significant assumption involved in his analysis is that the velocity component normal to the fin flat is negligible and can be assumed to be zero. This assumption, made on the basis of the experimental observations of earlier investigators, helped to simplify the problem considerably. The existence of only two velocity components made it possible to obtain a solution using the stream function and vorticity formulation.

However, subsequently, Shalaby [10,11] has solved the problem without neglecting the third velocity component. This involves the solution of the full set of Navier-Stokes equations governing the three components of velocity, pressure, and temperature. In such a case, the vorticity-stream function formulation cannot be used, and one has to attempt a solution of the equations governing the primitive variables directly.

2.2.1. Problem Formulation

The vertical fin array analysed is assumed to consist of a large number of vertical fins of height H, length L and a small negligible thickness 2t jutting out from a common vertical base plate. The spacing between two adjacent fin flats is S. Two adjacent fin flats and the fin base in between constitute one fin channel of the array. The following assumptions are also made:

1. The material of the fin array is assumed to have a high thermal conductivity. Thus the entire fin array is considered to be isothermal at a temperature T_h. It is assumed to be surrounded by a fluid (air) at a temperature T_∞.

2. Symmetry exists in the longitudinal direction and the problem therefore reduces to analysing one half of one fin channel domain (Fig. 1b).

The analytical solution of this problem requires the simultaneous solution of the equations of continuity, momentum, and energy. These equations expressed in dimensionless from are as follows:

$$\frac{\partial U}{\partial X} + \frac{\partial V}{\partial Y} + \frac{\partial W}{\partial Z} = 0 \tag{1}$$

$$U\frac{\partial U}{\partial X} + V\frac{\partial U}{\partial Y} + W\frac{\partial U}{\partial Z} = -\frac{\partial P}{\partial X} + \left(\frac{\partial^2 U}{\partial X^2} + \frac{\partial^2 U}{\partial Y^2} + \frac{\partial^2 U}{\partial Z^2}\right) + Gr_H\,\Theta \tag{2}$$

$$U\frac{\partial V}{\partial X} + V\frac{\partial V}{\partial Y} + W\frac{\partial V}{\partial Z} = -\frac{\partial P}{\partial Y} + \left(\frac{\partial^2 V}{\partial X^2} + \frac{\partial^2 V}{\partial Y^2} + \frac{\partial^2 V}{\partial Z^2}\right) \tag{3}$$

$$U\frac{\partial W}{\partial X} + V\frac{\partial W}{\partial Y} + W\frac{\partial W}{\partial Z} = -\frac{\partial P}{\partial Z} + \left(\frac{\partial^2 W}{\partial X^2} + \frac{\partial^2 W}{\partial Y^2} + \frac{\partial^2 W}{\partial Z^2}\right) \tag{4}$$

$$U\frac{\partial \Theta}{\partial X} + V\frac{\partial \Theta}{\partial Y} + W\frac{\partial \Theta}{\partial Z} = \frac{1}{Pr}\left(\frac{\partial^2 \Theta}{\partial X^2} + \frac{\partial^2 \Theta}{\partial Y^2} + \frac{\partial^2 \Theta}{\partial Z^2}\right) \tag{5}$$

2.2.2 Boundary Conditions

The state of open surfaces at the top, bottom, and the side of the fin channel (Fig.1b) poses a formidable problem. Obviously no definite boundary conditions can be assumed at these surfaces. In the present study an attempt has been made to accommodate these open boundaries by shifting them a certain distance away from the channel. At these extended top, bottom and side surfaces, ambient conditions are considered. Thus the problem is solved within the domain of an imaginary rectangular enclosure $A_1B_1B_2A_2D_1C_1C_2D_2$ as shown in Fig.2.

In terms of the dimensionless variables, the boundary conditions of the problem can now be stated.

At surface $A_1B_1C_1D_1$ Composed of ABCD, A_1B_1BA and DCC_1D_1 (Z = 0) :

The surfaces A_1B_1BA and DCC_1D_1 are respectively the top and bottom extensions of the fin base ABCD. On these imaginary surfaces, the following reasonable temperature boundary conditions are assumed. At the top of the fin array, the flow of air is essentially vertical and the thermal conductivity of air is small, hence the heat transfer by conduction in the Z direction is neglected. At the bottom, fresh air enters the fin channel at ambient temperature. On both these extended surfaces, the flow velocity in the Z direction (W) and the variation of U and V in the Z direction are neglected. These boundary conditions are considered to be physically realistic. Thus we have

Surface ABCD: $\Theta = 1,\ U = 0,\ V = 0,$ and $W = 0$ (6)

Surface A_1B_1BA : $\frac{\partial \Theta}{\partial Z} = 0,\ \frac{\partial U}{\partial Z} = 0,\ \frac{\partial V}{\partial Z} = 0,$

and $W = 0$ (7)

Surface DCC_1D_1 : $\Theta = 0,\ \frac{\partial U}{\partial Z} = 0,\ \frac{\partial V}{\partial Z} = 0,$

and $W = 0$ (8)

At surface $B_1B_2C_2C_1$ composed of the fin flat BCEF and $BFECC_1C_2B_2B_1$ $(Y = S^+/2)$:

Surface BCEF: $\Theta = 1,\ U = 0,\ V = 0,$ and $W = 0$ (9)

Surface $BFECC_1C_2B_2B_1$ is a plane of symmetry. Thus,

$$\frac{\partial \Theta}{\partial Y} = 0,\quad \frac{\partial U}{\partial Y} = 0,\quad V = 0 \text{ and } \frac{\partial W}{\partial Y} = 0 \qquad (10)$$

Surface $A_1A_2D_2D_1$ (Y = 0) : This is also a plane of symmetry. Hence

$$\frac{\partial \Theta}{\partial Y} = 0,\quad \frac{\partial U}{\partial Y} = 0,\quad V = 0 \text{ and } \frac{\partial W}{\partial Y} = 0 \qquad (11)$$

At the surfaces $A_1B_1B_2A_2$, $D_1C_1C_2D_2$ and $A_2B_2C_2D_2$: The top, bottom and side boundary surfaces $A_1B_1B_2A_2$, $D_1C_1C_2D_2$ and $A_2B_2C_2D_2$ should ideally be situated at infinite distances from the fin channel where the temperature Θ, and the velocities U, V and W tend to zero. However, it is not possible to solve the problem with the idealized boundary surfaces at infinity. Hence, these surfaces are assumed to be at some finite, but reasonably large distance away from the fin channel. Based on the flow visualization studies of air around the heated fin array, the side extension is kept equal to the length of the fin. The extended boundary at the top is fixed at a distance equal to double the height of the fin, while the bottom is fixed at a distance approximately equal to two-thirds the height of the fin. Thus, the conditions on the extended boundaries are:

At surface $A_1B_1B_2A_2$ $(X \simeq 3)$:

$$\Theta = 0,\ U = 0,\ V = 0, \text{ and } W = 0 \qquad (12)$$

At surface $C_1D_1D_2C_2$ $(X \simeq -0.67)$:

$$\Theta = 0,\ U = 0,\ V = 0, \text{ and } W = 0 \qquad (13)$$

At surface $A_2B_2C_2D_2$ $(Z \simeq 2L^+)$:

$$\Theta = 0,\ U = 0,\ V = 0, \text{ and } W = 0 \qquad (14)$$

The fresh air flowing in from the bottom and the side of the fin channel is assumed to be at the ambient temperature, neglecting the effect of back conduction. This simplification is introduced based on the physics of the problem and helps in overcoming the recirculation effect in the imaginary enclosure.

2.2.3 Local and Average Nusselt Numbers

The expressions for local Nusselt numbers are as follows:

Fin base

$$Nu_{bH} = -\left.\frac{\partial \Theta}{\partial Z}\right|_{Z=0} \qquad (15)$$

Fin flat

$$Nu_{fH} = -\left.\frac{\partial \Theta}{\partial Y}\right|_{Y=(\frac{S^+}{2})} \qquad (16)$$

The average Nusselt number is obtained by integrating the local Nusselt numbers. Thus

$$\overline{Nu}_{aH} = \frac{1}{(L^+ + \frac{S^+}{2})}\left[\int_0^{(\frac{S^+}{2})}\int_0^1 -\left[\frac{\partial \Theta}{\partial Z}\right]_{Z=0} dXdY + \int_0^{L^+}\int_0^1 \left[\frac{\partial \Theta}{\partial Y}\right]_{Y=(\frac{S^+}{2})} dXdZ\right] \qquad (17)$$

Previous investigators have also defined an average Nusselt number $(\overline{Nu}_{bH})$ using the fin base area alone. It is evident that

$$\overline{Nu}_{bH} = (1 + 2\frac{L^+}{S^+})\ \overline{Nu}_{aH} \qquad (18)$$

In Eqns. (17) and (18) the channel height (H) is used as the characteristic dimension. Later on, the spacing (S) is also used as the characteristic dimension to obtain an average Nusselt number $\overline{Nu}_S$ instead of $\overline{Nu}_{aH}$.

2.2.4 Results and Discussion

A numerical technique using finite-difference methods was adopted. The finite-difference formulation has been derived using the SIMPLE procedure of Patankar [12,13]. A uniform grid was used within the fin domain, while in the extended portions of the domain, the grid spacing was increased in a geometric progression.

The solution depends upon the dimensionless parameters L^+, S^+, Gr_H and Pr. Results have been obtained only for air (Pr=0.7). The values selected for the other parameters are given in Table 1. This table shows the thirteen sets of results obtained for the solution of the problem. The maximum values of the three velocity components, normalised by dividing by the square root of the Grashof number are tabulated along with the values of the average Nusselt numbers.

The variation of local Nusselt numbers (Nu_{bH} and Nu_{fH}) in the X direction on the fin flat and the fin base are plotted in Fig.3. It is seen that the values of Nu_{bH} and Nu_{fH} decrease as the distance X from the fin domain bottom increases. The maximum values of both local Nusselt numbers occur at the location where fresh air enters, as expected. The values decrease as one approaches the corner of the channel. It is also seen that the values of Nu_{bH} and Nu_{fH} vary widely over the fin base and the fin flat respectively and that the average values of Nu_{fH} are higher than those of Nu_{bH}.

The variation of the average Nusselt number with some of the dimensionless parameters will first be noted. In Table 1, serial numbers 1,3,6 and 9 show the variation of Nu_{aH} with Gr_H when the values of L^+ and S^+ are held constant at 0.25 and 0.2 respectively. It is seen that $\overline{Nu}_{aH}$ increases significantly from 9.70 to 28.99 as Gr_H increases from 2.5×10^3 to 1.25×10^7. On the other hand, inspection of serial numbers 11, 12, 13 and 6 in Table 1 ($Gr_H = 1.25 \times 10^6$, $S^+ = 0.2$) shows that $\overline{Nu}_{aH}$ remains essentially constant, around 15 as L^+ varies from 0.15 to 0.25.

The values obtained for the average Nusselt numbers $\overline{Nu}_{aH}$ are fitted by the least squares method to the following correlation, which should prove useful for design purposes.

$$\overline{Nu}_{aH} = 0.929\, Gr_H^{0.217}(S^+)^{0.015}(L^+)^{0.139} \qquad (19)$$

for $3.704 \times 10^4 \leq Gr_H \leq 10^8$, $0.1 \leq S^+ \leq 0.3$,

$0.15 \leq L^+ \leq 0.25$, $Pr = 0.7$. Excepting for two cases, correlation (19) agrees with the computational results to within $\pm$ 6.6%. However, in view of the fact that the correlation is based on only thirteen sets of data, it should be considered as tentative.

2.2.5 Comparison with Other Data

The values of the average Nusselt number $\overline{Nu}_s$ obtained in the present investigation are plotted in Fig.4 along with the experimental and theoretical data of previous investigators. It is seen that for $Ra_s^+ < 2 \times 10^2$, the results of the present investigation are higher than the experimental data of Welling and Wooldridge [3], Chaddock [4] and Aihara [5,7], while for higher values of Ra_s^+, the present results are in reasonably good agreement with the Aihara correlation for $2 \times 10^2 < Ra_s^+ < 2 \times 10^3$ and with the data of Welling and Wooldridge and the correlation of Chaddock for $2 \times 10^3 < Ra_s^+ < 2 \times 10^4$.

From the same figure, one may observe that the agreement between the present results and the previous results obtained by Saikhedkar [8] is poor for low values of Ra_s^+ upto 200. Nusselt number values are generally about 40-50% higher than the values obtained by neglecting the cross-flow velocity. However, for higher values of Ra_s^+ upto 20,000, the agreement between the values obtained by the two analyses is quite good, usually within $\pm$ 15%. It is thus evident that although the cross-flow velocity V is always small compared to U and W, it has significant effects on the heat transfer characteristics in some situations. Fig.4 also shows that Nu_s does not correlate well with Ra_s^+ alone. For this reason, in correlation (19), the Nusselt number is correlated with the parameters Gr_H, S^+ and L^+.

3. RADIATIVE HEAT TRANSFER FROM RECTANGULAR FIN ARRAYS

3.1 Previous Work

Problems of radiative heat transfer from fin arrays in general were studied initially without considering radiative interaction between the fins themselves and the base. Later on, the effects of fin-to-base and fin-to-adjacent fin interaction were considered by many investigators [14,15]. In these investigations, the variation of radiosity in one direction only was considered. Only Sparrow et al.[16] and Look et al. [17] have investigated radiative heat transfer from plates assuming a two-dimensional variation of radiosity.

The problem of radiative heat transfer from rectangular fin arrays has been solved by Saikhedkar [8, 18] with the radiosity being assumed to vary in both directions over the surfaces of the fin array. This analysis will now be discussed.

3.2 Mathematical Model

The array is the same as that shown in Fig.1(a). As in Section 2, the fin material is assumed to have a high thermal conductivity and consequently the surfaces of the fin array may be taken to be essentially isothermal at a temperature T_h. Further it is assumed that the fin surfaces are gray, the radiation is diffuse, the surrounding medium is non-participating and the transmissivity of the fin surfaces is zero.

The formulation is obtained by forming a fin channel into a enclosure with the help of imaginary surfaces. In the enclosure theory, the energy balance at a surface location takes account of all the radiant energy arriving at that location from all the directions in the enclosure. The summation over all directions leads to an integral equation for the distribution of the surface heat flux. Fig.5 shows two adjacent fin flats (surfaces 1 and 3) with the fin base (surface 2) in between. These three surfaces are assumed to form part of a six surface rectangular enclosure whose other three surfaces 4,5 and 6 are imaginary black isothermal surfaces maintained at the environment temperature T_∞. Considering the local heat balance at an elementary area dA_1 having position coordinates (x_1, y_1) on surface 1, the radiosity is equal to heat emitted plus the reflected irradiation, i.e.

$$J_1(x_1, y_1) = \varepsilon\ \sigma\ T_h^4 + \rho\ G_1(x_1, y_1) \tag{20}$$

The irradiation $G_1\ (x_1, y_1)$ is the sum of the radiant energy coming from all the surfaces of the enclosure and incident on dA_1.

On substituting for $G_1\ (x_1, y_1)$ in terms of shape factors in equation (20) and non-dimensionalizing, the following integral equation for dimensionless radiosity $J_1^*(X_1, Y_1)$ is obtained.

$$\begin{aligned} J_1^*(X_1, Y_1) &= \varepsilon + \frac{(1-\varepsilon)}{\pi} \Big[\int_0^{S^+}\int_0^1 J_2^*(X_2, Z_2)\, Z_2 Y_1\, dX_2\, dZ_2 / \{(X_1 - X_2)^2 + Y_1^2 + Z_2^2\}^2 \\ &+ (2S^+)^2 \int_0^{L^+}\int_0^1 J_3^*(X_3, Y_3)\, dX_3\, dY_3 / \{(X_1 - X_3)^2 + (Y_1 - Y_3)^2 + (2S^+)^2\}^2 \\ &+ (1/\Theta^4) \int_0^{S^+}\int_0^1 Z_4 \{(L^+) - Y_1\}\, dX_4\, dZ_4 / \{(X_1 - X_4)^2 + (Y_1 - L^+)^2 + Z_4^2\}^2 \\ &+ (1/\Theta)^4 \int_0^{S^+}\int_0^{L^+} Z_5(1 - X_1)\, dY_5\, dZ_5 / \{(X_1 - 1)^2 + (Y_1 - Y_5)^2 + Z_5^2\}^2 \\ &+ (1/\Theta^4) \int_0^{S^+}\int_0^{L^+} X_1 Z_6\, dY_6\, dZ_6 / \{(X_1^2 + (Y_1 - Y_6)^2 + Z_6^2\}^2 \Big) \end{aligned} \tag{21}$$

The integral equation for the dimensionless radiosity $J_2^*(X_2, Z_2)$ in the fin base can be obtained in a similar manner.

$$\begin{aligned} J_2^*(X_2, Z_2) &= \varepsilon + \frac{(1-\varepsilon)}{\pi} \Big[\int_0^{L^+}\int_0^1 J_1^*(X_1, Y_1)\, Y_1 Z_2\, dX_1 dY_1 / \{(X_2 - X_1)^2 + Y_1^2 + Z_2^2\}^2 \\ &+ \int_0^{L^+}\int_0^1 J_3^*(X_3, Y_3)\, Y_3 (2S^+ - Z_2)\, dX_3 dY_3 / \{(X_2 - X_3)^2 + Y_3^2 + (Z_2 - 2S^+)^2\}^2 \\ &+ (1/\Theta^4) \int_0^{S^+}\int_0^1 (L^+)^2\, dX_4\, dZ_4 / \{(X_2 - X_4)^2 + (L^+)^2 + (Z_2 - Z_4)^2\}^2 \\ &+ (1/\Theta^4) \int_0^{S^+}\int_0^{L^+} Y_5 (1 - X_2)\, dY_5\, dZ_5 / \{(X_2 - 1)^2 + Y_5^2 + (Z_2 - Z_5)^2\}^2 \\ &+ (1/\Theta^4) \int_0^{S^+}\int_0^{L^+} Y_6 X_2\, dY_6\, dZ_6 / \{(X_2^2 + Y_6^2 + (Z_2 - Z_6)^2\}^2 \Big] \end{aligned} \tag{22}$$

Since surfaces 1 and 3 are symmetrical,

$$J_1^*(X_1, Y_1) = J_3^*(X_3, Y_3)$$

Equations (21) and (22) have to be solved simultaneously for the unknown radiosities J_1^* and J_2^*. The dimensionless local heat flux and the average heat flux are then given by

$$q^* = \frac{\varepsilon}{1-\varepsilon}\ (1 - J^*) \tag{23}$$

$$\bar{q}^* = \Big[\int_{A_1^*} q_1^*(X_1, Y_1)\, dX_1\, dY_1 + \int_{A_2^*} \bar{q}_2^*(X_2, Z_2)\, dX_2\, dZ_2 + \int_{A_3^*} q_3^*(X_3, Y_3)\, dX_3\, dY_3 \Big] / (A_1^* + A_2^* + A_3^*) \tag{24}$$

The integral equations (21) and (22) have been solved numerically by adopting a finite difference scheme. The solution is seen to depend upon the four dimensionless parameters Θ, ε, L^+ and S^+.

3.3 Results and Discussion

Results have been obtained for the following values of the dimensionless parameters:

Θ = 1.1, 1.25, 1.4, 1.6

ε = 0.1, 0.5, 0.8

L^+ = 0.1, 0.175, 0.25

S^+ = 0.1, 0.2, 0.3.

A total number of 50 cases have been studied.

Local Radiosity: The variation of the dimensionless local radiosity J^* over the fin flat surface is plotted in Figs.6 and 7 for some typical situations in which ε = 0.8, S^+ = 0.2, L^+ = 0.25 and Θ is varied from 1.25 to 1.6. It is observed that the radiosity distribution on the fin flat is symmetrical in the x direction. The magnitude of the radiosity at the tips is less than the magnitude at the center. The results also show that the radiosity on the fin flat decreases in the y direction. It is largest at the common edge and lowest at the tip. The radiosity distribution on the fin base is also symmetrical both in the x and z directions. Its magnitude is lower at the tips and edges than at the center of the base.

Local Heat Transfer: The variation of local heat flux for the same cases is plotted in Figs. 6 and 7. It is seen that the heat transfer varies inversely as the radiosity. For the fin flat, q^* increases in the y direction. It is minimum at the common edge and maximum at the tip. This trend is similar to that observed by Sparrow et al. The local heat flux for the fin base is symmetrical both in the x and z direction.

Average Heat Transfer: Knowledge of the average radiative heat flux $\bar{q}^*$ is more important for design purposes. The variation of $\bar{q}$* has been plotted in Figs.8 and 9. It is observed that the average heat flux increases with emissivity, temperature ratio and spacing, but decreases with length.

Correlation: The results for average radiative heat flux have been fitted to the following two correlations which should prove useful for practical purposes.

1. For 0 < Θ 1.25

$$\bar{q}^* = [\,(0.33 - 3.0\,\varepsilon - 2.14\,\varepsilon^2) + (-0.38 + 2.99\,\varepsilon + 2.67\,\varepsilon^2)\,\Theta + (0.07 - 0.08\,\varepsilon - 0.59\,\varepsilon^2)\,\Theta^4\,]\,(S^+)^{0.64}\,(L^+)^{0.3} \tag{25}$$

2. For 1.25 < Θ < 1.60

$$\bar{q}^* = [\,[-0.06 + 0.23\,\varepsilon - 0.33\,\varepsilon^2) + (0.08 + 0.27\,\varepsilon + 0.05\,\varepsilon^2)\,\Theta + (-0.01 + 0.05\,\varepsilon - 0.04\,\varepsilon^2)\,\Theta^4]\,(S^+)^{0.71}\,(L^+)^{-0.29} \tag{26}$$

4. THE CONJUGATE PROBLEM

When the thermal conductivity of the fin material is assumed to be high (Sections 2 and 3), the fin can be treated as isothermal, and the natural convection and radiative heat transfer can be found independently of each other. However, if the fin material has a low thermal conductivity, then very often the fins cannot be treated as isothermal. There is then an interaction between the natural convection and radiative heat transfer modes and they cannot be solved independently of each other. The only way to analyse the problem in this situation consists in simultaneously solving the heat conduction equation within the fin, and the fluid flow and energy equations at the surface of the fin. Such a method in which conduction inside the fin is coupled with natural convection and radiation outside, is termed as a 'conjugate method'.

A few conjugate problems associated with rectangular fin arrays have been solved by Frost and Eraslan [19], Donovan and Rohrer [20] and Advani [21]. All these investigators have however made some simplifying assumptions. For example, in the first two investigations, the convective heat transfer coefficient has been assumed to be constant, while in the third, an integral method has been used for solving the momentum and energy equations at the surface. A more exact analysis has been done by Saikhedkar [8,22] This will now be described along with a few results.

4.1 Problem Formulation

The fin array to be analysed is again as shown in Fig.1 (a). However, now the fin base wall is assumed to be at a constant temperature T_b, while the fin flat temperature is assumed to vary both in the x and y directions. Radiation is diffuse. The surrounding air is treated as non-participating and at a temperature T_∞. The heat enters from the base into the fin flat by conduction and is dissipated to the surroundings by natural convection and radiation. Taking into account the heat conduction in two directions inside the fin, and the radiative and convective heat transfers from the surface, one obtains the following non-dimensional differential equation for the fin flat:

$$\frac{\partial^2\Theta_f}{\partial X^2} + \frac{\partial^2\Theta_f}{\partial Y^2} = C\,(Nu_{fS}\,\Theta_f/S^+) + [\,C\,\lambda/(1-\varepsilon)\,] \times [\,(\Theta_f + T_\infty^+)^4 - J_f^*\,] \tag{27}$$

The parameter C in equation (27) is called the conductivity parameter. It is a measure of the thermal resistance offered by the fin material. As C → 0, the fin tends to be isothermal. In order to solve equation (27), one needs the values of the local Nusselt numbers Nu_{fS} and Nu_{bS} which are obtained by solving the appropriate equations of momentum, energy and continuity+, equations (1) to (5) and the local radiosities J_f^* and J_b^* which are obtained by solving the radiosity integral equations (21) and (22).

The fin conduction equation is non-linear coupled and implicit. Since the unknowns are interdependent, an iterative procedure based on a modified Newton-Raphson method is used. After obtaining the values of Θ_f; Nu_{fS}, Nu_{bS}, J_f^* and J_b^*, the rates of heat transfer and the fin efficiency can be calculated.

+ With the cross velocity term neglected

The convective heat transfer rate from the fin array Q_C^+ is given by

$$Q_C^+ = \int_0^{S^+} \int_0^1 (Nu_{bS}/S^+)\, dX\, dX + 2 \int_0^{L^+} \int_0^1 (Nu_{fS}/S^+)\, \theta_f \quad dX \quad dY \qquad (28)$$

Similarly the radiative heat transfer rate from the fin array Q_R^+ is given by

$$Q_R^+ = \int_0^{S^+} \int_0^1 [\lambda/(1-\varepsilon)]\, [(T_\infty^+ + 1)^4 - J_b^*]\, dX\, dZ + 2 \int_0^{L^+} \int_0^1 [\lambda/(1-\varepsilon)]\, .\, [(\theta_f + T_\infty^+)^4 - J_f^*]\, dX\, dY \qquad (29)$$

The total heat flow rate from the fin array

$$Q_T^+ = Q_C^+ + Q_R^+ \qquad (30)$$

If the total heat flow rate from an identical isothermal fin array is Q_{TI}^+, then the efficiency η of the fin array is given by

$$\eta = Q_T^+ / Q_{TI}^+ \qquad (31)$$

4.2 Results and Discussion

An inspection of the governing differential and integral equations shows that the complete solution of the problem depends upon eight dimensionless parameters, L^+, S^+, Pr, Gr_S, C, ε, λ and T_∞^+. Since the computational effort required to obtain numerical results in large, only a limited number of solutions have been obtained. These results have been generated after thorough numerical experimentation to decide on the optimum grid size and relaxation parameter.

A few of the results obtained by varying the values of Gr_S and C will now be discussed for a typical situation for which $L^+ = 0.25$, $S^+ = 0.20$, $Pr = 0.7$, $\lambda = 0.04$ and $T_\infty^+ = 4.0$.

The temperature distributions in the fin flat obtained by varying the values of Gr_S and C are plotted in Fig.10. In all cases, θ_f decreases in the y direction, as expected. It is also seen that θ_f decreases with increasing values of Gr_S, because the convective heat transfer coefficient increases. θ_f also decreases with increasing values of C, because of the increasing thermal resistance inside the fin. For the case of $C = 0$, an isothermal distribution is obtained, as expected. From other results not presented here but discussed in reference (8), it is found that the value of θ_f decreases as the values of L^+ and S^+ increase.

The effect of varying the value of Gr_S from 10^3 to 10^5 on the total heat transfer rate from the fin array and on the fin efficiency is shown in Figs. 11 and 13a. It is seen that with increasing values of Gr_S, the convective heat transfer rate Q_C^+ increases rapidly, while the radiative heat transfer rate Q_R^+ decreases slightly. The total heat transfer rate therefore increases. These trends can be explained on the basis that the convective heat transfer coefficient increases significantly, while the temperature difference between the fin flats and the surroundings decreases a little. Because of increasing deviation from an isothermal fin flat distribution, the fin efficiency decreases.

The effect of varying the value of the conductivity parameter from 0 to 1 is shown in Figs.12 and 13b. The values of Q_C^+, Q_R^+ and Q_T^+ are all seen to decrease slightly as the value of C increases. These trends occur because of the fin flat temperature distribution discussed in Fig.10. As a result, the fin efficiency continuously decreases.

It is found from other results, not presented here but discussed in reference (8), that the fin efficiency increases with increasing values of S^+ and decreases with increasing values of L^+.

NOMENCLATURE

A	-	surface area
A^*	-	non-dimensional surface area
C	-	conductivity parameter, $kH/k_f t$
C_P	-	specific heat capacity
G	-	irradiation
Gr_H	-	Grashof number, $g\, \beta\, \Delta T\, H^3/\nu^2$
Gr_S	-	Grashof number, $g\, \beta\, \Delta T\, S^3/\nu^2$
H	-	fin height
J	-	local radiosity
J^*	-	local dimensionless radiosity $J/\sigma T_h^4$ or $J/\sigma (T_b - T_\infty)^4$
k	-	thermal conductivity of the fluid
k_f	-	thermal conductivity of fin material
L	-	length of fins
L^+	-	dimensionless fin length, L/H
$\overline{Nu}_{aH}$	-	average Nusselt number based on fin base and fin flat areas with H as the characteristic dimension
$\overline{Nu}_S$	-	average Nusselt number based on fin base and fin flat areas with S as the characteristic dimension

Nu_{bH} - local Nusselt number on fin base with H as the characteristic dimension

Nu_{bS} - local Nusselt number on fin base with S as the characteristic dimension

$\overline{Nu}_{bH}$ - average Nusselt number based on fin base areaa with H as the characteristic dimension

Nu_{fH} - local Nusselt number on fin flat with H as the characteristic dimension

Nu_{fS} - local Nusselt number on fin flat with S as the characteristic dimension

P - dimensionless pressure, $(p + \rho_\infty gx) H^2/\rho_\infty \nu^2$

p - pressure

Pr - Prandtl number, $C_p \mu / k$

Q - heat transfer rate

Q^+ - dimensionless heat transfer rate, $Q/kH(T_b - T_\infty)$

q - local radiative heat flux

q^* - dimensionless local radiative heat flux, $q/\sigma T_h^4$

$\overline{q}$ - average radiative heat flux from array

$\overline{q}^*$ - dimensionless average radiative heat flux from array, $q / \sigma T_h^4$

Ra_S^+ - modified Rayleigh number, $Gr_S\, Pr\, S^+$

S - spacing

S^+ - dimensionless spacing, S/H

T_b - temperature of fin base (Section 4)

T_h - temperature of isothermal fin array

T_∞ - ambient temperature

T_∞^+ - dimensionless surrounding air temperture, $T_\infty / (T_b - T_\infty)$

t - half-thickness of the fin

u,v,w - component of velocity in the x,y and z direction respectively

U,V,W - component of dimensionless velocity in the x,y and z direction respectively, uH/ν, vH/ν, wH/ν

x,y,z - rectangular Cartesian coordinates

X,Y,Z - dimensionless distances in x,y and z direction respectively, x/H, y/H, z/H

ΔT - temperature difference between fin array and environment, $(T_h - T_\infty)$

ε - emissivity

β - coefficient of thermal expansion

Θ - dimensionless temperature, $(T - T_\infty) / (T_h - T_\infty)$ or (T_h / T_∞)

Θ_f - dimensionless local fin flat temperature $(T - T_\infty) / (T_b - T_\infty)$

μ - dynamic viscosity

ν - kinematic viscosity, μ/ρ

ρ - density; reflectivity

η - fin array efficiency

λ - dimensionless parameter, $\sigma\, \varepsilon\, (T_b - T_\infty)^3 H/k$

REFERENCES

1. Sukhatme, S.P., Invited Lecture, Fifth National Heat and Mass Transfer Conference, Hyderabad, (1980).
2. Starner, K.E., and McManus, H.N., Journal of Heat Transfer, Trans ASME, 85 (1963) 273.
3. Welling J.R., and Wooldridge, C.B., Journal of Heat Transfer, Trans ASME, 87 (1965) 439.
4. Chaddock, J.B., ASHRAE Journal 12 (1970) 53.
5. Aihara T., Trans. JSME, 34 (1968) 915.
6. Aihara, T., Bulletin JSME, 13 (1970) 1192.
7. Aihara, T., Bulletin JSME , 13 (1970) 1182.
8. Saikhedkar, N.H., Heat Transfer from rectangular cross-sectioned vertical fin arrays, Ph.D., thesis, Indian Institute of Technology, Bombay, (1980).
9. Saikhdkar,N.H., and Sukhatme, S.P., Journal of Thermal Engineering 1 (1980) 125.
10. Shalaby, M.A.I., Natural convection heat transfer from rectangular fin arrays, Ph.D. thesis, Indian Institute of Technology, Bombay, (1983).
11. Shalaby, M.A.I., Gaitonde, U.N., and Sukhatme, S.P., Indian J. of Technology, 22 (1984) 321.
12. Patankar, S.V., Numerical heat transfer and fluid flow, McGraw-Hill, New York, (1980).
13. Patankar, S.V., Numerical Heat Transfer, 4 (1981) 409.
14. Eckert, E.R.G., Irvine, T.F., and Sparrow E.M., Journal of Aerospace Science 28 (1961) 763.
15. Sparrow, E.M., Gregg, J.L., Szel, J.V., and Manos, P., Journal of Heat Transfer, Trans ASME, Series C, 207 (1961)
16. Sparrow, E.M., and Haji-Sheikh, A., Journal of Heat Transfer, Trans ASME, Series C, 87 (1965) 103.
17. Look, D.C., and McKinney, S.E., AIAA Journal, 13 (1975) 1267.
18. Saikhedkar, N.H., Karlekar, B.V., and Sukhatme, S.P., Proceedings Fifth National Heat and Mass Transfer Conference, Hyderabad, (1980).
19. Frost, W., and Eraslan, A.H., Fluid Mechanics Institute, Stanfor University Press, (1968) 206.
20. Donovan, R.C., and Rohrer, W.M., Journal of Heat Transfer, Trans ASME, 93 (1971) 93.
21. Advani, C.G., Heat Transfer from vertical rectangular cross-sectioned fins losing heat

by free convection and radiation, Ph.D. thesis, Indian Institute of Technology, Bombay, (1971).

22. Saikhedkar, N.H., and Sukhatme, S.P., HMT-9-81, Proceedings, Sixth National Heat and Mass Transfer Conference, Madras, (1981).

TABLE 1 : COMPUTATIONAL RESULTS OF VERTICAL FIN ARRAY

Set. No.	Gr_H	Gr_S	S^+	L^+	$\frac{U_{max}}{\sqrt{Gr_H}}$	$\frac{V_{max}}{\sqrt{Gr_H}}$	$\frac{W_{max}}{\sqrt{Gr_H}}$	Nu_{bH}	Nu_{aH}	Nu_S
1.	2.5×10^3	5×10^2	0.2	0.25	0.77	0.012	-0.154	33.96	9.70	1.94
2.	10^6	10^3	0.1	0.25	0.97	0.012	-0.051	84.57	14.10	1.41
3.	1.25×10^5	10^3	0.2	0.25	0.244	- 0.016	-0.055	35.02	10.01	2.00
4.	3.704×10^4	10^3	0.3	0.25	0.181	0.004	-0.047	24.13	9.05	2.72
5.	10^7	10^4	0.1	0.25	0.777	0.005	-0.063	132.63	22.11	2.21
6.	1.25×10^6	10^4	0.2	0.25	0.502	0.007	-0.068	50.04	14.30	2.86
7.	3.704×10^5	10^4	0.3	0.25	0.333	0.015	-0.061	28.78	10.79	3.24
8.	10^8	10^5	0.1	0.25	0.73	0.013	-0.042	286.21	47.70	4.77
9.	1.25×10^7	10^5	0.2	0.25	0.668	0.010	-0.033	101.46	28.99	5.80
10.	3.704×10^6	10^5	0.3	0.25	0.586	0.011	-0.051	49.56	18.58	5.57
11.	1.25×10^6	10^4	0.2	0.15	0.142	0.005	0.007	36.60	14.64	2.93
12.	1.25×10^6	10^4	0.2	0.175	0.374	0.008	-0.042	42.51	15.46	3.09
13.	1.25×10^6	10^4	0.2	0.225	0.527	0.013	-0.059	47.75	14.69	2.94

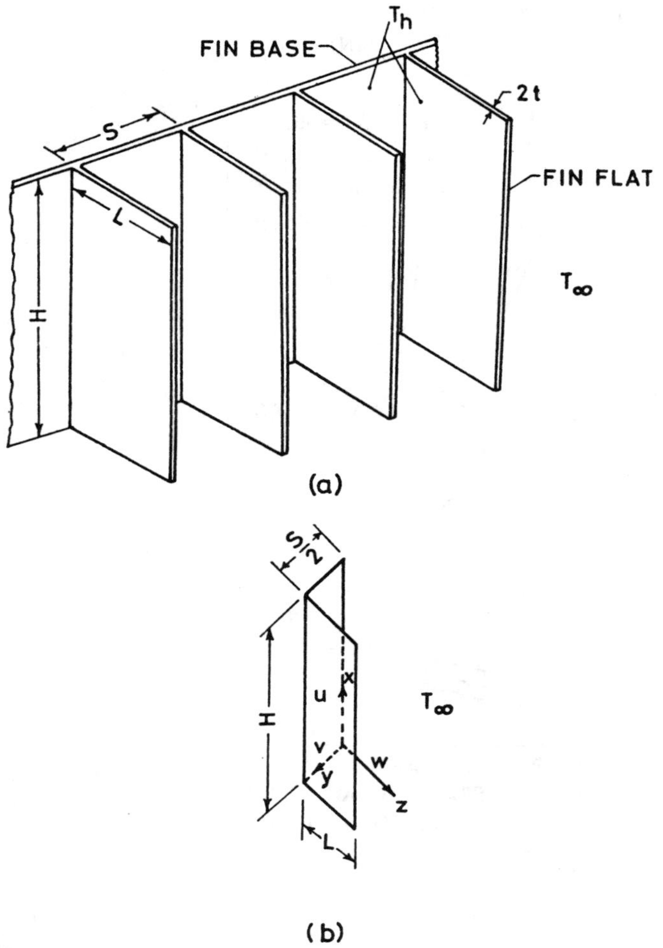

Fig.1 a) A Vertical Rectangular Fin Array
b) Fin Portion to be Analysed (Origin at Bottom Centre of Fin Base)

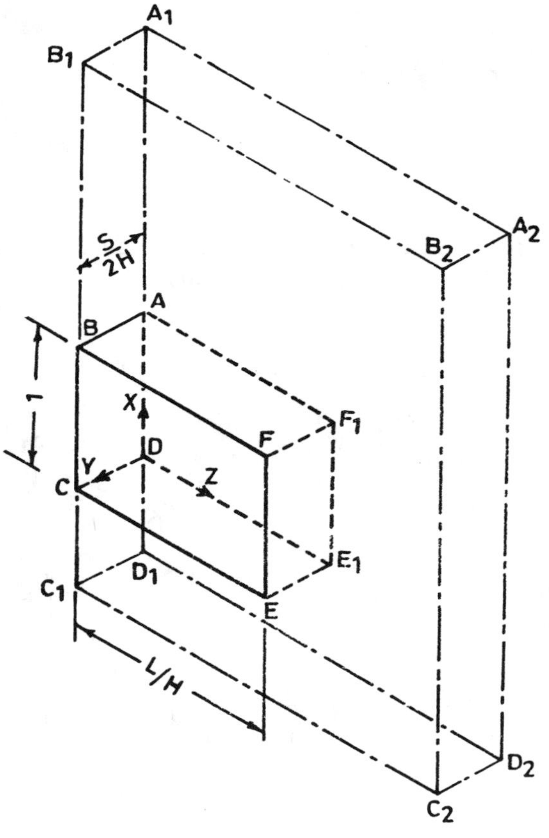

Fig.2 Domain of Computation

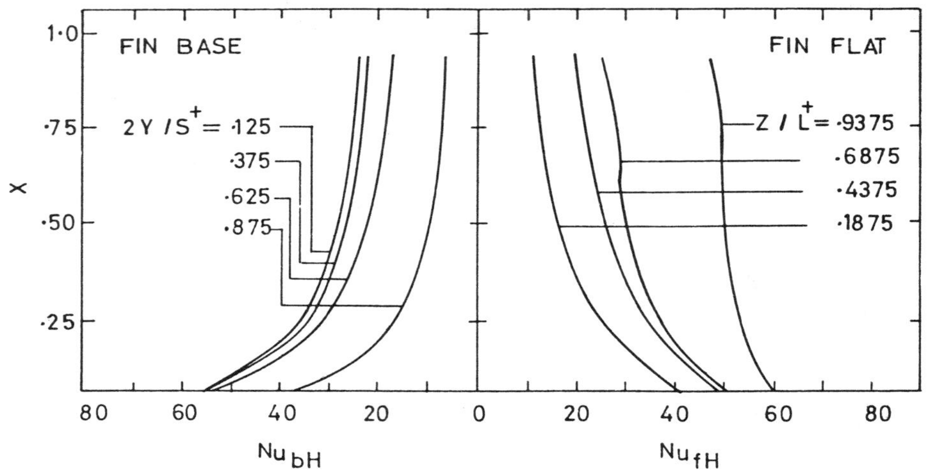

Fig.3 Local Nusselt Numbers - $Gr_S = 10^5$, $Pr = 0.7$, $L^+ = 0.25$, $S^+ = 0.2$

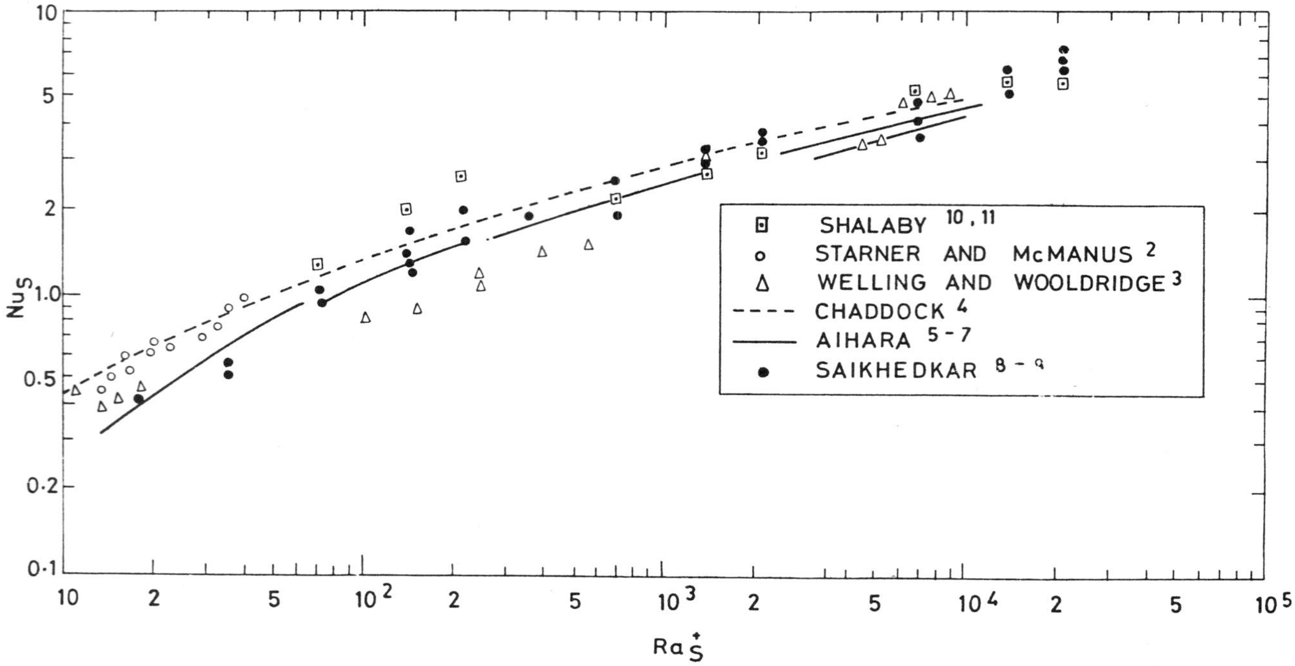

Fig.4 Average Nusselt Number Vs Modified Rayleigh Number

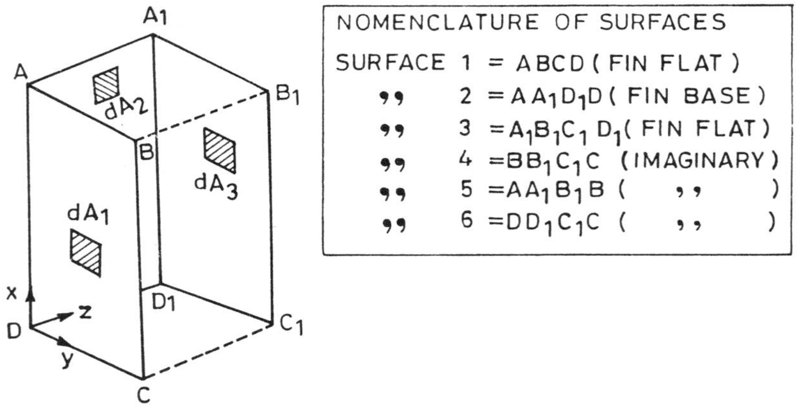

Fig.5 A Fin Channel Formed in to an Enclosure

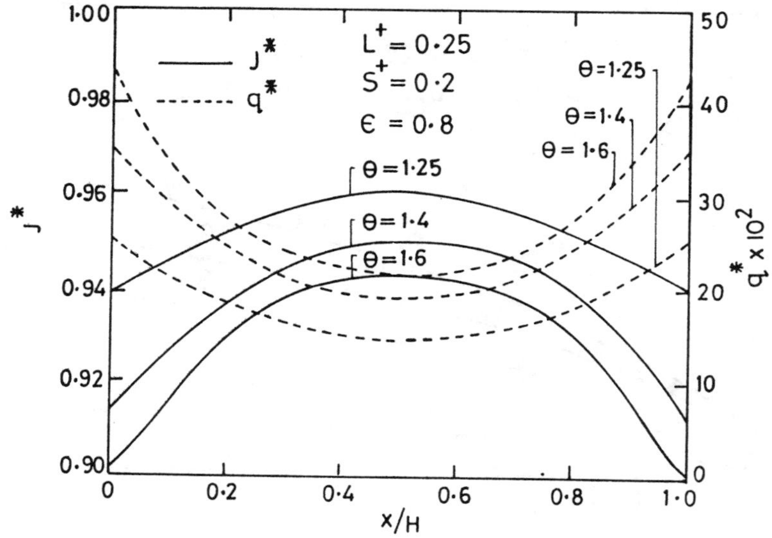

Fig.6 Variation of Local Radiosity and Heat Flux Along Fin Flat in X-Direction At y/L = 0.5

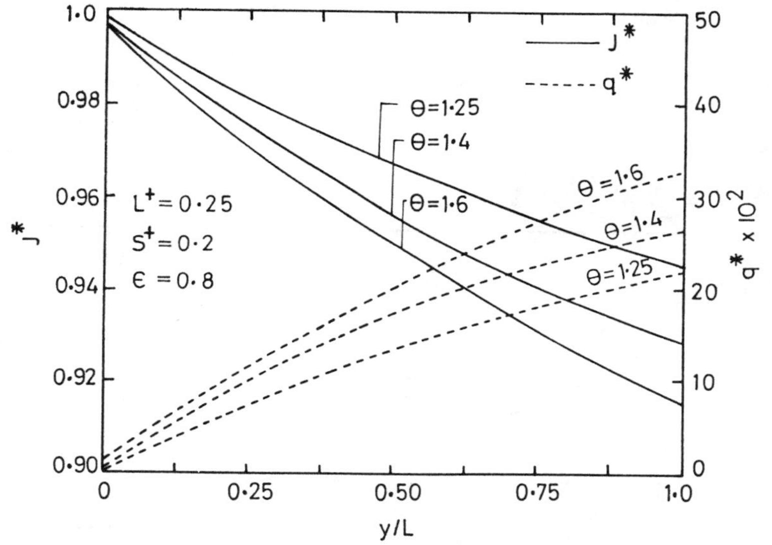

Fig.7 Variation of Local Radiosity and Heat Flux along Finflat in Y-Direction at x/H = 0.5

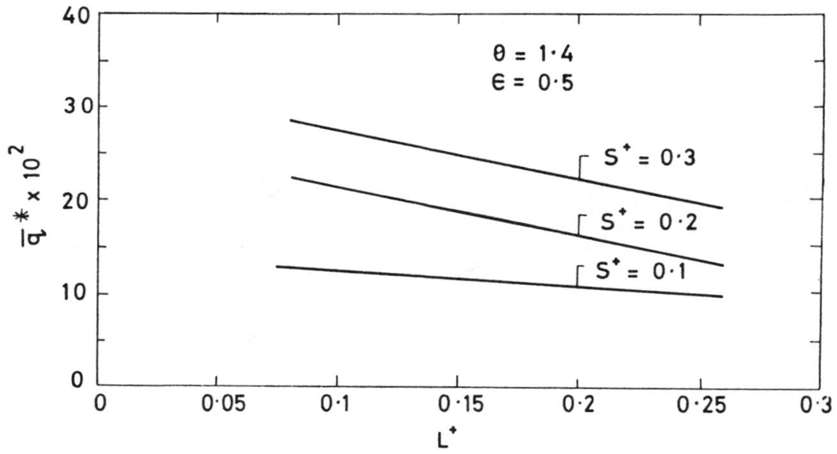

Fig.8 Effect of Fin Length and Spacing Average Radiative Heat Flux

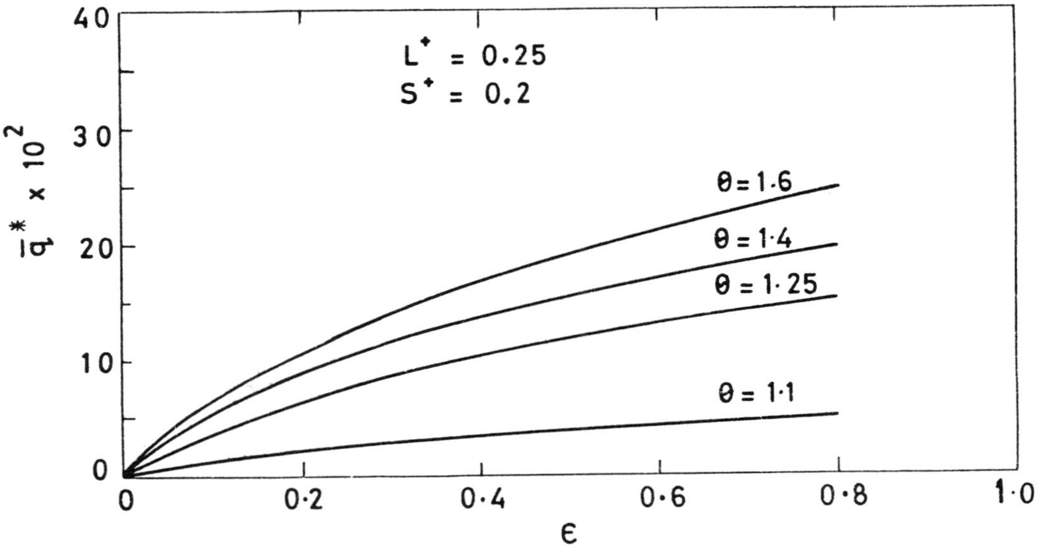

Fig.9 Effect of Emissivity and Temperature Ratio on Average Radiative Heat Flux

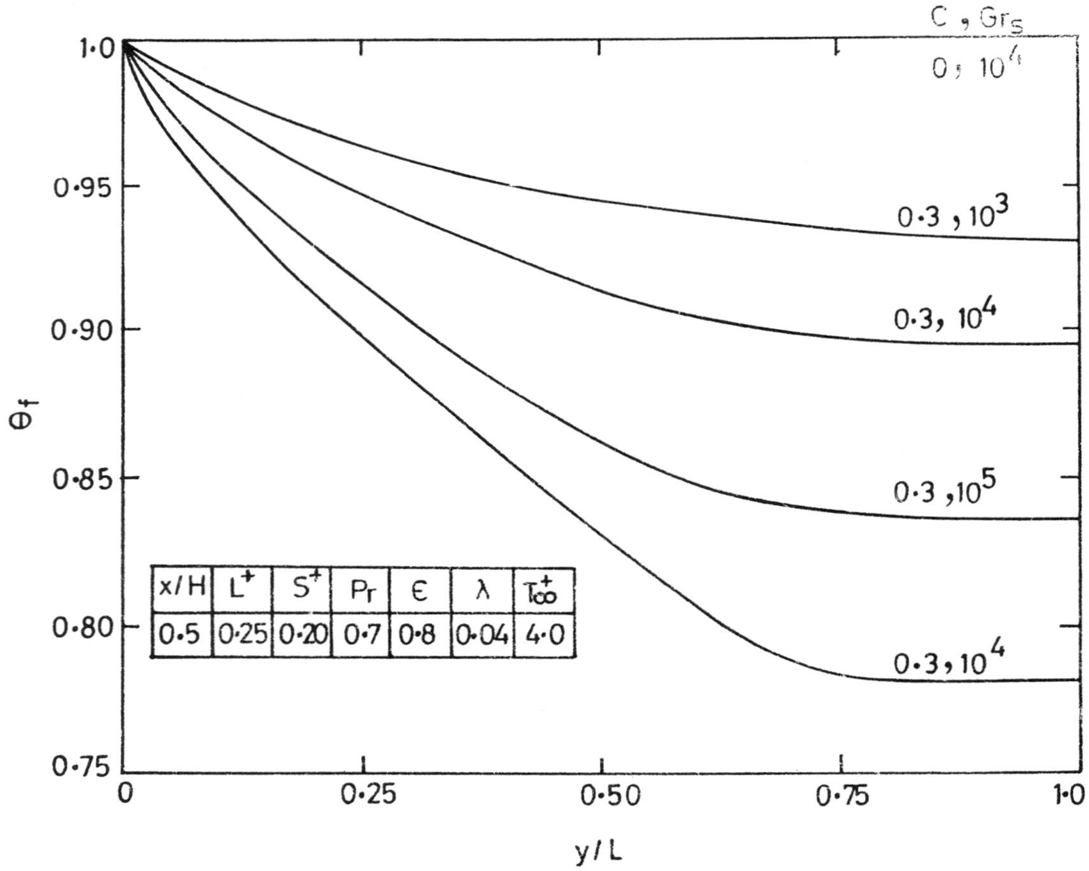

Fig.10 Effect of GrS and C on Fin Flat Temperature Distribution

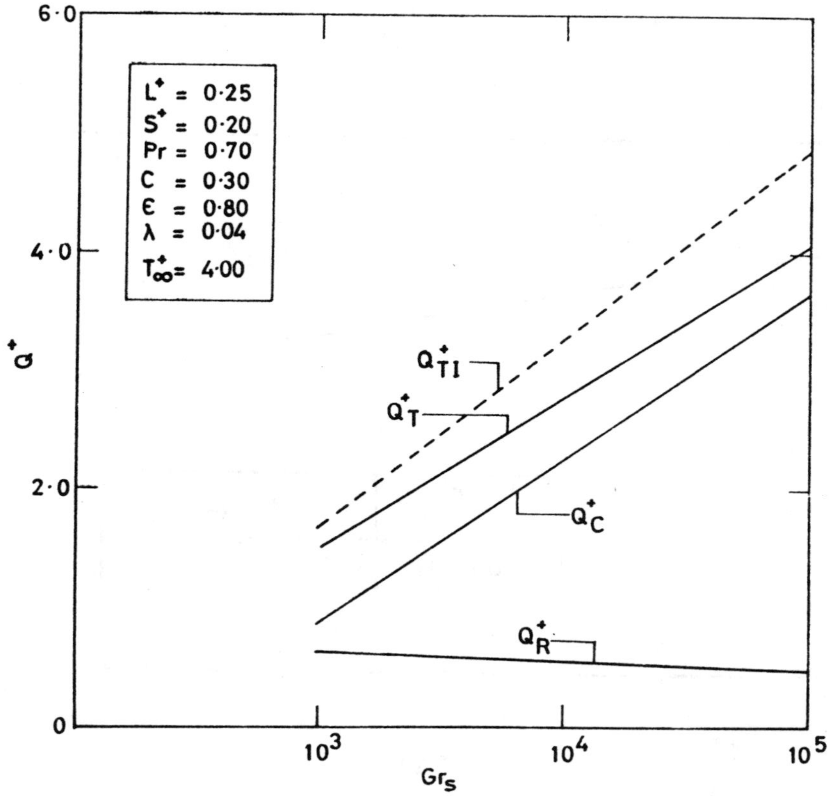

Fig.11 Effect of Grashof Number on Heat Transfer from Fin Array

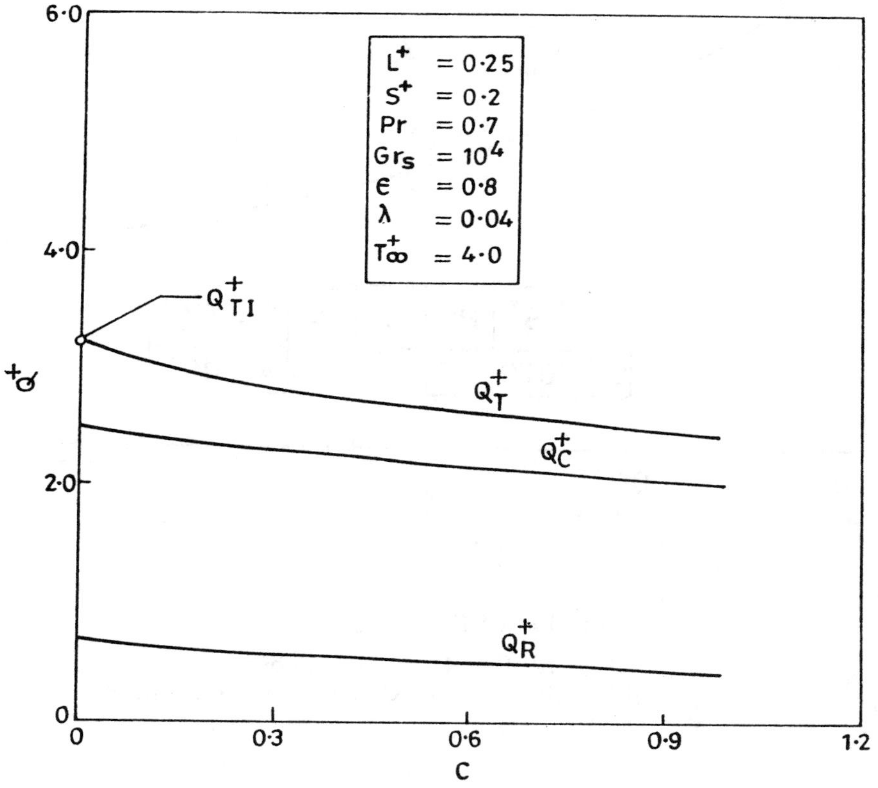

Fig.12 Effect of Conductivity Parameter on Heat Transfger From Fin Array

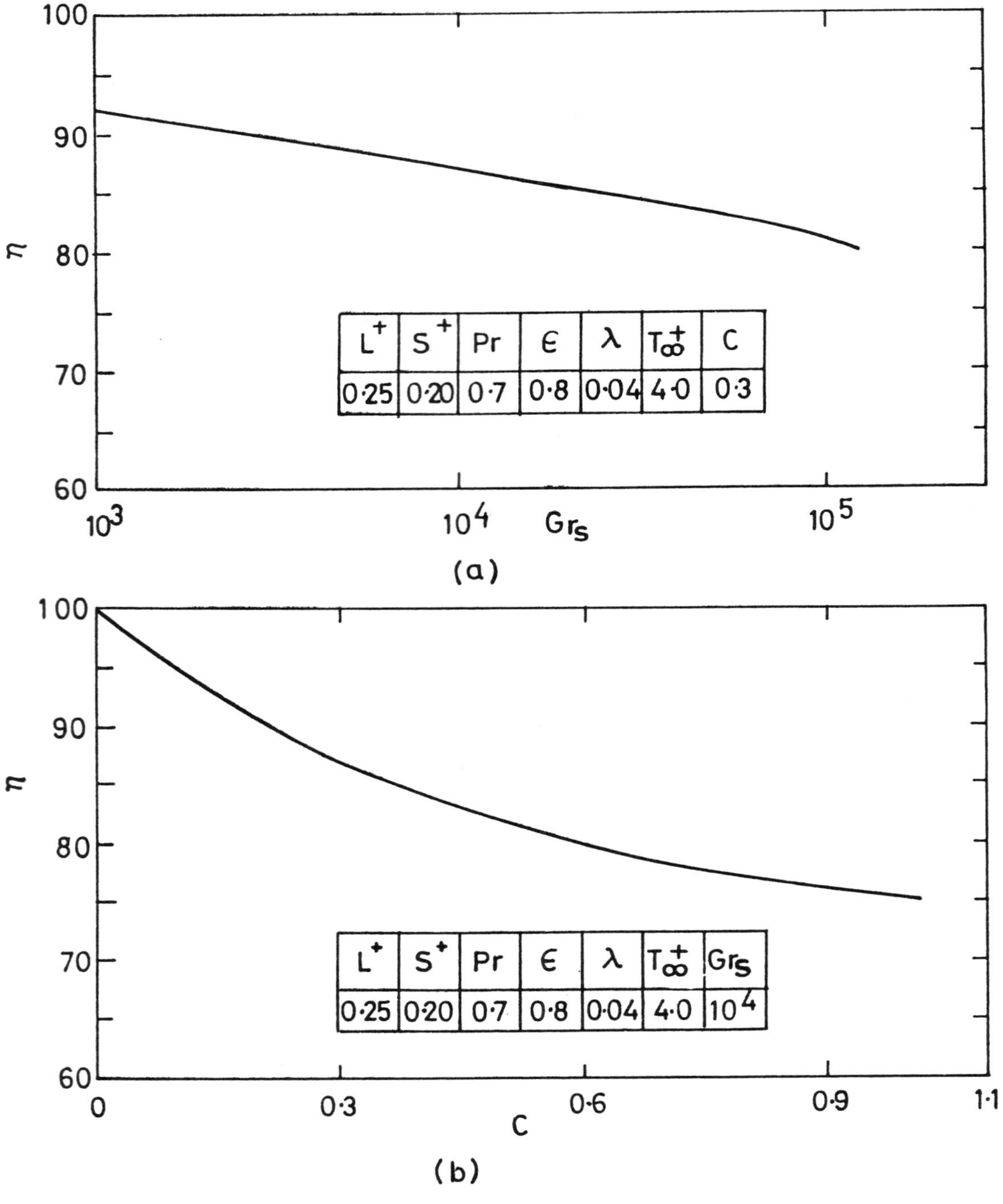

Fig.13 Variation of Fin Efficiency

a) Effect of Grashof Number
b) Effect of Fin Conductivity Parameter

Contact Drying of Polydispersed Free Flowing Granular Material

E. TSOTSAS and E. U. SCHLÜNDER
Institute für Thermische Verfahrenstechnik der Univ.
Karlsruhe, Kaiserstr. 12,7500 Karlsruhe 1, FRG

1. INTRODUCTION

In contact drying the heat needed to vaporize the water or other liquid is transferred to the particulate material through a wall. Contact drying is commonly carried out in disc dryers and in drum dryers. In order to achieve uniform drying the particle layers are agitated mechanically. The agitating elements in a disc dryer are stirrers, scrapers or paddles. There are rotating drum dryers with or without additional stirring elements as well as static drum dryers with flyers inside.

Contact dryers can be run without an inert gas atmosphere (vacuum contact drying). This is advantageous or imperative for drying duties as:

- Drying of material which decomposits at a certain critical temperature.
- Drying where the wetting fluid is valuable or toxic and therefore must be easily recovered.
- Drying of material which deteriorates or becomes explosive when exposed to atmosphere.

For the design of all kinds of contact dryers the 'drying rate curve', which correlates the drying rate $\dot{m}$ (in kg vapor per m^2 hot surface area per hour) with the average moisture content X of the product (in kg liquid per kg dry material) is needed.

In the beginning of this paper a method to predict the drying rate curve of a bed of equal particles will be briefly discussed. It will then be attempted to extend this method to multigranular packings. It is assumed that the material is free flowing and not hygroscopic.

2. RESISTANCES IN VACUUM CONTACT DRYING

The drying rate $\dot{m}$ is affected by both heat supply and vapor removal. In general, the heat supplied must overcome three heat transfer resistances: the contact resistance at the hot surface, the penetration resistance of the bulk, and the penetration resistance of the particle. On the other hand the vapor to be removed must overcome two mass transfer resistances: the permeation resistance of the porous particles and the permeation resistance of the bulk. In the absence of inert gas, the mass transfer resistances are flow resistances.

For a non agitated packed bed all these resistances lie in series (Fig.1). However, if the particulate material is mechanically agitated, we expect a random distribution of dry, partly wet and wet material, thus forming a random distribution of transport resistances partly in series and partly parallel.

In order to describe the rather complicated situation by a suitable model all authors take into consideration only one or two resistances which are assumed to be rate controlling, e.g. [1,2].

3. THE MODEL FOR MONODISPERSED PACKINGS

Mollekopf and Schlünder [3,4] regard vacuum contact drying of free flowing, mechanically agitated particulate material as a purely heat transfer controlled process. They consider the contact resistance at the hot surface and the penetration resistance of the bulk to be rate controlling.

The contact heat transfer coefficient α_{ws} can be predicted from first principles (s.Appendix A) and is a strong function of the particle diameter. Its value remains constant with good accuracy during a drying process at constant pressure and heating plate temperature and sets an upper limit for the drying rate.

Unlike $1/\alpha_{ws}$ which is completely independent of mechanical agitation, the bulk penetration resistance $1/\alpha_{sb}$ is a function of mixing intensity. Hence, in order to evaluate α_{sb} a model adequately describing the effect of random particle motion is needed. Mollekopf and Schlünder use for this purpose the so called penetration theory. The steady mixing process is replaced by a sequence of unsteady mixing steps. During a certain (fictitious) period of length t_R the bulk is static. A distinct drying front is penetrating from the hot surface into the bulk. Between the moving front and the hot surface all particles are dry, beyond the front all particles are wet. When the static period ends at $t = t_R$ the bulk is perfectly mixed. Thereafter the drying front moves again into the bulk. This time however, there are some particles which have already been dried during the first period. This situation is illustrated in Fig.2. Assuming that the packed bed may be considered as a quasi-continuum obeying the Fourier's law of heat transfer, one can make use of the Neumann's solution to calculate the location of the drying front, the transient temperature profiles, the bulk penetration resistance and finally the drying rate during each static period. The overall heat conductivity of the dry packed bed λ_{bed} needed for the calculation can be predicted from standard equations such as those summarized in Appendix B.

The last remaining problem is to estimate the length of the fictitious static period. This length t_R is postulated to be a function of the time scale of the mixing device (reciprocal stirrer speed in 1/min, drum revolutions per unit time, etc.) and has been defined as

$$t_R = t_{mix} \cdot N_{mix}. \qquad (1)$$

The quantity N_{mix}, the so called 'mixing number,' is a mechanical property of the mixing device in connection with the mechanical properties of the product, and a function of the time scale of the stirrer. It must be fitted to experimental results. N_{mix} simply says how often the mixing device must have turned around before the product has been ideally mixed once (within the scope of the penetration model).

The model briefly outlined above yielded a very good agreement between experiment and calculation for numerous drying rate curves measured in the various contact dryers shown in Fig.3. The pressure has been varied between 1 and 200 mbar, the overall temperature difference between 10 and 200°C, the particle diameter between 170 and 6600 µm and the speed of the stirrer between 0.2 and 130 revolutions per minute. The granulate material dried has been magnesium silicate, aluminium silicate, 'Perlkontakt' and PVC. The mixing number was found to lie between 2 and 25 and has been correlated for each dryer type to the Froude number. The same values of N_{mix} can be used to describe the heat transfer from the surface of immersed bodies to stirred beds of dry particulate material [5].

4. CONTACT DRYING OF MULTIGRANULAR PACKINGS

In order to apply the Neumann's solution to calculate the drying rate, the properties of the dry bed such as the density, ρ_{bed}, the specific heat capacity, c_{bed}, and the effective thermal conductivity λ_{bed}, must be constant (independent of temperature and length coordinate). The same must be valid for the product $X \cdot \Delta h_{ev}$, which describes the intensity of the latent heat sink in the wet bed. In the following it will be discussed, whether these assumptions are fulfilled in the case of multigranular packings. We distinguish between stochastic homogeneous and understochastic (inhomogeneous) packings.

5. STOCHASTIC HOMOGENEOUS MULTIGRANULAR PACKINGS

A multigranular packing is called to be stochastic homogeneous if the probability of a particle lying at a specific location in the bed is the same for all possible locations. From this definition follows that all physical properties of such a packing are independent of the length coordinate. The Neumann's solution can thus be used to predict the bulk penetration coefficient and the drying rate. One should however dispose of adequate equations for the calculation of the dispersity dependent properties of the packing. Such properties are the effective thermal conductivity λ_{bed}, the density ρ_{bed} or the corresponding bed porosity ψ, and the heat transfer coefficient wall-to-first particle layer, α_{ws}.

6. PREDICTION OF λ_{BED}, ψ AND α_{WS} HOMOGENEOUS MULTIGRANULAR PACKINGS

To predict the heat conductivity of the dry bulk material the method summarized in Appendix B has been used. Bauer [6] presents calculated and measured values of the overall heat conductivity of binary mixtures of spheres. For a constant volumetric composition of the mixture the number of contact points between spheres and consequently the heat conductivity of the bed increases with increasing diameter ratio of the involved fractions. For a constant diameter ratio the number of contact points changes with the composition of the mixture, so that in each case a maximal value is attained.

A model first proposed by Wise [7], see also [8], has been used for the calculation of the bed porosity. Assuming that all spheres touch their neighbours the structure of a random packing of spheres of different size can be reduced to a system composed of known tetrahedral subunits. The frequency distribution of these subunits can be calculated from the sphere size distribution using simple statistics and geometry. The overall porosity is then given by the sum of the porosities of the individual tetrahedra weighted by their relative frequencies. The absolute values obtained are unrealistically low, but can be corrected using a multiplying factor to simulate a moving apart of all the spheres. In this manner corrected, the model yields an at least qualitatively correct description of the relatively well known variation of the porosity with composition for binary mixtures of spheres. The porosity of the packing is maximal for equal spheres (ψ = 0.40), diminishing with any distribution in particle size. Plotting the porosity of a binary mixture over the mixture proportion yields nonsymmetric V-curves. The addition of few small spheres to a packing of large ones produces a large change in porosity. This change is all the more marked the greater the diameter ratio. The calculation for polydispersed packings renders to be cumbersome. In order to save calculating effort such packings have been simulated with bidispersed ones.

A very similar method has been used to predict the heat transfer coefficient α_{ws} of bidispersed, stochastic homogeneous packings. The elementary unit of the structure in this case is an orthogonal, triangular prism. The contact heat transfer coefficient calculated in this way is a monotonously increasing, convex function of the proportion of fine grained product in the mixture.

7. CALCULATED DRYING RATE CURVES FOR STOCHASTIC HOMOGENEOUS MULTIGRANULAR PACKINGS

Figure 4 shows calculated drying rate curves or a bidispersed stochastic homogeneous packing. Q_f is the mass proportion of fine product in the packed bed. For $Q_f = 0$ we have the drying rate curve of the coarse granulate. Adding fine product causes a rapid increase of drying rate, see curve for $Q_f = 0.05$. This can be explained with the increased density, thermal conductivity and contact coefficient of the bidispersed packing in comparison to the coarse monodispersed one. A maximal drying rate is attained for a Q_f of about 0.45. Further addition of small particles to the mixture has a decrease of the drying rate as result. This effect is due to the effective bed conductivity and the bed density decreasing after having reached a maximum. For $Q_f=1$ we obtain the drying rate curve of the fine product.

The behaviour illustrated in Fig.4 could be of great practical interest, for mixing of small particles in a bed of coarse material would drastically diminish the time or heating area required to accomplish the drying duty. It must however be borne in mind that this is the case only if the stirring device can provide and maintain a random distribution of particles in the bed. In the following sections it will be discussed, whether this is likely to happen in conventional technical apparatuses or not.

8. UNDERSTOCHASTIC MULTIGRANULAR PACKINGS

When the particles composing a mixture differ in size and/or density, mixing is accompanied by de-mixing. Continued operation of the mixer leads to an equilibrium state in which the components are in most cases partially separated. We will call such packings understochastic.

A multicomponent packing in a contact dryer is very likely to be understochastic. This would have the following consequences:

- The physical properties of the packed bed, such as density and thermal conductivity would be functions of the distance from the heating plate.

- The mixing at the end of the contact time assumed by the penetration model would no more be able to level the moisture profile. The moisture content and consequently the intensity of the heat sink would thus be a function of the distance from the heating plate.

Under such circumstances the solution of Neumann can not be applied to calculate the drying rate. In order to develop a model predicting the drying rate curve, information about the exact character of de-mixing is necessary. This will obviously depend on the type of dryer used. The following sections contain a detailed experimental and theoretical investigation of a disc dryer and a brief discussion about drum dryers.

9. EXPERIMENTAL SET UP

The experimental investigations reported have been made with the apparatus shown in Fig.5. The drying is carried out in a tubular container. The bottom of this container consists of an aluminium plate with a thickness of 5 mm and a diameter of 240 mm, which can be heated by an electrical resistor. In order to avoid heat losses a second heating plate is situated below the first one and is kept at the temperature of this. For the same reason the side walls of the container are insulated. The container is placed on a pneumatic device and can be set down onto a balance. The drying rate can in this manner be calculated from the mass change of the entire packing. This procedure excludes mistakes connected with collecting samples and is thus especially for the multigranular packings used in this investigation of great advantage. Container and balance lie in an autoclave, in order to maintain an atmosphere without inert gas. The granulated product can be mechanically agitated by a stirrer driven by an electrical motor through a vacuum tight coupling. The periphery mainly consists of two vacuum pumps and two methanol cooled condensors, to remove the vapor during the drying process. The methanol is pumped from a storage vessel through two cryostates to the condensors. The temperature of the heating plate and the pressure in the autoclave can be held constant for the duration of a run.

The stirrer shown in Fig.6, equipped with bristles directly contacting the heating surface, has been used as agitating unit. This provided a good mixing intensity and minimized the grinding of particles between stirrer blade and heating plate.

10. MATERIAL

The granulated material used for the experiments was porous aluminium silicate with a diameter range of 0.4 to 5.0 mm. Narrow fractions of this material gained by sieving have been regarded as monodispersed and used to produce bi- and poly-dispersed packings. The density of dry, monodispersed beds, slightly varying with the particle diameter, was about 1000 kg/m^3 and the specific heat capacity about 800 J/kgK. The particle porosity, being a weak function of particle diameter, was about 0.33, the thermal conductivity of the particles has been estimated as 1.0 W/mK and the particle roughness as 2.5 µm. The mean pore diameter, measured in means of mercury porosimetry was found to be about 0.6 µm. In order to guarantee the flowability of the heap the surface moisture of the particles was removed before each run.

11. MEASURED AND CALCULATED DRYING RATE CURVES OF MONODISPERSED PACKINGS

In the Figs. 7-9 some measured drying rate curves of monodispersed packings are compared with calculated ones.

The parameter in the runs shown in Fig. 7 is the particle diameter. The drying rate for fine material is, as expected, much higher than this for coarse material. This results from the contact heat transfer coefficient strongly increasing with decreasing particle size.

Fig.8 shows the effect of heating temperature on the drying rate curve.

The runs depicted in Fig.9 demonstrate how intensification of product mixing affects the drying rate. The effect of mixing on the drying rate is low when the actual drying rate comes close to the maximal one. This is the case for coarse grained material. For such material the contact resistance is rate controlling. The drying rate of fine product is on the contrary rather sensitive to variation of the mixing intensity. The bulk resistance of a fine grained packing tends to become drying rate controlling.

The values of N_{mix} used for the calculation are presented in Table 1.

n 1 / min \ d µm	4353	2413	1120	525
30	19.8	-	4.0	4.0
15	15.0	10.0	3.0	3.0
1	5.1	-	-	1.0

Table 1: Values of N_{mix} used to calculate drying rate curves of monodispersed material

The values for n = 15 1/min have been fitted to experimental data. In order to determine N_{mix} for other stirrer velocities the correlation given by Mollekopf [3] for disc dryers has been used. The mixing quality obviously depends on the particle diameter. This dependency is not due to flow or heat penetration resistances in the interior of particles. It is on the contrary a mechanical property of the stirring device used here, not being observed for other forms of agitators. The value of N_{mix} for fine grained material is almost the same to that used by Schlünder [5] in order to correlate heat transfer measurements made by Wunschmann [9] in an identical disc apparatus equipped with a bristlestirrer.

12. DRYING RATE CURVES OF BI- AND POLYDISPERSED PACKINGS

The upper curve depicted in Fig.10 is the drying rate curve of a fine grained monodispersed packing (d = 0.525 mm) and the lower the drying rate curve of a coarse grained one (d = 4. 353 mm). The remaining three curves are drying rate curves of bidispersed packings produced by mixing of the fine and the coarse monodispersed granulates in various mass proportions. The drying conditions were for all runs in Fig.10 the same. These results are typical for the form of drying rate curves of bidispersed material in the disc dryer. The drying rate, being at the beginning of drying almost equal to that of fine grained monodispersed product, decreases more or less steeply with decreasing moisture content of the bed. A bend-point can be observed in all three curves. The location of this point is obviously directly proportional to the proportion of fine product in the bed. The drying rate after the bend-point keeps on decreasing, at first slowly and then for small moisture contents more steeply, approaching null at the end of drying.

The form of measured drying rate curves is obviously gravely different from that expected for stochastic homogeneous packings. The mechanical agitation of multigranular packings in a disc dryer is evidently accompanied by de-mixing leading to a strongly understochastic packing. Visual observation of the packed bed revealed a far-going segregation of the two monodispersed fractions. The fine particles formed a layer lying on the heating plate and blocking it to coarse material. This observation can be used to explain the form of the drying rate curves, being divided to two rather distinct regions. In the first region the drying rate is dictated by the drying of fine product contacting the heating plate, in the second by the drying of coarse particles, the fine ones having already been dried out.

The assumption of drying taking place in two far-reaching distinct regions is supported by the data shown in Fig.11. For the runs depicted with the triangles dry fine product has been mixed to wet coarse material. The resulting drying rate curves are identical to the drying rate curve of the fully wetted packing after a moisture content of about 0.12. From the practical point of view the question arises, whether mixing of fine product to a coarse packing would increase the drying rate of the later or not. The presence of fine granulate would in all cases considerably diminish the contact resistance between wall and packing, but would also give rise to an additional heat transfer resistance (heat penetration through the fine layer) not existing for the monodispersed coarse material. The sum of these two resistances is to be compared to the contact resistance of the coarse material, and can in principle be lower or higher than this. For the system examined in this work the fine granulate causes generally a slight improvement of the drying rate of the coarse one, see Fig.10.

Fig.12 shows drying rates obtained for some polydispersed packings. The composition of the packings is depicted in Fig.13. The form of the curves is very similar to that of bidispersed material (compare the drying rate curve of packing 'a' to that of the bidispersed one).

13. CONTACT DRYING OF BIDISPERSED PACKINGS IN A DISC DRYER / MODEL

In order to calculate the drying rate curves of bidispersed packings in the disc dryer we idealise the existing de-mixing, and assume a complete stratification of the two fractions. We further subdivide the drying process in two parts,

- drying of the fine layer contacting the heating plate

drying of the coarse layer lying on the fine one,

in the following to be separately discussed. The boundary between fine and coarse product is assumed to be parallel to the heating plate, distinct and during the drying of fine granulate adiabatic.

1) Drying of the fine layer contacting the heating plate. The drying rate of the packing during this period will be equal to the drying rate of the fine layer. The process will proceed as if no coarse material would be existent. The solution of Mollekopf and Schlünder can thus be applied in unchanged form.

2) Drying of the coarse layer

The drying rate of the packing will now be equal to that of the coarse material. We continue to think in terms of the penetration theory and attempt to calculate the drying rate for the first static period after complete drying out of the fine product. At the beginning of this period the fine granulate will be at a temperature $T_{f,0}$ and the coarse one at T_S (saturation temperature). During the static period a drying front is assumed to penetrate into the bulk of coarse material. The situation is illustated in Fig.14. The general form of temperature profiles aside the layer boundary will be

$$T_f = A + B \operatorname{erf} \frac{z}{2\sqrt{\kappa_{bed,f}\, t}} \qquad (2a)$$

$$T_c = C + D \operatorname{erf} \frac{z}{2\sqrt{\kappa_{bed,c}\, t}} \qquad (2b)$$

Using the following boundary conditions

$$z \to -\infty \Longrightarrow \lim T_f = T_{f,0} \qquad (3a)$$

$$z = z_T \Longrightarrow T_c = T_S \qquad (3b)$$

$$-\lambda_{bed,c}\frac{\partial T_c}{\partial z} = \rho_{bed,c}\, X_c\, \Delta h_{ev}\frac{dz}{dt} \quad (3c)$$

$$z=0 \Rightarrow \lambda_{bed,c}\frac{\partial T_c}{\partial z} = \lambda_{bed,f}\frac{\partial T_f}{\partial z} \quad (3d)$$

$$T_c = T_f \quad (3e)$$

one can calculate the constants A,B,C and D:

$$A = C = \frac{T_S + F \operatorname{erf}\zeta\, T_{f,0}}{1 + F \operatorname{erf}\zeta} \quad (4a)$$

$$B = \frac{T_S - T_{f,0}}{1 + F \operatorname{erf}\zeta} \quad (4b)$$

$$D = \frac{F(T_S - T_{f,0})}{1 + F \operatorname{erf}\zeta} \quad (4c)$$

F is the ratio of the heat penetration coefficients of the two layers

$$F = \frac{\alpha^{dry}_{sb,f}}{\alpha^{dry}_{sb,c}} \quad (5)$$

with

$$\alpha^{dry}_{sb,f} = \frac{2}{\sqrt{\pi}}\frac{\sqrt{(\lambda\rho c)_{bed,f}}}{\sqrt{t_R}} \quad (6a)$$

$$\alpha^{dry}_{sb,c} = \frac{2}{\sqrt{\pi}}\frac{\sqrt{(\lambda\rho c)_{bed,f}}}{\sqrt{t_R}} \quad (6b)$$

ζ is the reduced instantaneous position of the drying front,

$$\zeta = \frac{z_T}{2\sqrt{\kappa_{bed,c}\, t}} \quad (7)$$

and can be calculated from the following equation

$$\sqrt{\pi}\,\zeta\, e^{\zeta^2}(1+F \operatorname{erf}\zeta) = \frac{c_{bed,c}\, F(T_{f,0} - T_S)}{X_c\, \Delta h_{ev}} \quad (8)$$

The. time averaged heat flux at the layer boundary as well as at the drying front can be calculated from the gradients of the temperature profile at the respective positions. It follws

$$\dot{q}_{Lo} = \left(\frac{1}{\alpha^{dry}_{sb,f}} + \frac{\operatorname{erf}\zeta}{\alpha^{dry}_{sb,c}}\right)^{-1}(T_{f,0} - T_S) \quad (9)$$

$$\dot{q}_{lat} = \dot{q}_{bo}\, e^{-\zeta^2} \quad (10)$$

The average drying rate during the fictitious static period observed can be calculated from $\dot{q}_{lat}$

$$\dot{m} = \frac{q_{lat}}{\Delta h_{ev}} \quad (11)$$

The decrease in moisture content of the packing is then

$$\Delta X = \frac{\dot{m}\, t_R\, A}{M_{dry}} \quad (12)$$

The change of the average bulk temperature of the (dry) fine product is

$$\Delta T_f = \frac{(\dot{q}_o - \dot{q}_{bo})\, t_R\, A}{Q_f\, M_{dry}\, c_{bed,f}} \quad (13)$$

with $\dot{q}_0$ the time averaged heat flux at the heating surface

$$\dot{q}_0 = \left(\frac{1}{\alpha_{ws}} + \frac{1}{\alpha^{dry}_{sb,f}}\right)^{-1}(T_W - T_{f,0}). \quad (14)$$

Assuming that, after perfect mixing, there is no heat transfer at all from dry and warm to wet and cold particles, see also [3], one can calculate the change of average bulk temperature of the coarse material

$$\Delta T_c = \frac{(\dot{q}_0 - \dot{q}_{bo})\, t_R\, A}{Q_c\, M_{dry}(c_{bed,c} + c_L x_c)} \quad (15)$$

Eqns.(12), (13) and (15) provide the values of moisture content and bulk temperature at the beginning of the second static period. The entire drying rate curve can thus be stepwise calculated.

The length of the static period t_R can be correlated with the time constant of the mixing device using the mixing number N_{mix}, see eqn. (1). In order to avoid renewed fitting on experimental data, we make the sweeping assumption that N_{mix} has for each layer the same value as for the corresponding monodispersed packing (see Table 1). It is then:

$$\alpha^{dry}_{sb,c} = \frac{2}{\sqrt{\pi}}\frac{\sqrt{(\lambda\rho c)_{bed,c}}}{\sqrt{t_{mix}\, N_{mix,c}}} \quad (16a)$$

$$\alpha^{dry}_{sb,f} = \frac{2}{\sqrt{\pi}}\frac{\sqrt{(\lambda\rho c)_{bed,f}}}{\sqrt{t_{mix}\, N_{mix,f}}}\, K_f \quad (16b)$$

K_f is a factor to correct the heat transfer coefficient of the fine layer for the deviation of reality from the assumed state of perfect stratification. By comparison of calculated and measured drying rate curves the value $K_f = 2$ has been found. This value has been successfully used to correlate all existing measurements and appears to be not only independent of heating temperature and pressure, but, for the investigated apparatus, also of packing composition and stirrer speed.

The calculation method presented above contains an inevitable discontinuity connected with transition from the fine-product-drying region to the coarse-product-drying region. This discontinuity can easily be smoothed away by exrapolating the drying rate curve for the coarse-product-drying region till this meets the one for the fine-product-drying-region.

The comparison between measurement and calculation is illustrated in Figs.15,16 and 17, using some representative runs. The same method has been used with fairly good results to predict drying rates of polydispersed packings. In order to do this the polydispersed packing had to be divided to two parts, each of them assumed to be stochastic homogeneous.

14. DRYING OF MULTIGRANULAR PACKINGS IN THE ROTARY DRUM DRYER

No measurements have been made in a drum dryer. It will however be attempted to discuss briefly the situation likely to exist in such an apparatus, the line of attacking the problem remaining mainly the same.

The de-mixing of multigranular packings in a rotary drum dryer need not and will not be the same as in the disc dryer. Possible de-mixing patterns are given by Donald and Roseman [10], s. Fig.18.

The radial de-mixing, depicted in Fig.18 as pattern A, can be interpretated as a sort of series arrangement of monodispersed or stochastic homogeneous polydispered regions in respect to the heating plate. It is the inverse case compared to the disc dryer, for the coarse particles are now contacting the heating wall blocking the fine granulate away. The coarse granulate should thus dry at first. Drying of the fine material would follow subsequently, the drying rate being gravely decreased in comparison to that of monodispersed fine product. The stratification model presented in this paper would be suitable for the prediction of the drying rate.

Pattern C is an axial de-mixing, dividing the packing in monodispersed or stochastic homogeneous polydispersed bands, each of them directly contacting the heating plate (arrangement in parallel). The drying rate could obviously be predicted as an average, weighted with the corresponding volume proportions or surface coverages, of the drying rate of fine and that of coarse product.

Pattern B is an hybrid situation between pattern A and pattern C. It could be handled as a combination of monodispersed regions arranged partly in series and partly parallel to each other.

According to Donald and Roseman the de-mixing pattern predominating depends on the static angles of repose of the components of the mixture.

15. CONCLUDING REMARKS

Contact drying of granular material is commonly carried out in agitated units. If the particles of the packing differ in size or density, the mechanical agitation is likely to lead to segregation (de-mixing) of the mixture components. Such occurances have a strong influence on the drying rate curve. Under certain circumstances mixing of dry, fine grained material to a bed of coarse grained product can improve the drying rate of the later. The model introduced by Mollekopf and Schlünder to predict the drying rate during vacuum contact drying of free flowing, mechanically agitated, particulate material can be extended in order to describe such cases.

NOMENCLATURE

A	hot surface are, m^2
A,B,C,D	constants, eqns. (2a,2b)
c	specific heat, J/kgK
d	particle diameter, m
F	ratio of heat penetration coefficients, eqn.(5)
Δh_{ev}	latent heat of evaporation, J/kg
K_f	correction coefficient, eqn. (16b)
$\dot{m}$	drying rate, kg/m^2s
M	mass, kg
n	number of revolutions of mixing device,1/s
N_{mix}	mixing number, eqn. (1)
P	pressure
$\dot{q}$	heat flux, W/m^2
Q	mass of component to total mass of packing (dry)
t	time, s
T	temperature, K
X	moisture content, (kg moisture) / (kg dry mat)
z	length normal to hot surface, m
z_T	penetration depth, m
α	heat transfer coefficient, W/m^2K
β	mass transfer coefficient, m/s
δ	surface roughness, m
ζ	reduced penetration depth
κ	thermal diffusivity, m^2/s
λ	thermal conductivity, W/mK
ρ	density, kg/m^2
ψ	porosity

Indices

bed	bed
bo	at the boundary
c	coarse grained product
dry	dry
f	fine grained product
lat	latent
L	liquid
mix	mixing
R	static period
S	saturation
sb	bulk
W	wall

W wall

ws wall-to-bed

0 at the beginning

0 at the hot surface

REFERENCES

1. Gunes, S., Ph.D. Thesis, Karlsruhe (1979)
2. Carn, R.M., and King, J.C., AIChE Symposium Series, No.163, 73, 103.
3. Mollekopf, N., Ph.D. Thesis, Karlsruhe (1983)
4. Schlünder, E.U., and Mollekopf, N., Chem. Eng. Process., 18 (1984)
5. Schlünder, E.U., Chem. Eng. Process., 18(1984) 31
6. Bauer, R., Int. Chem. Eng., 18(1978) 181
7. Wise, M.E., Philips Res. Rep., No.5,7 (1952) 321.
8. Dodds, J.A., Journal of Colloid and Interface Science, No.2, 77 (1980) 317
9. Wunschmann, J., Ph.D. Thesis, Karlsruhe (1974)
10. Donald, M.B. and Roseman, B., Brit. Chem. Eng., No.10, 7 (1962) 749.
11. Schlünder, E.U., Chem. - Ing.-Tech., 43(1971),651
12. Mollekopf, N., and Martin, H., VT-Verfahrenstechnik, 16(1982) 701
13. Zehner, P., VDI-Forschungsh., 558(1973)

APPENDIX A

Prediction of the contact heat transfer coefficient α_{ws}

The main reason for the contact heat transfer resistance is the fact that the heat conductivity of the gas in the wedge between particle and wall goes to zero when approaching the contact point, as set forth in ref.11. Later slight modifications of the original equation were published. The equation as recommended in ref.12 is used in this paper:

$$\alpha_{ws} = \phi_A \alpha_{wp} + (1-\phi_A) \frac{2\lambda_G / d}{\sqrt{2} + (2l + 2\delta)/d} + \alpha_{rad} \tag{A-1}$$

α_{wp} is the heat transfer coefficient for a single particle and is to be calculated by

$$\alpha_{wp} = \frac{4\lambda_G}{d}\left[\left(1 + \frac{2l + 2\delta}{d}\right).\ln\left(1 + \frac{d}{2l + 2\delta}\right) - 1\right]. \tag{A-2}$$

λ_G is the continuum heat conductivity of the gas. ϕ_A is a plate surface coverage factor and is of tne order of 0.80. d is the particle diameter and δ the roughness of the particle surface. l is the modified mean free path of the gas molecules and follows from

$$l = 2\,\frac{2-\gamma}{\gamma}\sqrt{\frac{2\pi\tilde{R}T}{\tilde{M}}}\;\frac{\lambda_G}{p(2c_{p,G} - \tilde{R}/\tilde{M})} \tag{A-3}$$

where γ is the accommodation coefficient, which is around 0.8 to 0.9 for normal gases at moderate temperatures.

α_{rad} accounts for heat transfer by radiation and can be calculated from the linearized Stefan-Boltzmann law:

$$\alpha_{rad} = 4C_{12}\;T^3 \tag{A-4}$$

with

$$C_{12} = \sigma \frac{1}{\frac{1}{\varepsilon_{wall}} + \frac{1}{\varepsilon_{bed}} - 1} \tag{A-5}$$

being the overall radiation exchange coefficient, which follows from the black body radiation coefficient $\sigma = 5.67 \times 10^{-8}$ W $m^{-2} K^{-4}$, and the emissivities of the wall surface ε_{wall} and the bed surface ε_{bed} as well.

APPENDIX B

Prediction of the dry bulk heat conductivity λ_{bed}

Various correlations for the prediction of λ_{bed} have been published over the last decades. The most elaborated and reliable one seems to be the correlation developed by Zehner [13] and Bauer [6]. It applies to monodispersed as well as to polydispersed packed beds of spherical and non-spherical particles of poor and good conductors within a wide temperature and pressure range ($100 < T < 1500$ K, $10^{-3} < p < 100$ bar). The calculation procedure has been summarized in ref. 3 and is repeated here. According to refs.13 and 6 the overall heat conductivity of packed beds λ_{bed} depends on the following parameters:

$$\lambda_{bed} = f(\lambda_S, \lambda_G, \lambda_R, \lambda_D, d, \psi, \phi_K, C_{Form}, f(\zeta_r)).$$

These parameters are:

λ_S	heat conductivity of particles
λ_G	heat conductivity of gas
λ_R	equivalent heat conductivity due to radiation
λ_D	equivalent heat conductivity due to molecular flow
d	particle diameter
ψ	void fraction
ϕ_K	relative flattened particle surface contact area
C_{Form}	particle shape factor
$f(\zeta_r)$	particle size distribution function

As set forth in refs. 13 and 6 the following correlations may be recommended for the prediction of λ_{bed}:

$$\frac{\lambda_{bed}}{\lambda_G} = (1 - \sqrt{1-\psi})\left[\frac{\psi}{\psi - 1 + \lambda_G/\lambda_D} + \psi\frac{\lambda_R}{\lambda_G}\right] \tag{B-1}$$

$$+ \sqrt{1-\psi}\left[\phi_K\frac{\lambda_S}{\lambda_G} + (1-\phi_K)\frac{\lambda'_{bed}}{\lambda_G}\right]$$

$$\frac{\lambda'_{bed}}{\lambda_G} = \frac{2}{K}\left\{\frac{B(\lambda_S/\lambda_G + \lambda_R/\lambda_G - 1)(\lambda_G/\lambda_D)(\lambda_G/\lambda_S)}{K^2} \times \ln\frac{(\lambda_S/\lambda_G + \lambda_R/\lambda_G)\,\lambda_G/\lambda_D}{B[1+(\lambda_G/\lambda_D - 1)(\lambda_S/\lambda_G + \lambda_R/\lambda_G)]} + \frac{B+1}{2B}\left(\frac{\lambda_R}{\lambda_G}\frac{\lambda_G}{\lambda_D} - B\left[1+\left(\frac{\lambda_G}{\lambda_D} - 1\right)\frac{\lambda_R}{\lambda_G}\right]\right) - \frac{B-1}{K}\frac{\lambda_G}{\lambda_D}\right\} \tag{B-2}$$

Where

$$K = \frac{\lambda_G}{\lambda_D}\left[1+\left(\frac{\lambda_R}{\lambda_G} - B\frac{\lambda_D}{\lambda_G}\right)\frac{\lambda_G}{\lambda_S}\right] - B\left(\frac{\lambda_G}{\lambda_D} - 1\right)\left(1+\frac{\lambda_R}{\lambda_G}\frac{\lambda_G}{\lambda_S}\right) \tag{B-3}$$

and

$$B = C_{Form}\left(\frac{1-\psi}{\psi}\right)^{10/9} f(\zeta_r) \tag{B-4}$$

Further,

$$\frac{\lambda_R}{\lambda_G} = \frac{4\sigma}{2/\varepsilon_p - 1}\,T^3\,\frac{\chi_R}{\lambda_G} \tag{B-5}$$

ε_p = 0.85 being the emissivity of the particle surface

$$\frac{\lambda_G}{\lambda_D} = 1 + \frac{1}{\chi_D} \tag{B-6}$$

$$\chi_R = R_{Form}\, d \tag{B-7}$$

and

$$\chi_D = D_{Form}\, d \tag{B-8}$$

d is the equivalent particle diameter,

$$d = \sqrt[3]{6\,V/\pi} \tag{B-9}$$

V being the particle volume. R_{Form} and D_{Form} are shape factors for the interstitial energy transport by radiation and molecular flow, respectively.

If the packed bed consists of mass fractions Δz_j with various particle diameters d_i, χ_R and χ_D must be calculated as follows:

$$\frac{1}{\chi_R} = \sum_{i=1}^{i=n}\frac{\Delta z_i}{R_{Form,i}\, d_i} \tag{B-10}$$

$$\frac{1}{\chi_D} = \sum_{i=1}^{i=n}\frac{\Delta z_i}{D_{Form,i}\, d_i} \tag{B-11}$$

The particle size distribution function $f(\zeta_r)$ was found to be [6].

$$f(\zeta_r) = 1 + 3\zeta_1 \tag{B-12}$$

where the distribution parameter ζ_1 is given by

$$\zeta_1 = \left[\frac{\sum_{i=1}^{i=n}\frac{\Delta z_i}{d_i^2}}{\left(\sum_{i=1}^{i=n}\frac{\Delta z_i}{d_i}\right)^2} - 1\right]^{1/2} \tag{B-13}$$

The set of eqns. (B-1) - (B-13) contains three shape factors, such as C_{Form}, R_{Form} and D_{Form}, as well as the relative particle contact surface area ϕ_K, which must be evaluated from experiments. Some of the paramets have been evaluated in ref.13 and are listed in Table B-1.

Table B-1: Shape factors and particle to particle contact areas ϕ_K according to Ref.13

Particle shape	C_{Form}	R_{Form}	D_{Form}	ϕ_K
Spheres	1.25	1	1	0.0077 (Remarks Ceramic) 0.0013 (Steel) 0.0253 (Copper)
Cylinders	2.50	1	?	?
Hollow Cylinders	$2.5[1+\left(\frac{d_i}{d_a}\right)^2]$	1	?	?
Arbitrary shaped	1.4	1	?	0.001 (Sand) ? (Slay)

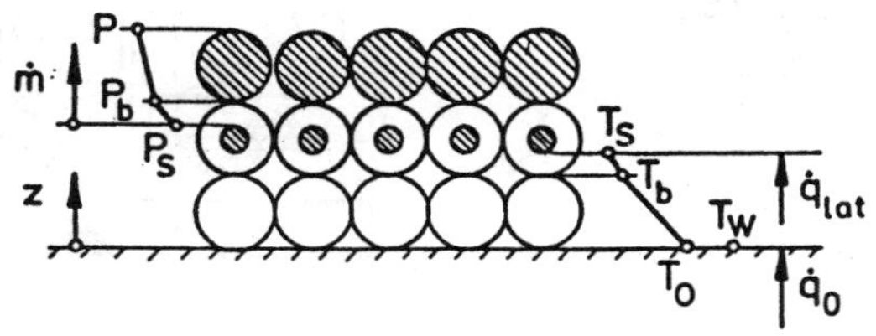

Heat in

Contact : $\frac{1}{\alpha_{WS}} = \frac{T_W - T_0}{\dot{q}_0}$

Bulk penetration : $\frac{1}{\alpha_{sb}} = \frac{T_0 - T_b}{\dot{q}_0}$

Particle penetrat.: $\frac{1}{\alpha_P} = \frac{T_b - T_s}{\dot{q}_{lat}}$

Mass out

Particle permeation: $\frac{1}{\beta_P} = \frac{P_S - P_b}{\dot{m}}$

Bulk permeation : $\frac{1}{\beta_b} = \frac{P_b - P}{\dot{m}}$

Fig.1 Heat and Mass Transfer resistances in contact drying

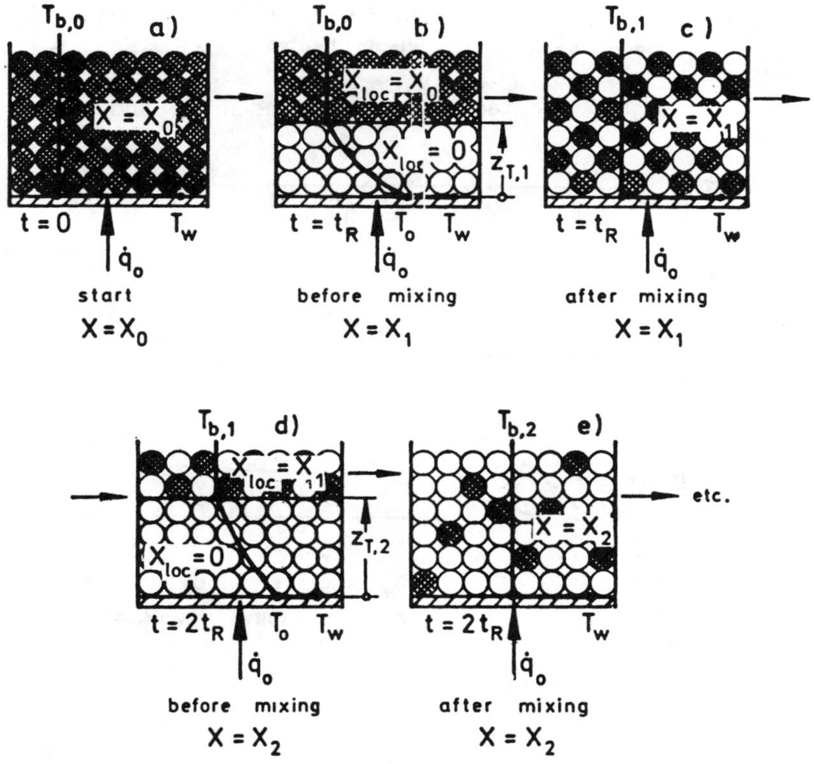

Fig.2 Illustration of the contact drying model for monodispersed material.

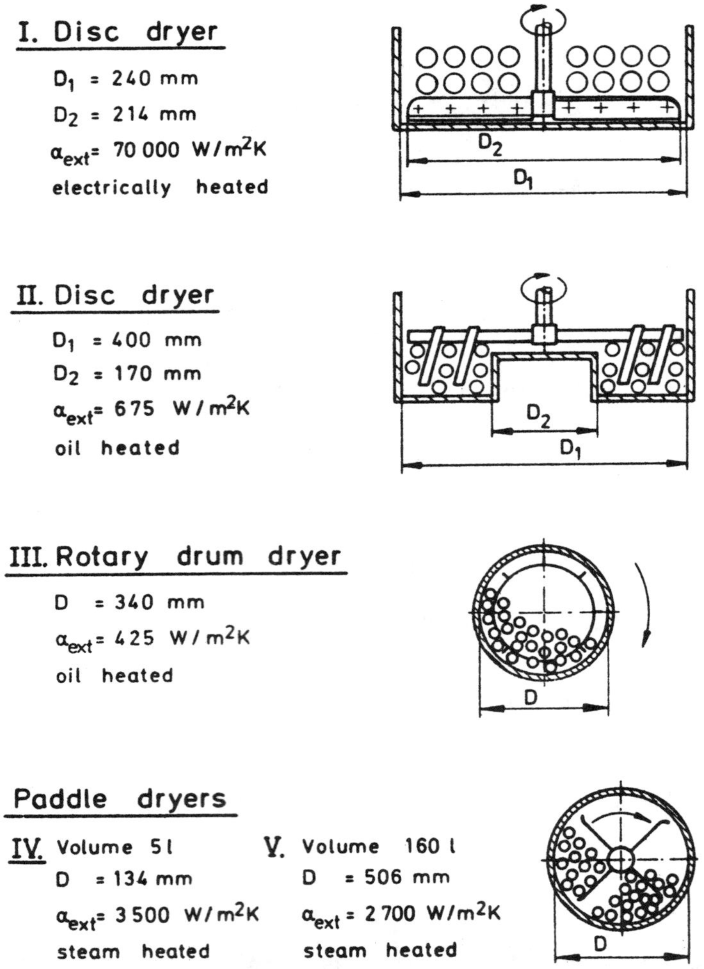

Fig.3 Contact dryers analyzed in [3] and [4]

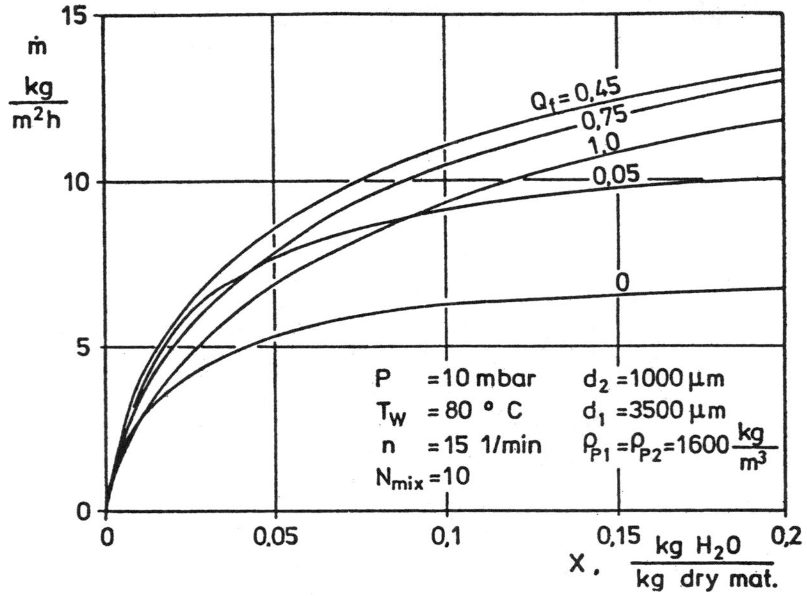

Fig.4 Calculated drying rate curves for several stochastic homogeneous, bidispersed packings.

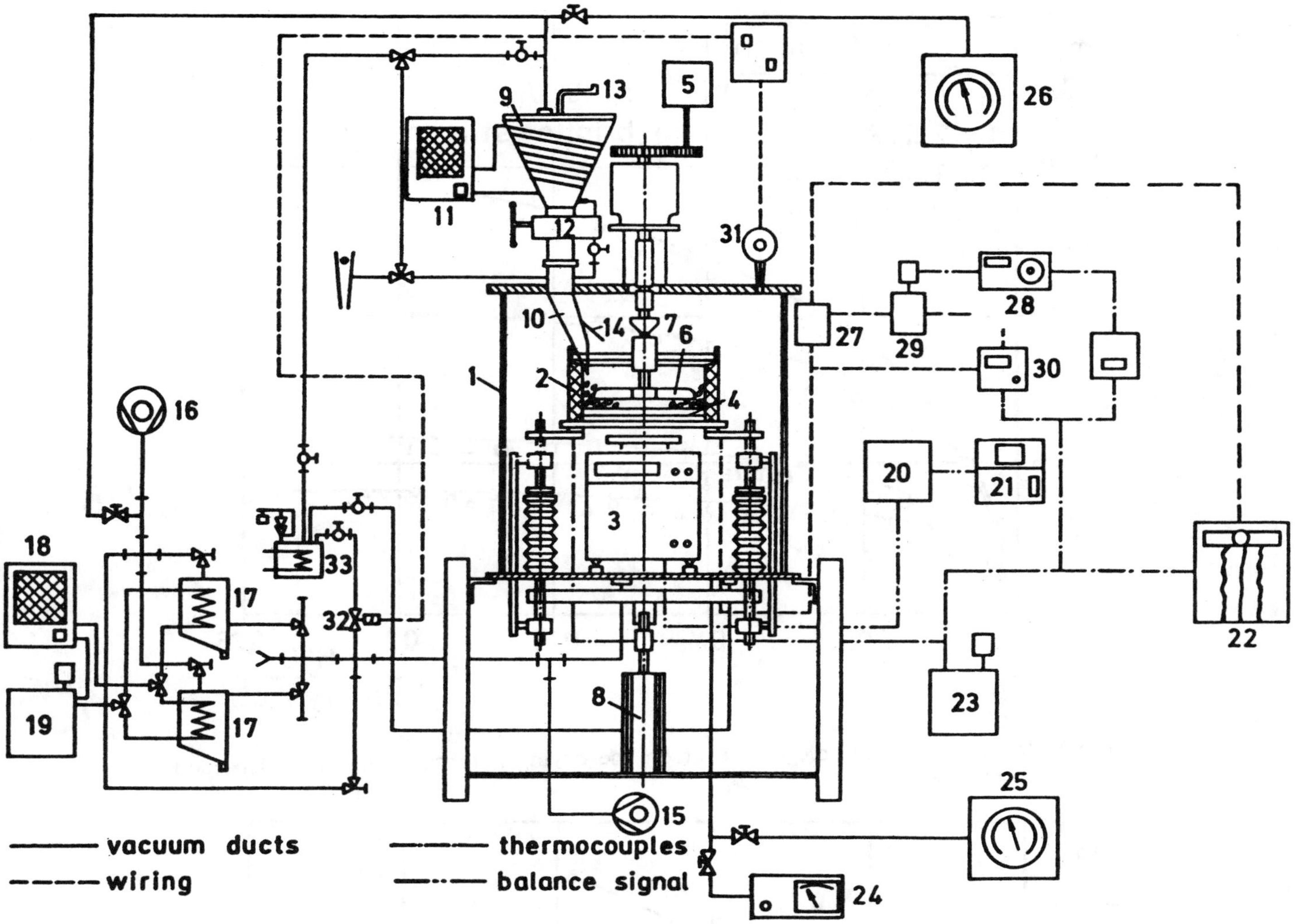

Fig.5 Experimental set up (disc dryer)

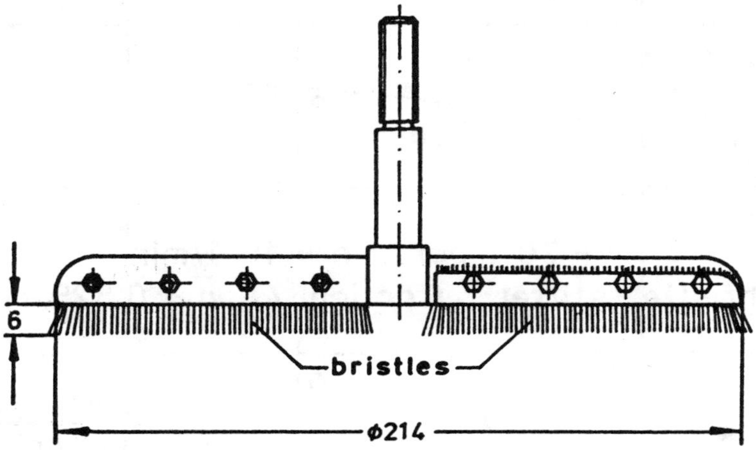

Fig.6 Stirring equipment (bristle-stirrer)

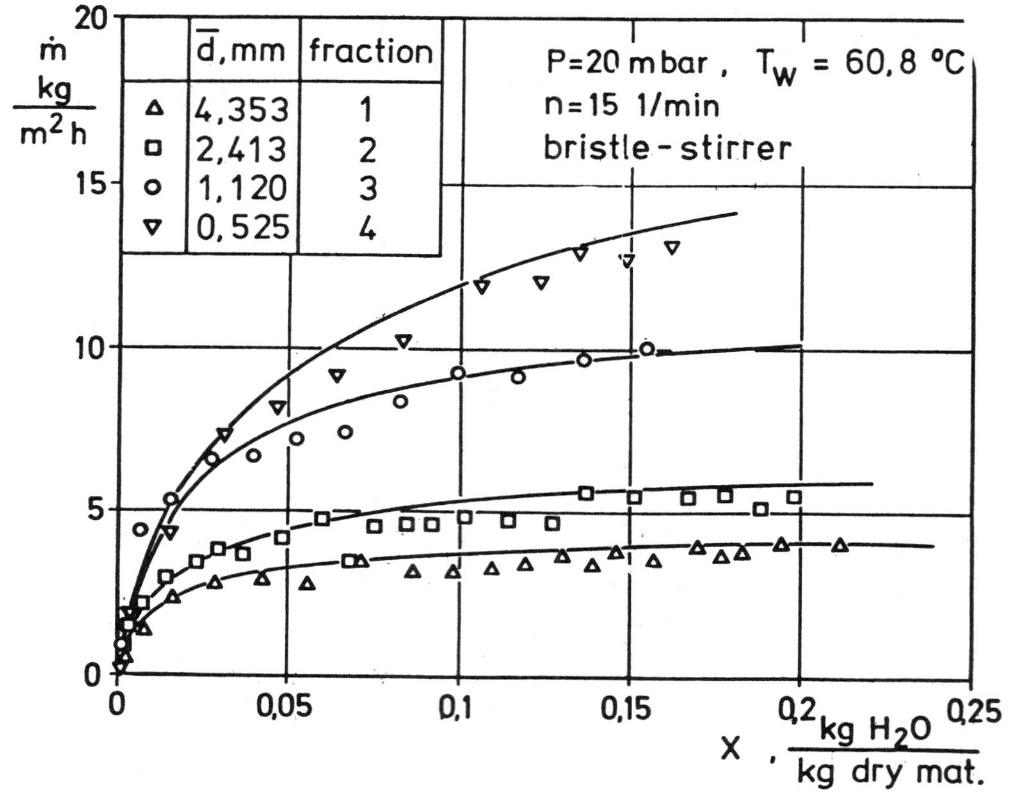

Fig.7 Influence of particle diameter on the drying rate curve of monodispersed packings.

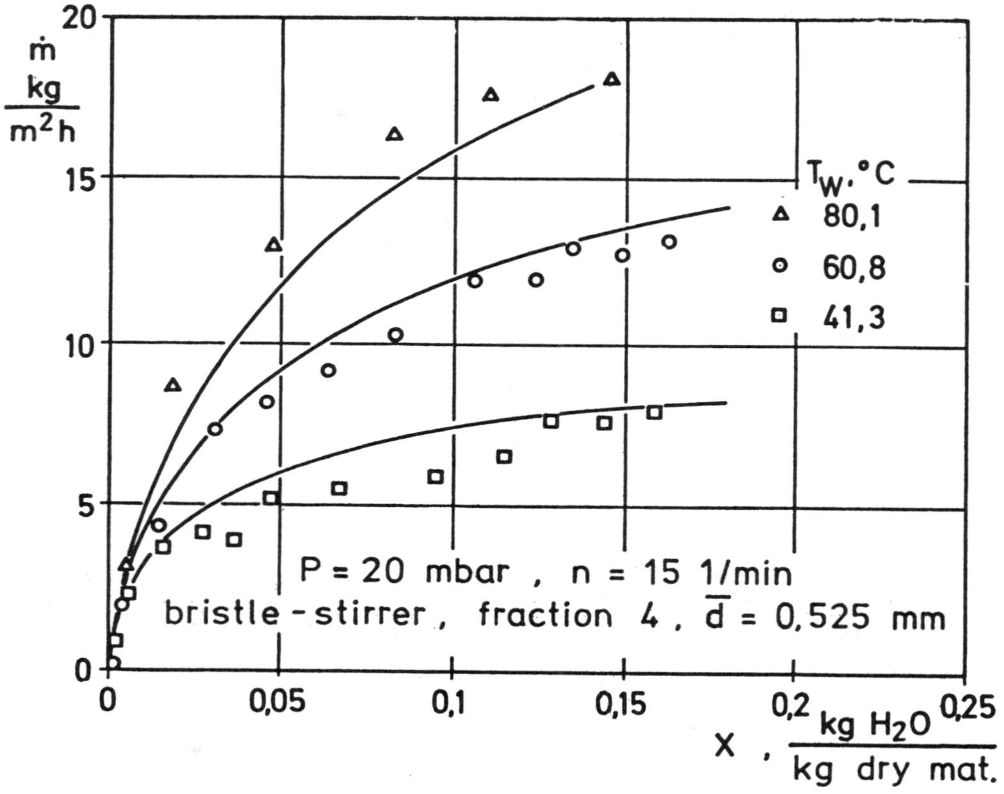

Fig.8 Influence of hot surface temperature on the drying rate curve of monodispersed packings.

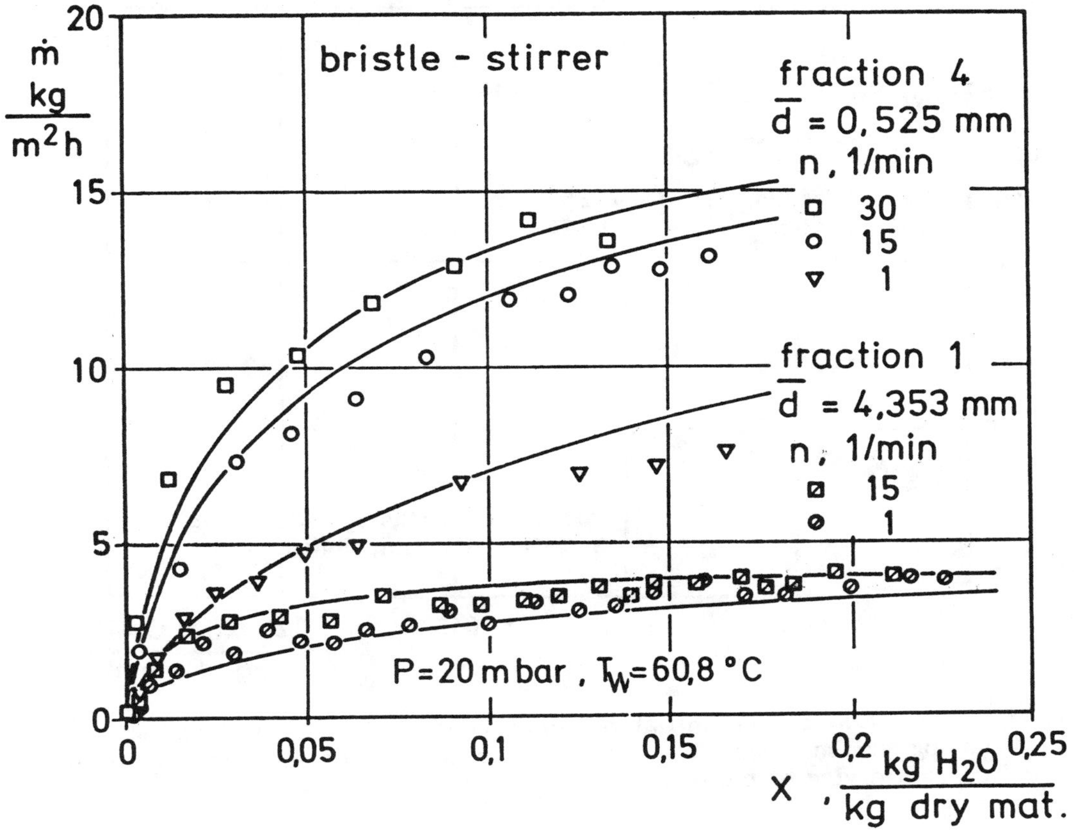

Fig.9 Influence of the number n of revolutions of the stirrer on the drying rate curve of monodispersed packings

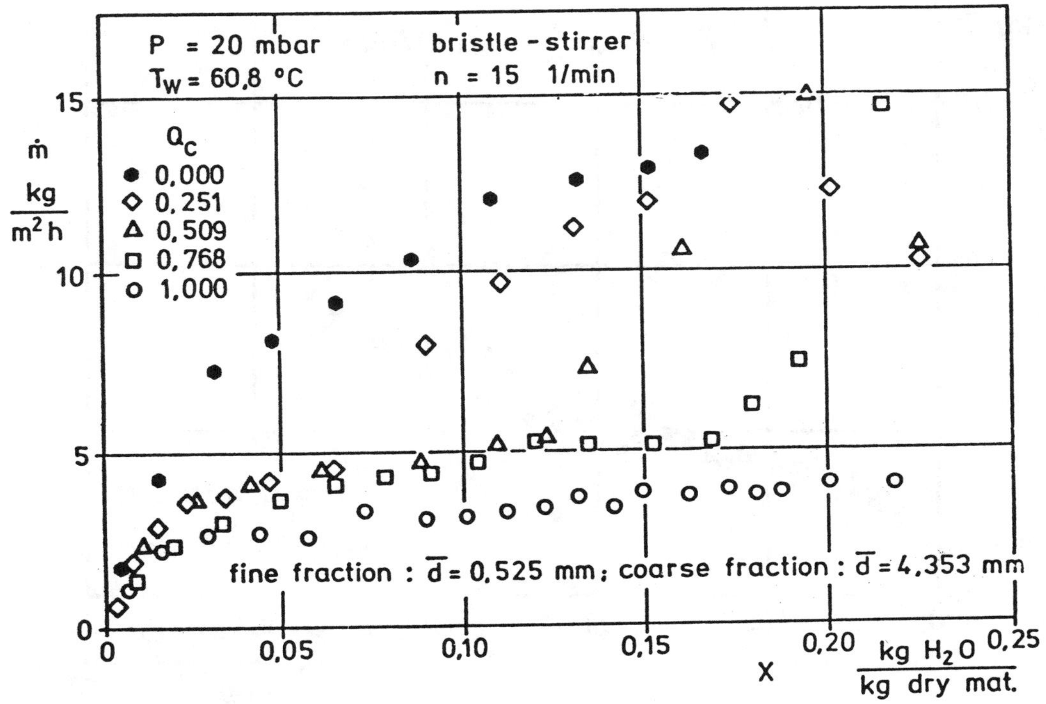

Fig.10 Measured drying rate curves of bidispersed packings. Variation of the composition of the packing.

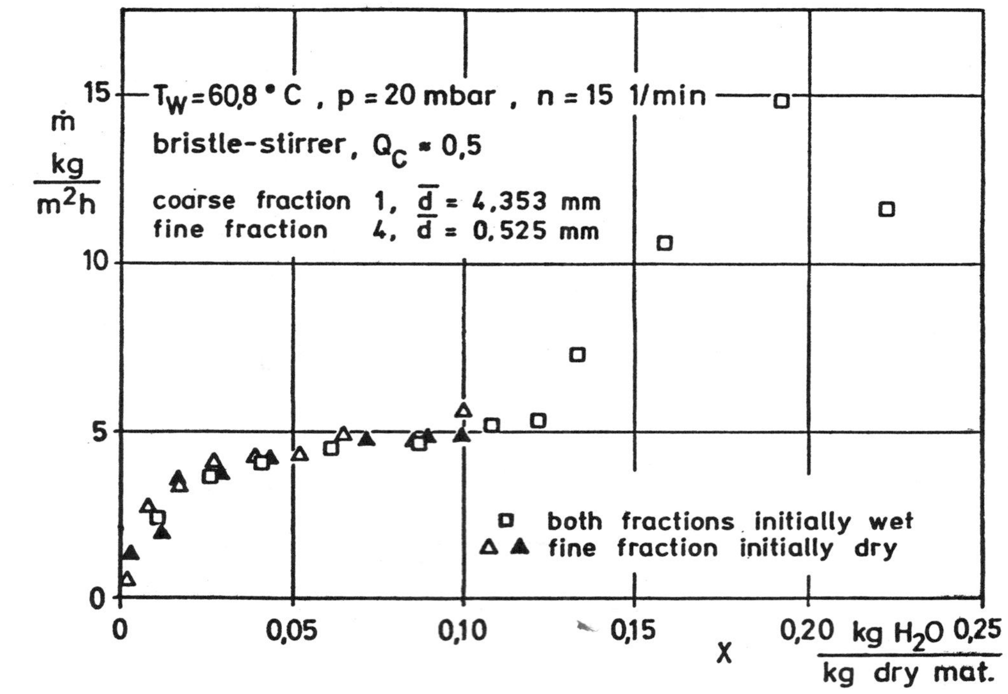

Fig.11 Illustration of the initial moisture content of fine grained product having no influence on the drying rate of coarse material, during contact drying of bidispersed packings in a disc dryer

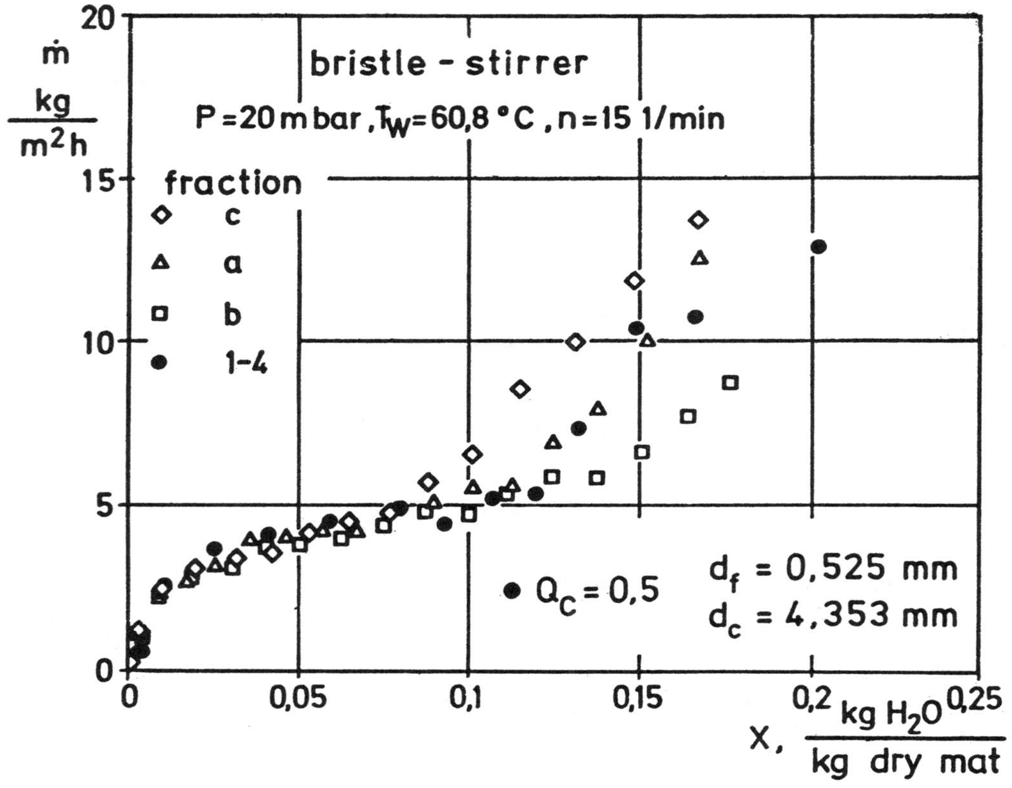

Fig.12 Measured drying rate curves of polydispersed particulate material in a disc dryer.

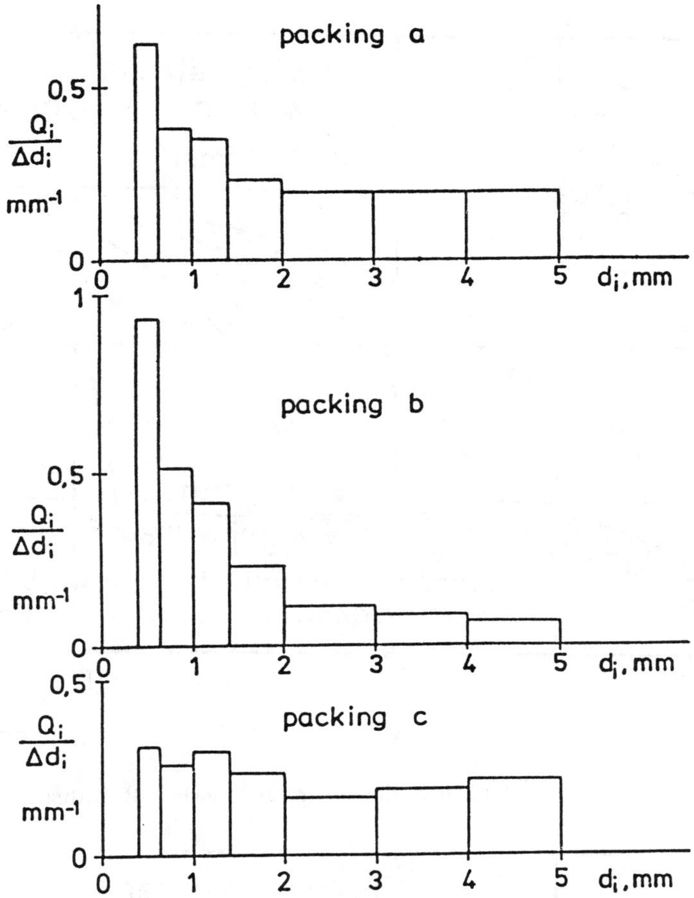

Fig.13 Composition of the polydispersed packings, whose drying rate curves are shown in Fig.12

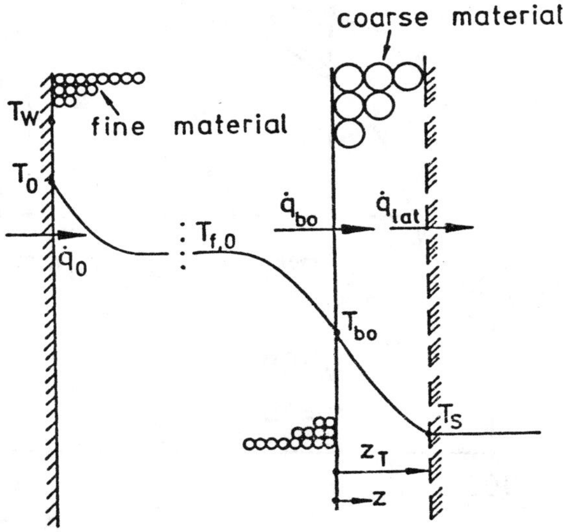

Fig.14 The stratification model for the description of contact drying of bidispersed packings in a disc dryer. Assumed temperature profiles during the first static period after drying out of the fine grained layer.

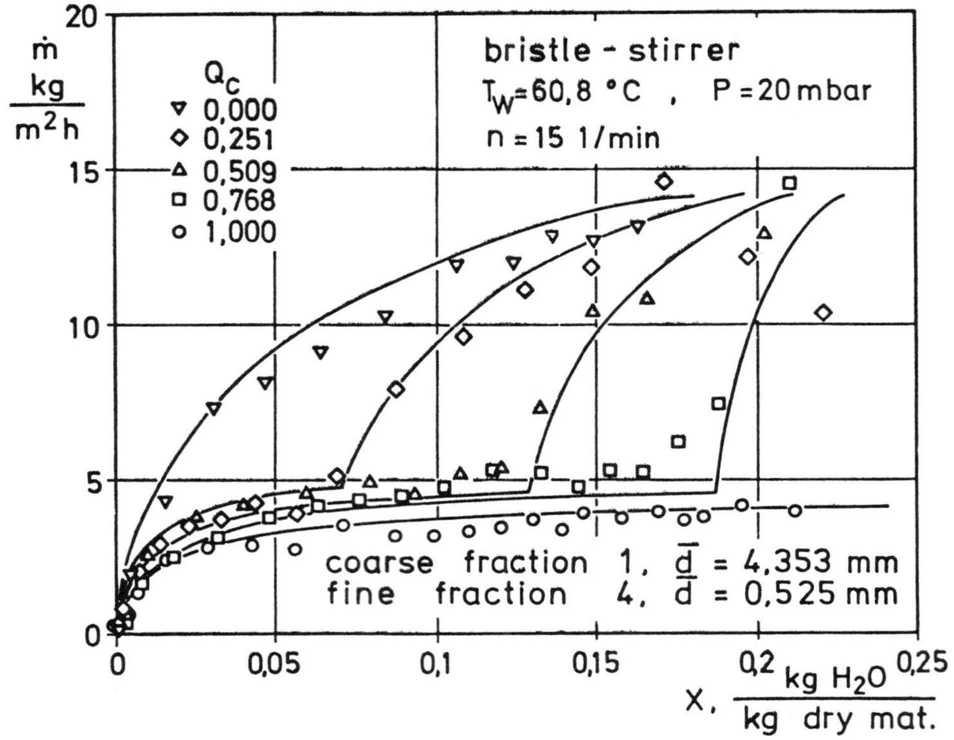

Fig.15 Measured and calculated drying rate curves of bidispersed packings in the disc dryer

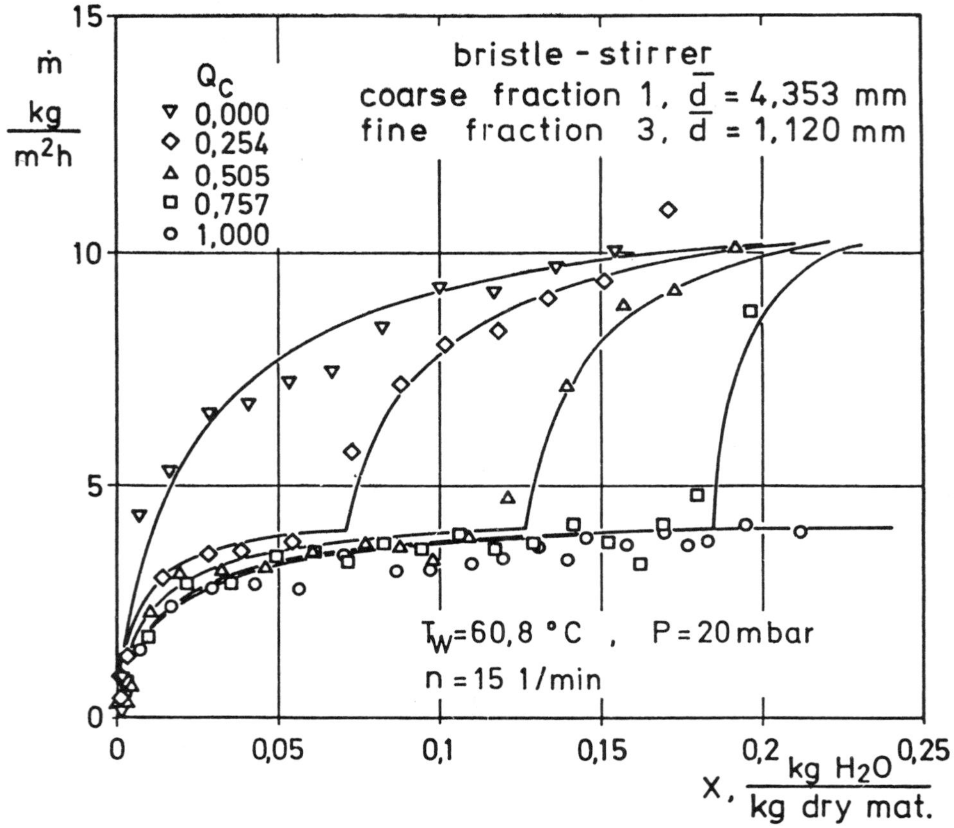

Fig.16 Measured and calculated drying rate curves of bidispersed packings in the disc dryer.

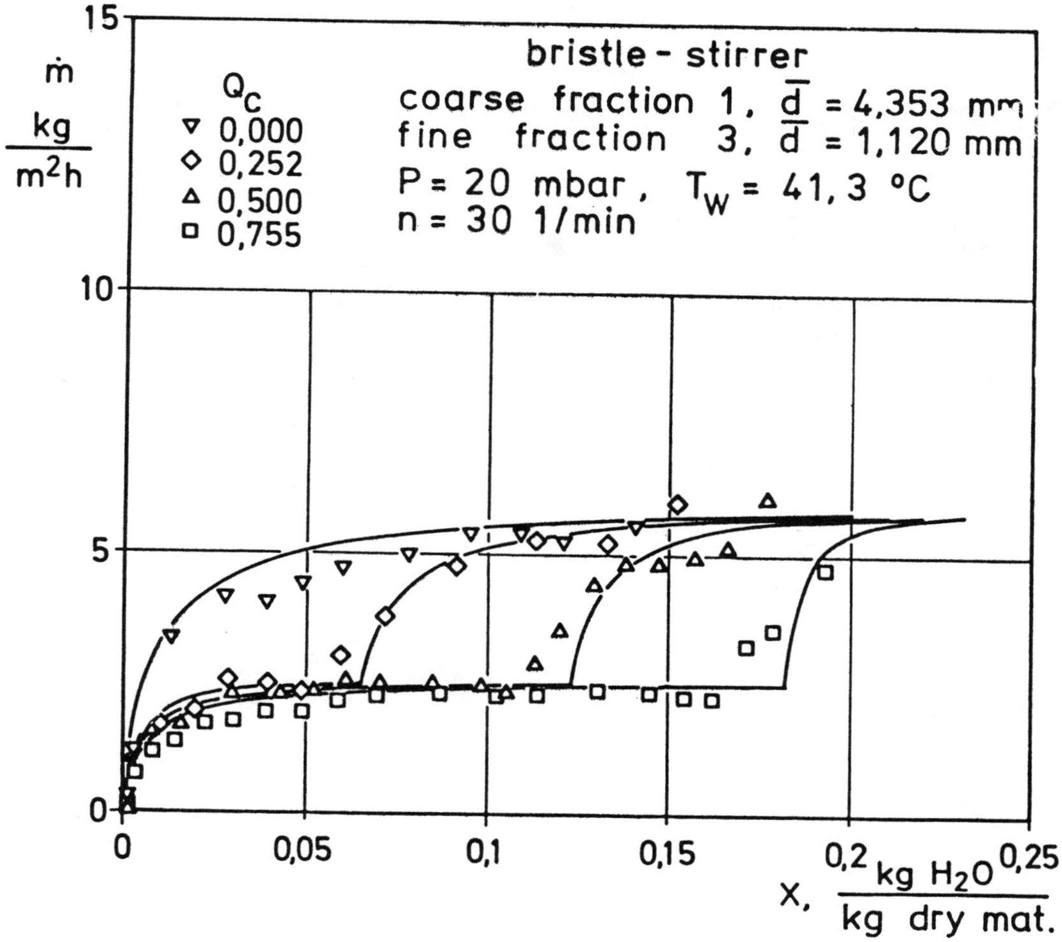

Fig.17 Measured and calculated drying rate curves of bidispersed packings in the disc dryer

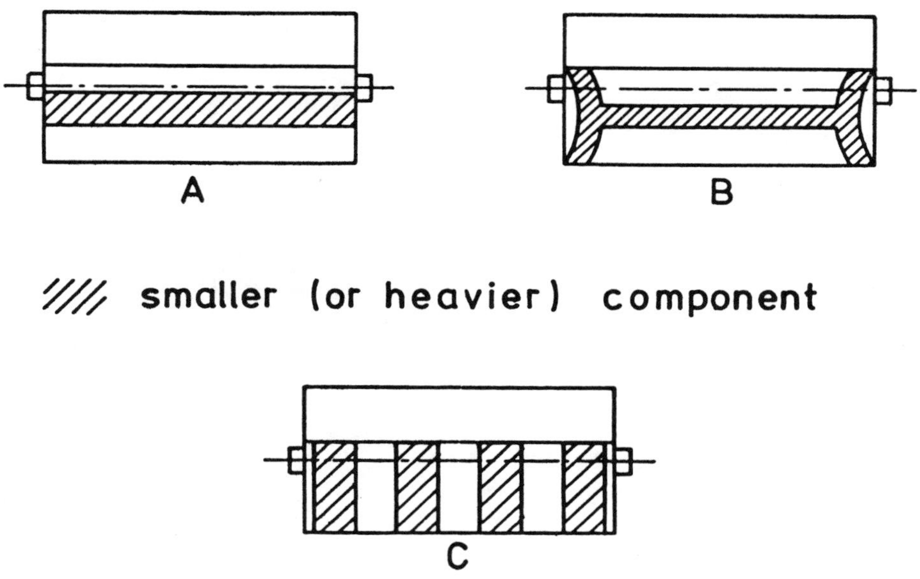

Fig.18 De-mixing patterns in a rotating drum dryer, as reported in [10]

Author Index